ST/SG/AC.10/11/Rev.6

Recommendations on the

TRANSPORT OF DANGEROUS GOODS

Manual of Tests and Criteria

Sixth revised edition

UNITED NATIONS
New York and Geneva, 2015

NOTE

The designations employed and the presentation of the material in this publication do not imply the expression of any opinion whatsoever on the part of the Secretariat of the United Nations concerning the legal status of any country, territory, city or area, or of its authorities, or concerning the delimitation of its frontiers or boundaries.

ST/SG/AC.10/11/Rev.6

UNITED NATIONS PUBLICATION
Sales No.: E.15.VIII.3
ISBN 978-92-1-139155-8
eISBN 978-92-1-057567-6
ISSN 1014-7160

FOREWORD

The "Recommendations on the Transport of Dangerous Goods, Manual of Tests and Criteria" supplement the "Recommendations on the Transport of Dangerous Goods, Model Regulations" and the "Globally Harmonized System of Classification and Labelling of Chemicals (GHS)". They contain criteria, test methods and procedures to be used for the classification of dangerous goods according to the provisions of Parts 2 and 3 of the Model Regulations, as well as of chemicals presenting physical hazards according to the GHS.

Originally developed by the Economic and Social Council's Committee of Experts on the Transport of Dangerous Goods, which adopted a first version in 1984, the Manual of Tests and Criteria has been regularly updated and amended. Presently, the updating is done under the auspices of the Committee of Experts on the Transport of Dangerous Goods and on the Globally Harmonized System of Classification and Labelling of Chemicals, which replaced the original committee in 2001.

This sixth revised edition includes all the amendments to the fifth revised edition adopted by the Committee at its fifth and sixth sessions in 2010 and 2012 (published under the symbols ST/SG/AC.10/11/Rev.5/Amend.1 and ST/SG/AC.10/11/Rev.5/Amend.2) and those adopted at its seventh session in 2014 (ST/SG/AC.10/42/Add.2).

The new amendments adopted in 2014 address in particular:

- A new test to measure the sensitiveness of the substance to drop-mass impact and to determine if the substance is too dangerous to transport in the form tested (Section 13, Test 3 (a) (vii): Modified Bureau of Mines impact machine test);

- Updated procedures for the classification of lithium metal and lithium ion cells and batteries (sub-section 38.3);

- A new test procedure for the classification of substances of Class 9 which may evolve flammable vapours (sub-section 38.4);

- A new Part V containing schemes for the classification of desensitized explosives for supply and use (including storage) according to the GHS;

- A new test for the classification of cartridges, small arms (UN No. 0012) (Appendix 9); and

- Updating of miscellaneous existing provisions.

GENERAL TABLE OF CONTENTS

GENERAL TABLE OF CONTENTS (continued)

GENERAL TABLE OF CONTENTS (continued)

SECTION 1

GENERAL INTRODUCTION

***NOTE:** This general introduction relates only to Parts I to III of the Manual of Tests and Criteria and its Appendices 1 to 9. At its second session (10 December 2004), the Committee of Experts on the Transport of Dangerous Goods and on the Globally Harmonized System of Classification and Labelling of Chemicals decided to add a new Part IV relating to tests methods concerning transport equipment. At its seventh session, the Committee of Experts on the Transport of Dangerous Goods and the Globally Harmonized System of Classification and Labelling of Chemicals decided to add a new Part V relating to classification procedures, test methods and criteria relating to sectors other than transport.*

1.1 Introduction

1.1.1 The purpose of this text is to present the United Nations schemes for the classification of certain types of dangerous goods and to give descriptions of the test methods and procedures considered to be the most useful for providing competent authorities with the necessary information to arrive at a proper classification of substances and articles for transport. The Manual of Tests and Criteria should be used in conjunction with the latest version of the Recommendations on the Transport of Dangerous Goods (hereafter referred to as the Recommendations) and of the Model Regulations on the Transport of Dangerous Goods annexed to these Recommendations (hereafter referred to as the Model Regulations).

1.1.2 It should be noted that the Manual of Tests and Criteria is not a concise formulation of testing procedures that will unerringly lead to a proper classification of products. It therefore assumes competence on the part of the testing authority and leaves responsibility for classification with them. The competent authority has discretion to dispense with certain tests, to vary the details of tests, and to require additional tests when this is justified to obtain a reliable and realistic assessment of the hazard of a product. In some cases, a small scale screening procedure may be used to decide whether or not it is necessary to perform larger scale classification tests. Suitable examples of procedures are given in the introductions to some test series and in Appendix 6. Examples which may be listed within various test procedures are for illustrative purposes and are provided for guidance only.

1.1.3 In situations where the proper classification of substances and articles of certain Hazard Classes or Divisions for transport is the responsibility of the Competent Authority, it is normal and accepted practice that due consideration will be given to testing or classification results of other Competent Authorities when provided.

1.2 Layout

1.2.1 The classification procedures, test methods and criteria are divided into three parts:

Part I: those relating to assignment of explosives to Class 1;

Part II: those relating to assignment of self-reactive substances to Division 4.1 and of organic peroxides to Division 5.2;

Part III: those relating to assignment of substances or articles to Class 2, Class 3, Class 4, Division 5.1, Class 8 or Class 9.

Part III contains some classification procedures, test methods and criteria which are also given in the Model Regulations. There are also a number of appendices which give information common to a number of different types of tests, on the National Contacts for Test Details, on an example method for emergency relief vent sizing of portable tanks for the transport of organic peroxides and self-reactive substances and on screening procedures.

1.2.2 The methods of test identification are given in Table 1.1.

Table 1.1: TEST IDENTIFICATION CODES

Part of Manual	Test series	Test type	Test number	Example of test identification code
I	1 - 8	(a), (b), etc.	(i), (ii), etc.[a]	2 (a) (i)
II	A - H	-	1, 2, etc.	A.1
III	L - T	-	1, 2, etc.	L.1

[a] *If only one test is given for a test type, the Roman numerals are not used.*

1.2.3 Each test is given a unique identification code and is edited as follows:

x.1 *Introduction*
x.2 *Apparatus and materials*
x.3 *Procedure* (including observations to be made and data to be collected)
x.4 *Test criteria and method of assessing results*
x.5 *Examples of results*

NOTE: *Examples of results are not normally given for tests on articles as these are too specific to the article tested and do not allow validation of the test procedure. Results on substances may vary from those given in the "Examples of results" if the physical form, composition, purity etc. of the substance is different. The results given should not be regarded as standard values.*

Figures x.1, x.2, x.3 etc. (i.e. diagrams of apparatus etc.)

NOTE: *Unless otherwise indicated, the dimensions given on the diagrams are in millimetres.*

1.3 Precedence of hazard characteristics

1.3.1 The table in 2.0.3.3 of Chapter 2.0 of the Model Regulations may be used as a guide in determining the class of a substance, mixture or solution having more than one risk, when it is not named in the Dangerous Goods List in Chapter 3.2 of the Model Regulations. For goods having multiple risks, which are not specifically listed by name in Chapter 3.2 of the Model Regulations, the most stringent packing group denoted to the respective hazard of the goods takes precedence over other packing groups, irrespective of the precedence of hazard table in 2.0.3.3 of Chapter 2.0 of the Model Regulations.

1.3.2 The precedence of hazard characteristics of the following are not dealt with in the Precedence of Hazard Table in Chapter 2.0 of the Model Regulations, since these primary characteristics always take precedence:

Substances and articles of Class 1;
Gases of Class 2;
Liquid desensitized explosives of Class 3;
Self-reactive substances and solid desensitized explosives of Division 4.1;
Pyrophoric substances of Division 4.2;
Substances of Division 5.2;
Substances of Division 6.1 with a packing group I inhalation toxicity;
Substances of Division 6.2; and
Material of Class 7.

1.3.3 Self-reactive substances, except for type G, giving a positive result in the self-heating test for Division 4.2, should not be classified in Division 4.2 but in Division 4.1 (see paragraph 2.4.2.3.1.1 of the Model Regulations). Organic peroxides of type G having properties of another class or division (e.g. UN 3149) should be classified according to the requirements of that class or division.

1.4 Safety

1.4.1 For the safety of laboratory personnel, the producer or other applicant for classification of a new product should provide all available safety data on the product e.g. the toxicity data.

1.4.2 Particularly when explosive properties are suspected, it is essential for the safety of workers that small scale preliminary tests are carried out before attempting to handle larger quantities. This involves tests for determining the sensitiveness of the substance to mechanical stimuli (impact and friction), and to heat and flame.

1.4.3 In tests involving initiation of potentially explosive substances or articles, a safe waiting period, prescribed by the test agency, should be observed after initiation.

1.4.4 Extra care should be taken when handling samples which have been tested since changes may have occurred rendering the substance more sensitive or unstable. Tested samples should be destroyed as soon as possible after the test.

1.5 General conditions for testing

1.5.1 The conditions given in the test prescriptions should be followed as closely as possible. If a parameter is not specified in the test prescription then the conditions given here should be applied. Where tolerances are not specified in the test prescription, it is implied that the accuracy is according to the number of decimal places given in any dimension e.g. 1.1 implies 1.05 to 1.15. In cases where conditions during a test deviate from those prescribed, the reason for the deviation should be stated in the report.

1.5.2 The composition of the test sample should be as close as possible to the concentration of the substance intended for transport. The contents of active substance(s) and diluent(s) should be specified in the test report with at least an accuracy of ± 2 % by mass. Components which can have a major effect on a test result, such as moisture, should be specified as accurately as possible in the test report.

1.5.3 All test materials in contact with the test substance should be such that, as far as possible, they do not affect the test results e.g. catalyse decomposition. In cases where such an effect cannot be excluded, special precautions should be taken to prevent the result being affected, e.g. passivation. The precautions taken should be specified in the test report.

1.5.4 The tests should be performed under the conditions (temperature, density etc.) which are representative of the expected circumstances of transport. If the transport conditions are not covered by the test conditions specified, supplementary tests may need to be performed which are specifically designed for the anticipated transport conditions e.g. elevated temperature. Where appropriate, e.g. when the result is particle size dependent, the physical conditions should be specified in the test report.

1.6 Recommended tests

1.6.1 The Manual gives descriptions of tests and criteria used to provide the necessary information to arrive at a proper classification. In some cases, there is more than one test for a particular property. As a result of comparative work with some of these tests, it has been possible to identify one test as the recommended test in a set of equivalent tests. The recommended tests for classifying explosive substances and articles (Part I of the Manual) are listed in Table 1.2 and for classifying self-reactive substances and organic peroxides (Part II of the Manual) in Table 1.3. Unless otherwise specified, all test methods given in Part III of the Manual are recommended tests as only one test is given for each property. The other tests in a set are considered to be alternative tests and may continue to be used for classification purposes.

1.6.2 As a result of comparative work, some tests have been deleted. However, as some countries maintain databases referenced by the test number, the tests currently given in the Manual have not been renumbered unless existing tests have been assigned to different test types.

1.6.3 The aim is to have only one United Nations test, or combination of tests, for each property. However, until the recommended tests have been used more widely, it is not possible to do this in all cases at present.

1.6.4 If new tests are proposed for inclusion in the Manual, the proposer should be able to provide justification that the new test is a significant improvement on the existing recommended test. In such cases, the new test may be included as an alternative test until it has been tried by laboratories of other countries.

Table 1.2: RECOMMENDED TESTS FOR EXPLOSIVES AND EXPLOSIVE ARTICLES

Test series	Test type	Test code	Test name
1	(a)	1 (a)	UN gap test
1	(b)	1 (b)	Koenen test
1	(c)	1 (c) (i)	Time / pressure test
2	(a)	2 (a)	UN gap test
2	(b)	2 (b)	Koenen test
2	(c)	2 (c) (i)	Time / pressure test
3	(a)	3 (a) (ii)	BAM Fallhammer
3	(b)	3 (b) (i)	BAM Friction apparatus
3	(c)	3 (c)	Thermal stability test at 75 °C
3	(d)	3 (d)	Small-scale burning test
4	(a)	4 (a)	Thermal stability test for unpackaged articles and packaged articles
4	(b)	4 (b) (i)	Steel tube drop test for liquids
4	(b)	4 (b) (ii)	Twelve metre drop test for unpackaged articles, packaged articles and packaged substances
5	(a)	5 (a)	Cap sensitivity test
5	(b)	5 (b) (ii)	USA DDT test
5	(c)	5 (c)	External fire test for Division 1.5
6	(a)	6 (a)	Single package test
6	(b)	6 (b)	Stack test
6	(c)	6 (c)	External fire (bonfire) test
6	(d)	6 (d)	Unconfined package test
7	(a)	7 (a)	EIS cap test
7	(b)	7 (b)	EIS gap test
7	(c)	7 (c) (ii)	Friability test
7	(d)	7 (d) (i)	EIS bullet impact test
7	(e)	7 (e)	EIS external fire test
7	(f)	7 (f)	EIS slow cook-off test
7	(g)	7 (g)	1.6 article external fire test
7	(h)	7 (h)	1.6 article slow cook-off test
7	(j)	7 (j)	1.6 article bullet impact test
7	(k)	7 (k)	1.6 article stack test
8	(a)	8 (a)	Thermal stability test for ANE
8	(b)	8 (b)	ANE gap test
8	(c)	8 (c)	Koenen test
8	(d)	8 (d)	Vented pipe tests[a]

[a] *These tests are intended for evaluating the suitability for transport in tanks.*

Table 1.3: RECOMMENDED TESTS FOR SELF-REACTIVE SUBSTANCES AND ORGANIC PEROXIDES

Test series	Test code	Test name
A	A.6	UN detonation test
B	B.1	Detonation test in package
C	C.1	Time/pressure test
C	C.2	Deflagration test
D	D.1	Deflagration test in the package
E	E.1	Koenen test
E	E.2	Dutch pressure vessel test
F	F.4	Modified Trauzl test
G	G.1	Thermal explosion test in package
H	H.1	United States SADT test (for packages)
H	H.2	Adiabatic storage test (for packages, IBCs and tanks)
H	H.4	Heat accumulation storage test (for packages, IBCs and small tanks)

1.7 Reporting

1.7.1 Classifications for Chapter 3.2 of the Model Regulations are made on the basis of consideration of data submitted to the Committee by governments, intergovernmental organisations and other international organisations in the form recommended in Figure 1 of the Recommendations. Supplementary data is required for the classification of:

Substances and articles of Class 1 (see 10.5);
Self-reactive substances of Division 4.1 (see 20.5); and
Organic peroxides of Division 5.2 (see 20.5).

1.7.2 Where tests are performed on packaged substances or articles, the test report should contain the quantity of substance or number of articles per package and the type and construction of the packaging.

PART I

CLASSIFICATION PROCEDURES, TEST METHODS AND CRITERIA RELATING TO EXPLOSIVES OF CLASS 1

CONTENTS OF PART I

NOTE 1: *The country or organization of origin of each test method is indicated in brackets after each test name.*

NOTE 2: *The test method recommended for use with each test type is indicated in* ***bold*** *and by an asterisk (see sub-section 1.6 of the General Introduction).*

CONTENTS OF PART I (continued)

CONTENTS OF PART I (continued)

CONTENTS OF PART I (continued)

SECTION 10

INTRODUCTION TO PART I

10.1 Purpose

10.1.1 Part I of the Manual presents the United Nations scheme for the classification of explosives. It includes a description of the procedures and test criteria considered to be the most useful for providing competent authorities with the necessary information to arrive at a proper classification of explosive substances and articles for transport. It should be used in conjunction with the classification flow charts in figures 10.1, 10.2, 10.3, 10.4 and 10.5, the general conditions for testing in sub-section 1.5 and the appropriate test prescriptions in sections 11 to 18 of this Manual.

10.1.2 Goods of Class 1 are assigned to one of six divisions, depending on the type of hazard they present (see paragraph 2.1.1.4 of the Model Regulations), and to one of thirteen compatibility groups which identify the kinds of explosive substances and articles that are deemed to be compatible. The general scheme for classifying a substance or article which is to be considered for inclusion in Class 1 is illustrated in Figure 10.1. The assessment is in two stages. In the first stage, the potential of a substance or article to explode should be ascertained and its stability and sensitivity, both chemical and physical, shown to be acceptable. In order to promote uniform assessments by competent authorities, it is recommended that, using the flow chart in Figure 10.2, data from suitable tests is analyzed systematically with respect to the appropriate test criteria. If the substance or article is provisionally accepted into Class 1, it is then necessary to proceed to the second stage and assign it to the correct division by use of the flow chart of Figure 10.3. With the exception of Compatibility Groups N and S, for which test data is necessary, assignment to a compatibility group is usually made without reference to testing. In the case of Compatibility Group S, the tests may be waived by the competent authority if classification by analogy is based on test results for a comparable article.

10.1.3 The test procedures allow assessment of the hazard of explosive substances and articles so that an appropriate classification for transport can be made by the competent authority.

10.2 Scope

10.2.1 New products which are considered as having explosive properties or are intended to function as explosives should first be considered for inclusion in Class 1. For substances such as self-reactive substances of Division 4.1 or organic peroxides of Division 5.2, refer to Part II of this Manual. In this context a new product is one which, in the opinion of the competent authority, involves any of the following:

(a) A new substance or a combination or mixture of substances, intended to function as an explosive or pyrotechnic, which is considered to be significantly different from other combinations or mixtures already classified;

(b) A new substance or article, not intended for explosive use, which has, or is suspected of having, explosive properties (see paragraph 2.1.1.5 of the Model Regulations);

(c) A new design of article containing an explosive substance or an article containing a new explosive substance or a new combination or mixture of explosive substances; or

(d) A new design of package for an explosive substance or article including a new type of inner packaging or a new arrangement of articles (a relatively minor change to an inner or outer packaging can be critical and can convert a lesser risk into a mass explosion risk).

The classification procedure should be undertaken before a new product is offered for transport.

10.2.2 The producer, or other applicant for the classification of a new product, should provide adequate information concerning the names and characteristics of all explosive substances in the product and should furnish the results of all relevant tests which have been done.

10.3 Acceptance procedure

10.3.1 *General description*

10.3.1.1 The acceptance procedure is used to determine whether or not a product as offered for transport is a candidate for Class 1. This is decided by determining whether a substance provisionally accepted for Class 1 is either too insensitive for inclusion in Class 1 or too dangerous for transport; or whether article(s) or packaged article(s) are too dangerous for transport.

10.3.2 *Test types*

10.3.2.1 The test methods used for deciding provisional acceptance into Class 1 are grouped into four series, numbered 1 to 4, and designed to provide the information necessary to answer the questions in Figure 10.2.

10.3.2.2 The question "Is it an explosive substance?" (box 4, Figure 10.2) is answered on the basis of national and international definitions of an explosive substance and the results of three types of Series 1 tests to assess possible explosive effects. The three types of test used are:

Type 1 (a): a shock test with defined booster and confinement to determine the ability of the substance to propagate a detonation;
Type 1 (b): a test to determine the effect of heating under confinement; and
Type 1 (c): a test to determine the effect of ignition under confinement.

10.3.2.3 Series 2 tests are used to answer the question "Is the substance too insensitive for acceptance into Class 1?" (box 6, Figure 10.2). In general the basic apparatus used is the same as that for Test Series 1 but with less stringent criteria, e.g. in the case of gap tests, the gap used is greater than zero. The following three types of test are used:

Type 2 (a): a shock test with defined initiation system and confinement to determine sensitivity to shock;
Type 2 (b): a test to determine the effect of heating under confinement; and
Type 2 (c): a test to determine the effect of ignition under confinement.

Figure 10.1: OVERALL SCHEME OF THE PROCEDURE FOR CLASSIFYING A SUBSTANCE OR ARTICLE IN CLASS 1

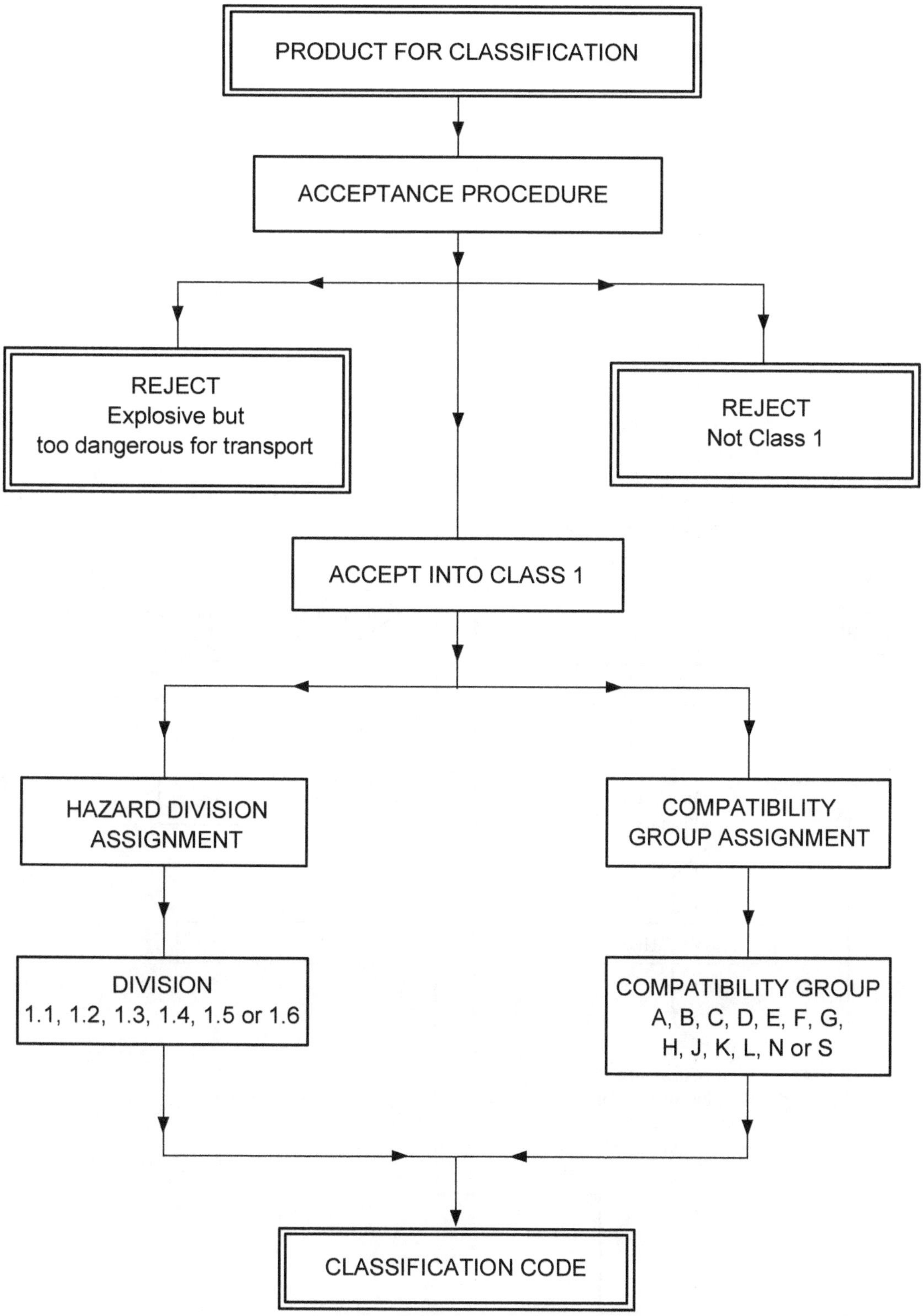

Figure 10.2: PROCEDURE FOR PROVISIONAL ACCEPTANCE OF A SUBSTANCE OR ARTICLE IN CLASS 1

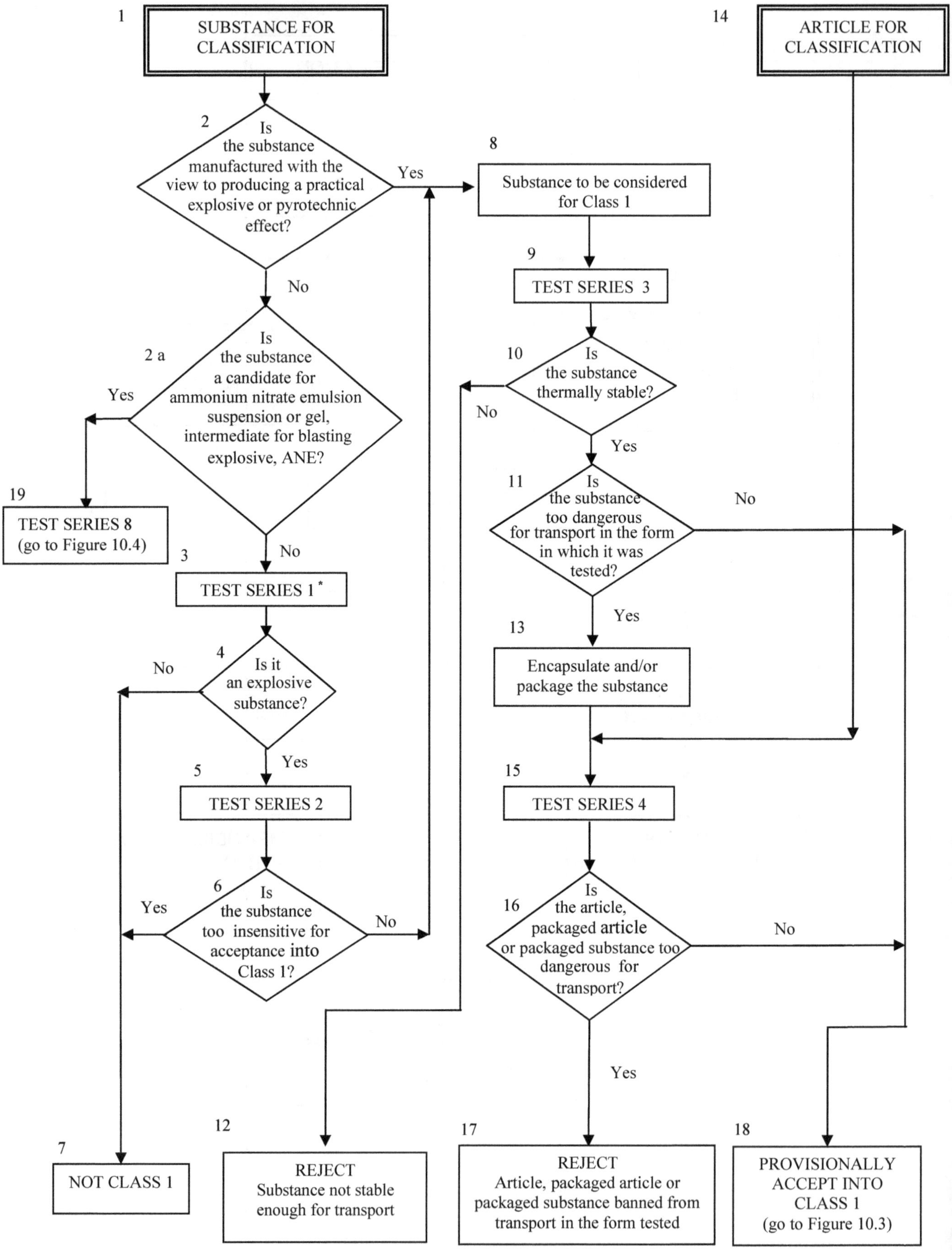

* *For classification purposes, start with test series 2.*

10.3.2.4 Test series 3 is used to answer the questions "Is the substance thermally stable?" (box 10, Figure 10.2) and "Is the substance too dangerous for transport in the form in which it was tested?" (box 11, Figure 10.2). This involves tests for determining the sensitiveness of the substance to mechanical stimuli (impact and friction), and to heat and flame. The following four types of test are used:

Type 3 (a): a falling weight test to determine sensitiveness to impact;
Type 3 (b): a friction, or impacted friction, test to determine sensitiveness to friction;
Type 3 (c): an elevated temperature test to determine thermal stability; and
Type 3 (d): an ignition test to determine the response of a substance to fire.

10.3.2.5 Series 4 tests are intended to answer the question "Is the article, packaged article or packaged substance too dangerous for transport?" (box 16, Figure 10.2). Conditions which may occur during transport include high temperature and high relative humidity, low temperature, vibration, bumping and dropping. The two types of test to be carried out are:

Type 4 (a): a test of thermal stability for articles; and
Type 4 (b): a test to determine the hazard from dropping.

10.3.3 ***Application of the test methods***

10.3.3.1 The numbering of test series 1 to 4 relates to the sequence of assessing the results rather than the order in which the tests are conducted. ***It may be important for the safety of experimenters that certain preliminary tests, using small amounts of material, be conducted first before proceeding to experiment with larger quantities.*** The results of these preliminary tests may also be used in the classification procedure.

10.3.3.2 The acceptance procedure for substances designed to have an explosive effect starts with the application of test types 3 (a), 3 (b), 3 (c) and 3 (d) to determine if the substance is too sensitive for transport in the form in which it is tested. If it proves to be thermally unstable, i.e. it fails test type 3 (c), it is not permitted to be transported. If it fails test types 3 (a), 3 (b) or 3 (d) it may either be encapsulated or otherwise desensitized or packaged to reduce its sensitiveness to external stimuli. Examples are water-wetted primary explosives and primary explosives which have been encapsulated in the form of detonators. The resulting new articles should be submitted to test series 4, and liquids or packaged solids to a test of type 4 (b), to determine whether or not their level of safety in transport is consistent with the requirements of Class 1. Desensitized substances should be re-examined under test series 3 for the same purpose. If a substance designed to have an explosive effect passes all tests in series 3 or an article designed to have an explosive effect passes all tests in series 4, the procedure for assignment to the appropriate division is applied.

10.3.3.3 Although test series 1 indicates whether a substance, not designed to have an explosive effect, has in fact potentially explosive properties, again it is more appropriate to start the testing procedure with test series 3. These tests involve relatively small sample sizes, which reduces the risk to test personnel. If test series 3 indicates that a substance is too sensitive for transport in the form in which it is tested, then the procedures for reducing its sensitiveness to external stimuli, outlined in 10.3.3.2, should be applied. If test series 3 indicates that the substance is not too sensitive for transport, the next step is the application of test series 2 which determines whether the substance is too insensitive for inclusion in Class 1. There is no real need to perform test series 1 at this point in the acceptance procedure since test series 2 answers the pertinent question regarding the degree of insensitiveness of the substance. Test series 1 is concerned with the resolution of questions relating to the explosive nature of the substance. The procedure for assignment to a division of Class 1 should be applied to substances which fail test series 2 but pass test series 3 i.e. they are not too insensitive for acceptance into Class 1 nor are they thermally unstable or too dangerous to transport in the form in which they are tested. It is important to note that a substance which fails test series 2 may still, if properly packaged, leave Class 1 provided that the product is not designed to have an explosive effect and does not exhibit any explosive hazard in test series 6 of the assignment procedure.

10.3.3.4 All articles or packaged articles containing substances which have failed test type 3 (a), 3 (b) or 3 (d) should be subjected to test series 4. If the article or packaged articles pass test type 4 (a), test type 4 (b) is performed. Packaged substances are subjected to test type 4 (b) only. If the product fails either test type 4 (a) or 4 (b), it should be rejected. However, the product may be modified and re-tested. If the competent authority suspects that the product may be subject to stimuli other than those specified in test type 4 (a) and 4 (b) resulting in potentially dangerous effects, additional information or tests may be required (see note under paragraph 2.1.3.3.1 of the Model Regulations).

10.3.3.5 If articles contain expensive, inert, control components, these may be replaced by inert components having a similar mass and volume.

10.4 Procedure for assignment to a division of Class 1

10.4.1 *General description*

10.4.1.1 Goods of Class 1 are assigned to one of six divisions, depending on the type of hazard they present (see paragraph 2.1.1.4 of the Model Regulations). The assignment procedure (Figure 10.3) applies to all substances and/or articles that are candidates for Class 1 except those declared from the outset to be in Division 1.1. A substance or article should be assigned to the division which corresponds to the results of the tests to which the substance or article, as offered for transport, has been subjected. Other test results, and data assembled from accidents which have occurred, may also be taken into account. As indicated in box 36 of Figure 10.3, there is authority to exclude an article from Class 1 by virtue of test results and the Class 1 definition.

10.4.2 *Test types*

10.4.2.1 The test methods used for assignment to a division are grouped into three series - numbered 5 to 7 - designed to provide the information necessary to answer the questions in Figure 10.3. The tests in series 5, 6 and 7 should not be varied unless the national authority is prepared to justify such action internationally.

10.4.2.2 The results from three types of series 5 tests are used to answer the question "Is it a very insensitive explosive substance with a mass explosion hazard?" (box 21, Figure 10.3). The test types are:

Type 5 (a): a shock test to determine the sensitivity to intense mechanical stimulus;

Type 5 (b): thermal tests to determine the tendency for transition from deflagration to detonation; and

Type 5 (c): a test to determine if a substance, when in large quantities, explodes when subjected to a large fire.

Figure 10.3: PROCEDURE FOR ASSIGNMENT TO A DIVISION OF CLASS 1

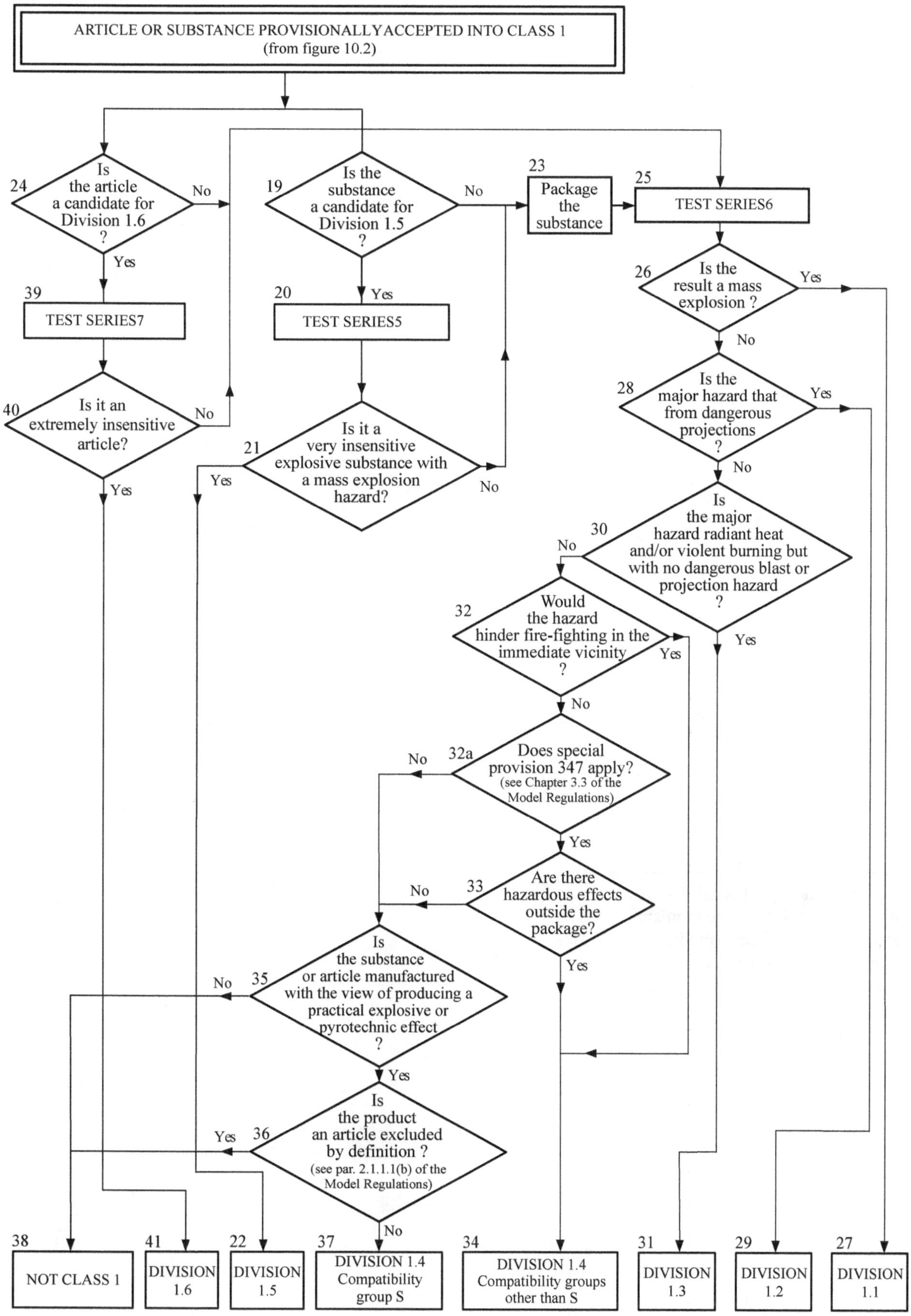

Figure 10.4: PROCEDURE FOR AMMONIUM NITRATE EMULSION, SUSPENSION OR GEL, INTERMEDIATE FOR BLASTING EXPLOSIVES

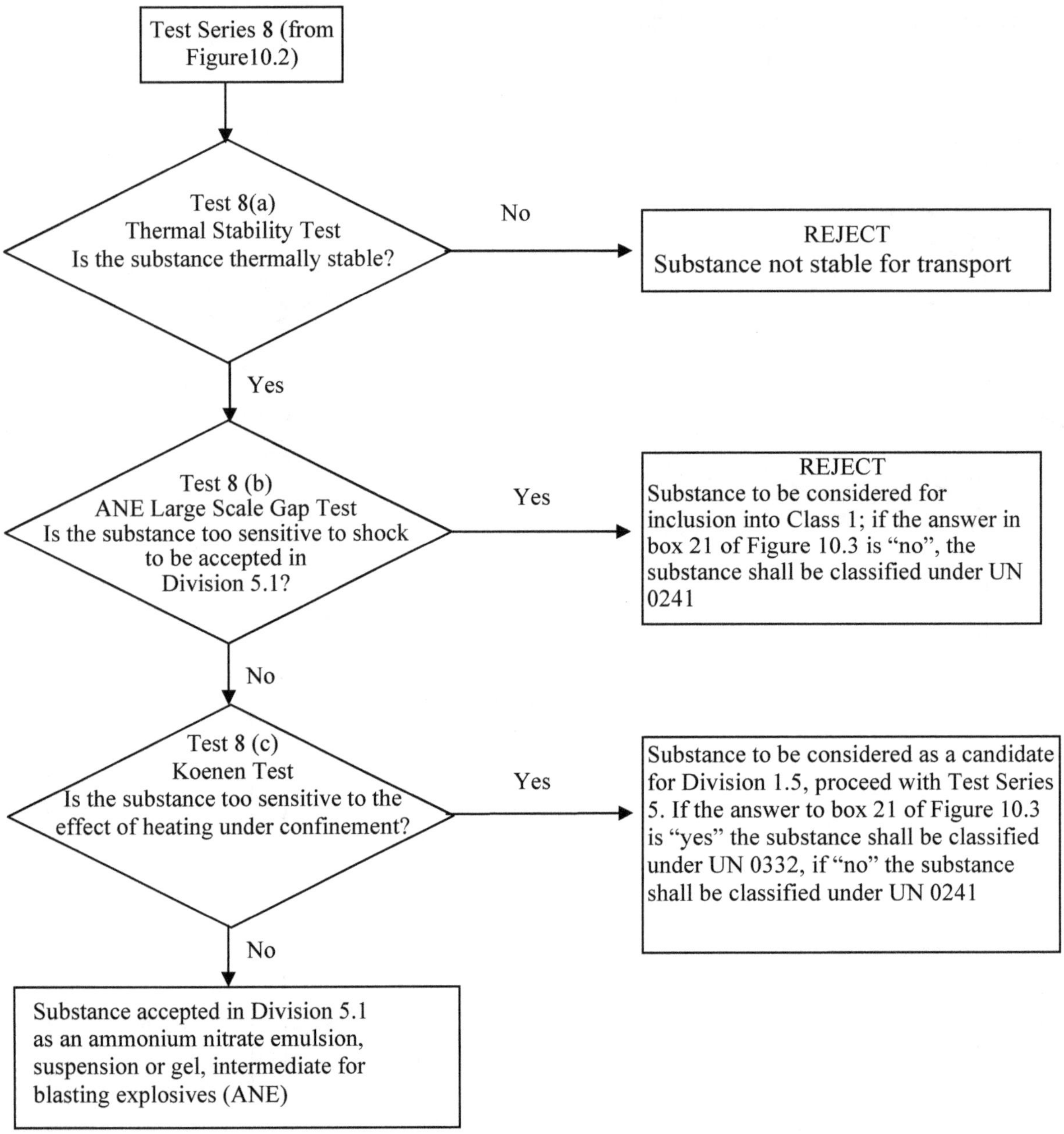

Figure 10.5: PROCEDURE TO DETERMINE REQUIRED SUBSTANCE TESTING FOR DIVISION 1.6

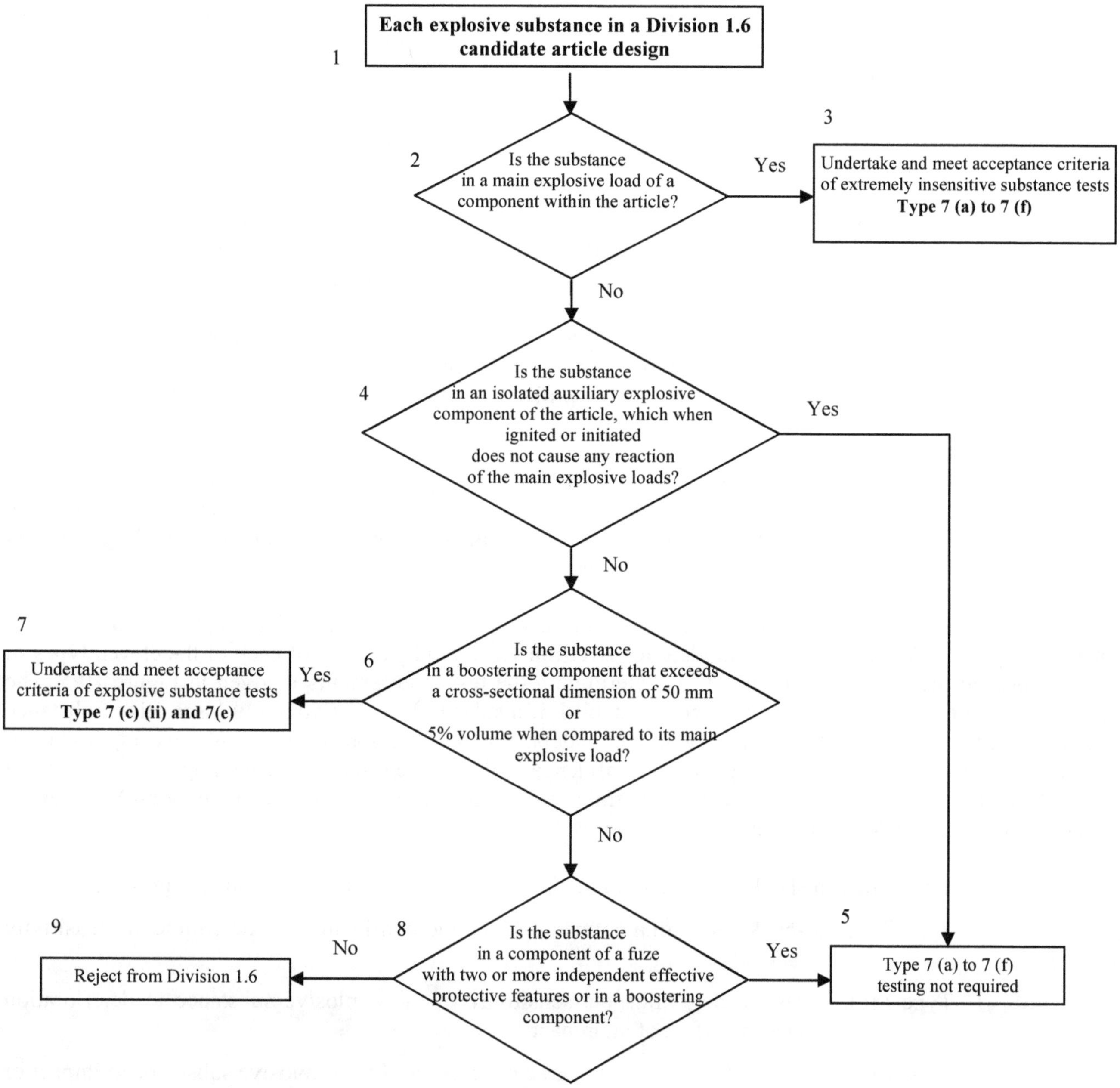

10.4.2.3 The results from four types of series 6 tests are used to determine which division, amongst Divisions 1.1, 1.2, 1.3 and 1.4, corresponds most closely to the behaviour of a product if a load is involved in a fire resulting from internal or external sources or an explosion from internal sources (boxes 26, 28, 30, 32 and 33 of Figure 10.3). The results are also necessary to assess whether a product can be assigned to Compatibility Group S of Division 1.4 and whether or not it should be excluded from Class 1 (boxes 35 and 36 of Figure 10.3). The four types of test are:

Type 6 (a): a test on a single package to determine if there is mass explosion of the contents;

Type 6 (b): a test on packages of an explosive substance or explosive articles, or non-packaged explosive articles, to determine whether an explosion is propagated from one package to another or from a non-packaged article to another;

Type 6 (c): a test on packages of an explosive substance or explosive articles, or non-packaged explosive articles, to determine whether there is a mass explosion or a hazard from dangerous projections, radiant heat and/or violent burning or any other dangerous effect when involved in a fire; and

Type 6 (d): a test on an unconfined package of explosive articles to which special provision 347 of Chapter 3.3 of the Model Regulations applies, to determine if there are hazardous effects outside the package arising from accidental ignition or initiation of the contents.

10.4.2.4 The question "Is it an extremely insensitive explosive article?" (box 40, Figure 10.3) is answered by series 7 tests and any candidate for Division 1.6 should pass one of each of the eleven types of test comprising the series. The protocol for determining the test requirements is given in Figure 10.5. The first six types of test (7(a) to 7(f)) are used to establish if a substance is an Extremely Insensitive Substance (EIS). The purpose of these tests is to develop an understanding of the sensitivity of substance(s) contained within the article, which informs and provides confidence in the article tests. The remaining five types of test (7(g), 7(h), 7(j), 7 (k) and 7 (l)) are used to determine if an article predominantly containing an EIS may be assigned to Division 1.6. The eleven test types are:

Type 7 (a): a shock test to determine the sensitivity to intense mechanical stimulus;

Type 7 (b): a shock test with a defined booster and confinement to determine the sensitivity to shock;

Type 7 (c): a test to determine the sensitivity of the explosive substance to deterioration under the effect of an impact;

Type 7 (d): a test to determine the degree of reaction of the explosive substance to impact or penetration resulting from a given energy source;

Type 7 (e): a test to determine the reaction of the explosive substance to an external fire when the material is confined;

Type 7 (f): a test to determine the reaction of the explosive substance in an environment in which the temperature is gradually increased to 365 °C;

Type 7 (g): a test to determine the reaction to an external fire of an article which is in the condition as presented for transport;

Type 7 (h): a test to determine the reaction of an article in an environment in which the temperature is gradually increased to 365 °C;

Type 7 (j): a test to determine the reaction of an article to impact or penetration resulting from a given energy source;

Type 7 (k): a test to determine whether the detonation of an article will initiate a detonation in an adjacent, like, article; and

Type 7 (l): a test to determine the sensitivity of an article to shock directed at vulnerable components.

10.4.2.5 The question "Is the substance a candidate for "ammonium nitrate emulsion or suspension or gel, intermediate for blasting explosives (ANE)?" (box 2(a), Figure 10.2) is answered by series 8 tests and any candidate should pass each of the three tests comprising the series. The three test types are:

Type 8 (a): a test to determine the thermal stability;
Type 8 (b): a shock test to determine sensitivity to intense shock;
Type 8 (c): a test to determine the effect of heating under confinement;

Test series 8 (d) has been included in this section as one method to evaluate the suitability of ANEs for the transport in portable tanks.

10.4.3 ***Application of the test methods***

10.4.3.1 Explanations of certain terms used in the assignment of divisions and compatibility groups are given in the Glossary in Appendix B of the Model Regulations (e.g. mass explosion, pyrotechnic substance, entire load, total contents, explode, explosion of the total contents).

10.4.3.2 Test series 5 should be used to determine whether a substance can be assigned to Division 1.5. Only those substances which pass all three types of test may be assigned to Division 1.5.

10.4.3.3 Test series 6 should be applied to packages of explosive substances and articles in the condition and form in which they are offered for transport. The geometrical arrangement of the products should be realistic in regard to the packing method, and the conditions of transport, and should be such as to produce the most disadvantageous test results. Where explosive articles are to be carried without packaging, the tests should be applied to the non-packaged articles. All types of packaging containing substances or articles should be subjected to the tests unless:

(a) The product, including any packaging, can be unambiguously assigned to a division by the competent authority on the basis of results from other tests or of available information; or

(b) The product, including any packaging, is assigned to Division 1.1.

10.4.3.4 Test types 6 (a), 6 (b), 6 (c) and 6 (d) are normally performed in alphabetical order. However, it is not always necessary to follow this order or to conduct all four types of test.

(a) Test type 6 (a) may be waived if explosive articles are carried without packaging or when only one article is in the package, (see also section 10.4.3.4.(d));

(b) Test type 6 (b) may be waived if, in each type of 6 (a) test, (see also section 10.4.3.4 (d))

(i) The exterior of the package is undamaged by internal detonation and/or ignition; or

(ii) The contents of the package fail to explode, or explode so feebly as would exclude propagation of the explosive effect from one package to another in test type 6 (b).

(c) Test type 6 (c) may be waived if, in a type 6 (b) test, there is a practically instantaneous explosion of virtually the total contents of the stack. In such cases the product is assigned to Division 1.1.

(d) Test type 6 (d) is a test used to determine whether a 1.4S classification is appropriate and is only used if special provision 347 of Chapter 3.3 of the Model Regulations applies. When testing articles to which special provision 347 applies, test type 6 (d) may be performed first. If the results of test type 6(d) indicate that a 1.4S classification is appropriate, then test types 6 (a) and 6 (b) may be waived.

10.4.3.5 If a substance gives a "-" result (no propagation of detonation) in the Series 1 type (a) test, the 6 (a) test with a detonator may be waived. If a substance gives a "-" result (no or slow deflagration) in a Series 2 type (c) test, the 6 (a) test with an igniter may be waived.

10.4.3.6 Tests types 7 (a) to 7 (f) should be used to establish that the explosive is an extremely insensitive detonating substance and then test types 7 (g), 7 (h), 7 (j), 7 (k) and 7 (l) used to establish that the articles containing EIS(s) may be assigned to Division 1.6.

10.4.3.7 Tests of types 7 (g), 7 (h), 7 (j), 7 (k) and 7(l) should be performed to determine if an article with EIS main explosive load(s) and appropriately insensitive boostering components may be assigned to Division 1.6. These tests are applied to articles in the condition and form in which they are offered for transport, except that non-explosive components may be omitted or simulated if the competent authority is satisfied that this does not invalidate the results of the tests. The procedure detailing testing requirements is given in Figure 10.5 and some points of explanation are given below.

(a) Complex articles may contain multiple substances and this procedure should be completed for all substances within the article to be classified.

(b) The question "Is the substance in a main explosive load of a component within the article?" (Box 2 of Figure 10.5) is answered by examining the design of the article. Main explosive load substances are those loaded into components within the article that are not fuze, boostering, or isolated auxiliary explosive components. All substances in main explosive loads must "Undertake and meet acceptance criteria of extremely insensitive substance tests, Type 7 (a) to 7 (f)" (Box 3 of Figure 10.5). If a '+' result is obtained for any main explosive load substance to any Type 7 (a) to 7 (f) test, the substance is not an EIS and the answer to the question in Box 24 of Figure 10.3 is "No". The article is not a candidate for Division 1.6.

(c) Answering the question "Is the substance in an isolated auxiliary explosive component of the article, which when ignited or initiated does not cause any reaction of the main explosive loads?" (Box 4 of Figure 10.5) requires knowledge of the design of the article plus the explosive effects that occur when such components are initiated or ignited, either in their design mode or accidentally. Typically these will be small explosive actuators or pyromechanical devices that produce movement, cutting or opening functions. If the answer is 'yes' to this question, Type 7 (a) to 7 (f) testing is not required for substances in isolated auxiliary explosive components and the article remains a candidate for Division 1.6.

(d) The question "Is the substance in a boostering component that exceeds a cross-sectional dimension of 50 mm or 5% volume when compared to its main explosive load?" (Box 6 of Figure 10.5) is answered by examining the design of the article. All substances in such larger boostering components, including those contained in explosive components of dual-protected fuzes in an article, must "Undertake and meet acceptance criteria of explosive substance tests, Type 7 (c) (ii) and 7 (e)" (box 7 of Figure 10.5). If a '+' result is obtained for any such larger boostering component substance to either Type 7 (c) (ii) and 7 (e) tests, the answer to the question in Box 24 of Figure 10.3 is "No". The article is not a candidate for Division 1.6.

(e) The question "Is the substance in a component of a fuze with two or more independent effective protective features or in a boostering component" (Box 8 of Figure 10.5) is answered by an understanding of the design and development of the article. If the answer is 'no', the article is not considered to have suitable intrinsic safety characteristics and the answer to the question in Box 24 of Figure 10.3 is 'No' the article is not a candidate for Division 1.6.

NOTE: *Knowledge of the design and explosive effects can be obtained by modelling or indicative tests etc.*

10.4.3.8 Test types 8 (a) to 8 (c) should be used to establish whether an ammonium nitrate emulsion or suspension or gel, intermediate for blasting explosives (ANE) may be assigned to Division 5.1. Substances failing any of the tests may be considered as a candidate for Class 1 in accordance with Figure 10.4.

10.4.3.9 If articles contain expensive, inert, control components, these may be replaced by inert components having a similar mass and volume.

10.5 Examples of test reports

10.5.1 Examples of test reports, with an illustration of the use of the flow charts on the application of the Class 1 acceptance and assignment procedures to musk xylene (UN 2956), are given in figures 10.6 to 10.9.

10.5.2 An example proforma for a test report on articles is given in Figure 10.10.

Figure 10.6: RESULTS FROM APPLICATION OF THE CLASS 1 ACCEPTANCE PROCEDURE

1.	**Name of substance**	:	5-tert-BUTYL-2,4,6-TRINITRO-m-XYLENE (MUSK XYLENE)
2.	**General data**		
2.1	Composition	:	99% tert-butyl-2,4,6-trinitro-m-xylene
2.2	Molecular formula	:	$C_{12}H_{15}N_3O_6$
2.3	Physical form	:	Fine crystalline powder
2.4	Colour	:	Pale yellow
2.5	Apparent density	:	840 kg /m^3
2.6	Particle size	:	< 1.7 mm
3.	**Box 2**	:	Is the substance manufactured with the view to producing a practical explosive or pyrotechnic effect?
3.1	Answer	:	No
3.2	Exit	:	Go to Box 3
4.	**Box 3**	:	Test Series 1
4.1	Propagation of Detonation	:	UN gap test (test 1(a))
4.2	Sample conditions	:	Ambient temperature
4.3	Observations	:	Fragmentation length 40 cm
4.4	Result	:	"+", propagation of detonation
4.5	Effect of heating under confinement	:	Koenen test (test 1(b))
4.6	Sample conditions	:	Mass 22.6 g
4.7	Observations	:	Limiting diameter 5.0 mm Fragmentation type "F" (time to reaction 52 s, duration of reaction 27 s)
4.8	Result	:	"+", shows some explosive effects on heating under confinement
4.9	Effect of ignition under confinement	:	Time/pressure test (test 1 (c) (i))
4.10	Sample conditions	:	Ambient temperature
4.11	Observations	:	No ignition
4.12	Result	:	"-", no effect on ignition under confinement
4.13	Exit	:	Go to Box 4
5.	**Box 4**	:	Is it an explosive substance?
5.1	Answer from Test Series 1	:	Yes
5.2	Exit	:	Go to box 5
6.	**Box 5**	:	Test Series 2
6.1	Sensitivity to shock	:	UN gap test (test 2(a))
6.2	Sample conditions	:	Ambient temperature
6.3	Observations	:	No propagation
6.4	Result	:	"-", not sensitive to shock
6.5	Effect of heating under confinement	:	Koenen test (test 2(b))
6.6	Sample conditions	:	Mass 22.6 g

6.7	Observations	:	Limiting diameter 5.0 mm Fragmentation type "F" (time to reaction 52 s, duration of reaction 27 s)
6.8	Result	:	"+", violent effect on heating under confinement
6.9	Effect of ignition under confinement	:	Time/pressure test (test 2 (c) (i))
6.10	Sample conditions	:	Ambient temperature
6.11	Observations	:	No ignition
6.12	Result	:	"-", no effect on ignition under confinement
6.13	Exit	:	Go to Box 6
7.	**Box 6**	:	Is the substance too insensitive for acceptance into Class 1?
7.1	Answer from Test Series 2	:	No
7.2	Conclusion	:	Substance to be considered for Class 1 (box 8)
7.3	Exit	:	Go to Box 9
8.	**Box 9**	:	Test Series 3
8.1	Thermal stability	:	75 °C/48 hour test (test 3 (c))
8.2	Sample conditions	:	100 g of substance at 75 °C
8.3	Observations	:	No ignition, explosion, self-heating or visible decomposition
8.4	Result	:	"-", thermally stable
8.5	Impact sensitivity	:	BAM fallhammer test (test 3 (a) (ii))
8.6	Sample conditions	:	as received
8.7	Observations	:	Limiting impact energy 25 J
8.8	Result	:	"-", not too dangerous to transport in form tested
8.9	Friction sensitivity	:	BAM friction test (test 3 (b) (i))
8.10	Sample conditions	:	as received
8.11	Observations	:	Limiting load > 360 N
8.12	Result	:	"-", not too dangerous to transport in form tested
8.13	Ease of deflagration to detonation transition	:	Small scale burning test (test 3 (d))
8.14	Sample conditions	:	Ambient temperature
8.15	Observations	:	Ignites and burns slowly
8.16	Result	:	"-", not too dangerous to transport in form tested
8.17	Exit	:	Go to box 10
9.	**Box 10**	:	Is the substance thermally stable?
9.1	Answer from test 3(c)	:	Yes
9.2	Exit	:	Go to box 11
10.	**Box 11**	:	Is the substance too dangerous for transport in the form in which it was tested?
10.1	Answer from Test Series 3	:	No
10.2	Exit	:	Go to box 18
11.	**Conclusion**	:	PROVISIONALLY ACCEPT INTO CLASS 1
11.1	Exit	:	Apply the Class 1 assignment procedure

Figure 10.7: PROCEDURE FOR PROVISIONAL ACCEPTANCE OF MUSK XYLENE IN CLASS 1

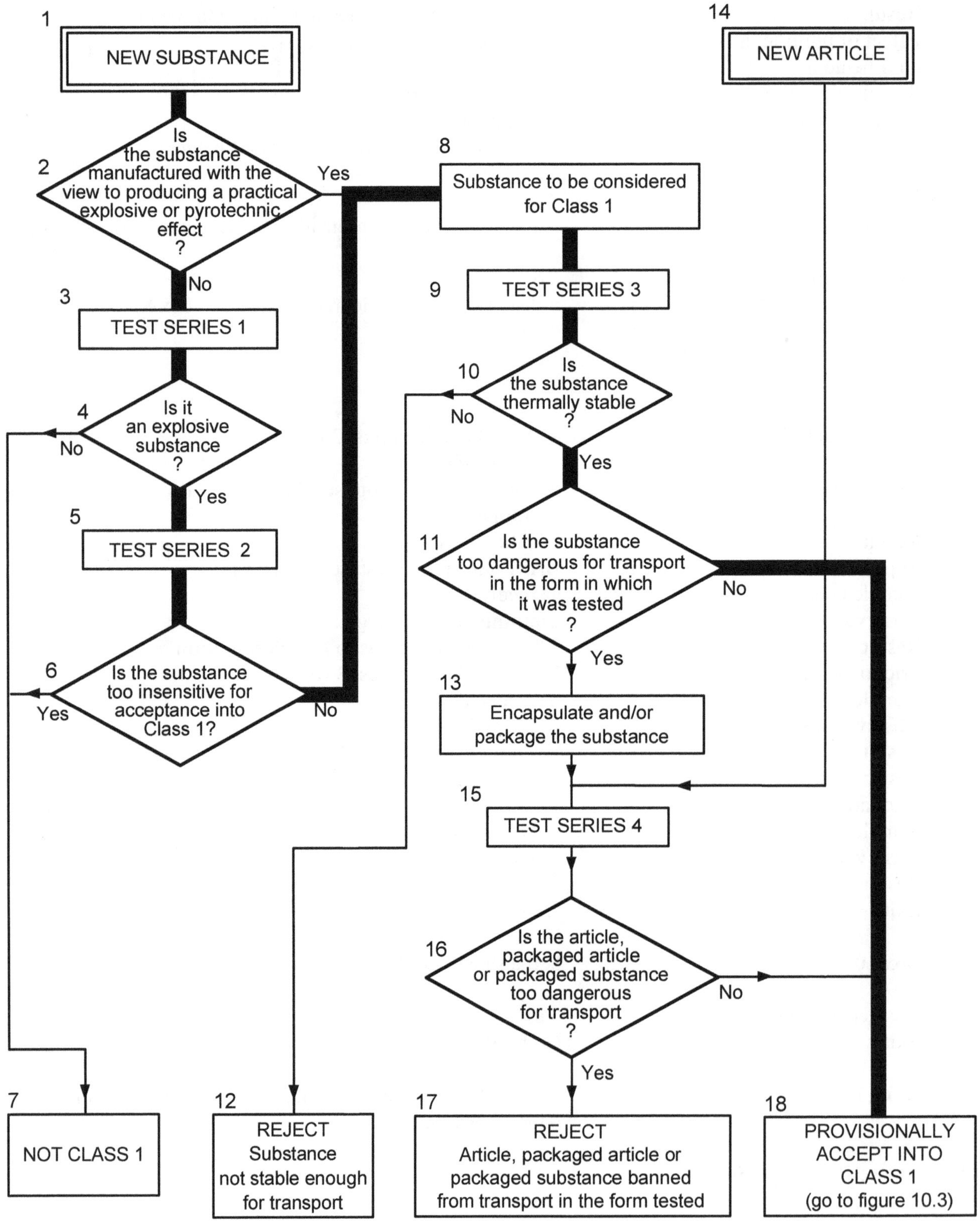

Figure 10.8: RESULTS FROM APPLICATION OF THE CLASS 1 ASSIGNMENT PROCEDURE

1.	**Box 19**	:	Is the substance a candidate for Division 1.5?
1.1	Answer	:	No
1.2	Result	:	Package the substance (box 23)
1.3	Exit	:	Go to box 25
2.	**Box 25**	:	Test Series 6
2.1	Effect of initiation in the package	:	Test 6(a) with detonator
2.2	Sample conditions	:	Ambient temperature, 50 kg fibreboard drum
2.3	Observations	:	Only localised decomposition around detonator
2.4	Result	:	No significant reaction
2.5	Effect of ignition in the package	:	Test 6(a) with igniter
2.6	Sample conditions	:	Ambient temperature, 50 kg fibreboard drum
2.7	Observations	:	Only localised decomposition around igniter
2.8	Result	:	No significant reaction
2.9	Effect of propagation between packages	:	Type 6(b) test not required as no effect outside package in 6(a) test
2.10	Effect of fire engulfment	:	Test 6(c)
2.11	Sample conditions	:	3 × 50 kg fibreboard drums mounted on steel frame above wooden crib fire
2.12	Observations	:	Only slow burning with black smoke occurred
2.13	Result	:	No effects which would hinder fire fighting
2.14	Exit	:	Go to box 26
3.	**Box 26**	:	Is the result a mass explosion?
3.1	Answer from Test Series 6	:	No
3.2	Exit	:	Go to box 28
4.	**Box 28**	:	Is the major hazard that from dangerous projections?
4.1	Answer from Test Series 6	:	No
4.2	Exit	:	Go to box 30
5.	**Box 30**	:	Is the major hazard radiant heat and/or violent burning but with no dangerous blast or projection hazard?
5.1	Answer from Test Series 6	:	No
5.2	Exit	:	Go to box 32
6.	**Box 32**	:	Is there nevertheless a small hazard in the event of ignition or initiation?
6.1	Answer from Test Series 6	:	No
6.2	Exit	:	Go to box 32a
7.	**Box 32a**	:	Does special provision 347 apply?
7.1.	Anwser	:	No
7.2.	Exit	:	Go to box 35
8.	**Box 35**	:	Is the substance or article manufactured with the view to producing a practical explosive or pyrotechnic effect?
8.1	Answer	:	No

8.2	Exit	:	Go to box 38
9.	**Conclusion**	:	NOT CLASS 1
9.1	Exit	:	Consider for another class/division

Figure 10.9: PROCEDURE FOR EXEMPTION OF MUSK XYLENE FROM CLASS 1

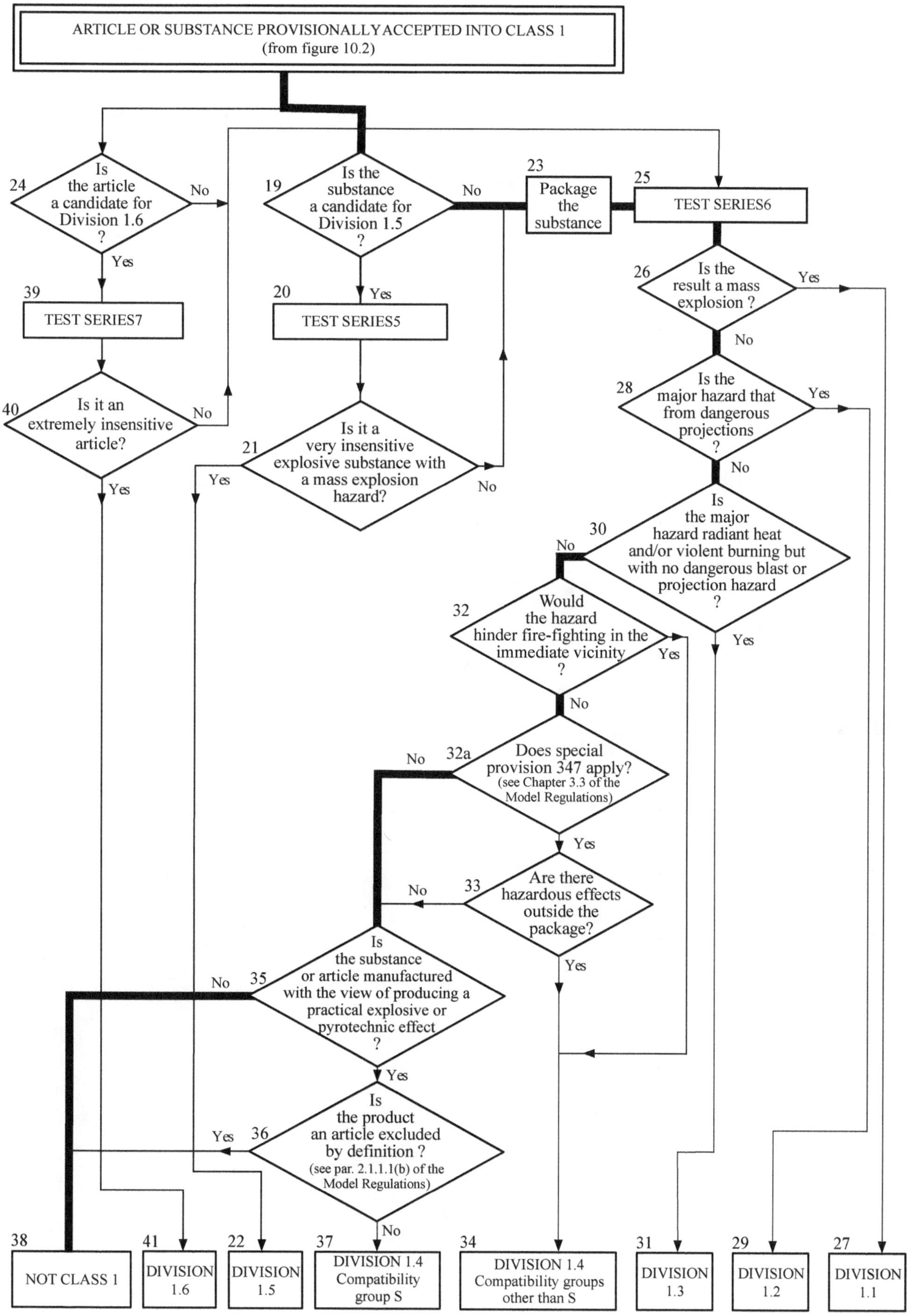

Figure 10.10: EXAMPLE OF A PROFORMA FOR A TEST REPORT FOR ARTICLES

Test method		Date of report		Data reference	
Product name		Lot number		Date of manufacture	

<u>CONSTRUCTION AND CONTENTS (attach drawings)</u>

<u>PACKAGING (if any)</u>

<u>PRETREATMENT OR CONDITIONING (if any)</u>

<u>TEST CONFIGURATION (including any variations or deviations from procedures described in the Manual)</u>

<u>TEST CONDITIONS</u>

Ambient temperature: °C Relative humidity: %

<u>OBSERVATIONS</u>

<u>TEST RESULT</u>

<u>CONCLUSION</u>

SECTION 11

TEST SERIES 1

11.1 Introduction

11.1.1 The question "Is it an explosive substance?" (box 4 of Figure 10.2) is answered on the basis of the results of three types of test to assess possible explosive effects. The question in box 4 is answered "yes" if a "+" is obtained in any of the three types of test.

11.2 Test methods

Test Series 1 is comprised of three types of test:

Type 1 (a): for determining propagation of detonation;
Type 1 (b): for determining the effect of heating under confinement; and
Type 1 (c): for determining the effect of ignition under confinement

The test methods currently used are listed in Table 11.1.

Table 11.1: TEST METHODS FOR TEST SERIES 1

Test code	Name of Test	Section
1 (a)	UN gap test [a]	11.4.1
1 (b)	Koenen test [a]	11.5.1
1 (c) (i)	Time/pressure test [a]	11.6.1
1 (c) (ii)	Internal ignition test	11.6.2

[a] *Recommended test*

11.3 Test conditions

11.3.1 As the apparent density of the substance has an important effect on the results from the type 1 (a) test, it should always be recorded. The apparent density of solids should be determined from measurement of the tube volume and sample mass.

11.3.2 If a mixture can separate out during transport, the test should be performed with the initiator in contact with the potentially most explosive part, if known.

11.3.3 The tests should be performed at ambient temperature unless the substance is to be transported under conditions where it may change its physical state or density.

11.3.4 If a liquid is being considered for transport in tank-containers, or intermediate bulk containers with a capacity exceeding 450 litres, a cavitated version of the type 1 (a) test should be performed (see special provision 26 of Chapter 3.3 of the Model Regulations).

11.3.5 For organic substances and mixtures of organic substances with a decomposition energy of 800 J/g or more, test 1 (a) need not be performed if the outcome of the ballistic mortar Mk.IIId test (F.1), or the ballistic mortar test (F.2) or the BAM Trauzl test (F.3) with initiation by a standard No. 8 detonator (see Appendix 1) is "No". In this case, the result of test 1 (a) is deemed to be "-". If the outcome of the F.1 or F.2 or F.3 test is "Low" or "Not low", the result of test 1 (a) shall be deemed "+". In this case, a "-" can only be obtained by performing test 1 (a).

11.4 Series 1 type (a) test prescription

11.4.1 *Test 1 (a): UN gap test*

11.4.1.1 *Introduction*

This test is used to measure the ability of a substance, under confinement in a steel tube, to propagate a detonation by subjecting it to the detonation from a booster charge.

11.4.1.2 *Apparatus and materials*

11.4.1.2.1 Solids

The apparatus for solids is shown in Figure 11.4.1.1. The test sample is contained in a cold-drawn, seamless, carbon steel tube with an external diameter of 48 ± 2 mm, a wall thickness of 4 mm and a length of 400 ± 5 mm. If the test substance may react with the steel, the inside of the tube may be coated with fluorocarbon resin. The bottom of the tube is closed with a plastics sheet pulled tightly (so that it plastically deforms) over the bottom of the tube and held tightly in place. The plastics sheet shall be compatible with the substance under test. The booster charge consists of 160 g RDX/wax (95/5) or PETN/TNT that has a minimum of 50% PETN in the mixture, 50 ± 1 mm in diameter with a density of 1 600 ± 50 kg/m^3. The charges may be pressed in one or more pieces, as long as the total charge is within the specifications, and the PETN/TNT charge is cast. A mild steel witness plate, 150 ± 10 mm square and 3 mm thick, may be mounted at the upper end of the steel tube and separated from it by spacers 1.6 ± 0.2 mm thick.

11.4.1.2.2 Liquids

The apparatus for liquids is the same as that for solids. When a cavitated version of the test is performed (see 11.3.4), one of the methods of cavitation given in Appendix 3 may be used.

11.4.1.3 *Procedure*

11.4.1.3.1 The sample is loaded to the top of the steel tube. Solid samples are loaded to the density attained by tapping the tube until further settling becomes imperceptible. The sample mass is determined and, if solid, the apparent density calculated using the measured internal volume of the tube.

11.4.1.3.2 The tube is placed in a vertical position and the booster charge is placed in direct contact with the sheet which seals the bottom of the tube. The detonator is fixed in place against the booster charge and initiated. Two tests should be performed unless detonation of the substance is observed.

11.4.1.4 *Test criteria and method of assessing results*

The test results are assessed on the basis of the type of fragmentation of the tube or on whether a hole is punched through the witness plate. The test giving the most severe assessment should be used for classification. The test result is considered "+" and the substance to propagate detonation if:

- The tube is fragmented completely; or
- A hole is punched through the witness plate.

Any other result is considered "-" and the substance is considered not to be able to propagate a detonation.

11.4.1.5 *Examples of results*

Substances	Apparent density (kg/m^3)	Fragmentation length (cm)	Witness plate	Result
Ammonium nitrate, prills	800	40	Domed	+
Ammonium nitrate, 200 µm	540	40	Holed	+
Ammonium nitrate/fuel oil, 94/6	880	40	Holed	+
Ammonium perchlorate, 200 µm	1 190	40	Holed	+
Nitromethane	1 130	40	Holed	+
Nitromethane/methanol, 55/45	970	20	Domed	-
PETN/lactose, 20/80	880	40	Holed	+
PETN/lactose, 10/90	830	17	No damage	-
TNT, cast	1 510	40	Holed	+
TNT, flaked	710	40	Holed	+
Water	1 000	<40	Domed	-

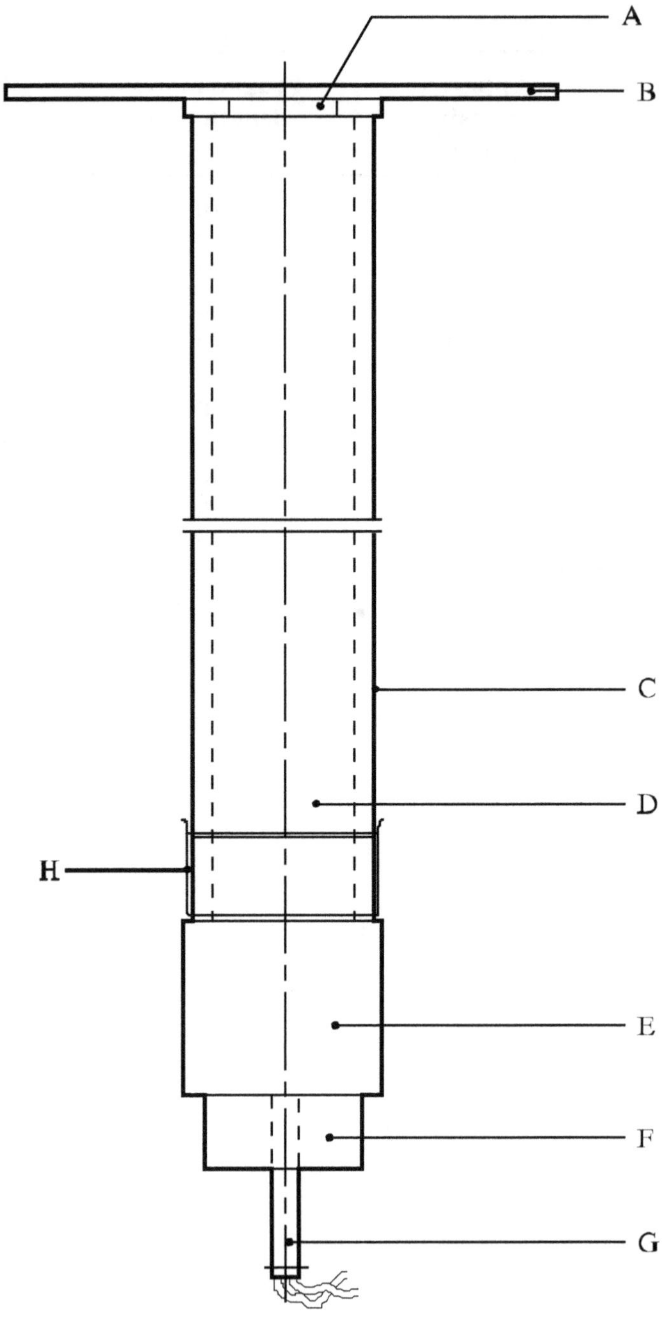

(A)	Spacers	(B)	Witness plate
(C)	Steel tube	(D)	Substance under investigation
(E)	RDX/wax or PETN/TNT booster charge	(F)	Detonator holder
(G)	Detonator	(H)	Plastics membrane

Figure 11.4.1.1: UN GAP TEST

11.5 Series 1 type (b) test prescription

11.5.1 *Test 1 (b): Koenen test*

11.5.1.1 *Introduction*

This test is used to determine the sensitiveness of solid and liquid substances to the effect of intense heat under high confinement.

11.5.1.2 *Apparatus and materials*

11.5.1.2.1 The apparatus consists of a non-reusable steel tube, with its re-usable closing device, installed in a heating and protective device. The tube is deep drawn from sheet steel conforming to specification DC04 (EN 10027-1), or equivalent A620 (AISI/SAE/ASTM), or equivalent SPCEN (JIS g 3141). The dimensions are given in Figure 11.5.1.1. The open end of the tube is flanged. The closing plate with an orifice, through which the gases from the decomposition of the test substance escape, is made from heat-resisting chrome steel and is available with the following diameter holes: 1.0 - 1.5 - 2.0 - 2.5 - 3.0 - 5.0 - 8.0 - 12.0 - 20.0 mm. The dimensions of the threaded collar and the nut (closing device) are given in Figure 11.5.1.1.

For quality control of the steel tubes, 1% of the tubes from each production lot shall be subjected to quality control and the following data shall be verified:

(a) The mass of the tubes shall be 26.5 ± 1.5 g, tubes to be used in one test sequence shall not differ in mass by more than 1 g;

(b) The length of the tubes shall be 75 ± 0.5 mm;

(c) The wall thickness of the tubes measured 20 mm from the bottom of the tube shall be 0.5 ± 0.05 mm; and

(d) The bursting pressure as determined by quasi-static load through an incompressible fluid shall be 30 ± 3 MPa.

11.5.1.2.2 Heating is provided by propane, from an industrial cylinder fitted with a pressure regulator, via a flow meter and distributed by a manifold to the four burners. Other fuel gases may be used provided the specified heating rate is obtained. The gas pressure is regulated to give a heating rate of 3.3 ± 0.3 K/s when measured by the calibration procedure. Calibration involves heating a tube (fitted with a 1.5 mm orifice plate) filled with 27 cm^3 of dibutyl phthalate. The time taken for the temperature of the liquid (measured with a 1 mm diameter thermocouple centrally placed 43 mm below the rim of the tube) to rise from 135 °C to 285 °C is recorded and the heating rate calculated.

11.5.1.2.3 Because the tube is likely to be destroyed in the test, heating is undertaken in a protective welded box, the construction and dimensions of which are given in Figure 11.5.1.2. The tube is suspended between two rods placed through holes drilled in opposite walls of the box. The arrangement of the burners is given in Figure 11.5.1.2. The burners are lit simultaneously by a pilot flame or an electrical ignition device. ***The test apparatus is placed in a protective area.*** Measures should be taken to ensure that the burner flames are not affected by any draughts. Provision should be made for extracting any gases or smoke resulting from the test.

11.5.1.3 *Procedure*

11.5.1.3.1 Normally substances are tested as received, although in certain cases it may be necessary to test the substance after crushing it. For solids, the mass of material to be used in each test is determined using a two-stage dry run procedure. A tared tube is filled with 9 cm^3 of substance and the substance tamped[1] with 80 N force applied to the total cross-section of the tube. If the material is compressible then more is added and tamped until the tube is filled to 55 mm from the top. The total mass used to fill the tube to the 55 mm level is determined and two further increments, each tamped with 80 N force, are added. Material is then either added, with tamping, or taken out as required to leave the tube filled to a level 15 mm from the top.

A second dry run is performed, starting with a tamped increment a third of the total mass found in the first dry run. Two more of these increments are added with 80 N tamping and the level of the substance in the tube adjusted to 15 mm from the top by addition or subtraction of material as required. The amount of solid determined in the second dry run is used for each trial filling being performed in three equal increments, each compressed to 9 cm^3. (This may be facilitated by the use of spacing rings.) Liquids and gels are loaded into the tube to a height of 60 mm taking particular care with gels to prevent the formation of voids. The threaded collar is slipped onto the tube from below, the appropriate orifice plate is inserted and the nut tightened by hand after applying some molybdenum disulphide based lubricant. It is essential to check that none of the substance is trapped between the flange and the plate, or in the threads.

11.5.1.3.2 With orifice plates from 1.0 mm to 8.0 mm diameter, nuts with an orifice of 10.0 mm diameter should be used; if the diameter of the orifice is above 8.0 mm, that of the nut should be 20.0 mm. Each tube is used for one trial only. The orifice plates, threaded collars and nuts may be used again provided they are undamaged.

11.5.1.3.3 The tube is placed in a rigidly mounted vice and the nut tightened with a spanner. The tube is then suspended between the two rods in the protective box. The test area is vacated, the gas supply turned on and the burners lit. The time to reaction and duration of reaction can provide additional information useful in interpreting the results. If rupture of the tube does not occur, heating is to be continued for at least five minutes before the trial is finished. After each trial the fragments of the tube, if any, should be collected and weighed.

11.5.1.3.4 The following effects are differentiated:

"O": Tube unchanged;
"A": Bottom of tube bulged out;
"B": Bottom and wall of the tube bulged out;
"C": Bottom of tube split;
"D": Wall of tube split;
"E": Tube split into two[2] fragments;
"F": Tube fragmented into three[2] or more mainly large pieces which in some cases may be connected with each other by a narrow strip;
"G": Tube fragmented into many mainly small pieces, closing device undamaged; and
"H": Tube fragmented into many very small pieces, closing device bulged out or fragmented.

Examples for the effect types "D", "E" and "F" are shown in Figure 11.5.1.3. If a trial results in any of the effects "O" to "E", the result is regarded as "no explosion". If a trial gives the effect "F", "G" or "H", the result is evaluated as "explosion".

---

1 ***For reasons of safety, e.g. the substance is friction sensitive, the substance need not be tamped.*** *In cases where the physical form of the sample can be changed by compression or compression of the sample is not related to the transport conditions, e.g. for fibrous materials, more representative filling procedures may be used.*

2 *The upper part of the tube remaining in the closing device is counted as one fragment.*

11.5.1.3.5 The series of trials is started with a single trial using an orifice plate of 20.0 mm. If, in this trial, the result "explosion" is observed, the series is continued with trials using tubes without orifice plates and nuts but with threaded collars (orifice 24.0 mm). If at 20.0 mm "no explosion" occurs, the series is continued with single trials using plates with the following orifices 12.0 - 8.0 - 5.0 - 3.0 - 2.0 - 1.5 and finally 1.0 mm until, at one of these diameters, the result "explosion" is obtained. Subsequently, trials are carried out at increasing diameters, according to the sequence given in 11.5.1.2.1, until only negative results in three tests at the same level are obtained. The limiting diameter of a substance is the largest diameter of the orifice at which the result "explosion" is obtained. If no "explosion" is obtained with a diameter of 1.0 mm, the limiting diameter is recorded as being less than 1.0 mm.

11.5.1.4 *Test criteria and method of assessing results*

The result is considered "+" and the substance to show some effect on heating under confinement if the limiting diameter is 1.0 mm or more. The result is considered "-" and the substance to show no effect on heating under confinement if the limiting diameter is less than 1.0 mm.

11.5.1.5 *Examples of results*

Substances	**Limiting diameter (mm)**	**Result**
Ammonium nitrate (crystalline)	1.0	+
Ammonium nitrate (high density prills)	1.0	+
Ammonium nitrate (low density prills)	1.0	+
Ammonium perchlorate	3.0	+
1,3-Dinitrobenzene (crystalline)	< 1.0	-
2,4-Dinitrotoluene (crystalline)	<1.0	-
Guanidine nitrate (crystalline)	1.5	+
Nitroguanidine (crystalline)	1.0	+
Nitromethane	<1.0	-
Urea nitrate (crystalline)	<1.0	-

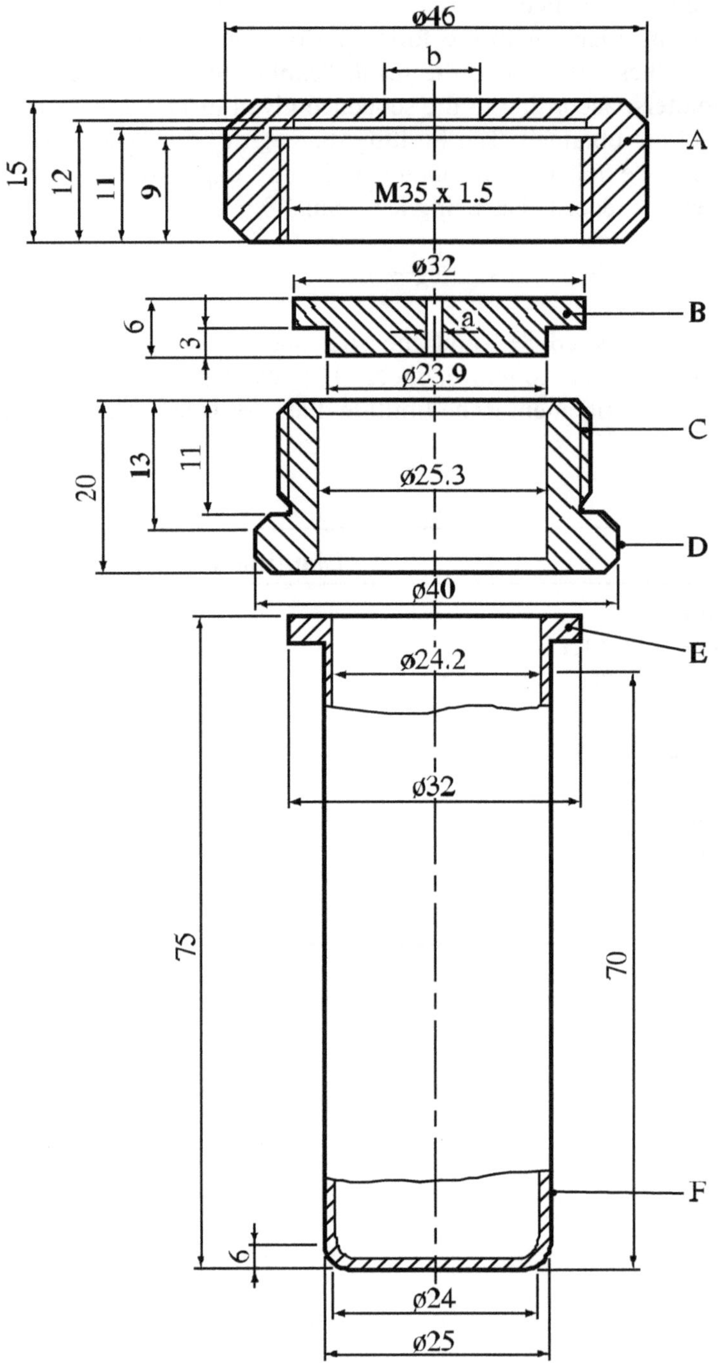

(A)	Nut (b = 10.0 or 20.0 mm) with flats for size 41 spanner	(B)	Orifice plate (a = 1.0 → 20.0 mm diameter)
(C)	Threaded collar	(D)	Flats for size 36 spanner
(E)	Flange	(F)	Tube

Figure 11.5.1.1: TEST TUBE ASSEMBLY

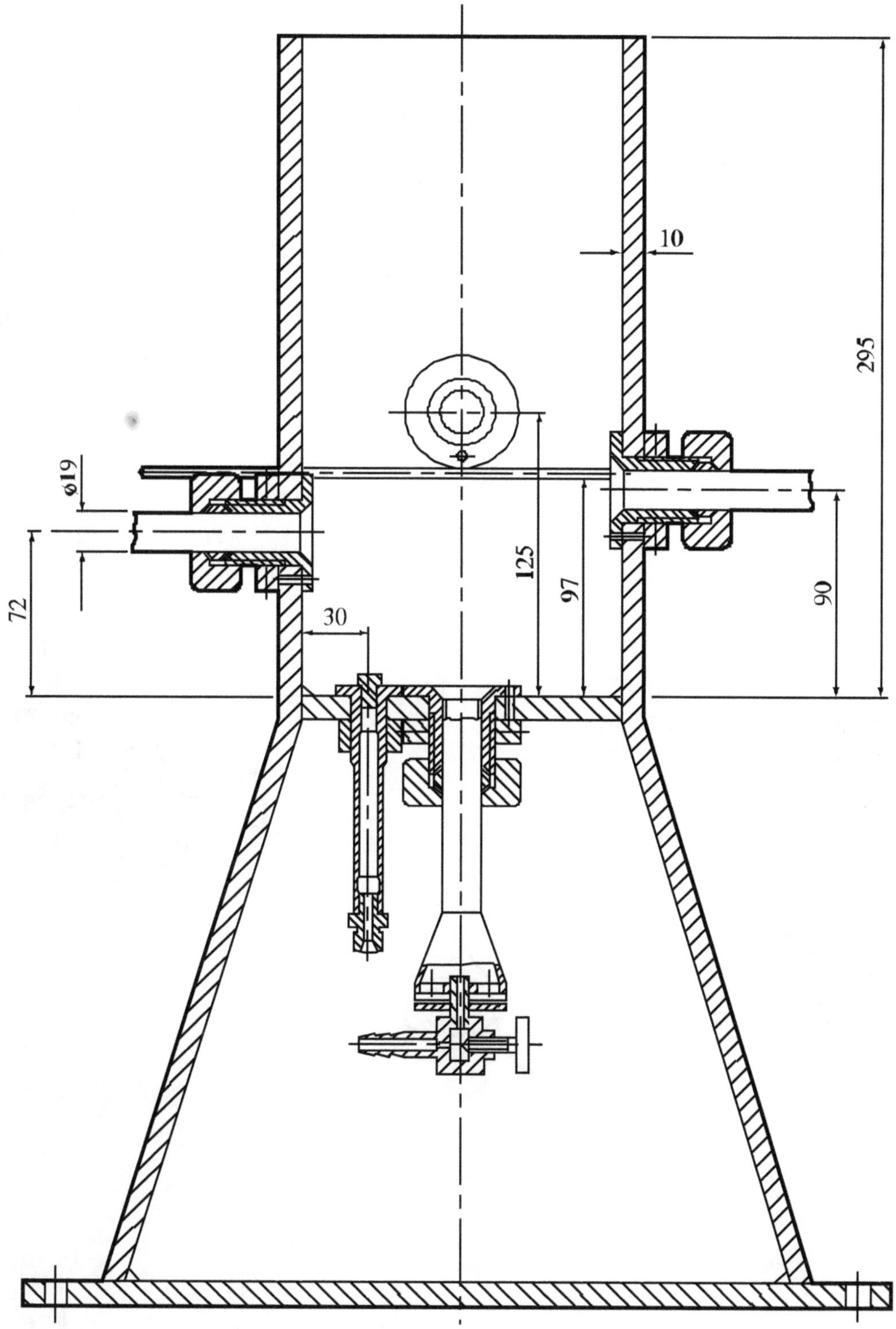

Figure 11.5.1.2: HEATING AND PROTECTIVE DEVICE

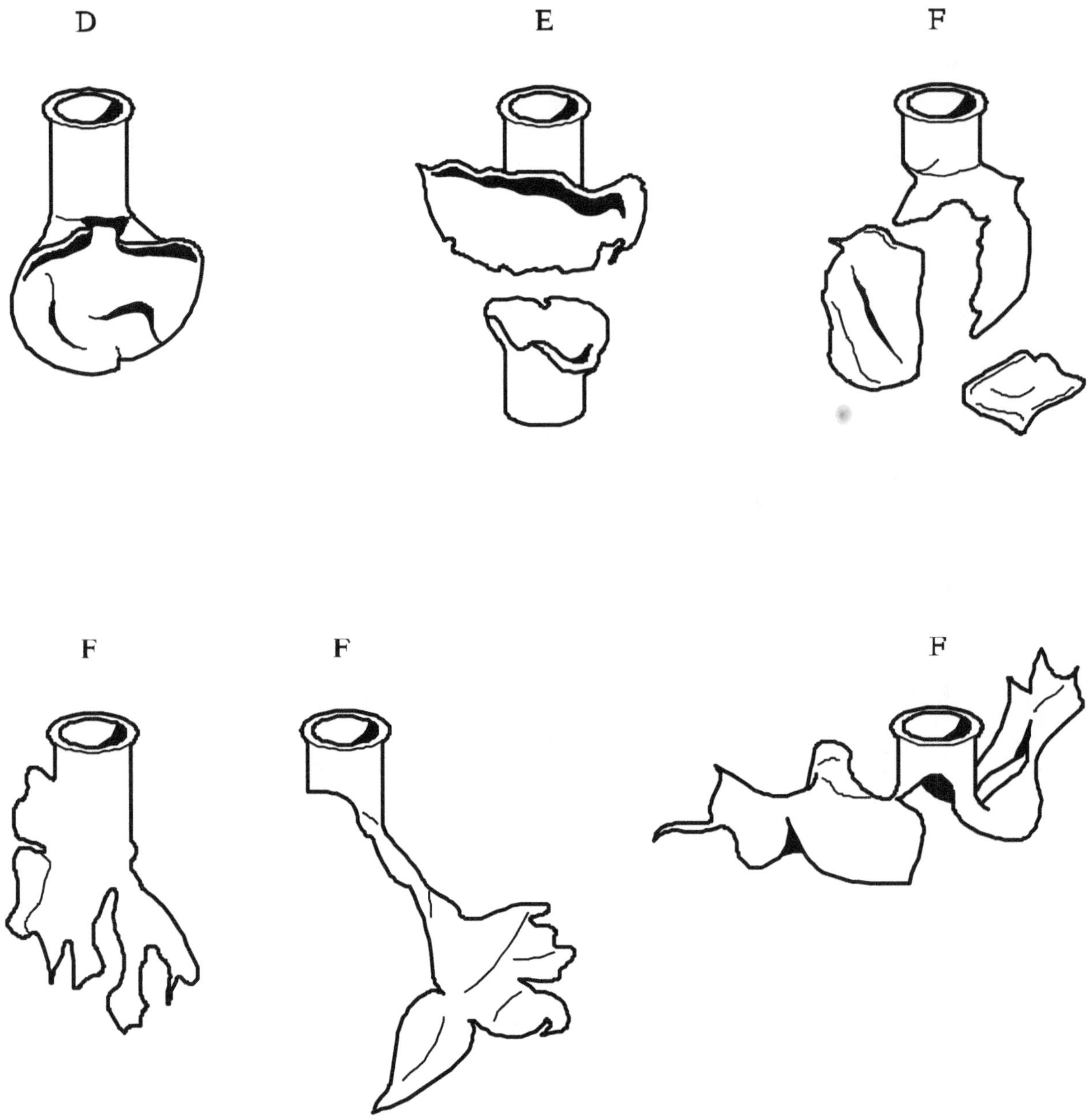

Figure 11.5.1.3: EXAMPLES OF EFFECT TYPES D, E AND F

11.6 Series 1 type (c) test prescriptions

11.6.1 *Test 1 (c) (i): Time/pressure test*

11.6.1.1 *Introduction*

This test is used to determine the effects of igniting the substance[3] under confinement in order to determine if ignition leads to a deflagration with explosive violence at pressures which can be attained with substances in normal commercial packages.

11.6.1.2 *Apparatus and materials*

11.6.1.2.1 The time/pressure apparatus (Figure 11.6.1.1) consists of a cylindrical steel pressure vessel 89 mm in length and 60 mm in external diameter. Two flats are machined on opposite sides (reducing the cross-section of the vessel to 50 mm) to facilitate holding whilst fitting the firing plug and vent plug. The vessel, which has a bore of 20 mm diameter, is internally rebated at either end to a depth of 19 mm and threaded to accept 1" British Standard Pipe (BSP). A pressure take-off, in the form of a side-arm, is screwed into the curved face of the pressure vessel 35 mm from one end and at 90° to the machined flats. The socket for this is bored to a depth of 12 mm and threaded to accept the 1/2" BSP thread on the end of the side-arm. A washer is fitted to ensure a gastight seal. The side-arm extends 55 mm beyond the pressure vessel body and has a bore of 6 mm. The end of the side-arm is rebated and threaded to accept a diaphragm type pressure transducer. Any pressure-measuring device may be used provided that it is not affected by the hot gases or decomposition products and is capable of responding to rates of pressure rise of 690-2 070 kPa in not more than 5 ms.

11.6.1.2.2 The end of the pressure vessel furthest from the side-arm is closed with a firing plug which is fitted with two electrodes, one insulated from, and the other earthed to, the plug body. The other end of the pressure vessel is closed by an aluminium bursting disc 0.2 mm thick (bursting pressure approximately 2 200 kPa) held in place with a retaining plug which has a 20 mm bore. A soft lead washer or a washer of a suitable deformable material (for example, polyoxymethylene) is used with both plugs to ensure a good seal. A support stand (Figure 11.6.1.2) holds the assembly in the correct attitude during use. This comprises a mild steel base plate measuring 235 mm × 184 mm × 6 mm and a 185 mm length of square hollow section (S.H.S.) 70 × 70 × 4 mm.

11.6.1.2.3 A section is cut from each of two opposite sides at one end of the length of S.H.S. so that a structure having two flat sided legs surmounted by an 86 mm length of intact box section results. The ends of these flat sides are cut to an angle of 60° to the horizontal and welded to the base plate.

11.6.1.2.4 A slot measuring 22 mm wide × 46 mm deep is machined in one side of the upper end of the base section such that when the pressure vessel assembly is lowered, firing plug end first, into the box section support, the side-arm is accommodated in this slot. A packing piece of steel 30 mm wide and 6 mm thick is welded to the lower internal face of the box section to act as a spacer. Two 7 mm thumb screws, tapped into the opposite face, serve to hold the pressure vessel firmly in place. Two 12 mm wide strips of 6 mm thick steel, welded to the side pieces abutting the base of the box section, support the pressure vessel from beneath.

11.6.1.2.5 The ignition system consists of an electric fusehead of the type commonly used in low tension detonators, together with a 13 mm square piece of primed cambric. Fuseheads with equivalent properties may be used. Primed cambric consists of a linen fabric coated on both sides with a potassium nitrate/silicon/sulphurless gunpowder pyrotechnic composition[4].

---

[3] *When testing thermally stable energetic liquids, such as nitromethane (UN 1261), variable results may be obtained because the substance may give two pressure peaks.*

[4] *Obtainable from the national contact for test details in United Kingdom (see Appendix 4).*

11.6.1.2.6 The procedure for the preparation of the ignition assembly for solids starts with separation of the brass foil contacts of an electric fusehead from its insulator, (see Figure 11.6.1.3). The exposed portion of insulation is then cut off. The fusehead is then fixed onto the terminals of the firing plug by means of the brass contacts such that the tip of the fusehead is 13 mm above the surface of the firing plug. A 13 mm square piece of primed cambric is pierced through the centre and positioned over the attached fusehead around which it is then folded and secured with fine cotton thread.

11.6.1.2.7 For liquids samples, leads are fixed onto the contact foils of the fusehead. The leads are then threaded through an 8 mm length of 5 mm outer diameter and 1 mm inner diameter silicone rubber tubing and the tubing is pushed up over the fusehead contact foils as shown in Figure 11.6.1.4. The primed cambric is then wrapped around the fusehead and a single piece of thin PVC sheathing, or equivalent, is used to cover the primed cambric and the silicone rubber tubing. The sheathing is sealed in position by twisting a length of thin wire tightly round the sheathing and rubber tubing. The leads are then fixed onto the terminals of the firing plug such that the tip of the fusehead is 13 mm above the surface of the firing plug.

11.6.1.3 *Procedure*

11.6.1.3.1 The apparatus, assembled complete with pressure transducer but without the aluminium bursting disc in position, is supported firing plug end down. 5.0 g[5] of the substance is introduced into the apparatus so as to be in contact with the ignition system. Normally no tamping is carried out when filling the apparatus unless it is necessary to use light tamping in order to get the 5.0 g charge into the vessel. If, even with light tamping, it is impossible to get all the 5.0 g of sample in, then the charge is fired after filling the vessel to capacity. Note should be taken of the charge weight used. The lead washer and aluminium bursting disc are placed in position and the retaining plug is screwed in tightly. The charged vessel is transferred to the firing support stand, bursting disc uppermost, which should be contained in a suitable, armoured fume cupboard or firing cell. An exploder dynamo is connected to the external terminals of the firing plug and the charge is fired. The signal produced by the pressure transducer is recorded on a suitable system which allows both evaluation and a permanent record of the time/pressure profile to be achieved (e.g. transient recorder coupled to a chart-recorder).

11.6.1.3.2 The test is carried out three times. The time taken for the pressure to rise from 690 kPa to 2 070 kPa above atmospheric is noted. The shortest time interval should be used for classification.

11.6.1.4 *Test criteria and method of assessing results*

The test results are interpreted in terms of whether a gauge pressure of 2 070 kPa is reached and, if so, the time taken for the pressure to rise from 690 kPa to 2 070 kPa gauge.

The result is considered "+" and the substance to show the ability to deflagrate if the maximum pressure reached is greater than or equal to 2 070 kPa. The result is considered "-" and the substance to show no likelihood of deflagration if the maximum pressure reached in any one test is less than 2 070 kPa gauge. Failure to ignite does not necessarily indicate that the substance has no explosive properties.

---

5 ***If preliminary safety-in-handling tests (e.g. heating in a flame) or unconfined burning tests (e.g. a Series 3 type (d) test) indicate that a rapid reaction is likely to occur, then the sample size should be reduced to 0.5 g until the severity of the confined reaction is known.*** *If it is necessary to use a 0.5 g sample, the sample size should be gradually increased until either a "+" result is obtained or the test is performed with a 5.0 g sample.*

11.6.1.5 *Examples of results*

Substances	**Maximum pressure (kPa)**	**Time for a pressure rise from 690 to 2 070 kPa (ms)**	**Result**
Ammonium nitrate (high density prills)	< 2 070	-	-
Ammonium nitrate (low density prills)	< 2 070	-	-
Ammonium perchlorate (2 μm)	> 2 070	5	+
Ammonium perchlorate (30 μm)	> 2 070	15	+
Barium azide	> 2 070	<5	+
Guanidine nitrate	> 2 070	606	+
Isobutyl nitrite	> 2 070	80	+
Isopropyl nitrate	> 2 070	10	+
Nitroguanidine	> 2 070	400	+
Picramic acid	> 2 070	500	+
Sodium picramate	> 2 070	15	+
Urea nitrate	> 2 070	400	+

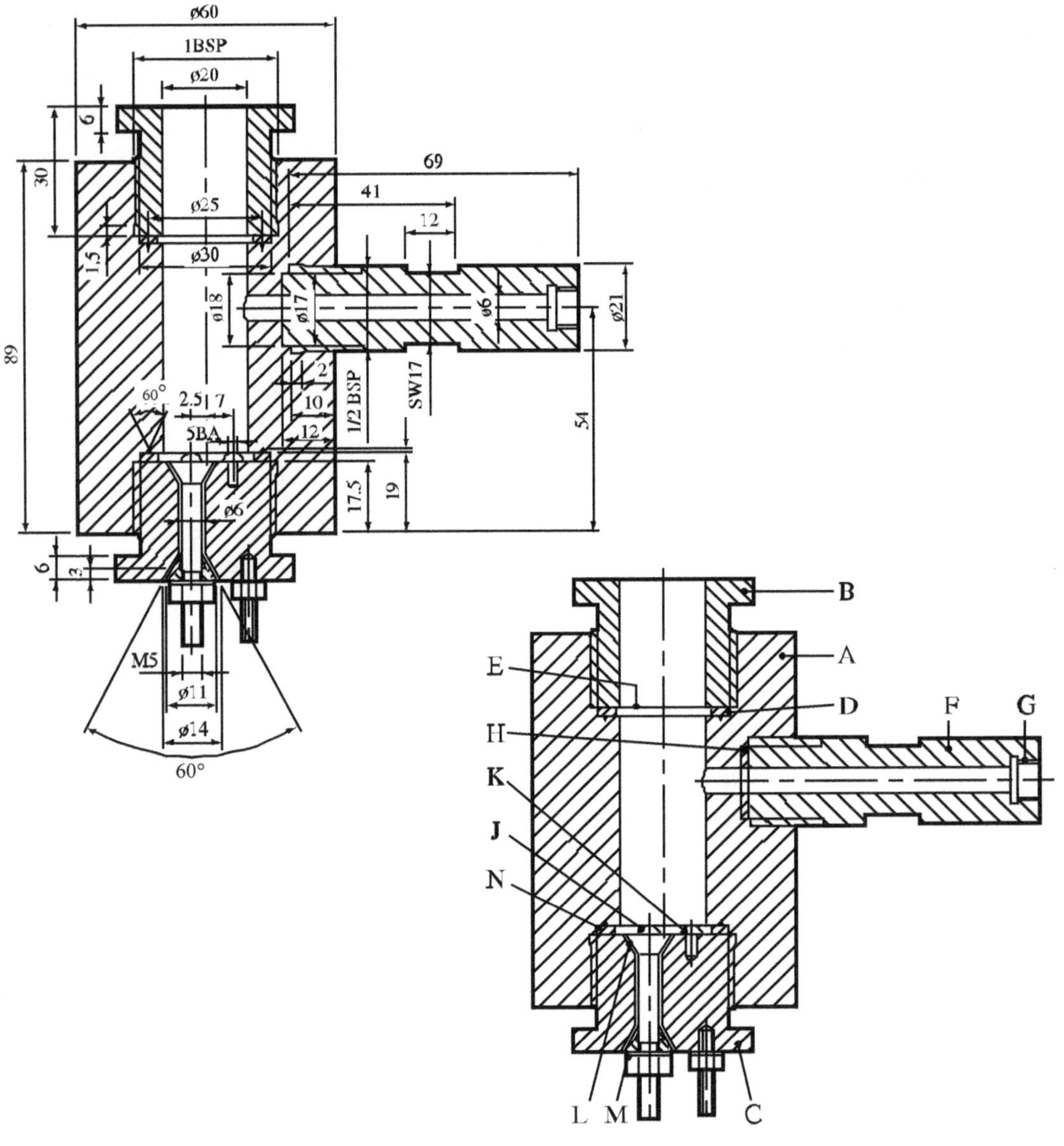

(A)	Pressure vessel body	(B)	Bursting disc retaining plug
(C)	Firing plug	(D)	Soft lead washer
(E)	Bursting disc	(F)	Side arm
(G)	Pressure transducer thread	(H)	Copper washer
(J)	Insulated Electrode	(K)	Earthed electrode
(L)	Insulation	(M)	Steel cone
(N)	Washer distorting groove		

Figure 11.6.1.1: APPARATUS

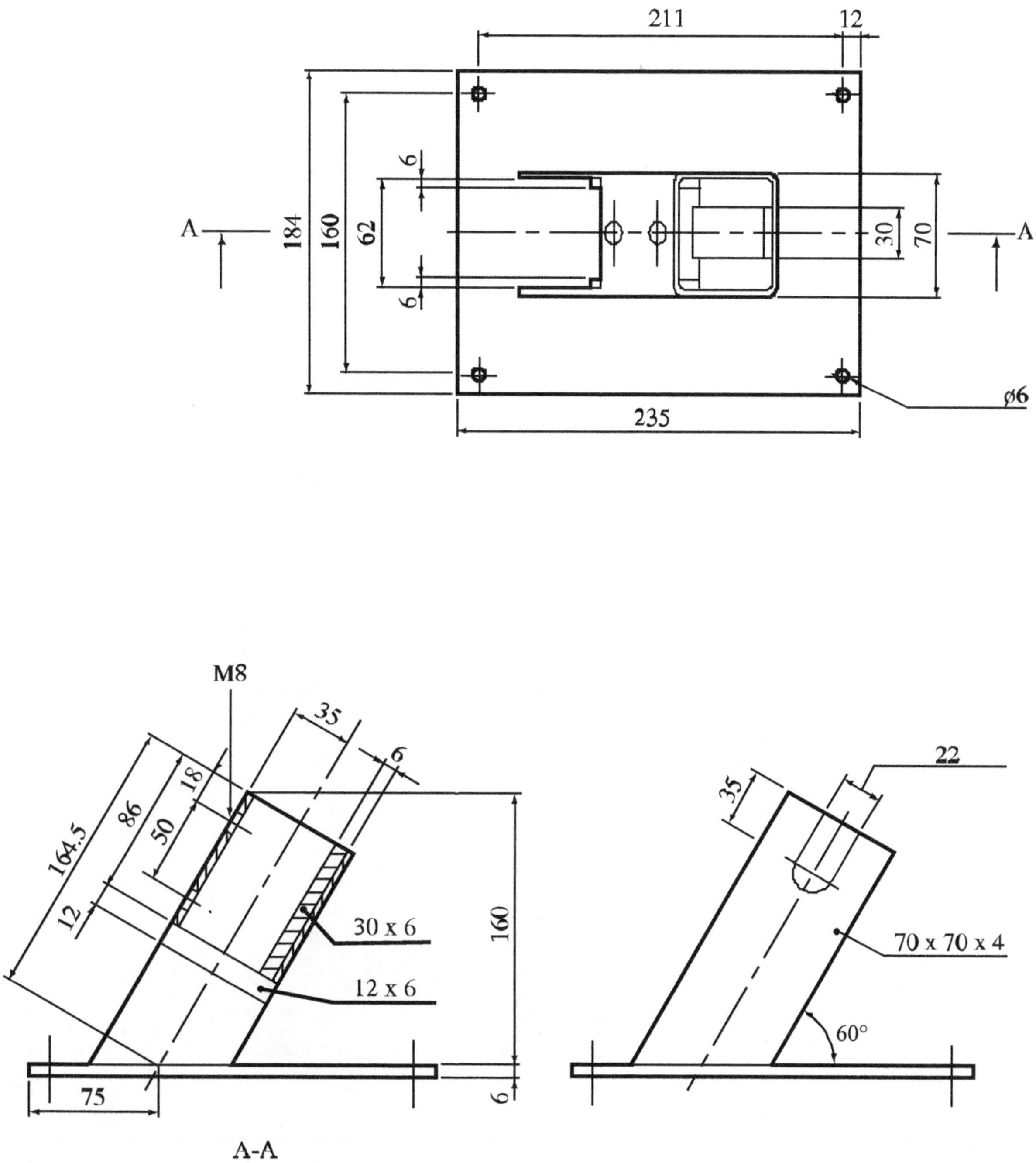

Figure 11.6.1.2: SUPPORT STAND

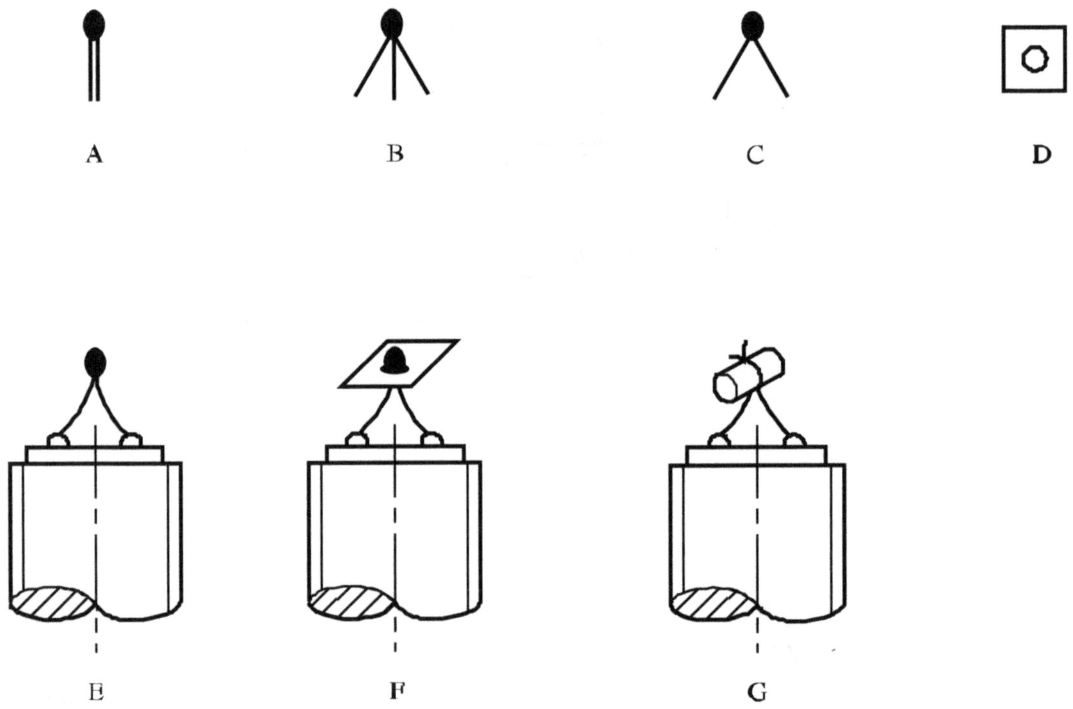

(A) Electrically ignited fusehead as manufactured
(B) Brass foil contacts parted from card insulator
(C) Insulating card cut off
(D) Primed cambric SR252 13 mm square with centre hole
(E) Fusehead fixed to pins on firing plug
(F) Cambric positioned on fusehead
(G) Cambric folded around and tied with thread

Figure 11.6.1.3: IGNITION SYSTEM FOR SOLIDS

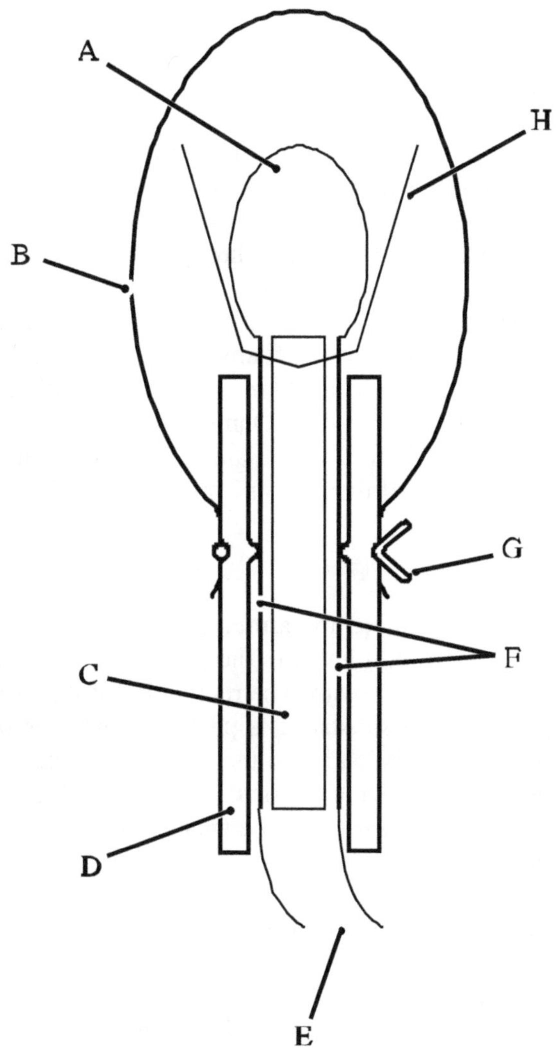

(A)	Fusehead
(B)	PVC sheath
(C)	Insulating card
(D)	Silicone rubber tubing
(E)	Firing leads
(F)	Foil contacts
(G)	Wire to make liquid-tight seal
(H)	Primed cambric

Figure 11.6.1.4: IGNITION SYSTEM FOR LIQUIDS

11.6.2 *Test 1 (c) (ii): Internal ignition test*

11.6.2.1 *Introduction*

This test is used to determine the tendency of a substance to undergo transition from deflagration to detonation.

11.6.2.2 *Apparatus and materials*

The experimental arrangement is shown in Figure 11.6.2.1. The sample of substance to be tested is contained in a 45.7 cm length of "3 inch schedule 80" carbon (A53 Grade B) steel pipe with inside diameter 74 mm, wall thickness 7.6 mm, capped at both ends with a "3 000 pound" forged steel pipe cap. An igniter consisting of 20 g of black powder (100% passed through No. 20 sieve, 0.84 mm, and 100% retained by No. 50 sieve, 0.297 mm) is located at the centre of the sample vessel. The igniter assembly consists of a cylindrical container 21 mm in diameter and 64 mm long which is made from 0.54 mm thick cellulose acetate which is held together by two layers of nylon filament reinforced cellulose acetate tape. The igniter capsule contains a small loop formed from a 25 mm length of nickel-chromium alloy resistance wire 0.30 mm in diameter having a resistance of 0.35 Ohms. This loop is attached to two insulated tinned copper lead wires 0.7 mm in diameter. The overall wire diameter, including insulation, is 1.3 mm. These lead wires are fed through small holes in the wall of the pipe and are sealed with epoxy resin.

11.6.2.3 *Procedure*

After the sample, at ambient temperature, is loaded into the pipe to a height of 23 cm, the igniter (with its leads inserted through small holes in the pipe wall) is inserted into the centre of the pipe and the leads pulled taut and sealed with epoxy resin. The remainder of the sample is then loaded and the top cap screwed on. For gelatinous samples, the substance is packed as near as possible to its normal shipping density. For granular samples, the substance is loaded to the density obtained by repeated tapping of the pipe against a hard surface. The tube is placed in a vertical position and the igniter is fired by a current of 15 amperes obtained from a 20-volt transformer. Three trials should be performed unless deflagration to detonation transition occurs earlier.

11.6.2.4 *Test criteria and method of assessing results*

The test result is considered "+" if either the pipe or at least one of the end caps is fragmented into at least two distinct pieces. Results in which the pipe is merely split or laid open, or in which the pipe or caps are distorted to the point at which the caps are blown off, are considered "-".

11.6.2.5 *Examples of results*

Substance	**Results**
Ammonium nitrate/fuel oil, aluminised	+
Ammonium nitrate prills, porous, low density	-
Ammonium perchlorate (45 µm)	+
Nitrocarbonitrate	-
TNT, granular	+
Watergel	+

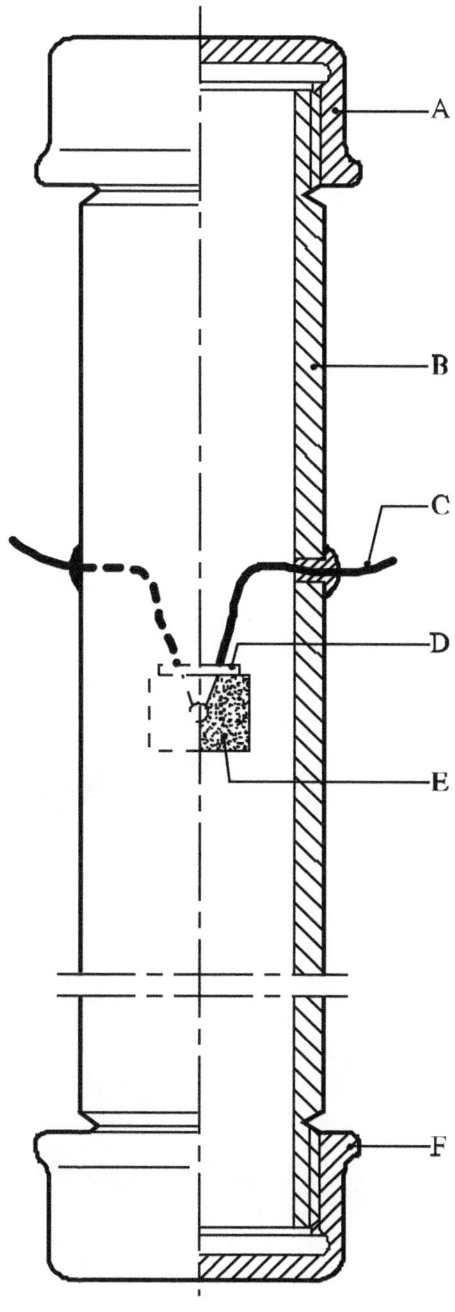

(A)	Forged steel cap	(B)	Steel pipe
(C)	Igniter leads	(D)	Seal
(E)	Igniter assembly	(F)	Forged steel cap

Figure 11.6.2.1: INTERNAL IGNITION TEST

SECTION 12

TEST SERIES 2

12.1 Introduction

12.1.1 The question "Is the substance too insensitive for inclusion in Class 1?" (box 6 of Figure 10.2) is answered on the basis of three types of test to assess possible explosive effects. The question in box 6 is answered "no" if a "+" is obtained in any of the three types of test.

12.2 Test methods

Test Series 2 comprises three types of test:

Type 2 (a): for determining sensitivity to shock;
Type 2 (b): for determining the effect of heating under confinement; and
Type 2 (c): for determining the effect of ignition under confinement

The test methods currently used are listed in table 12.1.

Table 12.1: TEST METHODS FOR TEST SERIES 2

Test code	Name of Test	Section
2 (a)	UN gap test[a]	12.4.1
2 (b)	Koenen test[a]	12.5.1
2 (c) (i)	Time/pressure test[a]	12.6.1
2 (c) (ii)	Internal ignition test	12.6.2

[a] *Recommended test.*

12.3 Test conditions

12.3.1 As the apparent density of the substance has an important effect on the results from the type 2 (a) test, it should always be recorded. The apparent density of solids should be determined from measurement of the tube volume and sample mass.

12.3.2 If a mixture can separate out during transport, the test should be performed with the initiator in contact with the potentially most explosive part, if known.

12.3.3 The tests should be performed at ambient temperature unless the substance is to be transported under conditions where it may change its physical state or density.

12.3.4 For organic substances and mixtures of organic substances with a decomposition energy of 800 J/g or more, test 2 (a) need not be performed if the outcome of the ballistic mortar Mk.IIId test (F.1), or the ballistic mortar test (F.2) or the BAM Trauzl test (F.3) with initiation by a standard No. 8 detonator (see Appendix 1) is "No". In this case, the result of test 2 (a) is deemed to be "-". If the outcome of the F.1 or F.2 or F.3 test is "Low" or "Not low", the result of test 2 (a) shall be deemed "+". In this case, a "-" can only be obtained by performing test 2 (a).

12.4 Series 2 type (a) test prescription

12.4.1 *Test 2 (a): UN gap test*

12.4.1.1 *Introduction*

This test is used to measure the sensitivity of a substance, under confinement in a steel tube, to detonative shock.

12.4.1.2 *Apparatus and materials*

The apparatus is shown in Figure 12.4.1.1. The test sample is contained in seamless, carbon steel tube with an external diameter of 48 ± 2 mm, a wall thickness of 4 mm and a length of 400 ± 5 mm. If the test substance may react with the steel, the inside of the tube may be coated with fluorocarbon resin. The bottom of the tube is closed a plastics sheet pulled tightly (so that it plastically deforms) over the bottom of the tube and held tightly in place. The plastics sheet shall be compatible with the substance under test. The booster charge consists of 160 g RDX/wax (95/5) or PETN/TNT that has a minimum of 50% PETN in the mixture, 50 ± 1 mm in diameter with a density of 1 600 ± 50 kg/m^3. The charges may be pressed in one or more pieces, as long as the total charge is within the specifications, and the PETN/TNT charge is cast. A polymethyl methacrylate (PMMA) spacer is required of diameter 50 ± 1 mm and length 50 ± 1 mm. A mild steel witness plate, 150 ± 10 mm square and 3 mm thick, may be mounted at the upper end of the steel tube and separated from it by spacers 1.6 ± 0.2 mm thick.

12.4.1.3 *Procedure*

12.4.1.3.1 The sample is loaded to the top of the steel tube. Solid samples are loaded to the density attained by tapping the tube until further settling becomes imperceptible. The sample mass is determined and, if solid, the apparent density calculated using the measured internal volume of the tube.

12.4.1.3.2 The tube is placed in a vertical position and the PMMA spacer placed in direct contact with the sheet which seals the bottom of the tube. After positioning the booster charge in contact with the PMMA spacer, the detonator is fixed in place against the bottom of the booster charge and initiated. Two tests should be performed unless detonation of the substance is observed.

12.4.1.4 *Test criteria and method of assessing results*

The test results are assessed on the basis of the type of fragmentation of the tube or whether the witness plate is holed. The test giving the most severe assessment should be used for classification. The test result is considered "+" and the substance to be sensitive to shock if:

(a) The tube is fragmented completely; or

(b) The witness plate is holed.

Any other result is considered "-" and the substance is considered to be not sensitive to detonative shock.

12.4.1.5 *Examples of results*

Substances	**Apparent density (kg/m^3)**	**Fragmentation length (cm)**	**Witness plate**	**Result**
Ammonium nitrate, prills	800	25	Domed	-
Ammonium nitrate, 200 μm	540	40	Holed	+
Ammonium nitrate/fuel oil, 94/6	880	40	Holed	+
Ammonium perchlorate, 200 μm	1 190	0	No damage	-
Nitromethane	1 130	0	No damage	-
PETN/lactose, 20/80	880	40	Holed	+
TNT, cast	1 510	20	No damage	-
TNT, flaked	710	40	Holed	+

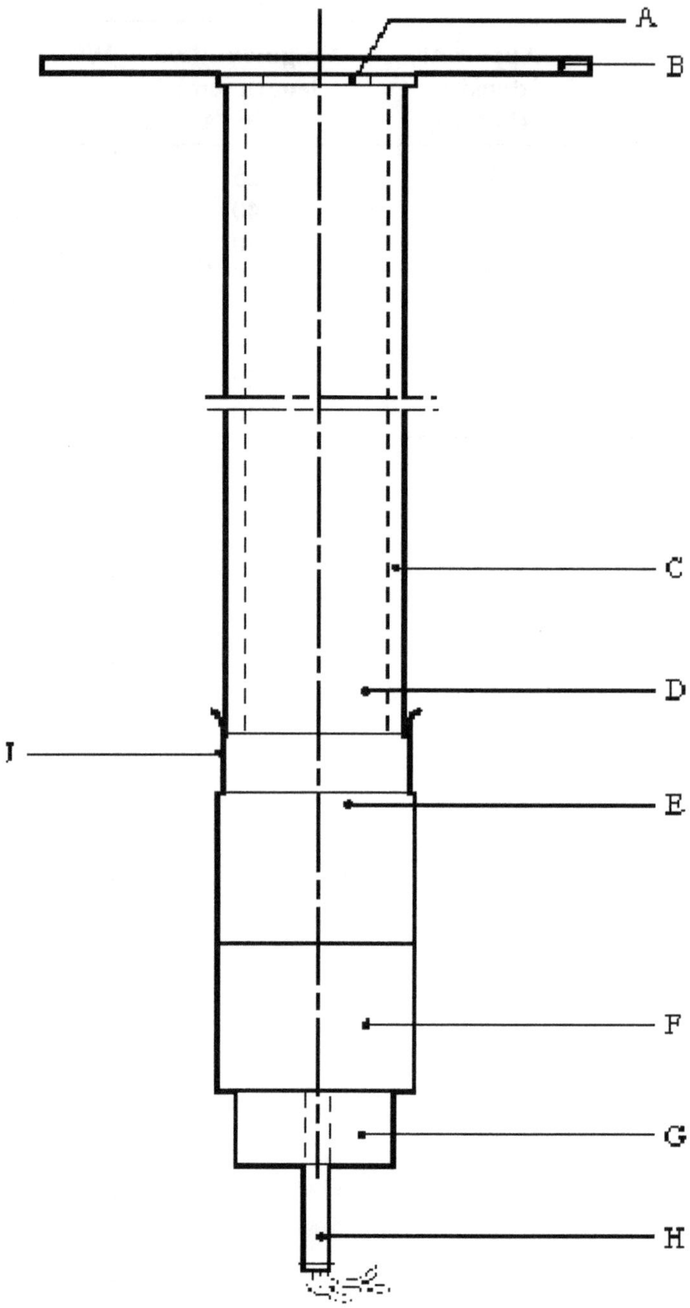

(A)	Spacers	(B)	Witness plate
(C)	Steel tube	(D)	Substance under investigation
(E)	PMMA spacer	(F)	RDX/wax or PETN/TNT booster charge
(G)	Detonator holder	(H)	Detonator
(J)	Plastics membrane		

Figure 12.4.1.1: UN GAP TEST

12.5 Series 2 type (b) test prescription

12.5.1 *Test 2 (b): Koenen test*

12.5.1.1 *Introduction*

This test is used to determine the sensitiveness of solid and liquid substances to the effect of intense heat under high confinement.

12.5.1.2 *Apparatus and materials*

12.5.1.2.1 The apparatus consists of a non-reusable steel tube, with its re-usable closing device, installed in a heating and protective device. The tube is deep drawn from sheet steel conforming to specification DC04 (EN 10027-1), or equivalent A620 (AISI/SAE/ASTM), or equivalent SPCEN (JIS g 3141). The dimensions are given in Figure 12.5.1.1. The open end of the tube is flanged. The closing plate with an orifice, through which the gases from the decomposition of the test substance escape, is made from heat-resisting chrome steel and is available with the following diameter holes: 1.0 - 1.5 - 2.0 - 2.5 - 3.0 - 5.0 - 8.0 - 12.0 - 20.0 mm. The dimensions of the threaded collar and the nut (closing device) are given in Figure 12.5.1.1.

For quality control of the steel tubes, 1% of the tubes from each production lot shall be subjected to quality control and the following data shall be verified:

(a) The mass of the tubes shall be 26.5 ± 1.5 g, tubes to be used in one test sequence shall not differ in mass by more than 1 g;

(b) The length of the tubes shall be 75 ± 0.5 mm;

(c) The wall thickness of the tubes measured 20 mm from the bottom of the tube shall be 0.5 ± 0.05 mm; and

(d) The bursting pressure as determined by quasi-static load through an incompressible fluid shall be 30 ± 3 MPa.

12.5.1.2.2 Heating is provided by propane, from an industrial cylinder fitted with a pressure regulator, via a flow meter and distributed by a manifold to the four burners. Other fuel gases may be used providing the specified heating rate is obtained. The gas pressure is regulated to give a heating rate of 3.3 ± 0.3 K/s when measured by the calibration procedure. Calibration involves heating a tube (fitted with a 1.5 mm orifice plate) filled with 27 cm^3 of dibutyl phthalate. The time taken for the temperature of the liquid (measured with a 1 mm diameter thermocouple centrally placed 43 mm below the rim of the tube) to rise from 135 °C to 285 °C is recorded and the heating rate calculated.

12.5.1.2.3 Because the tube is likely to be destroyed in the test, heating is undertaken in a protective welded box, the construction and dimensions of which are given in Figure 12.5.1.2. The tube is suspended between two rods placed through holes drilled in opposite walls of the box. The arrangement of the burners is given in Figure 12.5.1.2. The burners are lit simultaneously by a pilot flame or an electrical ignition device. ***The test apparatus is placed in a protective area.*** Measures should be taken to ensure that the burner flames are not affected by any draughts. Provision should be made for extracting any gases or smoke resulting from the test.

12.5.1.3 *Procedure*

12.5.1.3.1 Normally substances are tested as received, although in certain cases it may be necessary to test the substance after crushing it. For solids, the mass of material to be used in each test is determined using a two-stage dry run procedure. A tared tube is filled with 9 cm^3 of substance and the substance tamped[1] with 80 N force applied to the total cross-section of the tube. If the material is compressible then more is added and tamped until the tube is filled to 55 mm from the top. The total mass used to fill the tube to the 55 mm level is determined and two further increments, each tamped with 80 N force, are added. Material is then either added, with tamping, or taken out as required to leave the tube filled to a level 15 mm from the top.

A second dry run is performed, starting with a tamped increment a third of the total mass found in the first dry run. Two more of these increments are added with 80 N tamping and the level of the substance in the tube adjusted to 15 mm from the top by addition or subtraction of material as required. The amount of solid determined in the second dry run is used for each trial filling being performed in three equal increments, each compressed to 9 cm^3. (This may be facilitated by the use of spacing rings.) Liquids and gels are loaded into the tube to a height of 60 mm taking particular care with gels to prevent the formation of voids. The threaded collar is slipped onto the tube from below, the appropriate orifice plate is inserted and the nut tightened by hand after applying some molybdenum disulphide based lubricant. It is essential to check that none of the substance is trapped between the flange and the plate, or in the threads.

12.5.1.3.2 With orifice plates from 1.0 mm to 8.0 mm diameter, nuts with an orifice of 10.0 mm diameter should be used; if the diameter of the orifice is above 8.0 mm, that of the nut should be 20.0 mm. Each tube is used for one trial only. The orifice plates, threaded collars and nuts may be used again provided they are undamaged.

12.5.1.3.3 The tube is placed in a rigidly mounted vice and the nut tightened with a spanner. The tube is then suspended between the two rods in the protective box. The test area is vacated, the gas supply turned on and the burners lit. The time to reaction and duration of reaction can provide additional information useful in interpreting the results. If rupture of the tube does not occur, heating is to be continued for at least five minutes before the trial is finished. After each trial the fragments of the tube, if any, should be collected and weighed.

12.5.1.3.4 The following effects are differentiated:

"O": Tube unchanged;
"A": Bottom of tube bulged out;
"B": Bottom and wall of the tube bulged out;
"C": Bottom of tube split;
"D": Wall of tube split;
"E": Tube split into two[2] fragments;
"F": Tube fragmented into three[2] or more mainly large pieces which in some cases may be connected with each other by a narrow strip;
"G": Tube fragmented into many mainly small pieces, closing device undamaged; and
"H": Tube fragmented into many very small pieces, closing device bulged out or fragmented.

---

1 ***For reasons of safety, e.g. the substance is friction sensitive, the substance need not be tamped.*** *In cases where the physical form of the sample can be changed by compression or compression of the sample is not related to the transport conditions, e.g. for fibrous materials, more representative filling procedures may be used.*

2 *The upper part of the tube remaining in the closing device is counted as one fragment.*

Examples for the effect types "D", "E" and "F" are shown in Figure 12.5.1.3. If a trial results in any of the effects "O" to "E", the result is regarded as "no explosion". If a trial gives the effect "F", "G" or "H", the result is evaluated as "explosion".

12.5.1.3.5 The series of trials is started with a single trial using an orifice plate of 20.0 mm. If, in this trial, the result "explosion" is observed, the series is continued with trials using tubes without orifice plates and nuts but with threaded collars (orifice 24.0 mm). If at 20.0 mm "no explosion" occurs, the series is continued with single trials using plates with the following orifices 12.0 - 8.0 - 5.0 - 3.0 - 2.0 - 1.5 and finally 1.0 mm until, at one of these diameters, the result "explosion" is obtained. Subsequently, trials are carried out at increasing diameters, according to the sequence given in 12.5.1.2.1, until only negative results in three tests at the same level are obtained. The limiting diameter of a substance is the largest diameter of the orifice at which the result "explosion" is obtained. If no "explosion" is obtained with a diameter of 1.0 mm, the limiting diameter is recorded as being less than 1.0 mm.

12.5.1.4 *Test criteria and method of assessing results*

The result is considered "+" and the substance to show a violent effect on heating under confinement if the limiting diameter is 2.0 mm or more. The result is considered "-" and the substance to show no violent effect on heating under confinement if the limiting diameter is less than 2.0 mm.

12.5.1.5 *Examples of results*

Substances	**Limiting diameter (mm)**	**Result**
Ammonium nitrate (crystalline)	1.0	-
Ammonium perchlorate	3.0	+
Ammonium picrate (crystalline)	2.5	+
1,3-Dinitroresorcinol (crystalline)	2.5	+
Guanidine nitrate (crystalline)	1.5	-
Picric acid (crystalline)	4.0	+
PETN/wax (95/5)	5.0	+

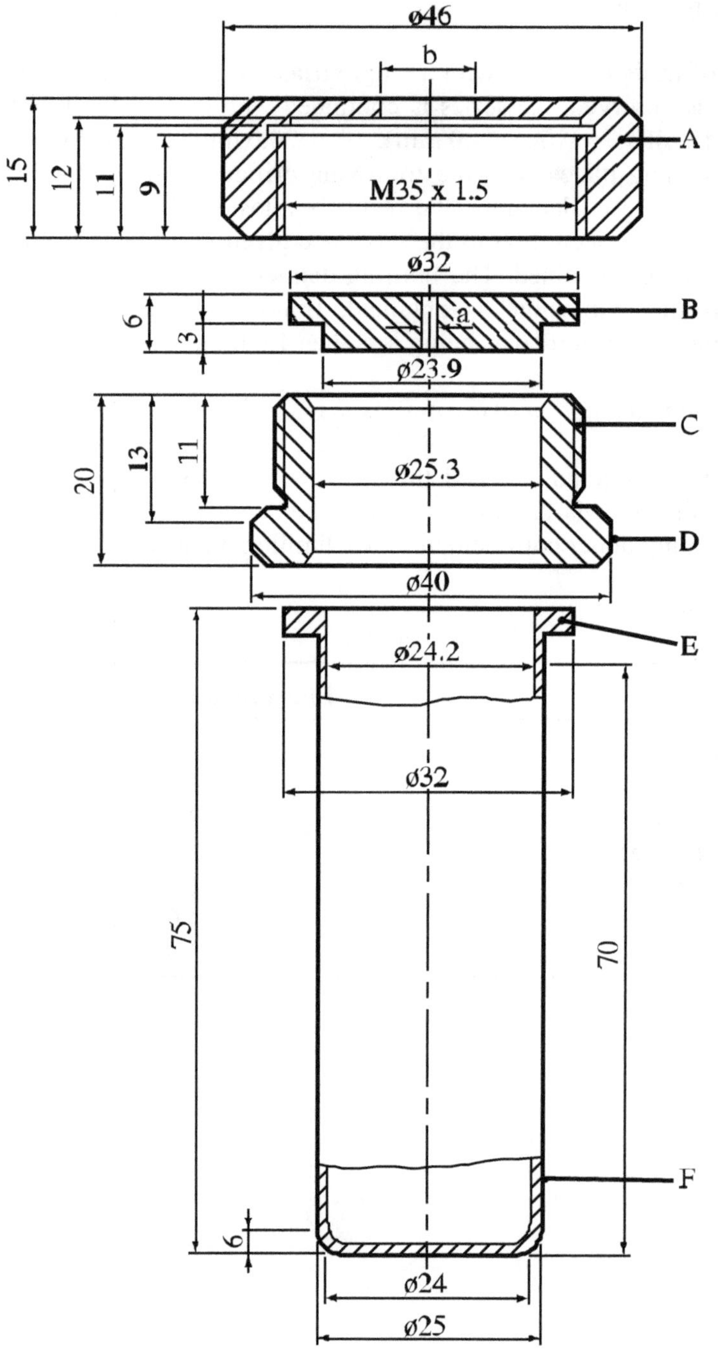

(A)	Nut (b = 10.0 or 20.0 mm) with flats for size 41 spanner	(B) (D)	Orifice plate (a = 1.0 → 20.0 mm diameter) Flats for size 36 spanner
(C)	Threaded collar	(F)	Tube
(E)	Flange		

Figure 12.5.1.1: TEST TUBE ASSEMBLY

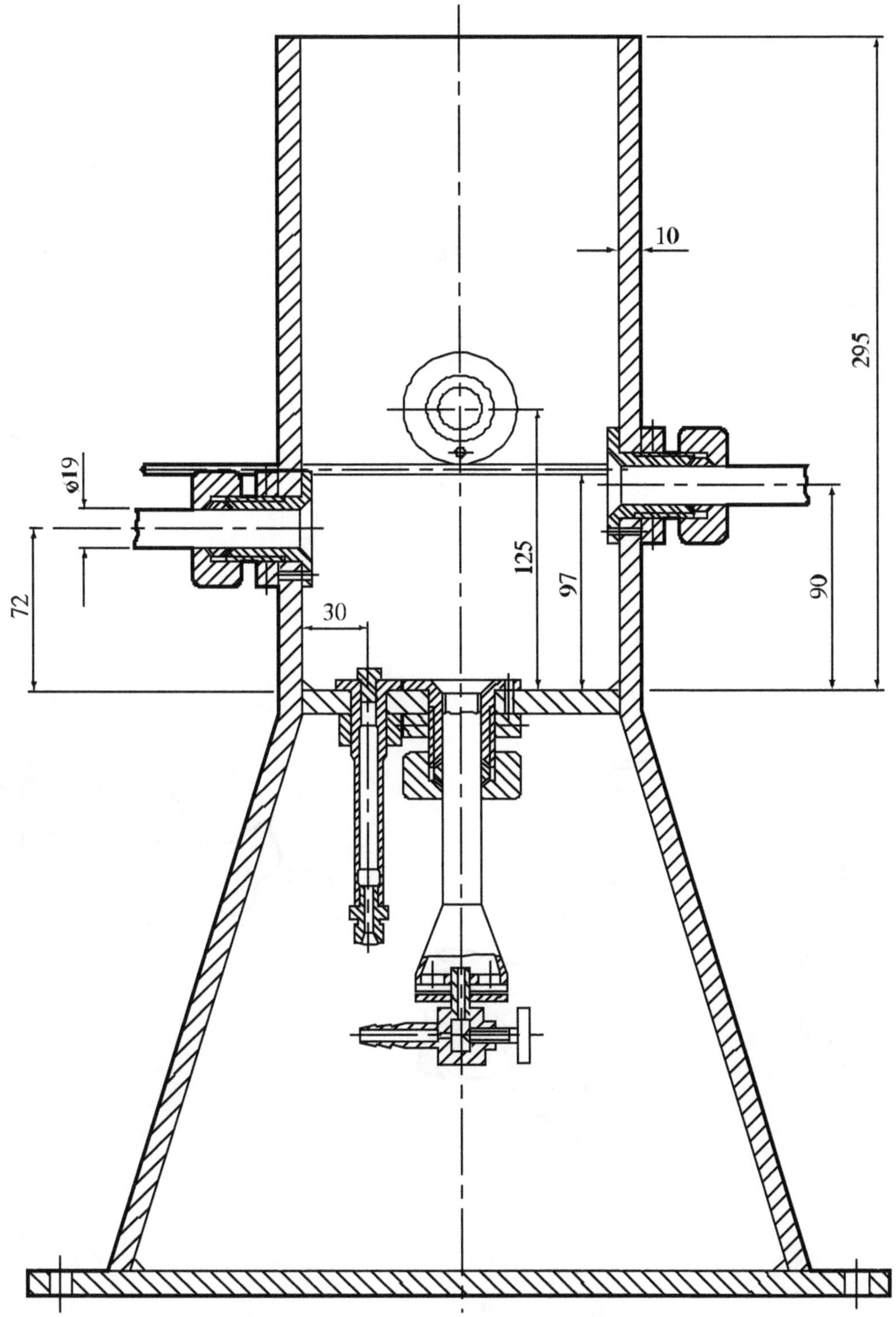

Figure 12.5.1.2: HEATING AND PROTECTIVE DEVICE

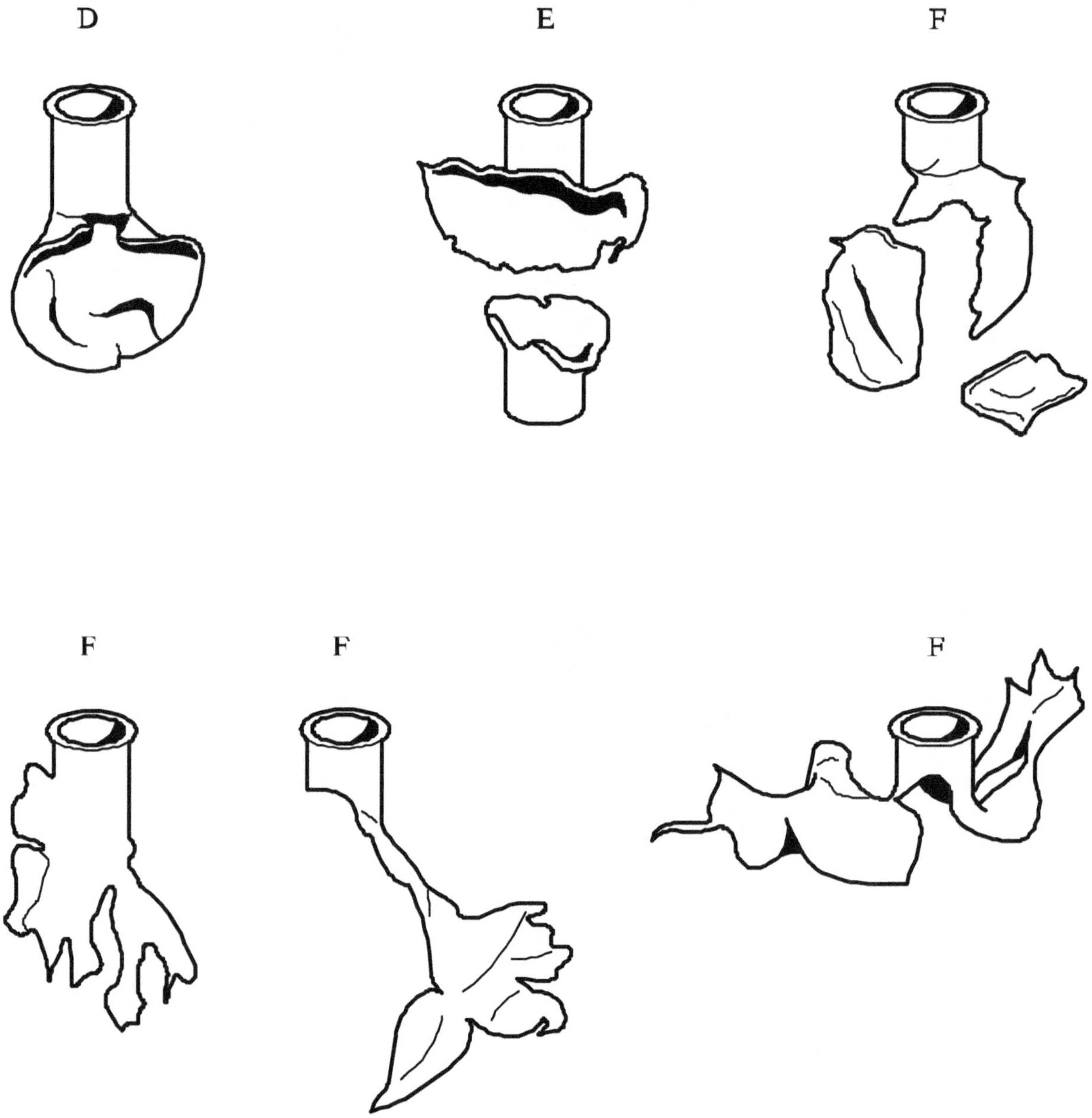

Figure 12.5.1.3 EXAMPLES OF EFFECT TYPES D, E AND F

12.6 Series 2 type (c) test prescriptions

12.6.1 *Test 2 (c) (i): Time/pressure test*

12.6.1.1 *Introduction*

This test is used to determine the effects of igniting the substance[3] under confinement in order to determine if ignition leads to a deflagration with explosive violence at pressures which can be attained with substances in normal commercial packages.

12.6.1.2 *Apparatus and materials*

12.6.1.2.1 The time/pressure apparatus (Figure 12.6.1.1) consists of a cylindrical steel pressure vessel 89 mm in length and 60 mm in external diameter. Two flats are machined on opposite sides (reducing the cross-section of the vessel to 50 mm) to facilitate holding whilst fitting the firing plug and vent plug. The vessel, which has a bore of 20 mm diameter, is internally rebated at either end to a depth of 19 mm and threaded to accept 1" British Standard Pipe (BSP). A pressure take-off, in the form of a side-arm, is screwed into the curved face of the pressure vessel 35 mm from one end and at 90° to the machined flats. The socket for this is bored to a depth of 12 mm and threaded to accept the 1/2" BSP thread on the end of the side-arm. A washer is fitted to ensure a gastight seal. The side-arm extends 55 mm beyond the pressure vessel body and has a bore of 6 mm. The end of the side-arm is rebated and threaded to accept a diaphragm type pressure transducer. Any pressure-measuring device may be used provided that it is not affected by the hot gases or decomposition products and is capable of responding to rates of pressure rise of 690-2 070 kPa in not more than 5 ms.

12.6.1.2.2 The end of the pressure vessel furthest from the side-arm is closed with a firing plug which is fitted with two electrodes, one insulated from and the other earthed to, the plug body. The other end of the pressure vessel is closed by an aluminium bursting disc 0.2 mm thick (bursting pressure approximately 2 200 kPa) held in place with a retaining plug which has a 20 mm bore. A soft lead washer or a washer of a suitable deformable material (for example, polyoxymethylene) is used with both plugs to ensure a good seal. A support stand (Figure 12.6.1.2) holds the assembly in the correct attitude during use. This comprises a mild steel base plate measuring 235 mm × 184 mm × 6 mm and a 185 mm length of square hollow section (S.H.S.) 70 × 70 × 4 mm.

12.6.1.2.3 A section is cut from each of two opposite sides at one end of the length of S.H.S. so that a structure having two flat sided legs surmounted by an 86 mm length of intact box section results. The ends of these flat sides are cut to an angle of 60° to the horizontal and welded to the base plate.

12.6.1.2.4 A slot measuring 22 mm wide × 46 mm deep is machined in one side of the upper end of the base section such that when the pressure vessel assembly is lowered, firing plug end first, into the box section support, the side-arm is accommodated in this slot. A packing piece of steel 30 mm wide and 6 mm thick is welded to the lower internal face of the box section to act as a spacer. Two 7 mm thumb screws, tapped into the opposite face, serve to hold the pressure vessel firmly in place. Two 12 mm wide strips of 6 mm thick steel, welded to the side pieces abutting the base of the box section, support the pressure vessel from beneath.

12.6.1.2.5 The ignition system consists of an electric fusehead of the type commonly used in low tension detonators, together with a 13 mm square piece of primed cambric. Fuseheads with equivalent properties may be used. Primed cambric consists of a linen fabric coated on both sides with a potassium nitrate/silicon/sulphurless gunpowder pyrotechnic composition[4].

---

[3] *When testing thermally stable energetic liquids, such as nitromethane (UN 1261), variable results may be obtained because the substance may give two pressure peaks.*

[4] *Obtainable from the national contact for test details in United Kingdom (see Appendix 4).*

12.6.1.2.6 The procedure for the preparation of the ignition assembly for solids starts with separation of the brass foil contacts of an electric fusehead from its insulator, (see Figure 12.6.1.3). The exposed portion of insulation is then cut off. The fusehead is then fixed onto the terminals of the firing plug by means of the brass contacts such that the tip of the fusehead is 13 mm above the surface of the firing plug. A 13 mm square piece of primed cambric is pierced through the centre and positioned over the attached fusehead around which it is then folded and secured with fine cotton thread.

12.6.1.2.7 For liquids samples, leads are fixed onto the contact foils of the fusehead. The leads are then threaded through an 8 mm length of 5 mm outer diameter and 1 mm inner diameter silicone rubber tubing and the tubing is pushed up over the fusehead contact foils as shown in Figure 12.6.1.4. The primed cambric is then wrapped around the fusehead and a single piece of thin PVC sheathing, or equivalent, is used to cover the primed cambric and the silicone rubber tubing. The sheathing is sealed in position by twisting a length of thin wire tightly round the sheathing and rubber tubing. The leads are then fixed onto the terminals of the firing plug such that the tip of the fusehead is 13 mm above the surface of the firing plug.

12.6.1.3 *Procedure*

12.6.1.3.1 The apparatus, assembled complete with pressure transducer but without the aluminium bursting disc in position, is supported firing plug end down. 5.0 g[5] of the substance is introduced into the apparatus so as to be in contact with the ignition system. Normally no tamping is carried out when filling the apparatus unless it is necessary to use light tamping in order to get the 5.0 g charge into the vessel. If, even with light tamping, it is impossible to get all the 5.0 g of sample in, then the charge is fired after filling the vessel to capacity. Note should be taken of the charge weight used. The lead washer and aluminium bursting disc are placed in position and the retaining plug is screwed in tightly. The charged vessel is transferred to the firing support stand, bursting disc uppermost, which should be contained in a suitable, armoured fume cupboard or firing cell. An exploder dynamo is connected to the external terminals of the firing plug and the charge is fired. The signal produced by the pressure transducer is recorded on a suitable system which allows both evaluation and a permanent record of the time/pressure profile to be achieved (e.g. transient recorder coupled to a chart-recorder).

12.6.1.3.2 The test is carried out three times. The time taken for the pressure to rise from 690 kPa to 2 070 kPa above atmospheric is noted. The shortest time interval should be used for classification.

12.6.1.4 *Test criteria and method of assessing results*

The test results are interpreted in terms of whether a gauge pressure of 2 070 kPa is reached and, if so, the time taken for the pressure to rise from 690 kPa to 2 070 kPa gauge.

The result is considered "+" and the substance to show the ability to deflagrate rapidly if the time for a pressure rise from 690 kPa to 2 070 kPa is less than 30 ms. The result is considered "-" and the substance to show no or slow deflagration if the rise time is 30 ms or more or a pressure of 2 070 kPa gauge is not reached. Failure to ignite does not necessarily indicate that the substance has no explosive properties.

---

[5] ***If preliminary safety-in-handling tests (e.g. heating in a flame) or unconfined burning tests (e.g. a Series 3 type (d) test) indicate that a rapid reaction is likely to occur, then the sample size should be reduced to 0.5 g until the severity of the confined reaction is known.*** *If it is necessary to use a 0.5 g sample, the sample size should be gradually increased until either a "+" result is obtained or the test is performed with a 5.0 g sample.*

12.6.1.5 *Examples of results*

Substances	**Maximum pressure (kPa)**	**Time for a pressure rise from 690 to 2 070 kPa (ms)**	**Result**
Ammonium nitrate (high density prills)	< 2 070	-	-
Ammonium nitrate (low density prills)	< 2 070	-	-
Ammonium perchlorate (2 µm)	> 2 070	5	+
Ammonium perchlorate (30 µm)	> 2 070	15	+
Barium azide	> 2 070	<5	+
Guanidine nitrate	> 2 070	606	-
Isobutyl nitrite	> 2 070	80	-
Isopropyl nitrate	> 2 070	10	+
Nitroguanidine	> 2 070	400	-
Picramic acid	> 2 070	500	-
Sodium picramate	> 2 070	15	+
Urea nitrate	> 2 070	400	-

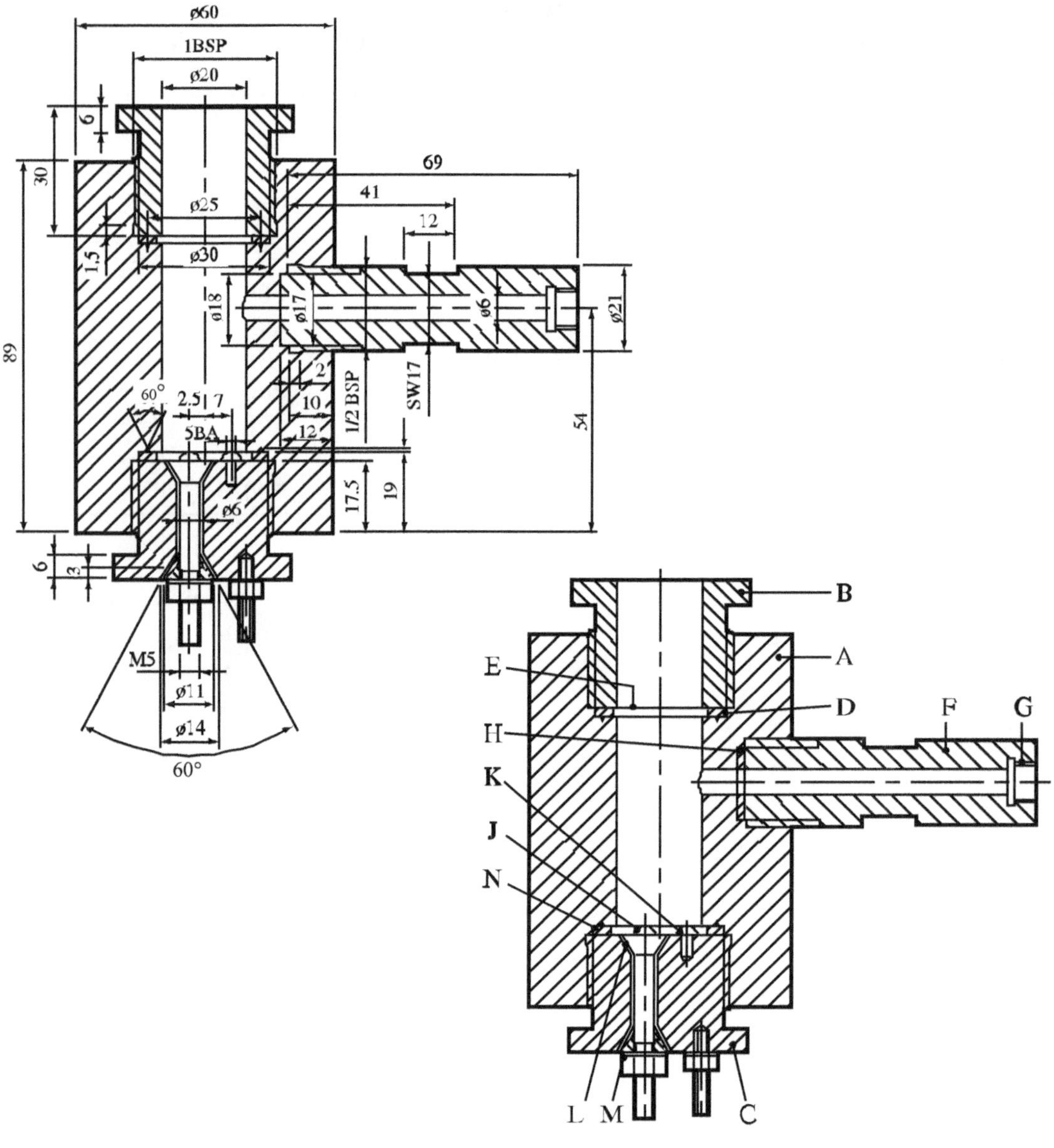

(A)	Pressure vessel body	(B)	Bursting disc retaining plug
(C)	Firing plug	(D)	Soft lead washer
(E)	Bursting disc	(F)	Side arm
(G)	Pressure transducer thread	(H)	Copper washer
(J)	Insulated electrode	(K)	Earthed electrode
(L)	Insulation	(M)	Steel cone
(N)	Washer distorting groove		

Figure 12.6.1.1: APPARATUS

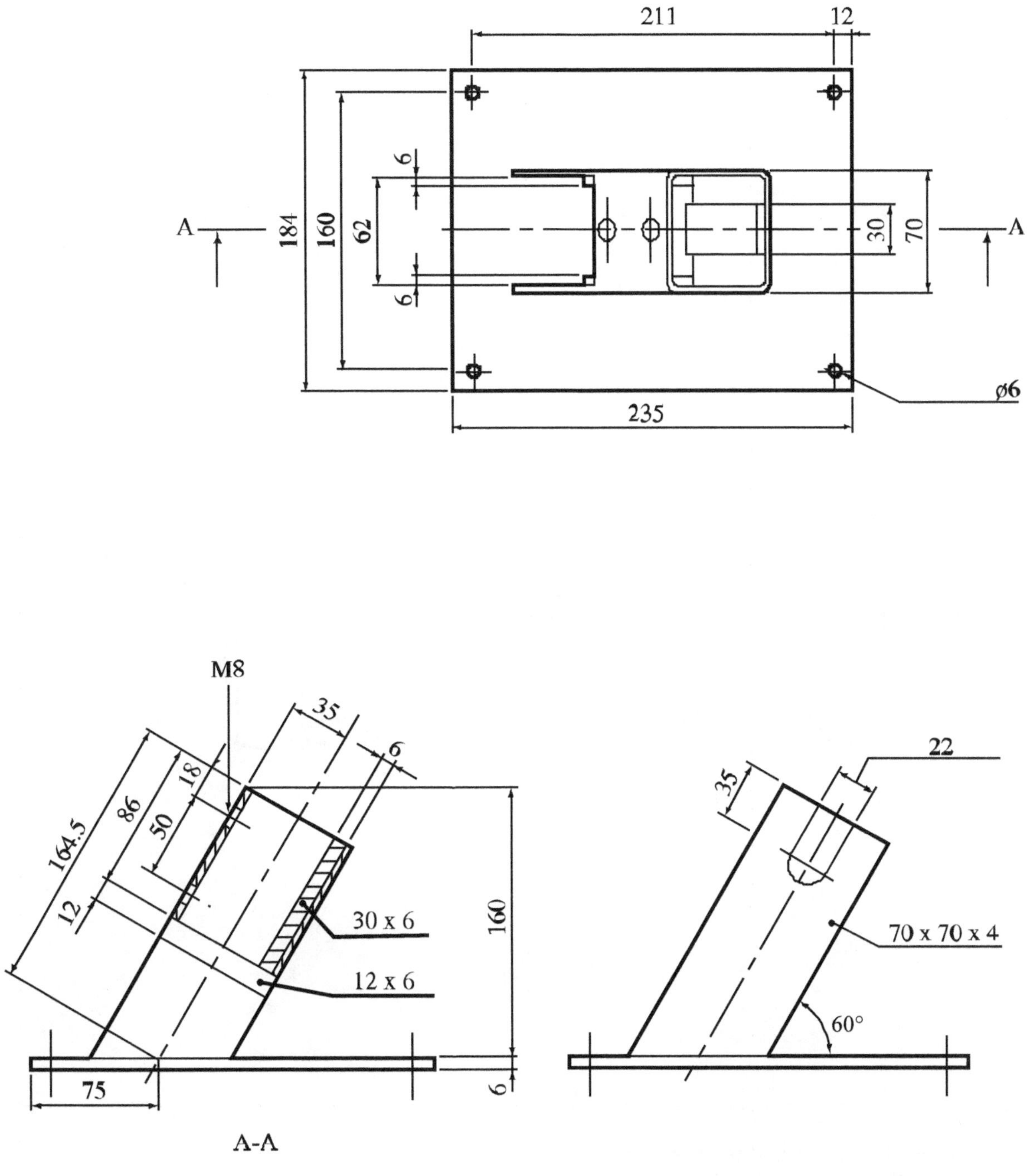

Figure 12.6.1.2: SUPPORT STAND

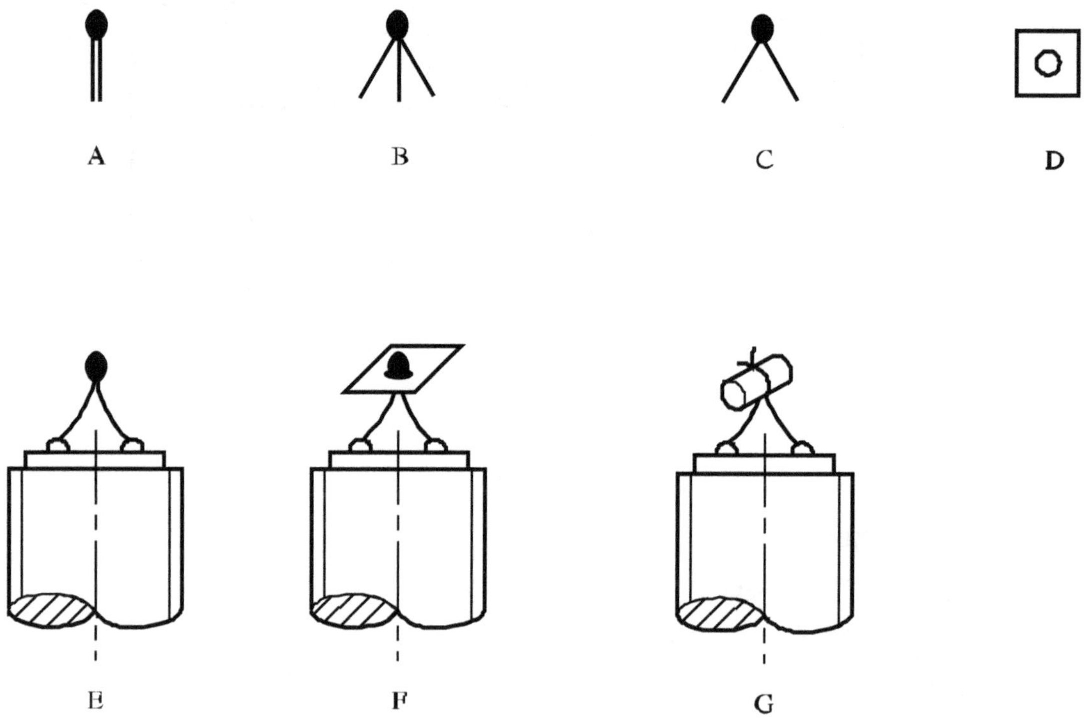

(A) Electrically ignited fusehead as manufactured
(B) Brass foil contacts parted from card insulator
(C) Insulating card cut off
(D) Primed cambric SR252 13 mm square with centre hole
(E) Fusehead fixed to pins on firing plug
(F) Cambric positioned on fusehead
(G) Cambric folded around and tied with thread

Figure 12.6.1.3: IGNITION SYSTEM FOR SOLIDS

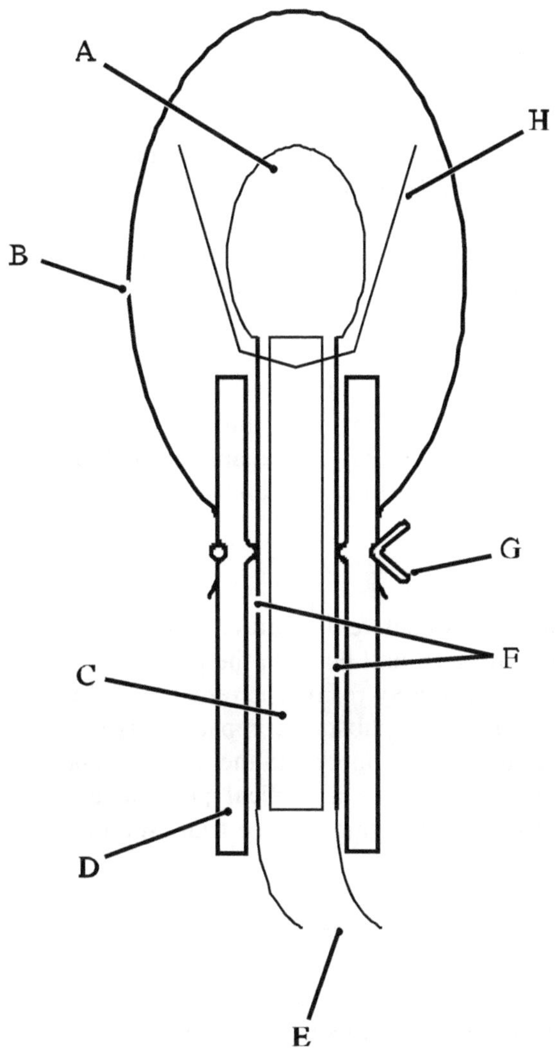

(A) Fusehead
(B) PVC sheath
(C) Insulating card
(D) Silicone rubber tubing
(E) Firing leads
(F) Foil contacts
(G) Wire to make liquid-tight seal
(H) Primed cambric

Figure 12.6.1.4: IGNITION SYSTEM FOR LIQUIDS

12.6.2 *Test 2 (c) (ii): Internal ignition test*

12.6.2.1 *Introduction*

This test is used to determine the tendency of a substance to undergo transition from deflagration to detonation.

12.6.2.2 *Apparatus and materials*

The experimental arrangement is shown in Figure 12.6.2.1. The sample of substance to be tested is contained in a 45.7 cm length of "3 inch schedule 80" carbon (A53 Grade B) steel pipe with inside diameter 74 mm, wall thickness 7.6 mm, capped at both ends with a "3 000 pound" forged steel pipe cap. An igniter consisting of 10 g of black powder (100% passed through No. 20 sieve, 0.84 mm, and 100% retained by No. 50 sieve, 0.297 mm) is located at the centre of the sample vessel. The igniter assembly consists of a cylindrical container 21 mm in diameter and 32 mm long which is made from 0.54 mm thick cellulose acetate which is held together by two layers of nylon filament reinforced cellulose acetate tape. The igniter capsule contains a small loop formed from a 25 mm length of nickel-chromium alloy resistance wire 0.30 mm in diameter having a resistance of 0.35 ohms. This loop is attached to two insulated tinned copper lead wires 0.7 mm in diameter. The overall wire diameter, including insulation, is 1.3 mm. These lead wires are fed through small holes in the wall of the pipe and are sealed with epoxy resin.

12.6.2.3 *Procedure*

After the sample, at ambient temperature, is loaded into the pipe to a height of 23 cm, the igniter (with its leads inserted through small holes in the pipe wall) is inserted into the centre of the pipe and the leads pulled taut and sealed with epoxy resin. The remainder of the sample is then loaded and the top cap screwed on. For gelatinous samples, the substance is packed as near as possible to its normal shipping density. For granular samples, the substance is loaded to the density obtained by repeated tapping of the pipe against a hard surface. The tube is placed in a vertical position and the igniter is fired by a current of 15 amperes obtained from a 20-volt transformer. Three trials should be performed unless deflagration to detonation transition occurs earlier.

12.6.2.4 *Test criteria and method of assessing results*

The test result is considered "+" if either the pipe or at least one of the end caps is fragmented into at least two distinct pieces. Results in which the pipe is merely split or laid open, or in which the pipe or caps are distorted to the point at which the caps are blown off, are considered "-".

12.6.2.5 *Examples of results*

Substance	**Results**
Ammonium nitrate/fuel oil, aluminised	-
Ammonium nitrate prills, porous, low density	-
Ammonium perchlorate (45 μm)	+
1,3-Dinitrobenzene, fine crystals	-
Nitrocarbonitrate	-
TNT, granular	+
Watergel	+

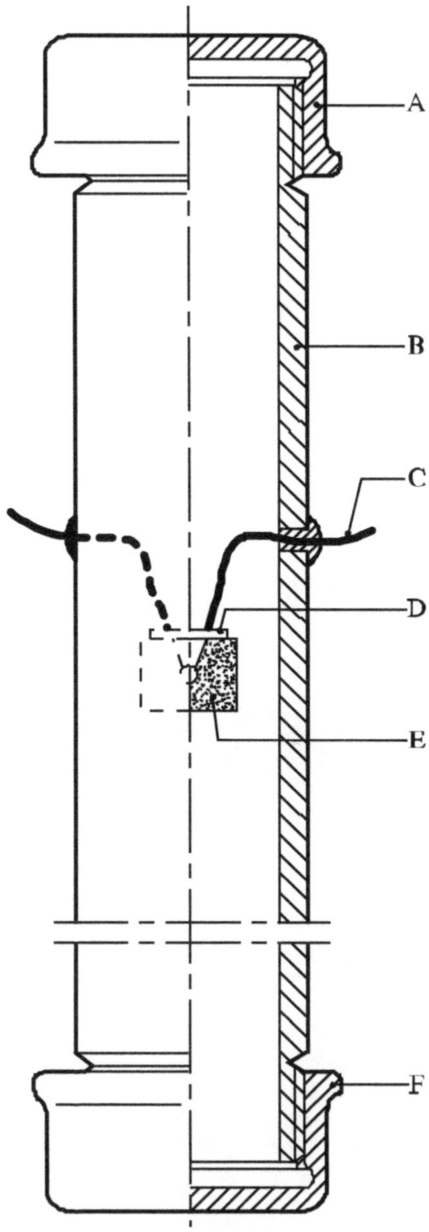

(A)	Forged steel cap	(B)	Steel pipe
(C)	Igniter leads	(D)	Seal
(E)	Igniter assembly	(F)	Forged steel cap

Figure 12.6.2.1: INTERNAL IGNITION TEST

SECTION 13

TEST SERIES 3

13.1 Introduction

The questions "Is the substance thermally stable?" (box 10 of Figure 10.2) and "Is the substance too dangerous for transport in the form in which it was tested?" (box 11 of Figure 10.2) are answered by determining the sensitiveness of the substance to mechanical stimuli (impact and friction), to heat and to flame. The question in box 10 is answered "no" if a "+" is obtained in test type 3(c) and the substance is considered too unstable to transport. The question in box 11 is answered "yes" if a "+" is obtained in any of the test types 3(a), 3(b) or 3(d). If a "+" is obtained, the substance may be encapsulated or otherwise desensitized or packaged to reduce its sensitiveness to external stimuli.

13.2 Test methods

Test Series 3 comprises four types of test:

Type 3 (a): for determining sensitiveness to impact;
Type 3 (b): for determining sensitiveness to friction (including impacted friction);
Type 3 (c): for determining the thermal stability of a substance; and
Type 3 (d): for determining the response of the substance to fire.

The test methods currently used are listed in Table 13.1.

Table 13.1: TEST METHODS FOR TEST SERIES 3

Test code	Name of Test	Section
3 (a) (i)	Bureau of Explosives impact machine	13.4.1
3 (a) (ii)	BAM Fallhammer[a]	13.4.2
3 (a) (iii)	Rotter test	13.4.3
3 (a) (iv)	30 kg Fallhammer test	13.4.4
3 (a) (v)	Modified type 12 impact tool	13.4.5
3 (a) (vi)	Impact sensitivity test	13.4.6
3 (a) (vii)	Modified Bureau of Mines impact machine test	13.4.7
3 (b) (i)	BAM friction apparatus[a]	13.5.1
3 (b) (ii)	Rotary friction test	13.5.2
3 (b) (iii)	Friction sensitivity test	13.5.3
3 (b) (iv)	ABL friction machine test	13.5.4
3 (c) (i)	Thermal stability test at 75 °C[a]	13.6.1
3 (c) (ii)	SBAT thermal stability test at 75 °C	13.6.2
3 (d)	Small-scale burning test[a]	13.7.1

[a] *Recommended test.*

13.3 Test conditions

13.3.1 ***Care must be taken if it is necessary to crush or cut explosive samples before use. Protective equipment, e.g. safety screens, should be used and the quantities kept to the minimum.***

13.3.2 For tests of types 3 (a) and (b), wetted substances should be tested with the minimum content of wetting agent provided for transport.

13.3.3 Test types 3 (a) and 3 (b) should be performed at ambient temperature unless otherwise specified or the substance is to be transported under conditions where it may change its physical state.

13.3.4 To obtain reproducible results, all factors of test types 3 (a) and 3 (b) should be carefully controlled and a suitable standard of known sensitiveness should be tested periodically.

13.3.5 Entrapped air bubbles render liquid substances much more sensitive to impact and hence test type 3 (a) methods for liquids employ special tools or procedures which allow the "adiabatic" compression of such bubbles in the liquid.

13.3.6 Tests of type 3 (b) need not be applied to liquids.

13.4 Series 3 type (a) test prescriptions

13.4.1 *Test 3 (a) (i): Bureau of Explosives impact machine*

13.4.1.1 *Introduction*

This test is used to measure the sensitiveness of a substance to drop-weight impact and to determine if the substance is too dangerous to transport in the form tested. It is applicable to solid and liquid substances by using two different sample assemblies.

13.4.1.2 *Apparatus and materials*

13.4.1.2.1 Solids

Drawings of the apparatus for the impact test of solids are shown in figures 13.4.1.1 and 13.4.1.2. The apparatus is designed so that a weight of mass 3.63 kg is free to fall between two parallel cylindrical guide rods, from pre-selected heights, onto a plunger-and-plug assembly. This assembly is in contact with the sample, which in turn is placed on a die-and-anvil assembly and confined in a cylindrical casing whose inside diameter is just sufficient to permit free movement of the plunger and plug. The plunger, plug, die, casing and anvil are hardened tool steel of hardness 50-55 on the Rockwell C scale and the mating surfaces and the surfaces in contact with the sample have a finish of 0.8 microns. The sample holder diameter is 5.1 mm.

13.4.1.2.2 Liquids

The apparatus for the impact test of liquid is similar to that for solids, except for the sample assembly. The sample assembly for testing liquids is shown in Figure 13.4.1.3.

13.4.1.3 *Procedure*

13.4.1.3.1 Solids

A 10 mg sample is loaded onto the die (C). The anvil (E) and die are placed in the sample housing (F) and the casing (D) screwed down over them. The plug (B) and plunger (A) are then inserted on top of the sample. The drop weight is raised to a height of 10.0 cm and released. Observations are made on whether an "explosion" occurs as evidenced by a flame or audible report. Ten trials are performed for each test sample.

13.4.1.3.2 Liquids

The rebound sleeve (A), the intermediate pin (B), and the striker (D) are assembled in the striker housing (C). A copper cup (E) is placed in the cup positioning block (not shown in Figure 13.4.1.3) and one drop of the liquid under test is placed in the cup (E). The housing (C) and its components (A, B and D) are placed over the top of the cup positioning block. The end of the striker (D) slips partway into the cup (E), but is prevented by the cup positioning block from actually touching the liquid in the cup. When the striker housing is lifted from the positioning block, the cup is held on the end of the striker by friction. The striker housing is then screwed down into the anvil housing and the dimensions of the tools are such that the bottom of the copper cup just touches the anvil when the striker housing is screwed in hand-tight. The whole unit is then placed in the same drop weight mechanism used for solids. The drop weight is raised to a height of 25.0 cm and released. Observations are made on whether an "explosion" occurs as evidenced by smoke, flame or audible report. Ten trials are performed for each test sample.

13.4.1.4 *Test criteria and method of assessing results*

13.4.1.4.1 Solids

The test result is considered "+" if a flame or an audible report is observed in at least 5 out of 10 trials at a drop height of 10 cm and the substance is considered too dangerous for transport in the form in which it was tested. Otherwise, the result is considered "-". Borderline cases may be resolved using the Bruceton method (see Appendix 2).

13.4.1.4.2 Liquids

The test result is considered "+" if smoke, flame or an audible report is observed in at least 1 out of 10 trials at a drop height of 25 cm and the substance is considered too dangerous for transport in the form in which it was tested. Otherwise, the result is considered "-".

13.4.1.5 *Examples of results*

13.4.1.5.1 Solids

Results from tests to determine the impact sensitiveness of solid:

Test sample	**Result**
Ammonium perchlorate	-
HMX, (dry)	+
NG Dynamite	-
PETN, (dry)	+
PETN/water, (75/25)	-
RDX, (dry)	+

13.4.1.5.2 Liquids

Results from tests to determine the impact sensitiveness of liquids:

Test sample	**Result**
NG	+
Nitromethane	-

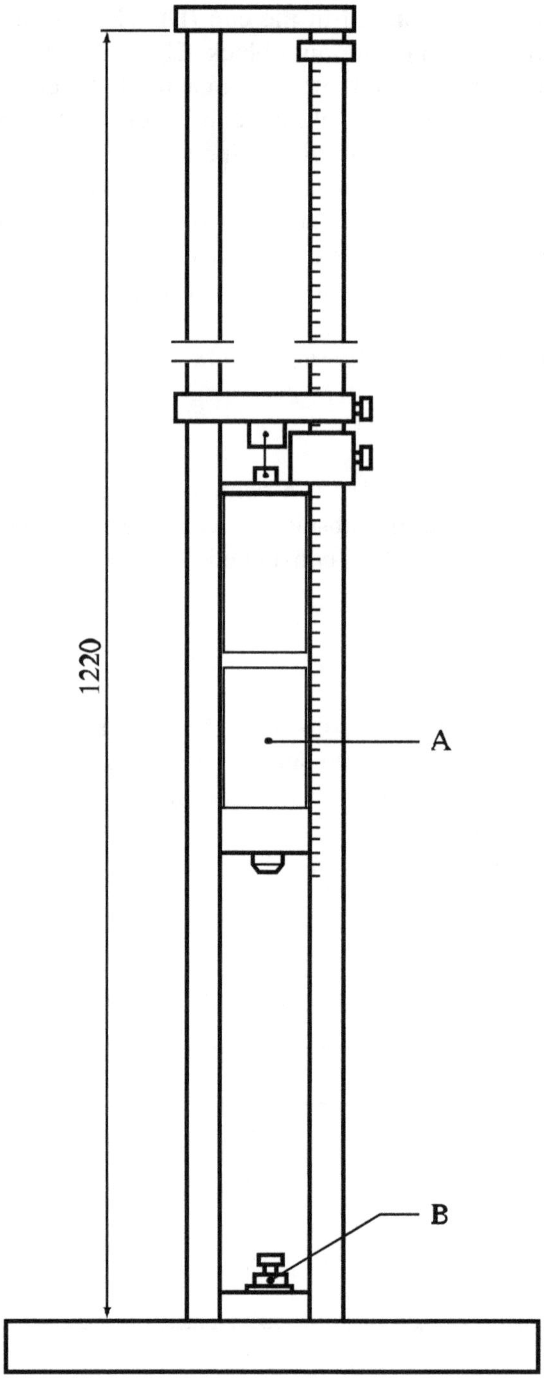

(A)	Drop weight
(B)	Sample assembly

Figure 13.4.1.1: BUREAU OF EXPLOSIVES IMPACT MACHINE

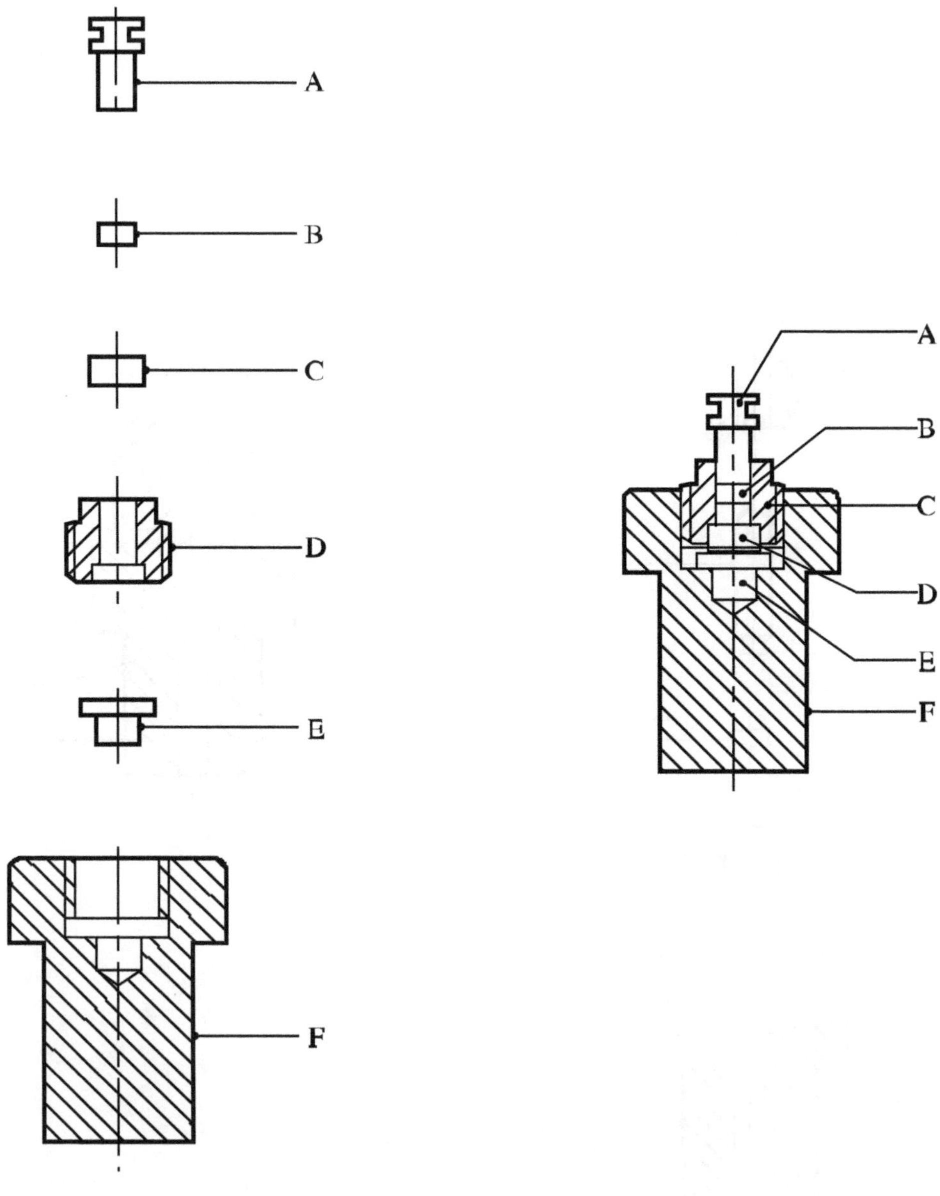

(A)	Plunger	(B)	Plug
(C)	Die	(D)	Casing
(E)	Anvil	(F)	Housing

Figure 13.4.1.2: SAMPLE ASSEMBLY FOR SOLIDS

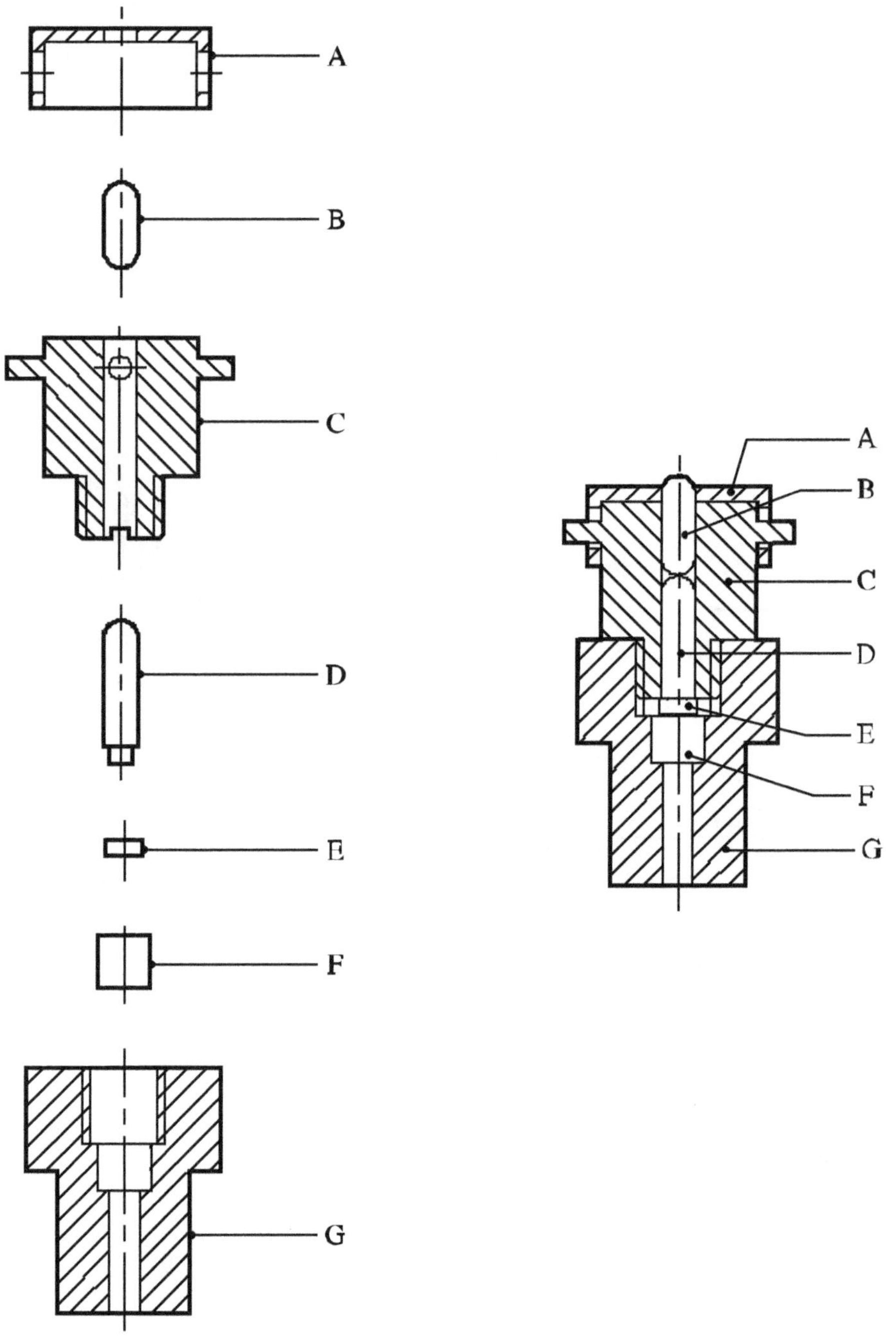

(A)	Rebound check sleeve	(B)	Intermediate pin
(C)	Striker housing	(D)	Striker
(E)	Copper cup	(F)	Anvil
(G)	Anvil housing		

Figure 13.4.1.3: SAMPLE ASSEMBLY FOR LIQUIDS

13.4.2 *Test 3 (a) (ii): BAM Fallhammer*

13.4.2.1 *Introduction*

This test is used to measure the sensitiveness of solids and liquids to drop-weight impact and to determine if the substance is too dangerous to transport in the form tested.

13.4.2.2 *Apparatus and materials*

13.4.2.2.1 The essential parts of the fallhammer are the cast steel block with base, the anvil, the column, the guides, the drop weights with release device and the impact device. A steel anvil is screwed onto the steel block and cast base. The support, into which is fixed the column (made from a seamless drawn steel tube), is bolted to the back of the steel block. The dimensions of the anvil, the steel block, the base and the column are given in Figure 13.4.2.1. The two guides which are fixed to the column by means of three cross-pieces are fitted with a toothed rack to limit the rebound of the drop weight and a movable graduated scale for adjusting the height of the fall. The drop weight release mechanism is adjustable between the guides and is clamped to them by the operation of a lever nut on two jaws. The apparatus is fixed onto a concrete block (600 × 600 mm) by means of four anchoring screws secured in the concrete, so that the base is in contact with the concrete over its whole area and the guides are exactly vertical. A wooden protective box with inner protective lining and which can be opened easily, surrounds the apparatus up to the level of the bottom cross-bar. An extraction system allows removal of any explosion gases or dust from the box.

13.4.2.2.2 The drop weights are shown in Figure 13.4.2.2. Each drop weight is provided with two locating grooves holding it between the guides as it drops, a suspension spigot, a removable cylindrical striking head and a rebound catch which are screwed on to the drop weight. The striking head is of hardened steel (HRC hardness of 60 to 63); its minimum diameter is 25 mm; it has a shoulder piece preventing it from being forced into the drop weight by the impact. Three drop weights are available with the following masses, 1.00 kg, 5.00 kg and 10.00 kg. The 1 kg-drop weight has a heavy steel centre fitted with the striking head. The 5 kg and 10 kg drop weights are of massive and compact steel, e.g. material specification at least St 37-1 in accordance with DIN 1700.

13.4.2.2.3 The sample of the substance under test is enclosed in an impact device consisting of two co-axial steel cylinders, one above the other in a hollow cylindrical steel guide ring. The cylinders are steel rollers from roller bearings with polished surfaces and rounded edges and a HRC hardness between 58 and 65. The dimensions of the cylinders and the ring are given in Figure 13.4.2.3. The impact device is placed on an intermediate anvil and centred by a locating ring with a ring of vent-holes to permit the escape of gases. The dimensions of the intermediate anvil are given in Figure 13.4.2.4 and those of the locating ring in Figure 13.4.2.3.

13.4.2.3 *Procedure*

13.4.2.3.1 For solid substances, other than paste-like or gel-types, the following points should be observed:

(a) Substances in powdered form are sieved (sieve mesh 0.5 mm), all that passes through the sieve is used for the test[1];

(b) Substances which have been compressed, cast or otherwise consolidated are broken into small pieces and sieved; the fraction passing a 1.0 mm sieve and retained on a 0.5 mm sieve is used for the test[1];

---

[1] *For substances which contain more than one constituent, the sieve fraction used for the test should be representative of the original sample.*

(c) Substances which are transported only in the form of charges are tested in the form of discs (chips) with a volume of 40 mm^3 (approximately 4 mm diameter and 3 mm thickness).

The cylinders and the guide ring should be degreased with acetone before use. The cylinders and guide ring should only be used once.

13.4.2.3.2 For substances in powdered form, a sample is taken with a cylindrical measure of 40 mm^3 capacity (3.7 mm diameter × 3.7 mm). For paste-like or gel-type substances, a cylindrical tube of the same capacity is inserted into the substance and after levelling off the surplus, the sample is removed from the tube by means of a wooden rod. For liquid substances, a fine-drawn pipette of 40 mm^3 capacity is used. The substance is placed in the open impact device, which is already in the locating ring on the intermediate anvil, and for powders or paste-like or gel-type substances, the upper steel cylinder is gently pressed until it touches the sample without flattening it. Liquid samples are placed in the open impact device in such a way that it fills the groove between the lower steel cylinder and the guide ring. The upper steel cylinder is lowered, with the aid of the depth gauge, until it is 2 mm from the lower cylinder (see Figure 13.4.2.5) and held in place by a rubber "O" ring. In some cases, capillary action causes the sample to exude from around the top of the sleeve. In these cases, the assembly should be cleaned and the sample re-applied. The filled impact device is placed centrally on the main anvil, the protective wooden box is closed and the appropriate drop weight, suspended at the required height, is released. In the interpretation of the results of the trial, distinction is made between "no reaction", "decomposition" (without flame or explosion) recognisable by change of colour or odour and "explosion" (with weak to strong report or inflammation). In some cases it is advisable to perform trials with appropriate inert reference substances to allow a better judgement of whether or not an audible report has occurred.

13.4.2.3.3 The limiting impact energy, characterising the impact sensitiveness of a substance, is defined as that lowest impact energy at which the result "explosion" is obtained from at least one out of at least six trials. The impact energy used is calculated from the mass of the drop weight and the fall height (e.g. 1 kg × 0.5 m ≈ 5 J). The 1 kg drop weight is used at fall heights of 10, 20, 30, 40 and 50 cm (impact energy 1 to 5 J); the 5 kg drop weight for fall heights of 15, 20, 30, 40, 50 and 60 cm (impact energy 7.5 to 30 J) and the 10 kg drop weight for fall heights of 35, 40 and 50 cm (impact energy 35 to 50 J). The series of trials is started with a single trial at 10 J. If at this trial the result "explosion" is observed, the series is continued with trials at stepwise lower impact energies until the result "decomposition" or "no reaction" is observed. At this impact energy-level, the trial is repeated up to the total number of six if no "explosion" occurs; otherwise the impact energy is reduced in steps until the limiting impact energy is determined. If at the impact energy level of 10 J the result "decomposition" or "no reaction" (i.e. no explosion) was observed, the test series is continued by trials at stepwise increased impact energies until for the first time the result "explosion" is obtained. Now the impact energy is lowered again until the limiting impact energy is determined.

13.4.2.4 *Test criteria and method of assessing results*

The test results are assessed on the basis of:

(a) Whether an "explosion" occurs in any of up to six trials at a particular impact energy; and

(b) The lowest impact energy at which at least one "explosion" occurs in six trials.

The test result is considered "+" if the lowest impact energy at which at least one "explosion" occurs in six trials is 2 J or less and the substance is considered too dangerous for transport in the form in which it was tested. Otherwise, the result is considered "-".

13.4.2.5 *Examples of results*

Substance	Limiting impact engergy (J)	Result
Ethyl nitrate (liquid)	1	+
Hexal 70/30	10	-
Hydrazine perchlorate (dry)	2	+
Lead azide (dry)	2.5	-
Lead styphnate	5	-
Mannitol hexanitrate (dry)	1	+
Mercury fulminate (dry)	1	+
NG (liquid)	1	+
PETN (dry)	3	-
PETN/wax 95/5	3	-
PETN/Wax 93/7	5	-
PETN/wax 90/10	4	-
PETN/water 75/25	5	-
PETN/lactose 85/15	3	-
RDX/water 74/26	30	-
RDX (dry)	5	-
Tetryl (dry)	4	-

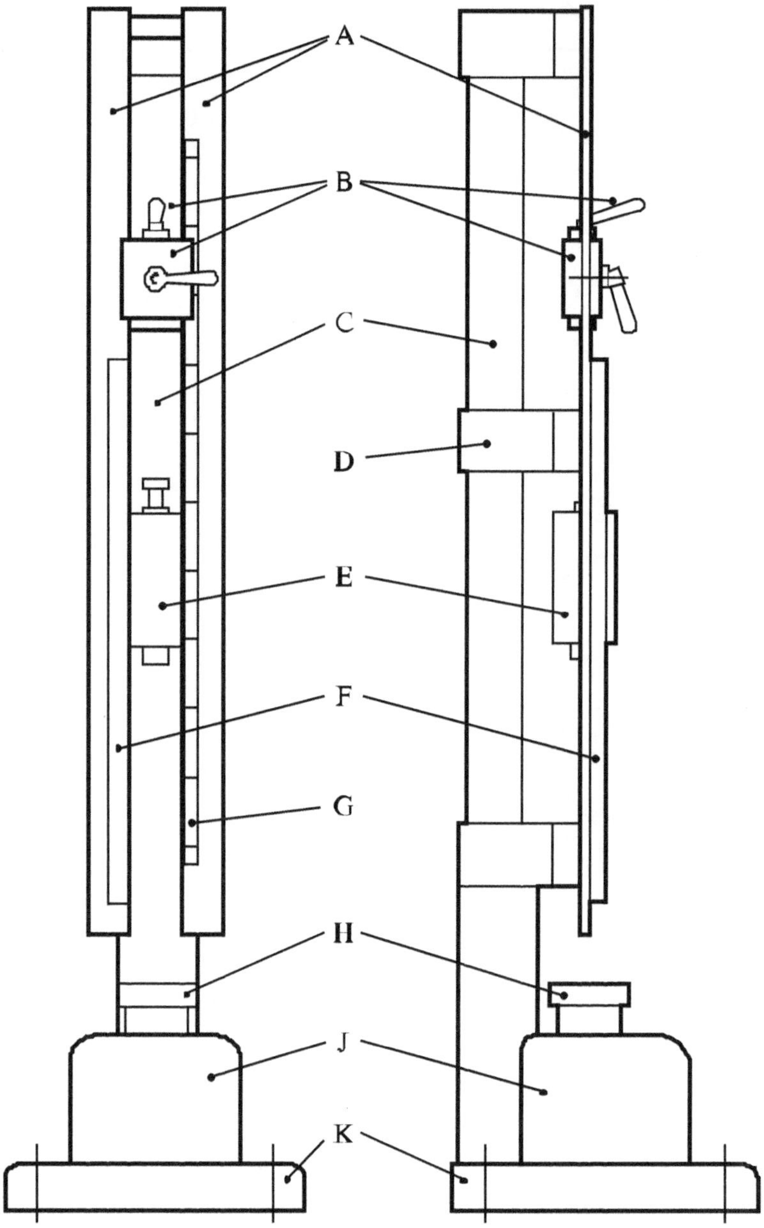

(A)	Two guides	(B)	Holding and releasing device
(C)	Column	(D)	Middle cross-piece
(E)	Drop weight	(F)	Toothed rack
(G)	Graduated scale	(H)	Anvil 100 mm diameter × 70 mm
(J)	Steel block 230 × 250 × 200 mm	(K)	Base 450 × 450 × 60 mm

Figure 13.4.2.1: BAM FALLHAMMER GENERAL VIEW, FRONT AND SIDE DIMENSIONS

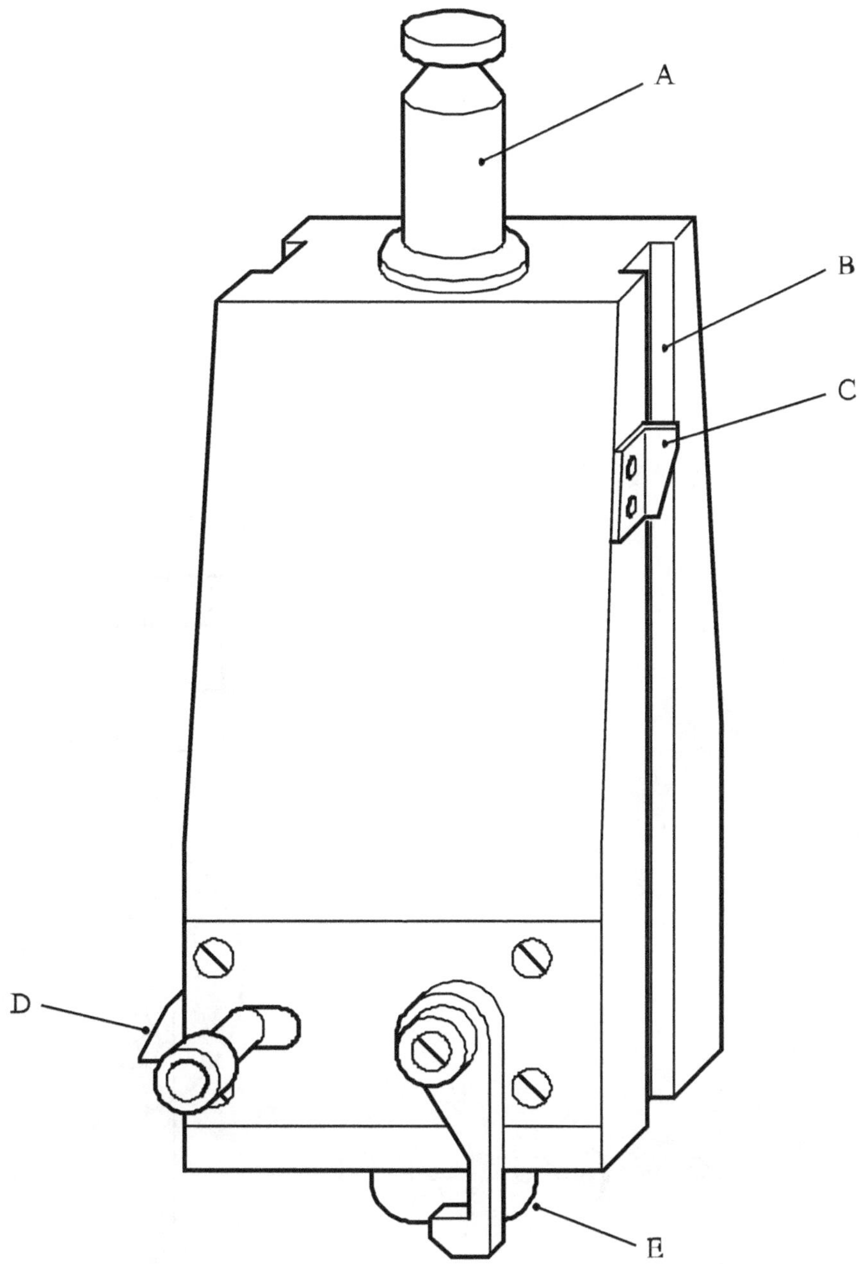

(A) Suspension spigot
(B) Positioning groove
(C) Height marker
(D) Rebound catch
(E) Cylindrical striking head

Figure 13.4.2.2: DROP WEIGHT

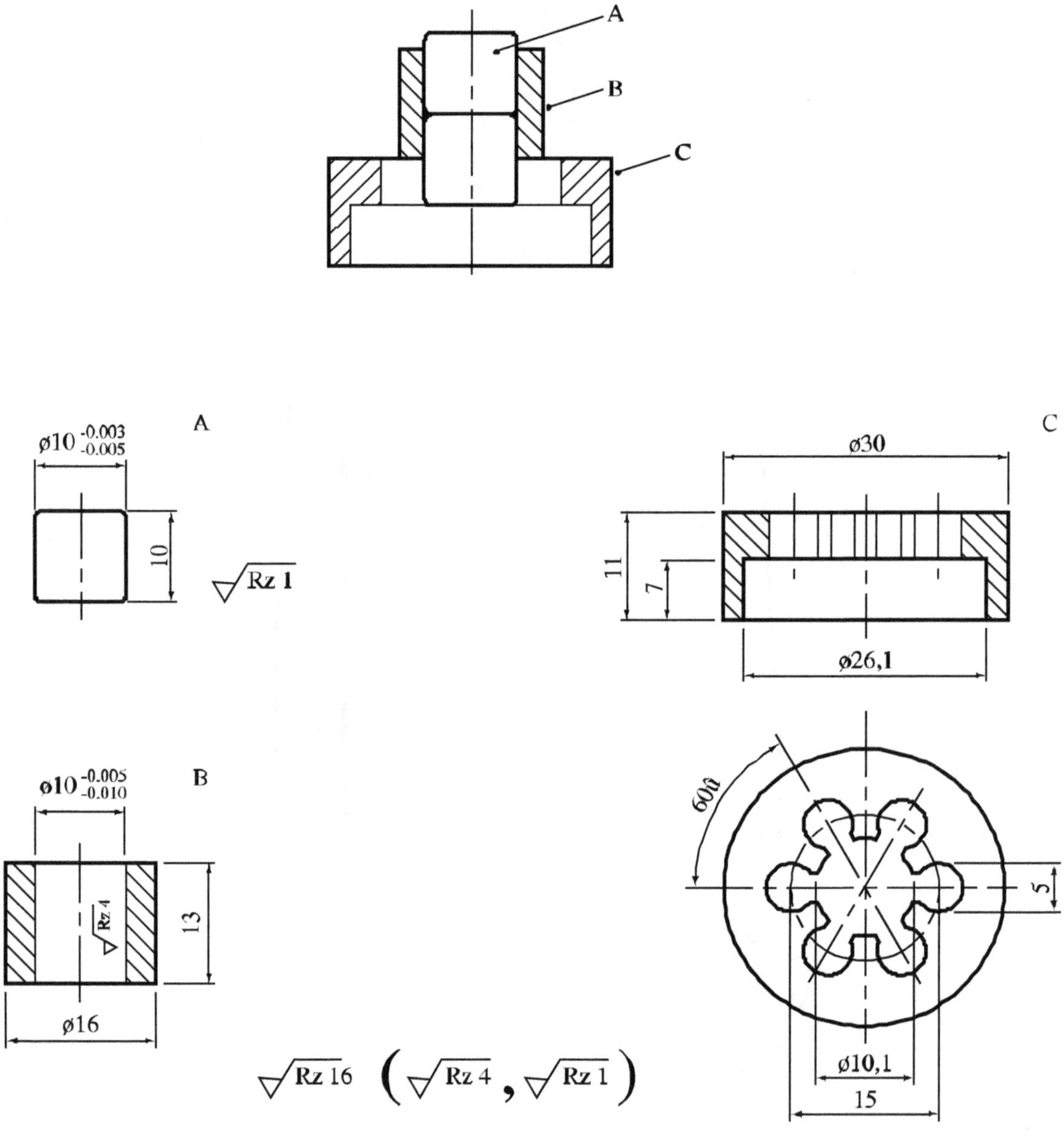

(A) Steel cylinder(s)
(B) Guide collar
(C) Locating ring

Figure 13.4.2.3: IMPACT DEVICE FOR SUBSTANCES IN POWDERED, PASTE-LIKE OR GEL TYPE FORM AND LOCATING RING

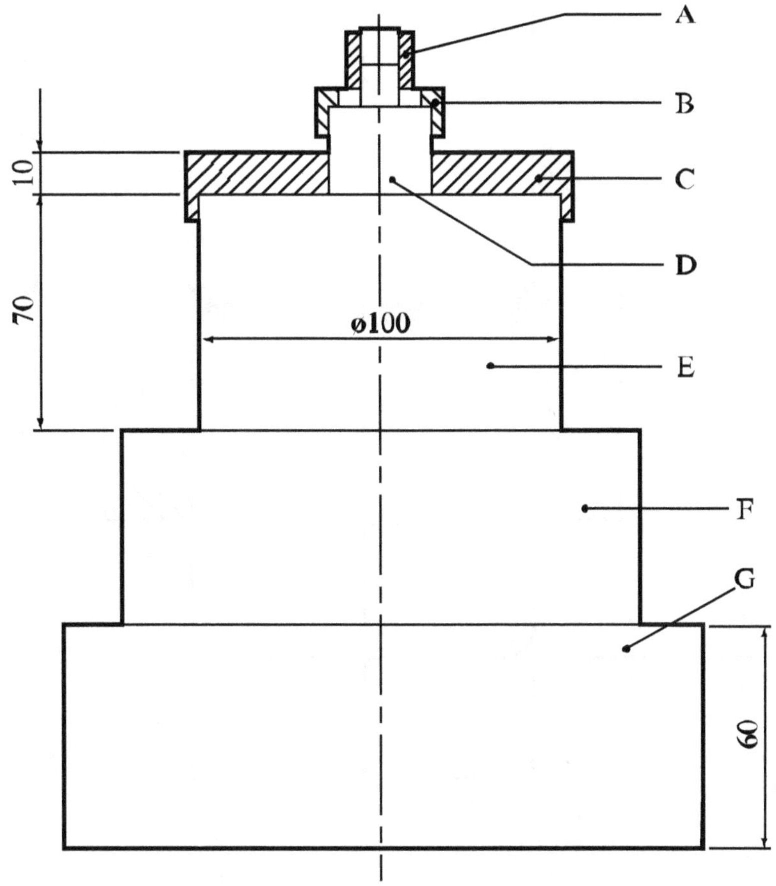

(A)	Impact device
(B)	Locating ring
(C)	Locating plate
(D)	Intermediate anvil 26 mm diameter × 26 mm
(E)	Anvil 100 mm diameter × 70 mm
(F)	Steel block 230 × 250 × 200 mm
(G)	Base 450 × 450 × 60 mm

Figure 13.4.2.4: LOWER PART

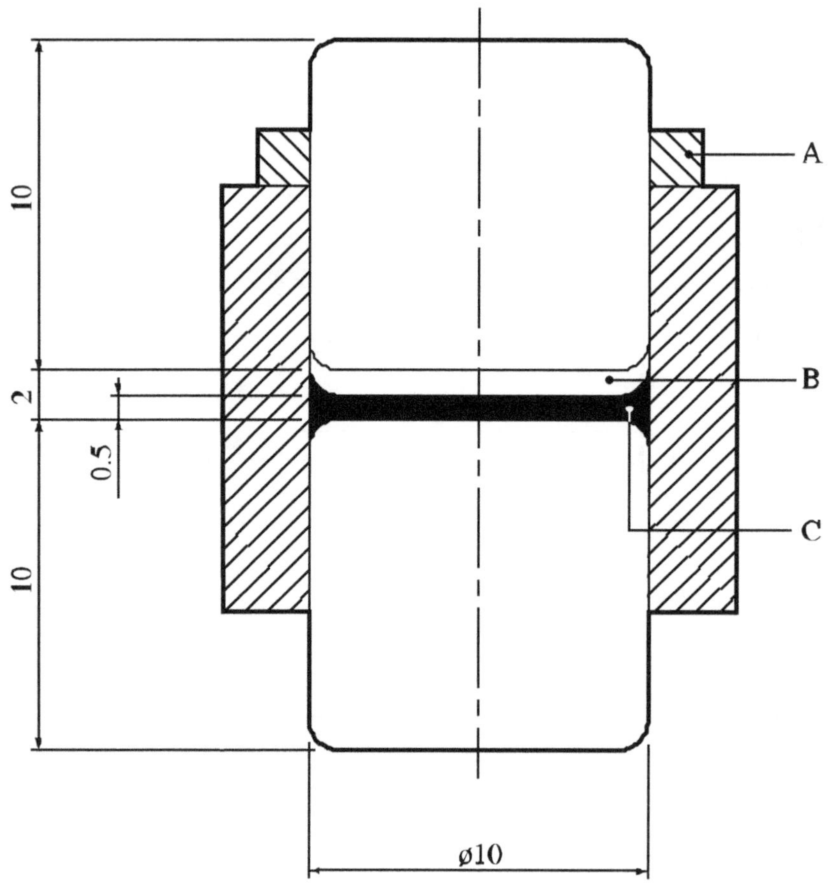

(A) Rubber ring (can sometimes be omitted)
(B) Space free from liquid
(C) Liquid sample spread round circumference of steel cylinder

Figure 13.4.2.5: IMPACT DEVICE FOR LIQUIDS

13.4.3 *Test 3 (a) (iii): Rotter test*

13.4.3.1 *Introduction*

This test is used to measure the sensitiveness of the substance to drop-weight impact and to determine if the substance is too dangerous to transport in the form tested. It is applicable to solid and liquid substances by using two different sample assemblies. The procedure can involve a direct comparison with a standard explosive, the median drop heights (50 per cent probability of ignition) being determined by the Bruceton method.

13.4.3.2 *Apparatus and materials*

13.4.3.2.1 Solids

The Rotter type impact machine (5 kg weight) and tools, are shown in outline in Figure 13.4.3.1 with an enlargement of the chamber in Figure 13.4.3.2. The hardened steel anvils, brass caps, measuring device (0.03 cm^3 volume), tamping device and gas measuring burette (50 cm^3) are manufactured to standard drawings. The standard explosive is RDX, recrystallized from cyclohexanone and dried according to a standard procedure.

13.4.3.2.2 Liquids

The equipment used for testing liquids is the Rotter type impact machine but with a different type of impact assembly (Figure 13.4.3.3) and drift (Figure 13.4.3.4), and without the burette. A 2 kg weight is employed. The various items shown in figures 13.4.3.2 to 13.4.3.4 are manufactured to standard drawings, as is a hardened steel disc for calibration of the cup and plunger assemblies.

13.4.3.3 *Procedure*

13.4.3.3.1 Solids

For solids, other than pastes or gels, the following points should be observed:

- If necessary, substances in coarse powdered form are crushed to pass through a 850 µm sieve; and
- Cast substances are either crushed and passed through a 850 µm sieve or 0.03 cm^3 discs are cut from the solid with nominal dimensions of 4 mm diameter and 2 mm thickness.

13.4.3.3.2 Powdered substances are measured into the caps with the measuring device and low bulk density substances are tamped with the tamping device. The loaded cap is placed upon the anvil in such a way as to avoid inversion of the cap before the substance makes contact with the anvil tip. The cap is then rotated to spread the explosive evenly, the chamber is closed, the striker is adjusted so that it is in contact with the cap and the chamber placed in position in the machine. The logarithms of the standard drop heights are arranged on a linear scale. The initial heights for starting the Bruceton (see Appendix 2) runs for the sample under test and the standard are determined by interpolation between the nearest "go" (ignition) and "no go" (non ignition) until these occur at adjacent levels. In a normal test, 50-shot Bruceton runs are performed. If the Sample Comparison Test (SCT) procedure (see Appendix 2) is used, caps of the standard and the sample under test are fired alternately, a separate Bruceton run being operated for each. When testing any explosive substance, a "go" is said to have occurred if 1 cm^3 or more of gaseous products are recorded on the manometer or if so indicated by non-standard transient movement of the manometer fluid which is confirmed by the presence of smoke on opening the anvil housing. For some pyrotechnics a lesser effect, e.g. colour change, is accepted as evidence of a "go". After testing each cap, the anvil and the interior of the chamber must be thoroughly cleaned and dried; the anvil is examined and changed if visibly damaged. Dropping the weight from heights much in excess of 200 cm can itself damage the anvil. Unless obtained from the SCT procedure, data for the standard is obtained from 50-shot running mean determinations.

13.4.3.3.3 Liquids

Individual cups and plungers for use with liquids are paired off before testing begins. The calibration disc is inserted into each cup in turn, its plunger added and the assembly located in the impact chamber. After the ball-bearing has been placed on top of the plunger, the upper part of the housing is fitted to the cap retainer and locked in position. The cap is then inserted and screwed down until the ball seating contacts the ball. This initial setting is read off a circular scale of 100 divisions on top of the upper part of the housing, and the reading is specific to the individual combination of cup and plunger used. Each division on the circular scale corresponds to a vertical displacement of 0.02 mm. For the test itself, an O-ring is placed in the cup. 0.025 cm^3 of the liquid under test is metered into the cavity, a suitable dispenser being a 0.5 cm^3 gas-tight syringe used in conjunction with a ratchet and a finely tipped plastic nozzle. A stainless steel disc is then dropped on to the O-ring, this seals off 0.025 cm^3 of air. The plunger is then located on top. The assembly is placed in the impact chamber, the ball-bearing on top of the plunger, and the upper part of the housing fitted and locked into position. The cap is then screwed down by hand until it contacts the ball (Figure 13.4.3.3). A standard pre-compression is applied to the sample chamber by screwing down the cap to the initial calibration position for the particular cup and plunger used, and for a further standard number of divisions on the circular scale. The housing is placed under the drop weight machine with the spherically recessed drift (Figure 13.4.3.4) resting on top of the ball-bearing. The testing procedure is similar to that for solids, the same Bruceton scale being used. A "go" is said to have occurred if a "bang" louder than that from a drop of equal height on to an inert liquid is heard, or if there is residual pressure in the sample chamber, or if on dismantling decomposition products are seen or smelt. Following "no goes" unchanged liquid remains in the sample chamber. After testing, the cup and plunger are thoroughly cleaned and if either shows signs of damage (usually pitting) it is replaced, when recalibration using the calibration disc is required. In any case, the O-ring and stainless steel disc are replaced by new components after each test.

13.4.3.4 *Test criteria and method of assessing results*

13.4.3.4.1 Solids

The test results are assessed on the basis of:

(a) Whether a "go" is observed in a trial;

(b) Determination of the median drop height for the RDX reference standard and the sample by the Bruceton method (see Appendix 2); and

(c) Comparison of the running mean median drop height of the standard (H_1) with the median drop height of the sample (H_2) using the equation:

Figure of Insensitiveness (F of I) = $80 \times H_2/H_1$

(if $H_2 \geq 200$ cm then the F of I is given as > 200)

The test result is considered "+" if the F of I is less then or equal to 80 and the substance is considered too dangerous for transport in the form in which it was tested. The test result is considered "-" if the F of I is greater than 80. If the F of I obtained for the substance under test is less than 80, a direct comparison with the standard RDX may be made by using the Sample Comparison Test (SCT - see Appendix 2) procedure with 100 shots on each substance. If there is 95% or greater confidence that the substance under test is not more sensitive than RDX, the substance under test is not too dangerous for transport in the form tested.

13.4.3.4.2 Liquids

The test results are assessed on the basis of:

- Whether a "go" is observed in a test; and
- Determination of the median drop height for the sample by the Bruceton method.

The median drop height for liquids is calculated as for solids, and the result quoted directly. For samples which do not give "goes" at drop heights of approximately 125 cm, the median height is quoted as "> 125 cm". The test result is considered "+" and the liquid too dangerous for transport in the form tested if it is more sensitive in this test than iso-propyl nitrate. This is normally determined from the median height value, but if the median height obtained for the substance under test is less than that quoted for iso-propyl nitrate, 14.0 cm, a direct comparison with iso-propyl nitrate may be made by using the SCT procedure with 100 shots on each substance. If there is 95% or greater confidence that the substance under test is not more sensitive than iso-propyl nitrate, the substance under test is not too dangerous for transport in the form tested. The result is considered "-" if the median height is greater than or equal to that of iso-propyl nitrate.

13.4.3.5 *Examples of results*

13.4.3.5.1 Solids

Substance	F of I	Result
Blasting Gelatine-geophex	15	+
Blasting Gelatine-Submarine	15	+
Cordite	20	+
1,3-Dinitrobenzene	>200	-
Guanidine nitrate	>200	-
HMX	60	+
Lead azide (service)	30	+
PETN	50	+
PETN/wax 90/10	90	-
RDX	80	+
Tetryl	90	-
TNT	140	-

13.4.3.5.2 Liquids

Substance	Median height(cm)	Result
Diethylene glycol dinitrate	12	+
Diethylene glycol mononitrate	46	-
1,1-Dinitroethane	21	-
Dinitroethylbenzene	87	-
Glyceryl trinitrate (nitroglycerine, NG)	5	+
Isopropyl nitrate	14	+
Nitrobenzene	>125	-
Nitromethane	62	-
Triethylene glycol dinitrate	10	+
Triethylene glycol mononitrate	64	-

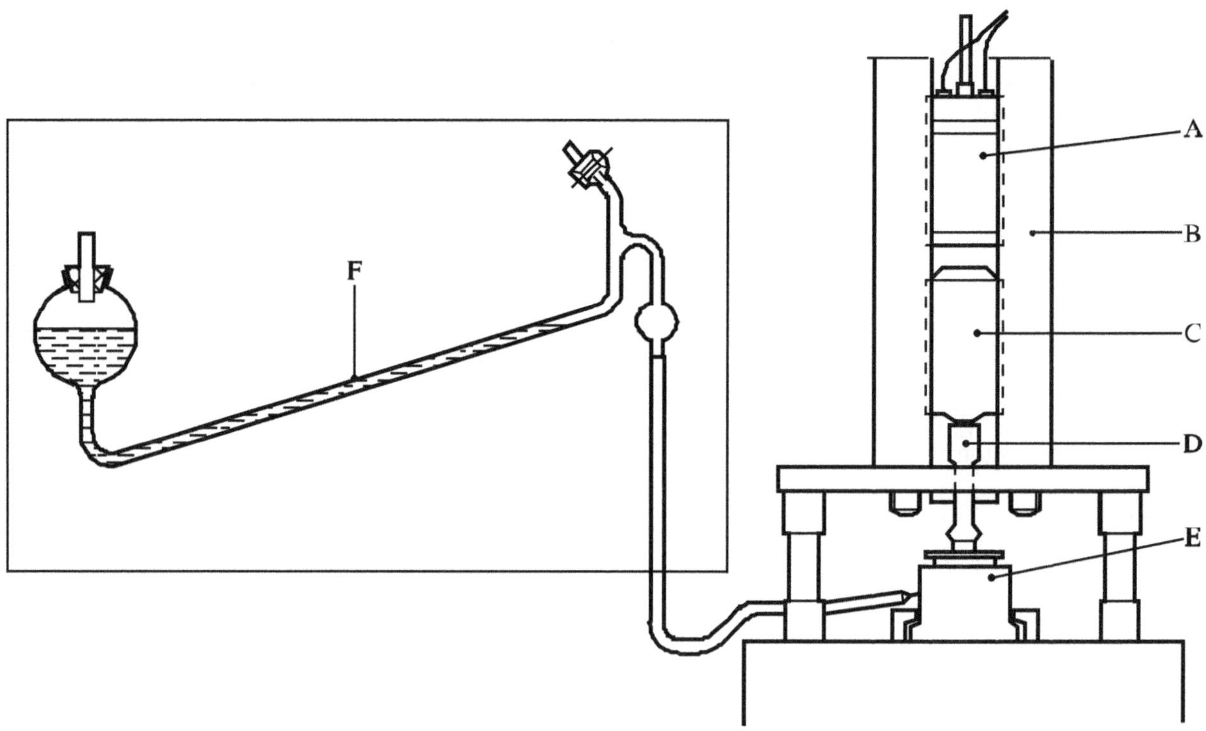

(A)	Magnet
(B)	Tubular guides
(C)	Weight
(D)	Drift
(E)	Chamber
(F)	Manometer containing dyed paraffin oil

Figure 13.4.3.1: ROTTER TEST

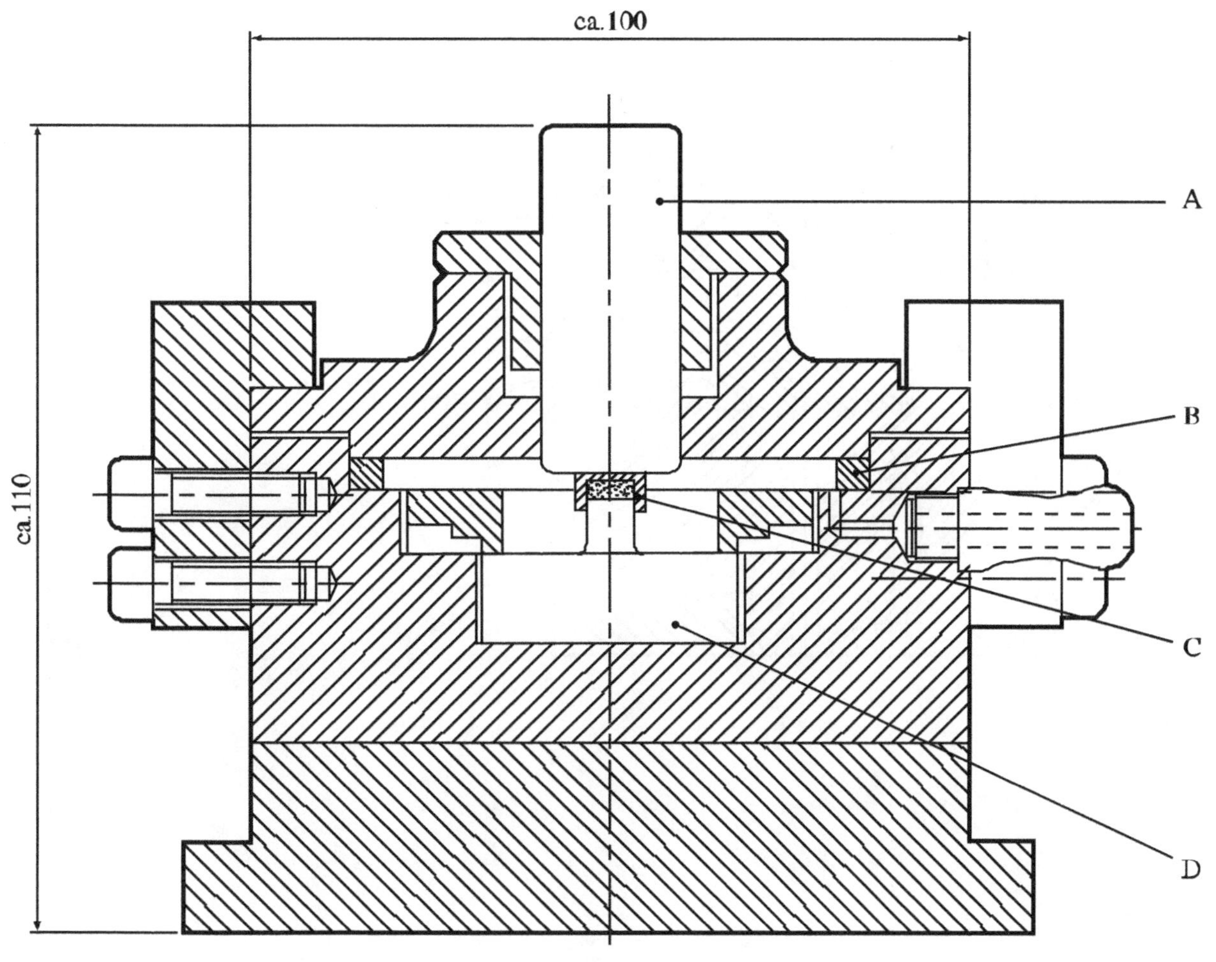

(A) Striker
(B) Sealing ring
(C) Cap
(D) Anvil

Figure 13.4.3.2: EXPLOSION CHAMBER

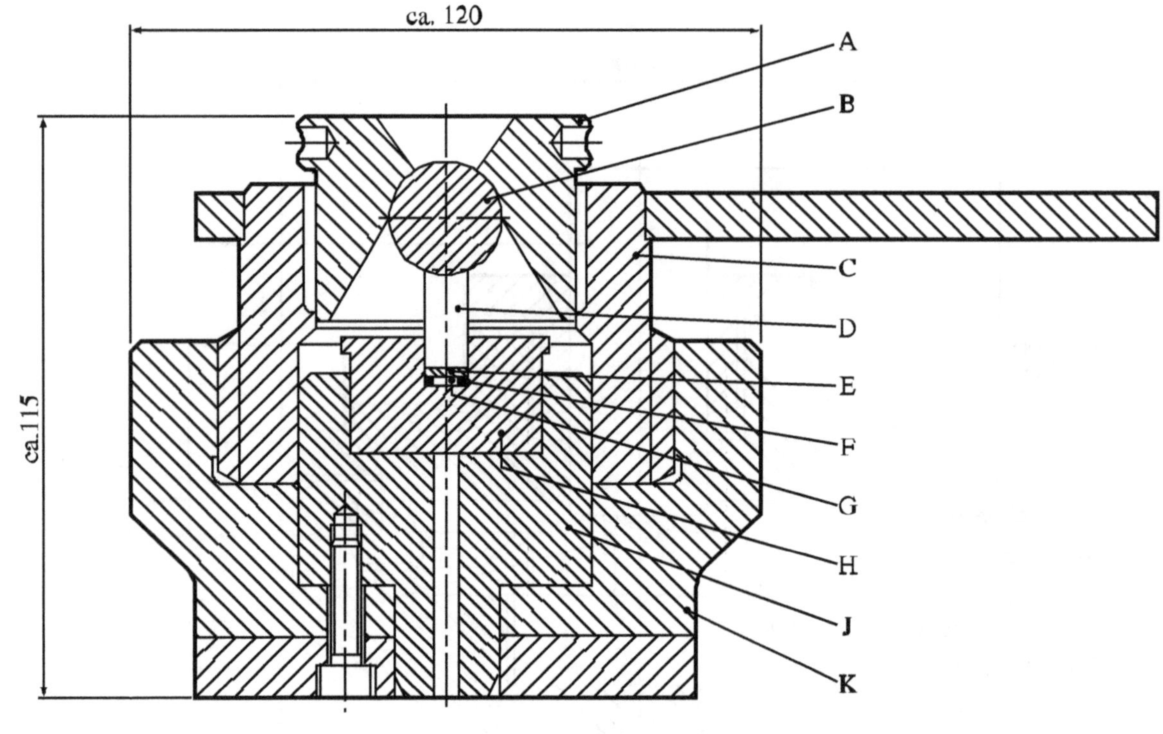

(A)	Cap
(B)	7/8 inch (22.2 mm) ball bearing
(C)	Upper part of housing
(D)	Plunger made of hardened tool steel
(E)	Stainless steel disc
(F)	Rubber O-ring
(G)	Test sample
(H)	Cup made of hardened tool steel
(J)	Impact chamber
(K)	Cap retainer

Figure 13.4.3.3: CUP AND PLUNGER ASSEMBLY AND HOUSING FOR LIQUIDS

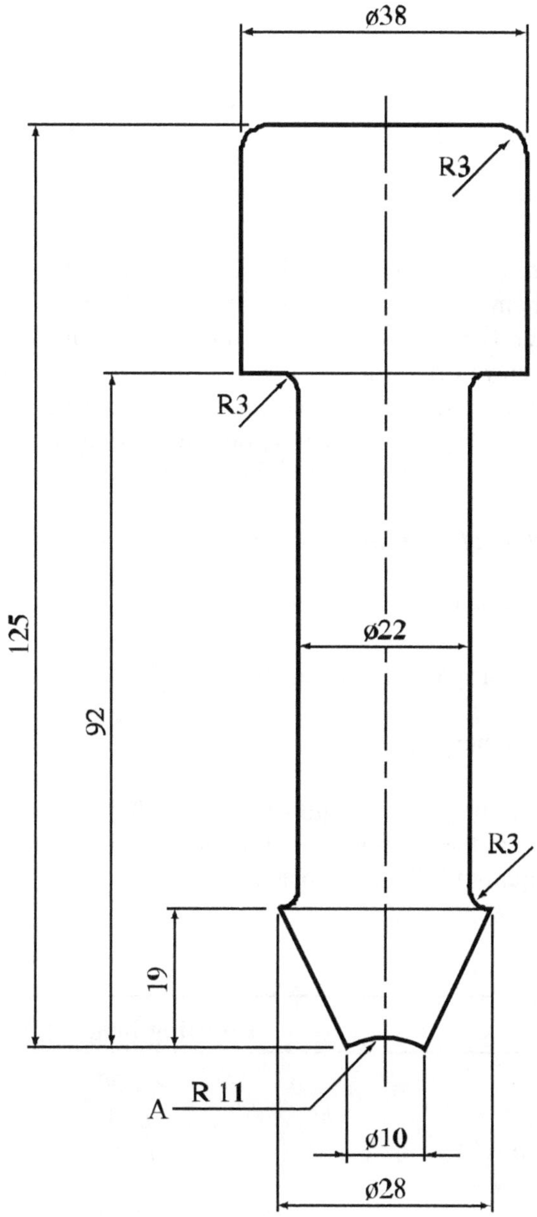

(A)	Spherical

Figure 13.4.3.4: LIQUID IMPACT TEST INTERMEDIATE DRIFT

13.4.4 *Test 3 (a) (iv): 30 kg Fallhammer test*

13.4.4.1 *Introduction*

This test is used to measure the sensitiveness of solids and liquids to drop-weight impact and to determine if the substance is too dangerous to transport in the form tested.

13.4.4.2 *Apparatus and materials*

The test apparatus and materials are shown in detail in figures 13.4.4.1 and 13.4.4.2. The steel sample tray (wall thickness ca. 0.4 mm) is 8 mm deep, 50 mm wide and 150 mm long.

13.4.4.3 *Procedure*

The sample tray is uniformly filled with the substance to a depth of 8 mm. The tray is placed on the anvil in such a way that the hammer falls onto a point 25 mm from one end, on the axis of the tray (see Figure 13.4.4.1). The fallhammer is released from a height of 4.00 m to 0.25 m in steps of 0.25 m. Propagation is considered to have occurred if explosive effects, primarily deformation of the tray walls, are observed at least 100 mm from the impacted point in the sample. Three trials are performed at each height. The limiting drop height is the highest height at which no propagation occurs in three trials. If there is no propagation using a drop height of 4.00 m, the limiting height is recorded as "≥ 4.00 m".

13.4.4.4 *Test criteria and method of assessing results*

The test results are assessed on the basis of:

(a) Whether there is propagation of reaction; and

(b) The limiting drop height.

The test result is considered "+" if a limiting drop height of less than 0.75 m is observed and the substance is considered too dangerous for transport in the form in which it was tested. The test result is considered "-" if a limiting drop height greater than or equal to 0.75 m is observed.

13.4.4.5 *Examples of results*

Substances	Limiting height (m)	Result
Ammonium perchlorate	≥ 4.00	-
HMX 0 - 100 μm (70% min ≤40 μm) [a]	0.50	+
HMX 80 - 800 μm (50% min ≥315 μm) [a, b]	1.75	-
Hydrazine nitrate, melted [c]	0.25	+
Mining explosive [d]	≥ 4.00	-
Nitroglycerine	0.50	+
Nitroguanidine	≥ 4.00	-
PETN fine (40% min ≤ 40 μm)	0.50	+
RDX 0 - 100 μm (55% min ≤ 40 μm) [a]	1.00	-
RDX mean size 125 - 200 μm	2.00	-
TNT flakes [e]	≥ 4.00	-
TNT cast	≥ 4.00	-

[a] Recrystallized from cyclohexanone

[b] RDX content : max. 3%

[c] 60-80 °C

[d] AN based, with Pentolite 11.5% and Al 8.5%

[e] Melting point ≥ 80.1 °C

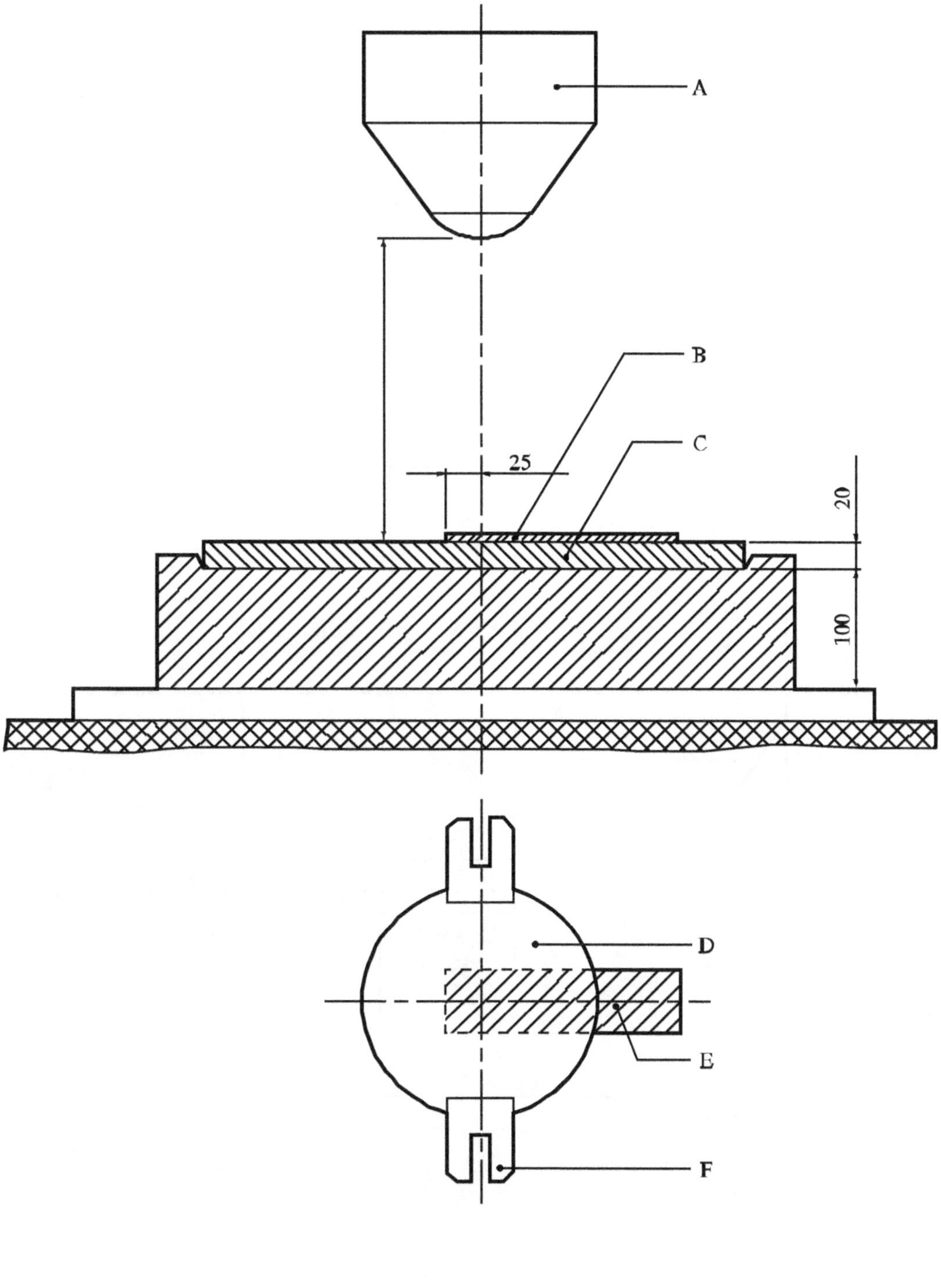

(A)	30 Kg weight	(B)	Sample
(C)	removable anvil	(D)	30 Kg weight
(E)	Sample	(F)	Guide - lug

Figure 13.4.4.1: 30 kg FALLHAMMER TEST

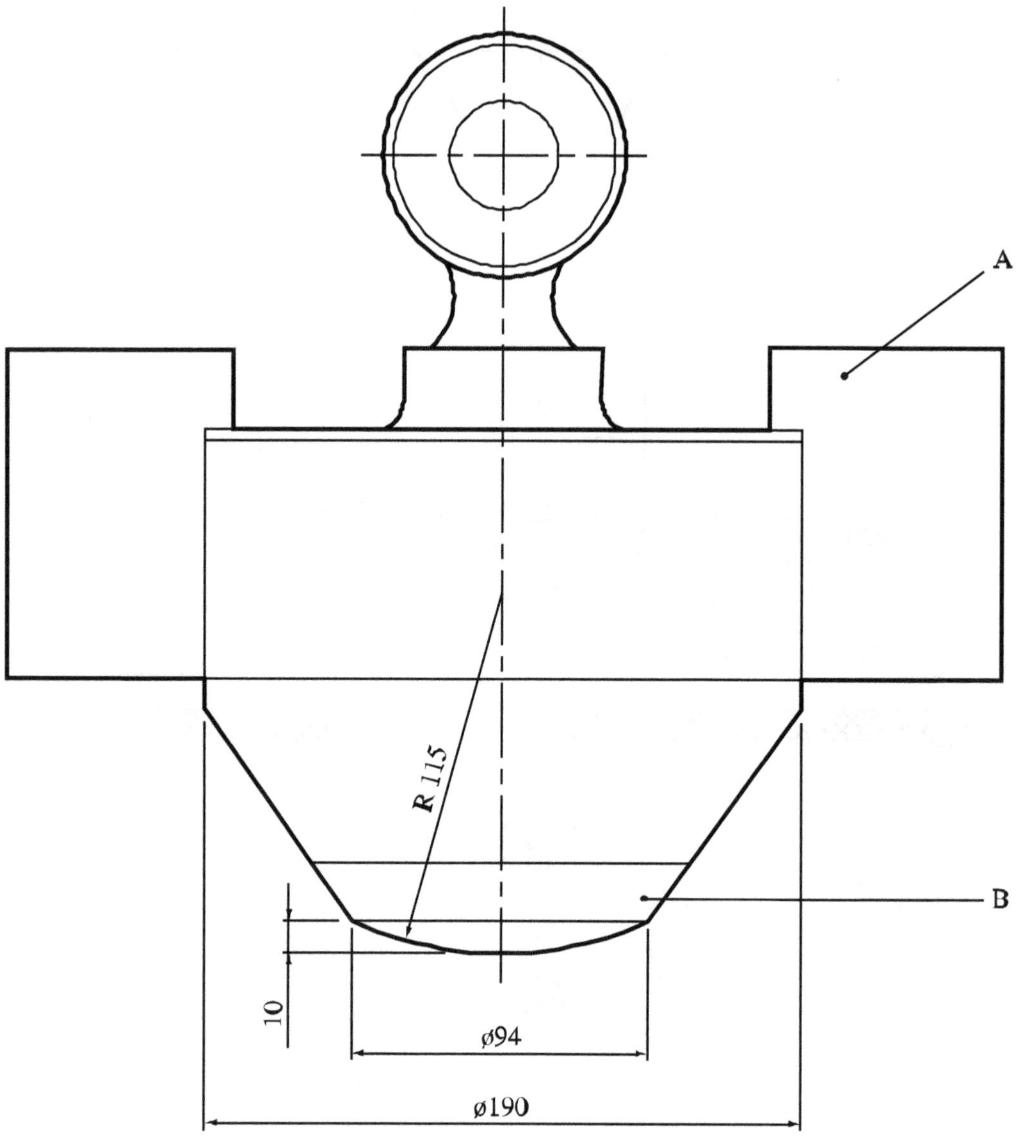

(A) Guide - lug
(B) Removable nose

Figure 13.4.4.2: DROP WEIGHT

13.4.5 *Test 3 (a) (v): Modified type 12 impact tool*

13.4.5.1 *Introduction*

This test is used to measure the sensitiveness of substances to drop-weight impact and to determine if the substance is too dangerous to transport in the form tested. It is applicable to solid and liquid substances by using two different sample assemblies.

13.4.5.2 *Apparatus and materials*

The following apparatus and materials are required:

(a) A drop mechanism capable of dropping, via three guides, a mass of 1.0, 1.5, 1.8, 2.0, 2.5 or 5.0 kg through a vertical distance of up to 3.0 m onto an intermediate mass resting on a sample placed on an anvil. The drop and intermediate masses are used in the following combinations:

(i) 1.5 kg intermediate mass with a 1.0, 1.5, 1.8 or 2.0 kg drop mass;

(ii) 2.0 kg intermediate mass with a 1.0 or 2.0 kg drop mass; and

(iii) 2.5 kg intermediate mass with 2.5 or 5.0 kg drop mass;

(b) A target assembly (modified Type 12 tool) consisting of an anvil (impact surface 32 mm diameter) and a guide for the intermediate mass;

(c) Garnet paper cut into 25 ± 2 mm squares;

(d) A balance with an accuracy of ± 1 mg;

(e) Brass caps with 10.0 mm diameter, 4.8 mm height and 0.5 mm wall thickness;

(f) Stainless steel discs 8.4 mm diameter and 0.4 mm thick;

(g) Neoprene O-rings 8.4 mm diameter and 1.3 mm thick;

(h) 50 μl syringe;

(i) Micro-spatula.

13.4.5.3 *Procedure*

13.4.5.3.1 Solids

The intermediate mass is raised. 30 ± 5 mg of the test substance is placed in a loose pile centrally on the anvil (for less sensitive substances, 30 ± 5 mg of the test substance is placed on a square of garnet paper and the garnet paper with the test substance is placed on the anvil). The intermediate mass is then carefully lowered onto the substance on the anvil. The drop mass is raised to 36.0 cm (the height in the middle of the logarithmic series of drop heights) and released to drop onto the intermediate mass. The intermediate mass is raised. The trial is assessed as positive if the sample reacts with an audible report, if there is production of smoke or odour, or there is visual evidence of an ignition. Note the type of reaction that occurs. The surfaces are cleaned with a cloth. The initial drop height for application of the Bruceton method (see Appendix 2) is determined by interpolation between the nearest drop heights giving positive and negative results until these occur on adjacent levels. 25 trials are then performed selecting the heights using the Bruceton method with a base 10 logarithmic interval of 0.093 giving the following series of drop heights: 6.5, 8, 10, 12, 15, 19, 24, 29, 36, 45, 55, 69, 85, 105, 131, 162 and 200 cm. The median height calculated from the results using the procedure given in Appendix 2. The combination of a 1.8 kg drop mass and a 1.5 kg intermediate mass, without the use of garnet paper, has been found to be optimal for determining whether substances are more or less sensitive than RDX.

13.4.5.3.2 Liquids

An O-ring is inserted into a cap and tamped down to the bottom. 25 μl[2] of the test substance is placed in the cap using a syringe. A stainless steel disc is placed on the O-ring. The intermediate mass is raised and the cap assembly is placed on the anvil. The intermediate mass is lowered carefully so that it fits into the cap and compresses the O-ring. The drop mass is raised and released to drop onto the intermediate mass. The intermediate mass is raised. The trial is assessed as positive if the sample reacts with an audible report, if there is production of smoke or odour, or if there is visual evidence of an ignition. Note the type of reaction that occurred. The initial height is selected using the procedure given in 13.4.5.3.1. 25 trials are performed and the median height calculated using the procedure given in Appendix 2. The combination of a 1.0 kg drop mass and a 1.5 kg intermediate mass (designed for testing liquids) has been found to be optimal for determining whether substances are more or less sensitive than isopropyl nitrate.

13.4.5.4 *Test criteria and method of assessing results*

13.4.5.4.1 The test results are assessed on the basis of:

(a) Whether a positive reaction is obtained in a trial; and

(b) Determination of the median drop height (H_{50}) for the sample by the Bruceton method.

Details of the statistics used to determine H_{50} and the standard deviation can be found in Appendix 2.

13.4.5.4.2 Solids

The test result is considered "+" if the median drop height (H_{50}) is less than or equal to that of dry RDX and the substance is considered too dangerous for transport in the form in which it was tested. The test result is considered "-" if the medium drop height (H_{50}) is greater than that of dry RDX.

13.4.5.4.3 Liquids

The test result is considered "+" if the median drop height (H_{50}) is less than that of isopropyl nitrate and the substance is considered too dangerous for transport in the form in which it was tested. The test result is considered "-" if the medium drop height (H_{50}) is equal to or greater than that of isopropyl nitrate.

---

2 *The relationship between the sample volume and the sensitiveness of the liquid is a function specific to the liquid. The volume selected in this procedure is suitable for determining relative sensitiveness. A determination of the relationship between the sensitiveness and the sample volume should be carried out when more detailed information is required about the substance.*

13.4.5.5 *Examples of results*

13.4.5.5.1 Solids

Substances	Median height (cm)	Result
1.8 kg drop mass, 1.5 kg intermediate mass, no garnet paper		
PETN (superfine)	15	+
RDX Grade 1	38	+
RDX/water (75/25)	> 200	–
Tetryl	> 200	–
TNT (200 mesh)	> 200	–
2.5 kg drop mass, 2.5 kg intermediate mass, with garnet paper		
PETN (superfine)	5	+
RDX (Cal 767)	12	+
Tetryl	13	-
TNT (200 mesh)	25	-

13.4.5.5.2 Liquids

Substances	Median height (cm)	Result
1.0 kg drop mass, 2.0 kg intermediate mass		
Isopropyl nitrate (99%, boiling point. 101-102 °C)	18	-
Nitromethane	26	-
TEGDN	14	+
TMETN	10	+
TEGDN/TMETN (50/50)	13	+

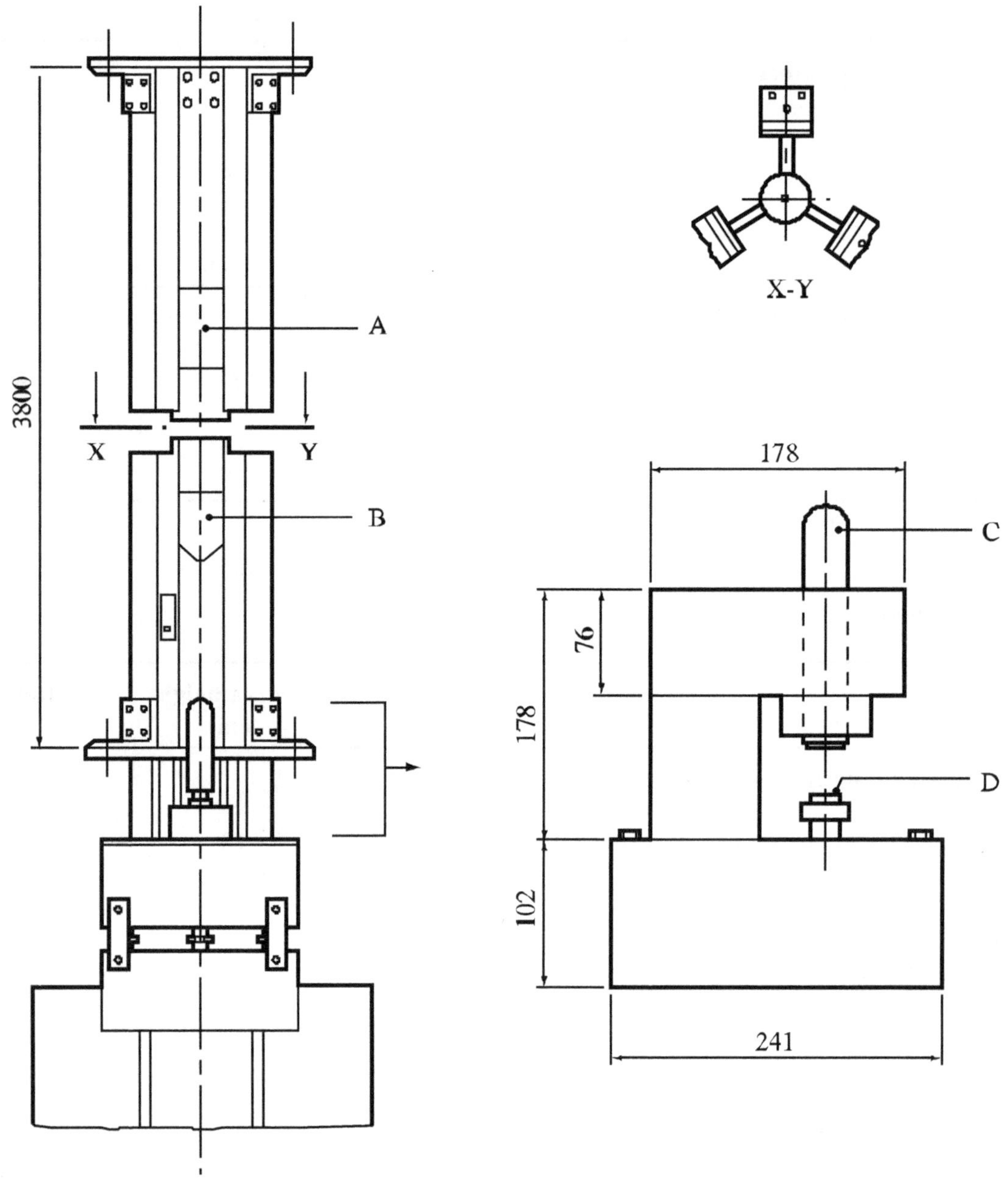

(A)	Electromagnet
(B)	Drop mass (e.g. 2.5 kg)
(C)	Intermediate mass (e.g. 2.5 kg , diameter 32 mm)
(D)	Anvil (impact surface 32 mm diameter)

Figure 13.4.5.1: MODIFIED TYPE 12 IMPACT TOOL (Full, top and expanded side view)

13.4.6 *Test 3 (a) (vi): Impact sensitivity test*

13.4.6.1 *Introduction*

This test is used to measure the sensitiveness of the substance to drop-weight impact and to determine if the substance is too dangerous to transport in the form tested. It is applicable to solid and liquid substances by using two different sample assemblies.

13.4.6.2 *Apparatus and materials*

13.4.6.2.1 Figure 13.4.6.1 shows a diagram of the impact apparatus. The main components of the apparatus are:

(a) An anvil made of seamless steel;

(b) Vertical parallel guide columns for a falling weight;

(c) A steel weight (10 kg) with a limit stop - the weight impact head is made of hardened steel (Rockwell Chardness 60-63);

(d) A catching and dropping device;

(e) A rack which prevents the weight from repeatedly falling and impacting on the sample on the anvil; and

(f) A measuring rule with 1 mm scale divisions.

13.4.6.2.2 The test sample is placed into the roller assembly 2 (solids) or 3 (liquids). Dimensions and requirements for these roller assemblies are shown in figures 13.4.6.2 and 13.4.6.3. The following additional equipment is also required:

(a) A laboratory balance with a weighing error of no more than 0.005 g;

(b) A hydraulic press which provides a compression pressure of 290 MPa; and

(c) A standard explosive, tetryl (recrystallized from acetone) with crystal sizes 0.200-0.270 mm.

13.4.6.3 *Procedure*

13.4.6.3.1 Solids

13.4.6.3.1.1 As a rule substances are tested in the form in which they are received. Wetted substances should be tested with the minimum quantity of wetting agent required for transport. Depending on their physical form, the substances should then be subjected to the following procedures:

(a) Granulated, flaky, pressed, cast and similar substances are milled and sieved; particles should pass through a sieve with 0.9-1.0 mm mesh sizes;

(b) Elastic substances are cut with a sharpened knife on a wooden surface into pieces not more than 1 mm in size. Elastic substances samples are not sieved; and

(c) Powdered and plastic explosives samples are not ground and sieved.

Roller assemblies for solid samples are degreased with acetone or ethyl alcohol. Prepared test assemblies should have 0.02-0.03 mm difference between the diameters of the sleeves and the rollers. The components may be re-used if they remain within specification.

13.4.6.3.1.2 To determine the lower sensitivity limit of the explosive under test, a sample with mass of 100 ± 5 mg is placed on the roller surface in the opened roller of assembly 2. The sleeve should be aligned with the direction of the groove downwards. The second roller is placed on the explosive sample and the upper roller is used to smooth it out by pressing and rotating. The assembly containing the explosive is placed on a hydraulic press, where it is compressed to a pressure of 290 MPa. For plastic, elastic and pasty explosives, the pressure is preselected so that the explosive is not pressed out beyond the faces of the rollers. Wetted explosives are not compressed. The sleeve with the rollers and the explosive is then turned over in a tray and placed against the rollers as far as they will go. This ensures that the explosive is in contact with the sleeve groove. The assembly containing the explosive is placed on the anvil of the impact apparatus. The weight (10 kg) falls and impacts the sample.

13.4.6.3.1.3 The lower limit of explosive impact sensitivity is defined as the maximum drop height of the 10 kg steel weight which does not give positive results in 25 trials. The drop height is selected from the following range: 50, 70, 100, 120, 150, 200, 250, 300, 400, 500 mm. Tests are carried out beginning with 150 mm drop height. A sound effect, flash or burn traces on the rollers and sleeve are taken as a positive reaction. Discolouration of the sample is not considered as a sign of explosion. If positive results are obtained at this height the test is repeated with the next lower drop height. Conversely, when negative results are obtained, the next higher drop height is used. Thus the maximum drop height for a 10 kg mass is obtained when no positive reaction occurs in 25 tests. If, at the 50 mm drop height, positive results are obtained in 25 tests, it is established that the lower sensitivity limit of explosive tested in roller assembly 2 is < 50 mm. When no positive reaction occurs in 25 tests for the 500 mm drop height, then the lower impact sensitivity limit for the explosive tested in roller assembly 2 is expressed as 500 mm or more.

13.4.6.3.2 Liquids

13.4.6.3.2.1 Roller assemblies 3 are degreased with acetone or ethyl alcohol. Usually 35-40 roller assemblies are prepared. The roller assemblies should have a 0.02-0.03 mm difference between the diameters of sleeves and rollers.

13.4.6.3.2.2 To determine the lower sensitivity limit, the liquid substance is placed in the cap with a dropping tube or pipette. The cap is positioned in the centre of the lower roller and is completely filled with the liquid substance. The second roller is carefully placed on the cap containing the liquid substance, the roller assembly is placed on the impact apparatus anvil and the steel weight is dropped. The result is noted.

13.4.6.3.2.3 The lower limit of an explosive impact sensitivity is defined as the maximum drop height of the 10 kg steel weight which does not give positive results in 25 trials. The drop height is selected from the following range: 50, 70, 100, 120, 150, 200, 250, 300, 400, 500 mm. Tests are carried out beginning with a 150 mm height. If positive results are obtained at this height, the test is repeated with the next lower drop height. Conversely, when negative results are obtained, the next higher drop height is used. Thus the maximum drop height for 10 kg mass is obtained when no positive reaction occurs in 25 tests. If, at the 50 mm drop height, one or more positive results are obtained in 25 tests then it is established that the lower sensitivity limit of explosive tested in roller assembly 3 is < 50 mm. If no positive reaction occurs in 25 tests for the 500 mm drop height, the lower impact sensitivity limit for the explosive tested in roller assembly 3 is expressed as 500 mm or more.

13.4.6.4 *Test criteria and method of assessing results*

13.4.6.4.1 Solids

The test results are assessed on the basis of:

(a) Whether one or more positive results are obtained in 25 trials at a particular height; and

(b) The lowest height at which a positive result is obtained.

The test result is considered "+" if the lowest height at which a positive result is obtained with assembly 2 is less than 100 mm and the substance is considered too dangerous for transport in the form in which it was tested. The test result is considered "-" if the lowest impact height for a positive result with assembly 2 is equal to or greater than 100 mm.

13.4.6.4.2 Liquids

The test results are assessed on the basis of:

(a) Whether one or more positive results are obtained in 25 trials at a particular height; and

(b) The lowest height at which a positive result is obtained.

The test result is considered "+" if the lowest height at which a positive result is obtained with assembly 3 is less than 100 mm and the substance is considered too dangerous for transport in the form in which it was tested. The test result is considered "-" if the lowest impact height for a positive result with assembly 3 is equal to or greater than 100 mm.

13.4.6.5 *Examples of results*

13.4.6.5.1 Solids

Substances	**Lower limit in assembly 2 (mm)**	**Result**
Ammonal (80.5% ammonium nitrate, 15% trotyl, 4.5% aluminium)	150	-
Ammonal, splitting (66% ammonium nitrate, 24% hexogen, 5% aluminium)	120	-
Ammonite 6ZhV (79% ammonium nitrate, 21% Trotyl)	200	-
Ammonite T-19 (61% ammonium nitrate, 19% Trotyl, 20% sodium chloride)	300	-
Cyclotrimethylenetrinitramine (dry)	70	+
Cyclotrimethylenetrinitramine / wax 95/5	120	-
Cyclotrimethylenetrinitramine / water 85/15	150	-
Granulite AS-8 (91.8% ammonium nitrate, 4.2% machine oil, 4% aluminium)	>500	-
Pentaerythritol tetranitrate (dry)	50	+
Pentaerythritol tetranitrate / paraffin 95/5	70	+
Pentaerythritol tetranitrate / paraffin 90/10	100	-
Pentaerythritol tetranitrate / water 75/25	100	-
Picric acid	>500	-
Tetryl	100	-
Trinitrotoluene	>500	-

13.4.6.5.2 Liquids

Substances	Lower limit in assembly 2 (mm)	Result
Di-(2,2-dinitro-2-fluoro-ethyl)formal/methylene chloride 65/35	400	-
Isopropyl nitrate	>500	-
Nitroglycerine	<50	+
Nitromethane	>500	-

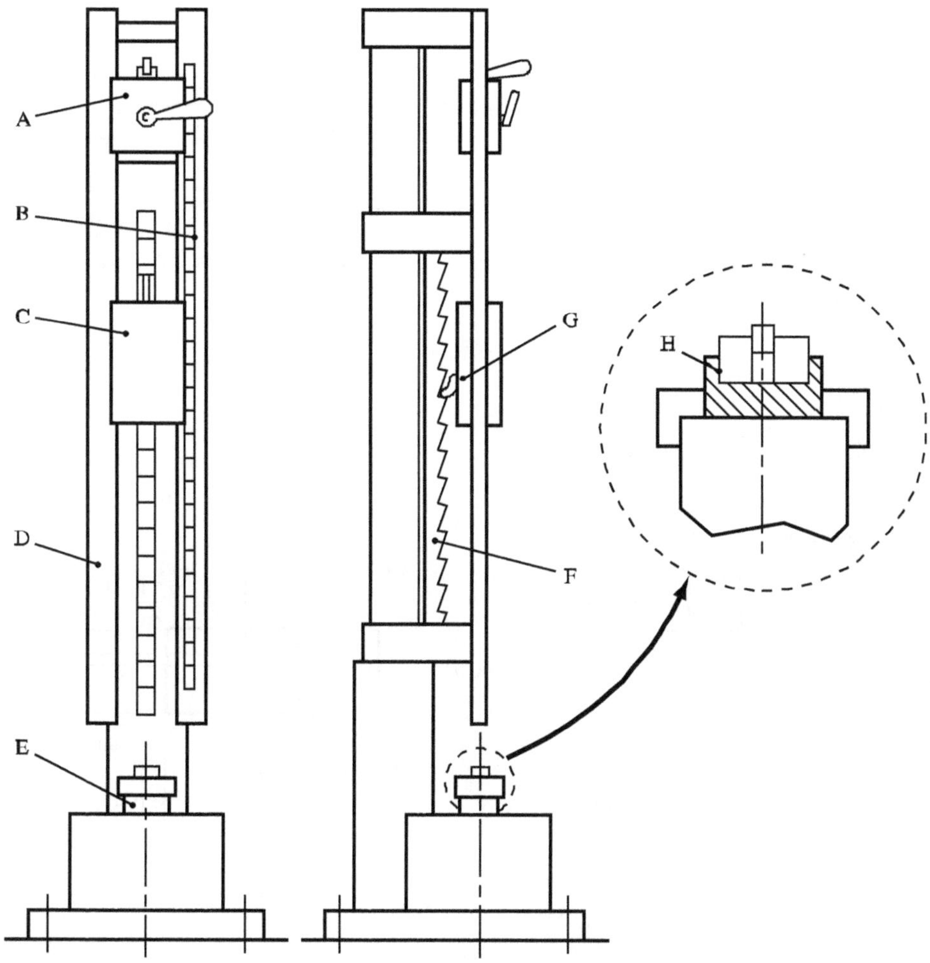

(A)	Catching and dropping device	(B)	Graduated scale
(C)	Drop weight	(D)	Guide column
(E)	Anvil	(F)	Rack
(G)	Ratchet to catch rebound	(H)	Magnified view of roller assembly

Figure 13.4.6.1: IMPACT APPARATUS

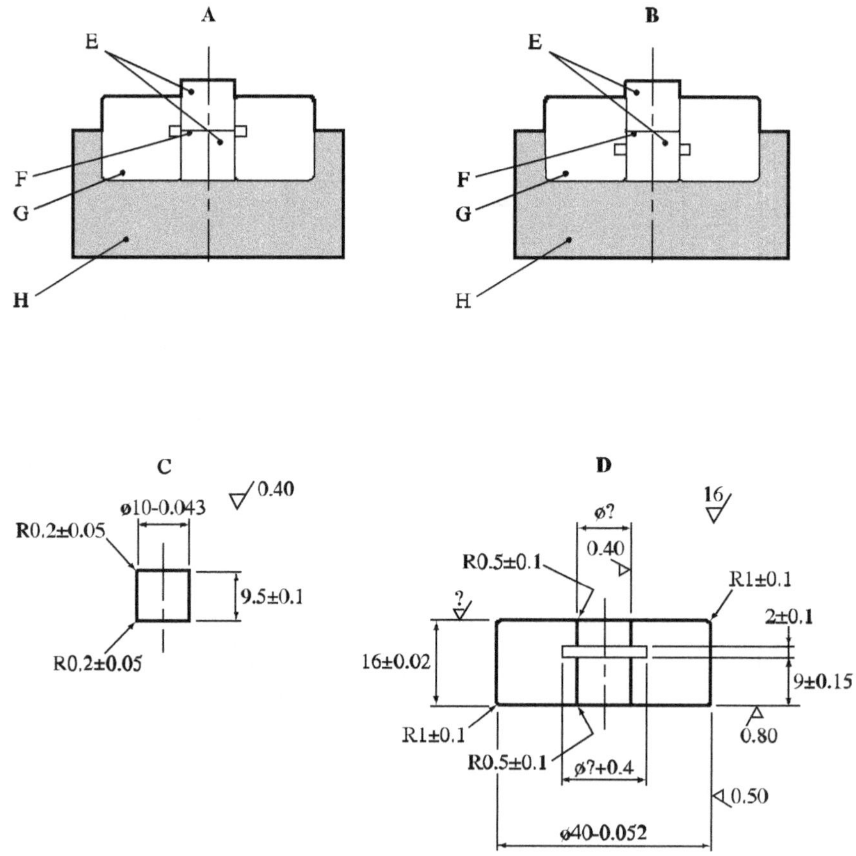

(A)	Sleeve position "groove facing upwards"	(B)	Sleeve position "groove facing downwards"
(C)	Roller of ball bearing steel to HRC 63 - 66	(D)	Sleeve of tool carbon steel to HRC 57 - 61
(E)	Rollers	(F)	Sample
(G)	Sleeve	(H)	Tray

Figure 13.4.6.2: ROLLER ASSEMBLY 2

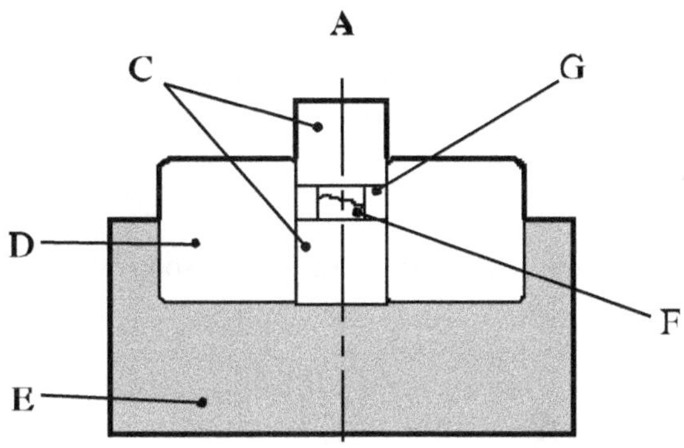

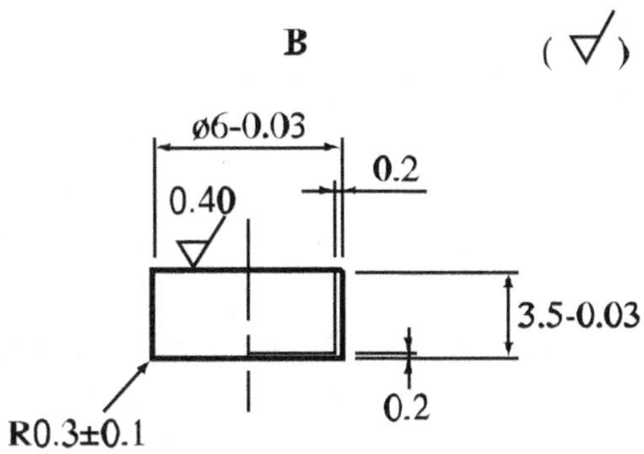

(A)	Roller assembly 3
(B)	3 micron nickel plated copper (M2) cap
(C)	Rollers
(D)	Sleeve
(E)	Tray
(F)	Sample
(G)	Cap

Figure 13.4.6.3: ROLLER ASSEMBLY 3

13.4.7 *Test 3 (a) (vii): Modified Bureau of Mines impact machine test*

13.4.7.1 *Introduction*

This test is used to measure the sensitiveness of the substance to drop-mass impact and to determine if the substance is too dangerous to transport in the form tested. The test substance is subjected to a vertical impact force through an intermediate hammer via a drop mass. It is applicable to solid, semisolid, liquid, and powder substances.

13.4.7.2 *Apparatus and materials*

13.4.7.2.1 The general design of the MBOM impact test apparatus is given in Figure 13.4.7.1. The following components are required:

A mechanism containing a drop mass of 2.0 kg , two drop mass guide rails, a drop mass holding, lifting, and dropping mechanism, and a 1.0 kg intermediate hammer containing a 1.27 cm diameter steel insert with a surface roughness of 1.3 – 1.8 μm that is resting on a sample placed on a steel anvil (impact surface 3.8 cm diameter) with a surface roughness of 1.3 – 1.8 μm. Details of the target area are given in Figure 13.4.7.2.

13.4.7.3 *Procedure*

13.4.7.3.1 Solid sample placement

As a rule substances are tested in the form in which they are received. Wetted substances should be tested with the minimum quantity of wetting agent required for transport. Depending on the physical form, the substances should then be subjected to the following procedures:

(a) Powders are to be tested on the anvil in a monolayer; i.e., the thickness of the granular material. Place enough granules on the anvil to cover an area in excess of the 1.3 cm^2 area of the insert.

(b) Solid propellants are tested in the form of thin, uniform slices. The slices are usually square, having a minimum edge length of 1.6 cm and a thickness of 0.08 ± 0.01 cm. This thickness is easily obtainable with the use of a microtome cutting tool.

The intermediate hammer is raised. The test substance is placed centrally on the anvil. The intermediate hammer is then carefully lowered onto the substance on the anvil.

13.4.7.3.2 Liquids and semisolids sample placement

Depending on the physical form, the substances should then be subjected to the following procedures:

(a) Liquids are tested with a controlled thickness and a fixed gap of 0.05 cm above the liquid level using a spring between the hammer collar and the guide collar (adjustable tension). The thickness of the liquid sample is controlled by putting a piece of 0.015 cm thick tape (compatible with the substance) with a 1.6 cm diameter hole in it on the anvil. The intermediate hammer is raised. The hole in the tape is centrally positioned on the anvil such that the intermediate hammer insert does not touch the tape. A 0.05 cm feeler gauge is used to set the proper gap above the liquid. The tape hole is filled with the liquid substance and levelled-off using a straight-edge ensuring that no air gaps are present in the sample. The intermediate hammer is then carefully lowered to 0.05 cm above the substance on the anvil.

(b) Semisolids (slurries, gels, etc.) are prepared and tested in much the same way as liquid samples; however, the sample thickness is governed by the largest particle size. If the largest particle size is greater than the 0.015 cm thickness then a monolayer sample is spread on the anvil in a monolayer; i.e., the thickness of the granular material. If the cohesive properties of the semisolid are not practical for a 0.015 cm thickness, then the minimum attainable thickness is used. Place enough granules on the anvil to cover an area in excess of the 1.3 cm^2 area of the intermediate hammer insert.

13.4.7.3.3 Machine operation

The drop mass is raised to the desired height (17 cm for solids and semisolids and 11 cm for liquids) and released to drop onto the intermediate hammer. Observations are made on whether a "reaction" occurs as evidenced by audible report or production of smoke, fire, charring or visible light as observed by human senses. The type of reaction that occurs is documented. The surfaces are cleaned with a cloth or light abrasive pad to remove any residual material from the anvil or intermediate hammer insert. The anvil and intermediate hammer insert are inspected for scratches, scoring, divots, or other damage which may affect the surface roughness. If damaged these items should be replaced before use on the next trial. Six trials are performed for each test sample.

13.4.7.4 *Maintenance and calibration*

Moving parts should be inspected to ensure that they are freely moving and that friction between them is minimal. The distance between the drop mass and the intermediate hammer that is resting on the anvil should be verified. The contact area between the intermediate hammer insert and anvil should be uniform. The test machine should be periodically cleaned and calibrated according to a schedule based on the amount of usage. At a minimum, the machine should be calibrated on an annual basis.

13.4.7.5 *Test criteria and method of assessing results*

13.4.7.5.1 Solids

The test result is considered "+" if a reaction (see 13.4.7.3.3) is observed in at least 1 out of 6 trials at a drop height of 17 cm and the substance is considered too dangerous for transport in the form in which it was tested. Otherwise, the result is considered "–". Borderline cases may be resolved using the Bruceton method (see Appendix 2).

13.4.7.5.2 Liquids

The test result is considered "+" if a reaction (see 13.4.7.3.3) is observed in at least 1 out of 6 trials at a drop height of 11 cm and the substance is considered too dangerous for transport in the form in which it was tested. Otherwise, the result is considered "–". Borderline cases may be resolved using the Bruceton method (see Appendix 2).

13.4.7.6 *Examples of results*

Substances[1]	**Result**
RDX (dry)	+
PBXN-8	–
Nitrocellulose/DNT (90/10)	–
PETN (dry)	+
Nitroglycerin	+

1 *Data acquired at relative humidity of 10-30% and temperature of 16-24 °C*

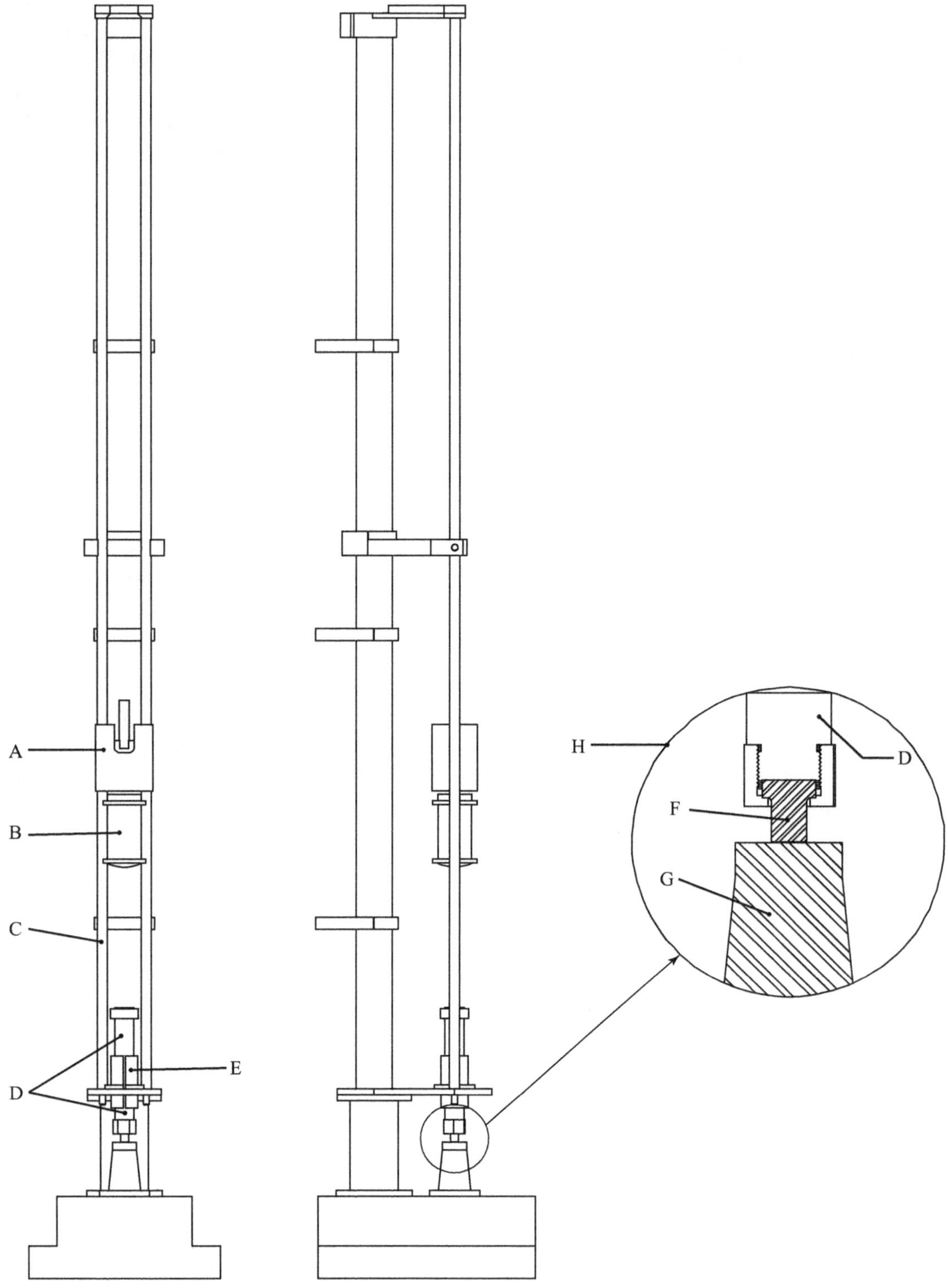

(A)	Drop mass lifting, holding, and dropping device	(B)	Drop mass
(C)	Drop mass guild rails	(D)	Intermediate hammer
(E)	Intermediate hammer guide	(F)	Intermediate hammer insert
(G)	Anvil	(H)	Magnified view of target area

Figure 13.4.7.1: MBOM Impact machine

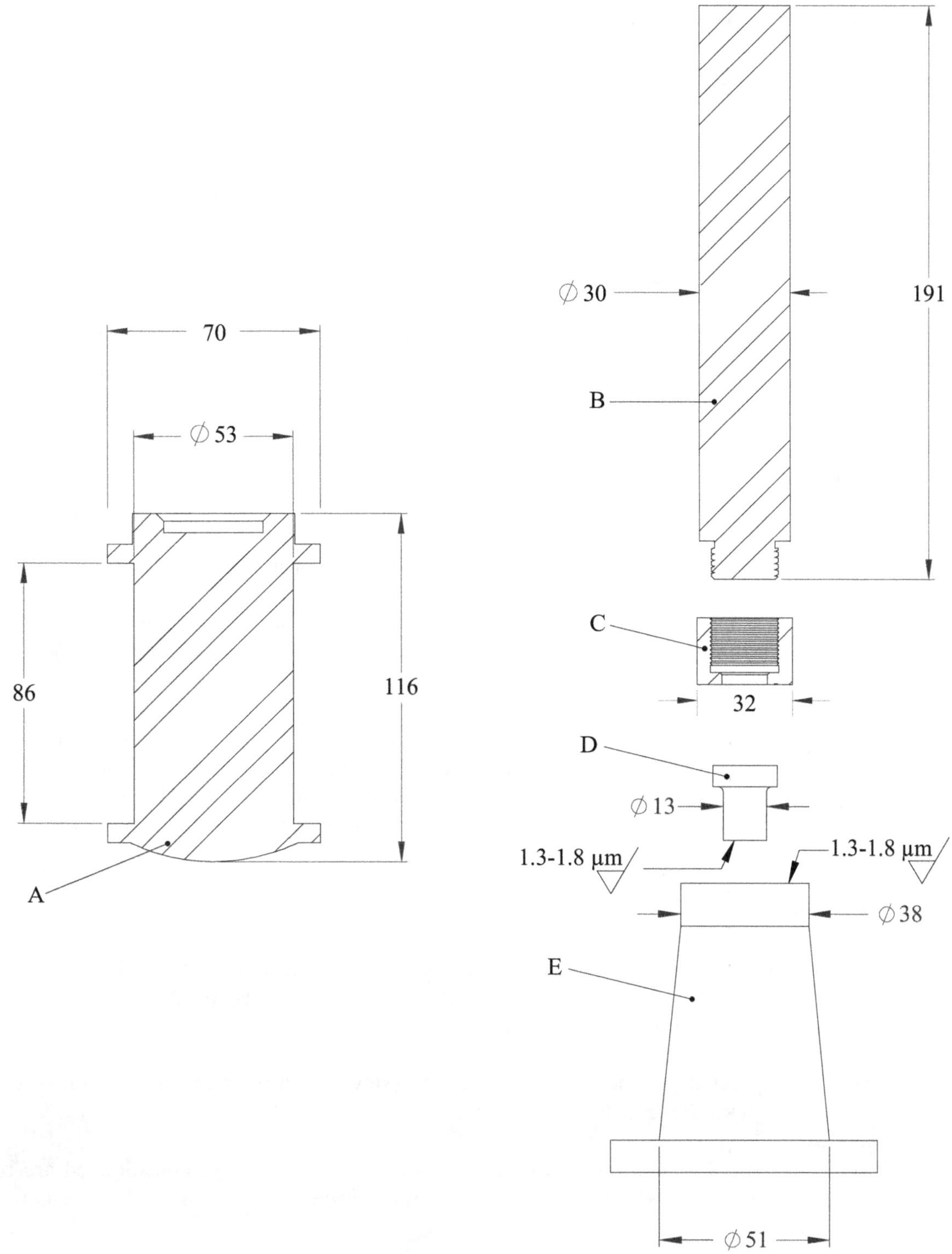

(A)	Drop mass
(B)	Intermediate hammer
(C)	Intermediate hammer insert retaining nut
(D)	Intermediate hammer insert
(E)	Anvil

Figure 13.4.7.2: MBOM Impact machine drop mass and target area detail

13.5 Series 3 type (b) test prescriptions

13.5.1 *Test 3 (b) (i): BAM friction apparatus*

13.5.1.1 *Introduction*

This test is used to measure the sensitiveness of the substance to frictional stimuli and to determine if the substance is too dangerous to transport in the form tested.

13.5.1.2 *Apparatus and materials*

13.5.1.2.1 The friction apparatus (see Figure 13.5.1.1) consists of a cast steel base, on which is mounted the friction device proper. This comprises a fixed porcelain peg and a moving porcelain plate (see paragraph 13.5.1.2.2). The porcelain plate is held in a carriage which runs in two guides. The carriage is connected to an electric motor via a connecting rod, an eccentric cam and suitable gearing such that the porcelain plate is moved, once only, backwards and forwards beneath the porcelain peg a distance of 10 mm. The loading device pivots on an axis so that the porcelain peg can be changed; it is extended by a loading arm which is fitted with 6 notches for the attachment of a weight. Zero load is obtained by adjusting a counterweight. When the loading device is lowered onto the porcelain plate, the longitudinal axis of the porcelain peg is perpendicular to the plate. There are different weights of masses up to 10 kg . The loading arm is fitted with 6 notches of distances of 11 cm, 16 cm, 21 cm, 26 cm, 31 cm and 36 cm from the axis of the porcelain peg. A weight is hung into a notch on the loading arm by means of a ring and hook. The use of different weights in different notches results in loads on the peg of 5 - 10 - 20 - 40 - 60 -80 - 120 - 160 - 240 - 360 N. If necessary, intermediate loads may be used.

13.5.1.2.2 The flat porcelain plates are made from technical white porcelain and, before being fired, their two rubbing surfaces (roughness 9 - 32 microns) are thoroughly roughened by being rubbed with a sponge. The sponge marks are clearly visible. The cylindrical porcelain pegs are also in technical white porcelain and their roughened ends are rounded. The dimensions of plate and peg are given in Figure 13.5.1.2.

13.5.1.3 *Procedure*

13.5.1.3.1 Normally the substances are tested as received. Wetted substances should be tested with the minimum content of wetting agent provided for transport. In addition, for solids substances other than paste-like or gel-types, the following points should be observed:

(a) Substances in powdered form are sieved (sieve mesh 0.5 mm); everything that passes through the sieve is used for the test[3];

(b) Substances which have been compressed, cast or otherwise consolidated are broken into small pieces and sieved; all that passes through a 0.5 mm mesh sieve is used for the test[3];

(c) Substances which are transported only in the form of charges are tested in the form of discs or chips of 10 mm^3 (minimum diameter: 4 mm).

Each part of the surface of the plate and peg must only be used once; the two ends of each peg will serve for two trials and the two friction surfaces of a plate will each serve for three trials.

13.5.1.3.2 A porcelain plate is fixed on the carriage of the friction apparatus so that the grooves of the sponge-marks on it run transversely to the direction of movement. The quantity to be tested, about 10 mm^3, is taken from substances in powdered form by means of a cylindrical measure (2.3 mm diameter × 2.4 mm

---

[3] *For substances which contain more than one constituent, the sieved sample should be representative of the original substance.*

deep); for paste-like or gel-type substances, a rectangular 0.5 mm thick gauge with a 2 × 10 mm window is used; the window is filled with the substance to be tested on the plate, and the gauge is removed carefully. The firmly clamped porcelain peg is placed onto the sample as in Figure 13.5.1.2; the loading arm is loaded with the required weights and the switch is operated. Care must be taken to ensure that the peg rests on the sample, and that there is enough of the substance to come under the peg when the porcelain plate moves in front of the peg.

13.5.1.3.3 The series of trials is started with a single trial at a load of 360 N. The results of each trial are interpreted in terms of "no reaction", "decomposition" (change of colour or odour) or "explosion" (report, crackling, sparking or flame). If in the first trial the result "explosion" is observed, the series is continued with trials at stepwise lower loads until the result "decomposition" or "no reaction" is observed. At this friction load level the trial is repeated up to the total number of six if no "explosion" occurs; otherwise the friction load is reduced in steps until the lowest load is determined at which no "explosion" occurs in six trials. If in the first trial at 360 N the result "decomposition" or "no reaction" occurs, up to five further trials are performed. If in all six trials at the highest load, the result "decomposition" or "no reaction" occurs, the substance is deemed to be insensitive to friction. If an "explosion" is obtained, the load is reduced as above. The limiting load is defined as the lowest load at which the result "explosion" is obtained from at least one out of at least six trials.

13.5.1.3.4 Test criteria and method of assessing results

The test results are assessed on the basis of:

(a) Whether an "explosion" occurs in any of up to six trials at a particular friction load;

(b) The lowest friction load at which at least one "explosion" occurs in six trials.

The test result is considered "+" if the lowest friction load at which one "explosion" occurs in six trials is less than 80 N and the substance is considered too dangerous for transport in the form in which it was tested. Otherwise, the test result is considered "-".

13.5.1.5 *Examples of results*

Substances	**Limiting load (N)**	**Result**
Blasting gelatine (75% NG)	80	-
Hexanitrostilbene	240	-
HMX (dry)	80	-
Hydrazine perchlorate (dry)	10	+
Lead azide (dry)	10	+
Lead styphnate	2	+
Mercury fulminate (dry)	10	+
Nitrocellulose 13.4% N (dry)	240	-
Octol 70/30 (dry)	240	-
PETN (dry)	60	+
PETN/wax 95/5	60	+
PETN/wax 93/7	80	-
PETN/wax 90/10	120	-
PETN/water 75/25	160	-
PETN/lactose 85/15	60	+
Picric acid (dry)	360	-
RDX (dry)	120	-
RDX (water wet)	160	-
TNT	360	-

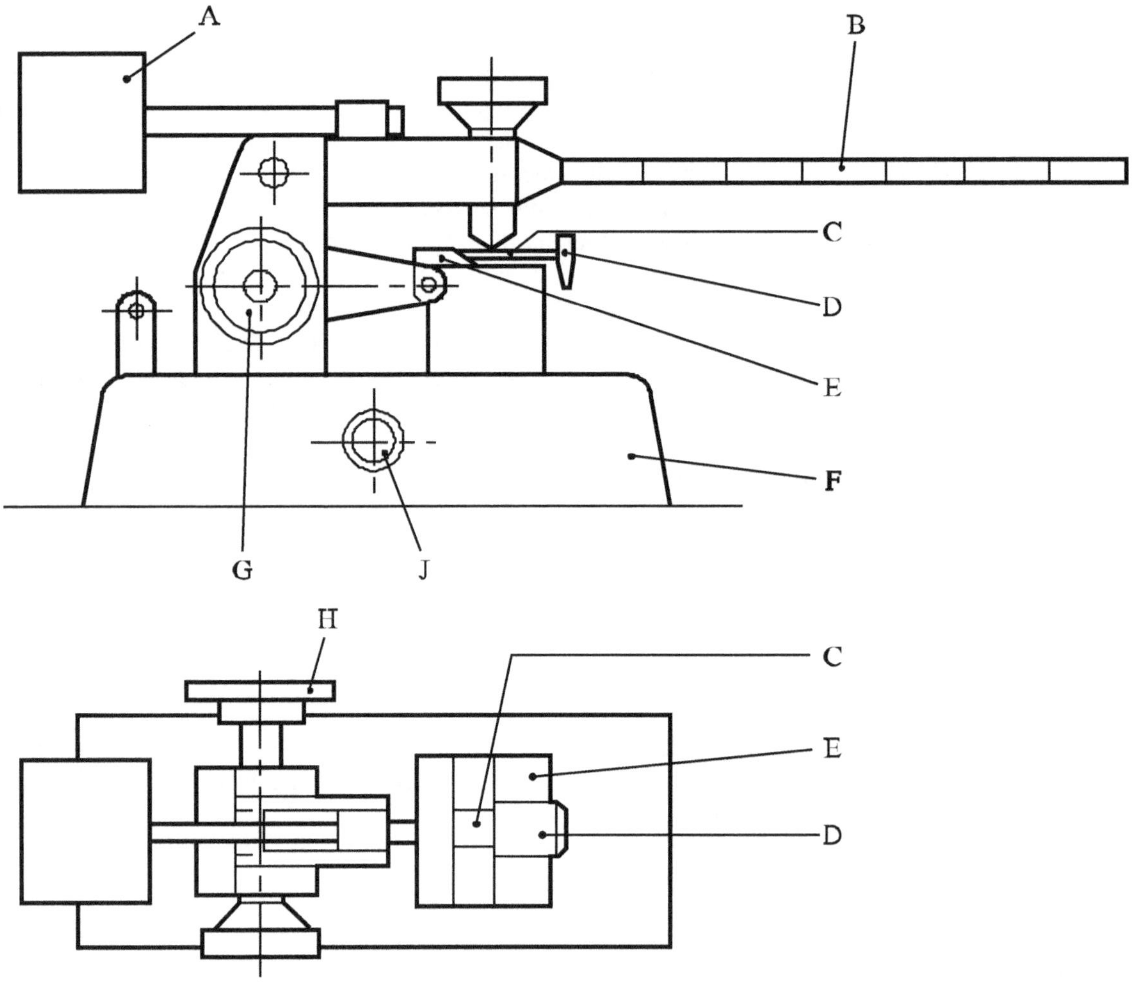

(A) Counter weight
(B) Loading arm
(C) Porcelain plate held on carriage
(D) Adjusting rod
(E) Movable carriage
(F) Steel base
(G) Handle for setting the carriage at the starting position
(H) Points the direction to electric motor drive
(J) Switch

Figure 13.5.1.1: BAM FRICTION APPARATUS

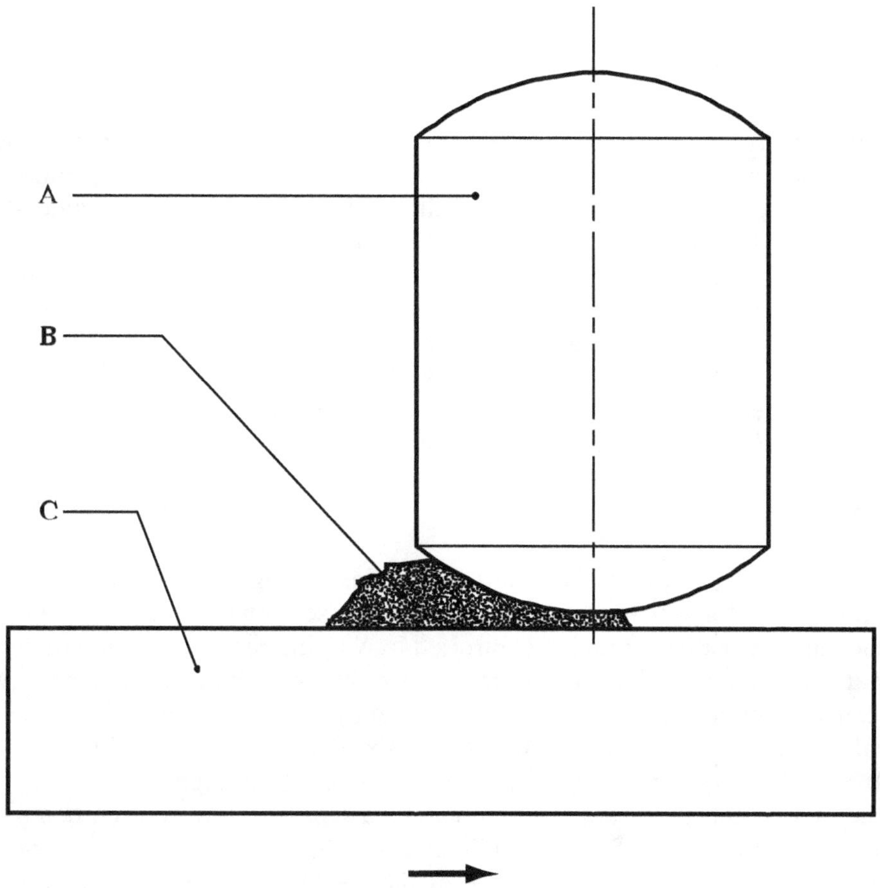

(A)	Porcelain peg 10 mm diameter × 15 mm
(B)	Sample under test
(C)	Porcelain plate 25 × 25 × 5 mm

Figure 13.5.1.2: PORCELAIN PLATE AND PEG

13.5.2 *Test 3 (b) (ii): Rotary friction test*

13.5.2.1 *Introduction*

This test is used to measure the sensitiveness of the substance to mechanical frictional stimuli and to determine if the substance is too dangerous to transport in the form tested. The principle of operation is to hold a thin sample under a load between the prepared surfaces of a flat bar and the periphery of a wheel of specified diameter.

13.5.2.2 *Apparatus and materials*

The apparatus is shown diagrammatically in Figure 13.5.2.1. The bar (A) is made of general purpose mild steel whose surface has been prepared by grit-blasting to a finish of 3.2 μm ± 0.4 μm. A similar grit-blasting treatment is applied to the periphery of the wheel (C), made of the same steel and is 70 mm diameter by 10 mm thick. The sample under test is either cut as a sliver or spread as a powder so that its thickness is no more than approximately 0.1 mm on the bar. The wheel is mounted on spigots on the end of a rotor, the other end of which holds a pivoted latch operated by a relay switching mechanism in the circuit of a solenoid. The load is applied by compressed air (B) to a predetermined pressure. When the firing switch is operated, this latch is moved into the path of a striker on the periphery of a heavy flywheel which drives the rotor and, therefore, the wheel through 60°, after which the friction surfaces are separated by means of a cam on the rotor and a push rod operated by the loading cylinder.

13.5.2.3 *Procedure*

In the normal procedure, the load is maintained using an air pressure of 0.275 MPa, except for very sensitive explosives, for which it may be necessary to use a smaller load. The angular velocity of the wheel is used as the variable parameter and is controlled by varying the speed of the motor which drives the flywheel. The initial velocity for starting a run is determined by testing at the step velocity which is closest to the mean of the nearest ignition and non-ignition, and repeating the process until these occur at adjacent step levels. In a normal test, a 50-shot Bruceton (see Appendix 2) run with a logarithmic step size of 0.10 is employed. If the Sample Comparison Test (SCT) procedure (see Appendix 2) is used, samples of the standard and the sample under test are fired alternately, a separate Bruceton run being operated for each. An ignition is usually identified by a flash or audible report, but even a little smoke or blackening of the sample is considered as an ignition for the purpose of the test. Each sample is used only once, as are the contacting surfaces of the bar and the wheel. In order to monitor the long-term behaviour of the equipment, regular measurements are carried out on a standard explosive viz. RDX, recrystallized from cyclohexanone and dried according to a standard method. Unless obtained from the SCT procedure, data for the standard is obtained from 50-shot running mean determinations.

13.5.2.4 *Test criteria and method of assessing results*

The test results are assessed on the basis of:

(a) Whether an ignition is observed in a trial;

(b) Determination of the median striking velocity for the RDX reference standard and the sample by the Bruceton method (see Appendix 2);

(c) Comparison of the running mean median striking velocity of the standard (V_1) with that of the sample (V_2) using the equation:

Figure of Friction (F of F) = $3.0\ V_2/V_1$

Standard RDX is assigned a F of F of 3.0.

The test result is considered "+" if the F of F is less than or equal to 3.0 and the substance is considered too dangerous for transport in the form in which it was tested. The test result is considered "-" if the F of F is greater than 3.0. If the F of F obtained for the substance under test is less than 3.0, a direct comparison with the standard RDX may be made by using the SCT procedure with 100 shots on each substance. If there is 95% or greater confidence that the substance under test is not more sensitive than RDX, the substance under test is not too dangerous for transport in the form tested.

13.5.2.5 *Examples of results*

Substances	**Figure of Friction**	**Result**
Blasting gelatine-geophex	2.0	+
Blasting gelatine-submarine	1.3	+
Lead azide	0.84	+
PETN/wax 90/10	4.0	-
RDX	3.4	-
Tetryl	4.5	-
TNT	5.8	-

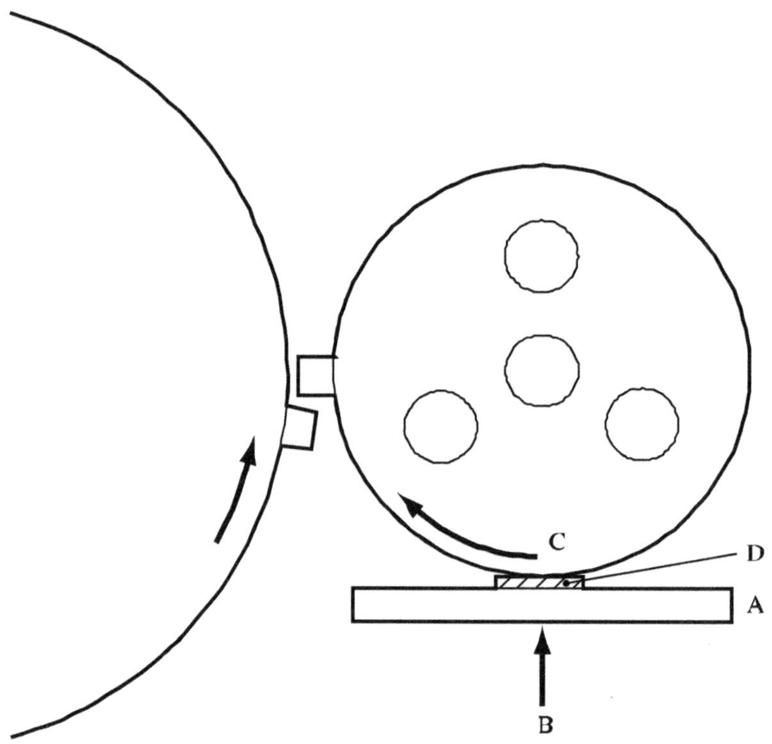

(A)	Mild steel bar
(B)	Compressed air load
(C)	Rotating wheel in contact with sample
(D)	Sample

Figure 13.5.2.1: ROTARY FRICTION TEST

13.5.3 *Test 3 (b) (iii): Friction sensitivity test*

13.5.3.1 *Introduction*

This test is used to measure the sensitiveness of the substance to mechanical frictional stimuli and to determine if the substance is too dangerous to transport in the form tested.

13.5.3.2 *Apparatus and materials*

13.5.3.2.1 The general design of the friction test apparatus is given in Figure 13.5.3.1. The apparatus consists of four main components: pendulum, pendulum holder, apparatus body and hydraulic press. The apparatus is mounted on a concrete base. The roller assembly 1 with explosive under test is placed in the apparatus body. The compression of the explosive sample, placed between two rollers, up to the prescribed pressure is carried out by means of a hydraulic press. The movement of the upper roller along the substance by 1.5 mm is carried out using impact from the pendulum weight.

13.5.3.2.2 The roller assembly 1 consists of a sleeve and two rollers. Its dimensions and requirements are shown in Figure 13.5.3.2.

13.5.3.3 *Procedure*

13.5.3.3.1 As a rule, substances should be tested as received. Wetted substances are to be tested with the minimum quantity of the wetting agent required for transport. The substances should be subjected to the following procedures:

(a) Granulated, flaky, pressed, cast and in other ways packed substances are milled and sieved; the substance particles for test should pass through the sieves with 0.50 ± 0.05 mm mesh sides;

(b) Elastic substances are cut into pieces not more than 1 mm in size with a sharpened knife on a wooden surface. Elastic substances samples are not sieved;

(c) Powdered, plastic and pasty explosives samples are not ground and sieved.

Before use, the roller assemblies are degreased. The devices may be used again if they remain within specification.

13.5.3.3.2 To determine the lower friction sensitivity limit of tested explosive, the 20 mg of sample is placed into the open roller assembly. By gently pressing and turning the upper roller, the explosive sample is evenly smoothed between the rollers. The roller assembly containing the explosive sample is placed into the chamber of the apparatus body where it is compressed up to the chosen pressure. The sleeve is lowered by maintaining the pressure so that the explosive sample is pressed between roller faces and is raised above the sleeve. A striking pin is then moved so that its impact end touches the roller. The striking pin is impacted by the pendulum weight causing friction between the upper roller and the sample. The roller movement is 1.5 mm. The throw-off angle for the pendulum is chosen according to the following table, depending upon the retaining pressure of the sample. The tests are performed until the maximum retaining pressure is found which does not give an explosion in 25 trials. An explosion is considered to have occurred if there is a sound effect, a flash or burn traces on the rollers. The lower friction sensitivity limit is considered as the maximum retaining pressure which gives no explosion in 25 tests and which distinguishes from the pressure which still gives explosions but differ by not more than:

10 MPa - at test pressure up to 100 MPa

20 MPa - at test pressure 100 to 400 MPa

50 MPa - at test pressure above 400 MPa.

If no explosion is obtained in 25 tests at a pressure of 1 200 MPa, the lower friction sensitivity limit is expressed as "1 200 MPa or more". If one or more explosions are obtained in 25 tests at a pressure of 30 MPa, the lower friction sensitivity is expressed as "less than 30 MPa".

RELATIONSHIP BETWEEN RETAINING PRESSURE OF EXPLOSIVE SAMPLE AND THROW-OFF ANGLE OF PENDULUM PROVIDING THE CONSTANT VALUE OF ROLLER SHIFT

Retaining pressure of explosive sample (MPa)	Throw-off angle of pendulum (° from vertical)	Retaining pressure of explosive sample (MPa)	Throw-off angle of pendulum (° from vertical)
30	28	40	32
50	35	60	38
70	42	80	43
90	46	100	47
120	54	140	58
160	61	180	64
200	67	220	70
240	73	260	76
280	78	300	80
320	82	340	83
360	84	380	85
400	86	450	88
500	91	550	93
600	95	650	97
700	100	750	101
800	103	850	106
900	107	950	108
1 000	110	1 100	115
1 200	118		

13.5.3.4 *Test criteria and method of assessing results*

The test results are assessed on the basis of:

(a) Whether an "explosion" occurs in any one of 25 trials; and

(b) The maximum retaining pressure at which there is no explosion in any of 25 trials.

The test result is considered "+" if the lower impact friction sensitivity limit is less than 200 MPa and the substance is considered too dangerous for transport in the form in which it was tested. The test result is considered "-" if the lower impact friction sensitivity limit is greater than or equal to 200 MPa.

13.5.3.5 *Examples of results*

Substances	**Lower limit (MPa)**	**Result**
Ammonium nitrate	1 200	-
Lead Azide	30	+
PETN (dry)	150	+
PETN/paraffin (95/5)	350	-
PETN/TNT (90/10)	350	-
PETN/water (75/25)	200	-
Picric acid	450	-
RDX (dry)	200	-
RDX/water (85/15)	350	-
TATB	900	-
TNT	600	-

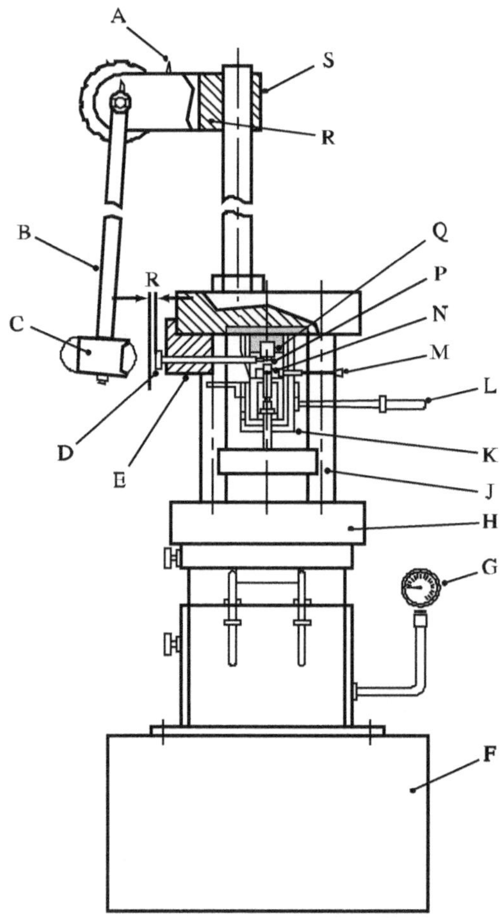

(A)	Trigger	(B)	Pendulum lever
(C)	Pendulum weight	(D)	Striking pin
(E)	Guide for striking pin	(F)	Base
(G)	Pressure gauge	(H)	Hydraulic press
(J)	Apparatus support	(K)	Apparatus body
(L)	Handle for lowering of roller assembly sleeve	(M)	Roller assembly pusher
(N)	Sleeve	(P)	Roller
(Q)	Chamber	(R)	Pendulum holder
(S)	Pendulum holder support		

Figure 13.5.3.1: APPARATUS FOR IMPACT FRICTION TEST

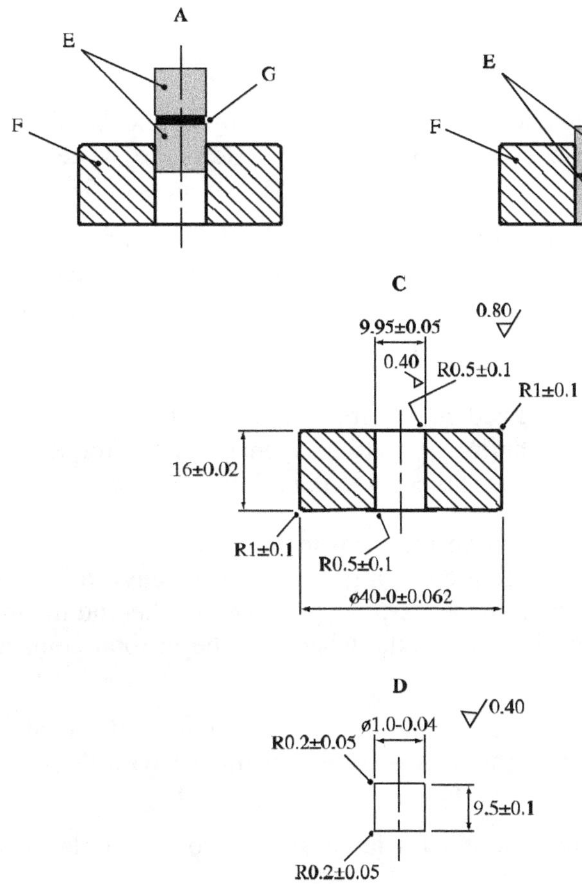

(A) Initial position of rollers
(B) Rollers in position for test
(C) Sleeve of tool carbon steel HRC 57 - 61
(D) Roller of ball bearing steel HRC 63 - 66
(E) Rollers
(F) Sleeve
(G) Substance under test

Figure 13.5.3.2: ROLLER ASSEMBLY 1

13.5.4 *Test 3 (b) (iv): ABL friction machine test*

13.5.4.1 *Introduction*

This test is used to measure the sensitiveness of the substance to friction stimuli and to determine if the substance is too dangerous to transport in the form tested. The test substance is subjected to a vertical compression force under a non-rotating wheel, while the substance is moved in a horizontal direction on a sliding anvil. It is applicable to solid, semisolid, and powder substances.

13.5.4.2 *Apparatus and materials*

13.5.4.2.1 The following apparatus and materials are required:

(a) A mechanism capable of applying a force hydraulically through a non-rotating steel wheel to a sample placed on steel anvil. Both the wheel and anvil have a surface roughness of 1.3 – 1. 8 μm and a Rockwell C hardness of 55 – 62.

(b) A pendulum system that is capable of being positioned and released at an angle that will impart a predetermined velocity to the sliding anvil. A travel distance of approximately 2.5 cm perpendicular to the applied force on the wheel is achieved with this system.

13.5.4.3 *Procedure*

13.5.4.3.1 As a rule, substances are tested in the form in which they are received. Wetted substances should be tested with the minimum quantity of wetting agent required for transport. Depending on the physical form, the substances should then be subjected to the following procedures:

(a) Powders are to be tested on the anvil in a monolayer; i.e., the thickness of the granular material, if possible. Place enough granules on the anvil to approximately cover an area 1.3 cm long by 0.65 cm wide starting about 0.65 cm behind the initial contact point of the wheel with the anvil such that the wheel will be in total contact with the sample when lowered onto it.

(b) Solid propellants are tested in the form of thin, uniform slices with a thickness of 0.08 + 0.01 cm. This thickness is easily obtainable with the use of a microtome cutting tool.

(c) Semisolids will be smoothed with a spatula to a thin layer with uniform thickness approximately 0.015 cm.

With the friction wheel raised, the test substance is placed on the anvil below the wheel such that the wheel will be in total contact with the sample when lowered onto it. The friction wheel is then carefully lowered onto the substance on the anvil and the desired normal force is applied to the wheel [250 N at 2.4 m/s or 445 N at 1.2 m/s]. The pendulum is raised to the desired angle to achieve the appropriate test velocity and released. Observations are made on whether a “reaction” occurs as evidenced by audible report or production of smoke, fire, charring or visible light as observed by human senses. The type of reaction that occurs is documented. The force on the wheel is removed and any excess test substance is cleaned from the area. The wheel is indexed and shifted across the anvil in order to ensure that fresh surfaces are used for each trial.

13.5.4.4 *Maintenance and calibration*

The maximum speed of the anvil should be calibrated to 2.4 m/s and 1.2 m/s. The downward force on the wheel should be verified. The test machine should be periodically cleaned and calibrated according to a schedule based on the amount of usage. At a minimum, the machine should be calibrated on an annual basis.

13.5.4.5 *Test criteria and method of assessing results*

The test result is considered "+" if the lowest friction load at which at least one reaction occurs in six trials is 250 N at 2.4 m/s or 445 N at 1.2 m/s or less and the substance is considered too dangerous for transport in the form in which it was tested. Otherwise, the result is considered "–".

13.5.4.6 *Examples of results*

Substances[1]	**Result**
RDX (class 5)	–
RDX (class 7)	–
PBXN-8	–
PBXN-10	–
Aluminium/TNT (80/20 Mixture)	–
PETN (dry)[2]	+

[1] *Data acquired at 2.4 m/s, relative humidity of 10-30%, and temperature of 16-24 °C unless noted otherwise.*

[2] *Data acquired at 2.4 m/s and 1.2 m/s.*

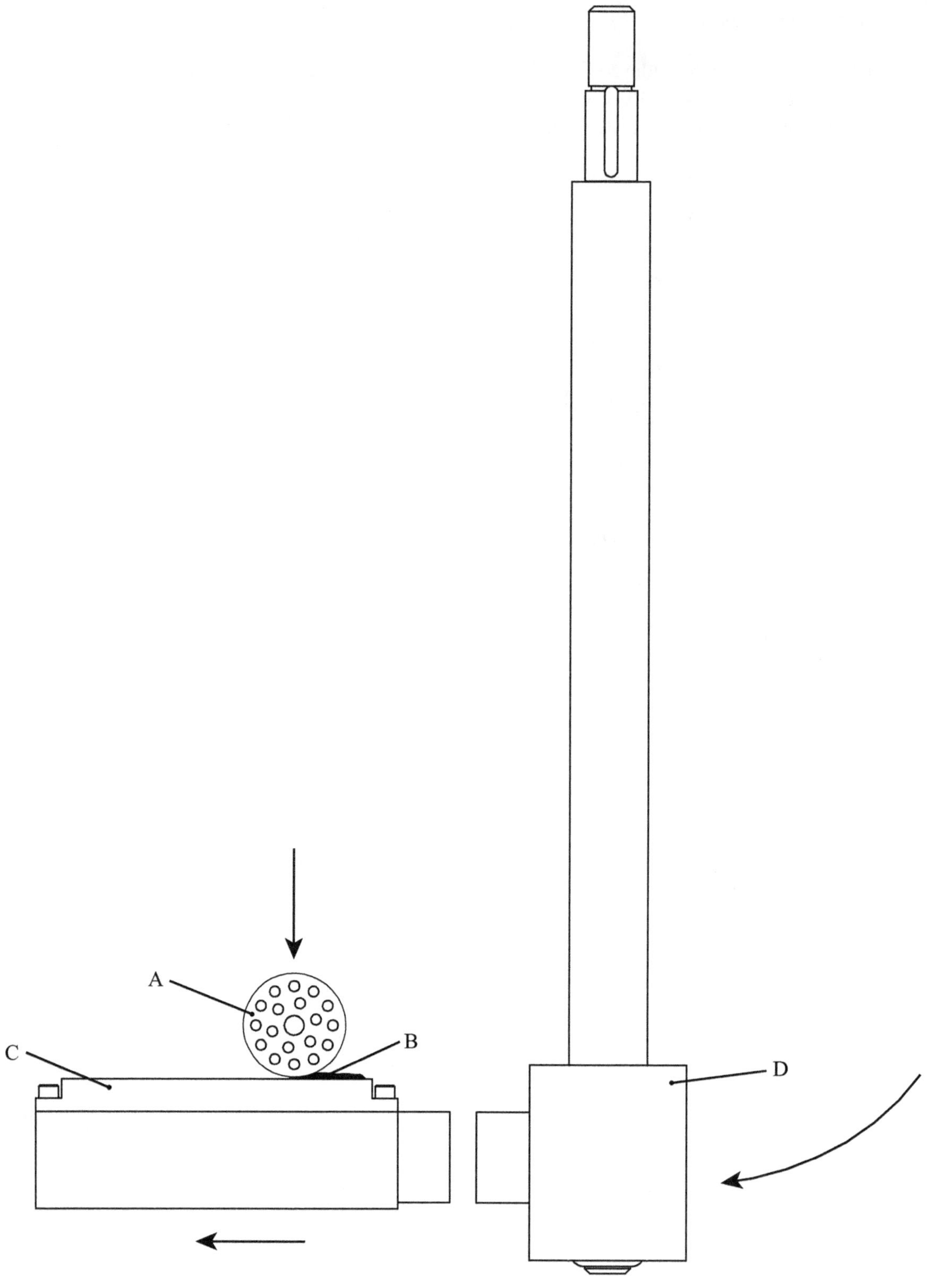

(A)	Non-rotating wheel	(B)	Sample
(C)	Anvil	(D)	Pendulum

Figure 13.5.4.1: ABL Friction machine

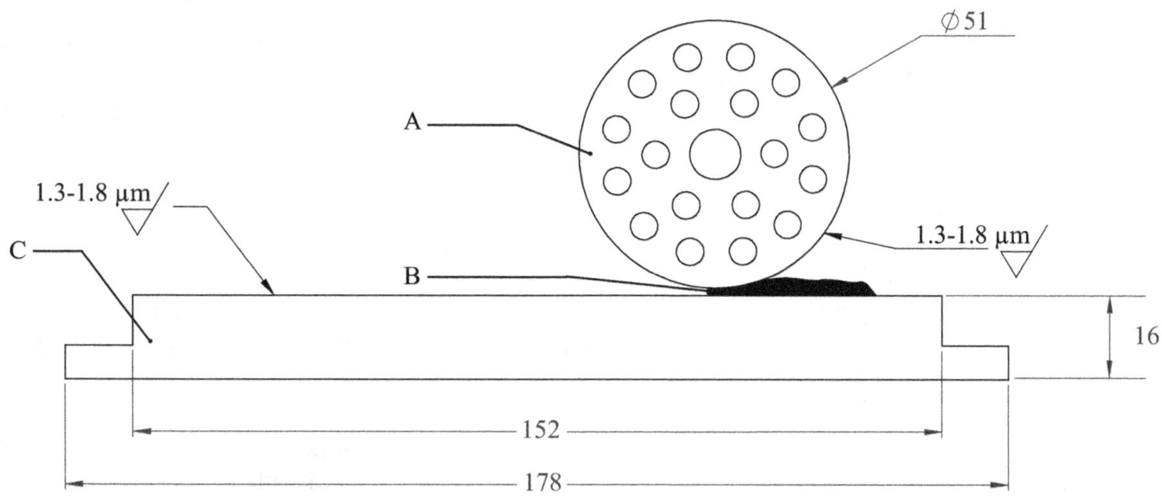

(A)	Non-rotating wheel 5.1 cm diameter maximum × 3.18 cm
(B)	Sample under test
(C)	Anvil 18 cm × 5.7 cm × 1.6 cm maximum

Figure 13.5.4.2: ABL Friction machine wheel and anvil detail

13.6 Series 3 type (c) test prescription

13.6.1 *Test 3 (c) (i): Thermal stability test at 75 °C*

13.6.1.1 *Introduction*

This test is used to measure the stability of the substance when subjected to elevated thermal conditions to determine if the substance is too dangerous to transport.

13.6.1.2 *Apparatus and materials*

13.6.1.2.1 The following apparatus is required:

(a) An electric oven equipped with ventilation, explosion-proof electrical features, and thermostatic control adequate to maintain and record the temperature at 75 ± 2 °C. The oven should have dual thermostats or some kind of protection against thermal run-away if the thermostat malfunctions;

(b) A lipless beaker of 35 mm diameter and 50 mm high and a watch-glass of 40 mm diameter;

(c) A balance capable of determining the sample weight to ± 0.1 g;

(d) Three thermocouples and a recording system;

(e) Two flat-bottomed glass tubes of 50 ± 1 mm diameter and 150 mm length and two 0.6 bar (60 kPa) pressure resisting stoppers.

13.6.1.2.2 An inert substance, whose physical and thermal properties are similar to the test substance, should be used as the reference substance.

13.6.1.3 *Procedure*

13.6.1.3.1 ***In dealing with a new substance, several screening tests, involving heating small samples at 75 °C for 48 hours, are performed to explore its behaviour.*** If no explosive reaction occurs using a small amount of substance, the procedure given in 13.6.1.3.2 or 13.6.1.3.3 should be used. If explosion or ignition occurs then the substance is too thermally unstable for transport.

13.6.1.3.2 Uninstrumented test: A 50 g sample is weighed into a beaker, covered and placed in an oven. The oven is heated to 75 °C and the sample left at oven temperature for 48 hours or until ignition or explosion occurs, whichever is sooner. If ignition or explosion does not occur but there is evidence, e.g. fuming or decomposition, that some self-heating has occurred, the procedure given in should be performed. However, if the substance shows no evidence of thermal instability, it may be regarded as thermally stable and no further testing of this property is necessary.

13.6.1.3.3 Instrumented test: A 100 g (or 100 cm^3 if the density is less than 1 000 kg /m^3) sample is placed in one tube and the same quantity of reference substance is placed in the other. Thermocouples T_1 and T_2 are inserted into the tubes at half-height of the substances. If the thermocouples are not inert with respect to both the substance being tested and the reference substance, they should be enclosed in sheaths which are inert. Thermocouple T_3 and the covered tubes are placed in the oven as shown in Figure 13.6.1.1. The temperature difference (if any) between test sample and reference is measured for 48 hours after the sample and any reference substance reach 75 °C. Evidence of decomposition of the sample is noted.

13.6.1.4 *Test criteria and method of assessing results*

13.6.1.4.1 The result from an uninstrumented test is considered "+" if ignition or explosion occurs and "-" if no changes are observed. The result of a instrumented test is considered "+" if an ignition or explosion occurs or if a temperature difference (i.e. self-heating) of 3 °C or greater is recorded. If no ignition or explosion occurs but self-heating of less than 3 °C is noted, additional tests and/or evaluation may be required to determine if the sample is thermally unstable.

13.6.1.4.2 If the test result is "+", the substance should be considered too thermally unstable for transport.

13.6.1.5 *Examples of results*

Substances	Observations	Result
70% AP, 16% A1, 2.5% Catocene, 11.5% binder	Oxidation reaction took place on Catocene (burning rate catalyst). Sample discoloured on the surface, but there was no chemical decomposition.	-
PETN/Wax 90/10	Negligible mass loss	-
RDX, 22% water wetted	Mass loss < 1%	-
Gelatine dynamite (NG 22%, DNT 8%, Al 3%)	Negligible mass loss	-
ANFO explosive	Mass loss < 1%	-
Slurry explosives[a]	Negligible mass loss, slight (eventual) swelling	-

[a] *Various types.*

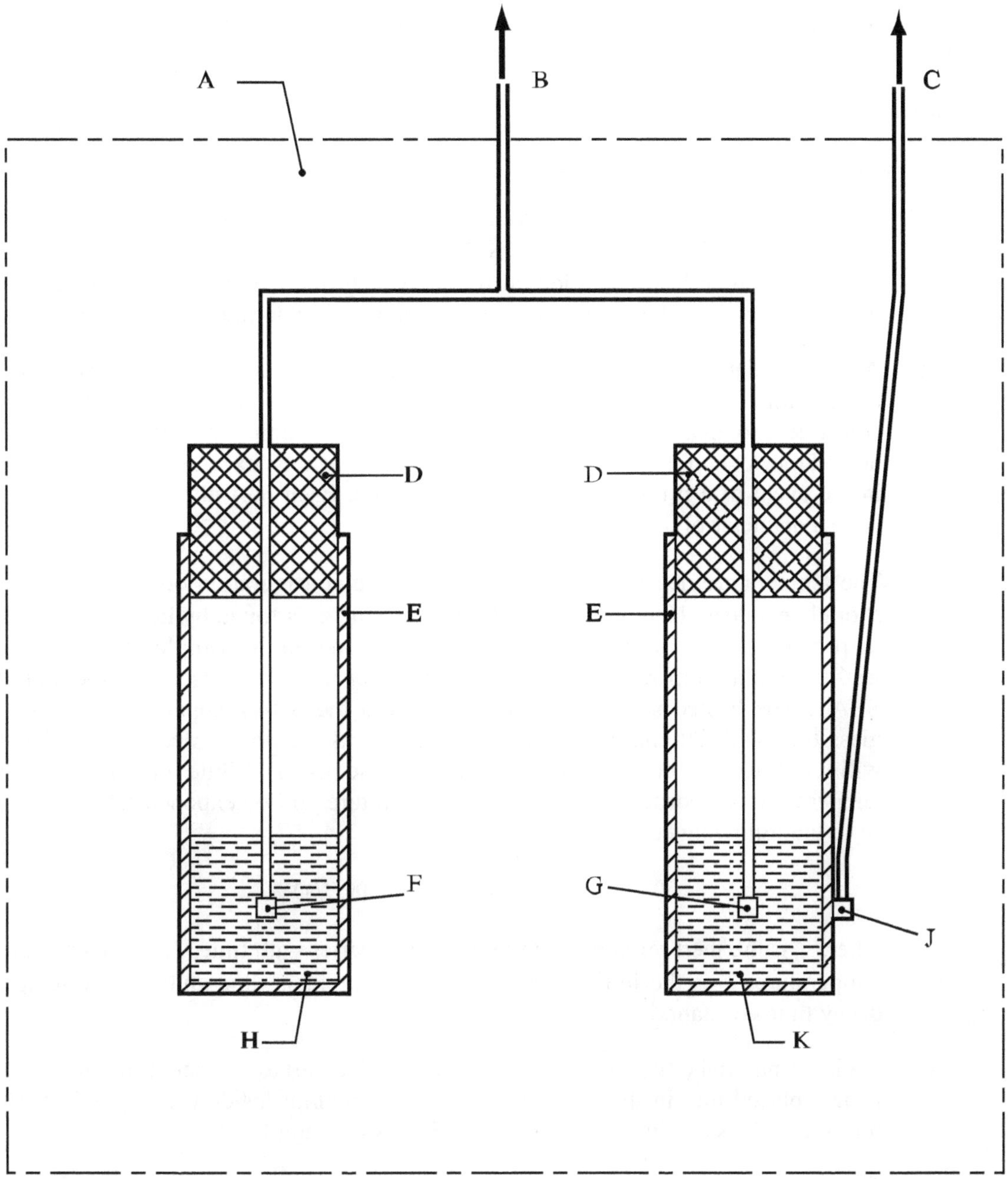

(A)	Heating oven	(B)	To millivoltmeter (T_1 - T_2)
(C)	To millivoltmeter (T_3)	(D)	Stoppers
(E)	Glass tubes	(F)	Thermocouple No 1 (T_1)
(G)	Thermocouple No 2 (T_2)	(H)	100 cm^3 of sample
(J)	Thermocouple No 3 (T_3)	(K)	100 cm^3 of reference substance

Figure 13.6.1.1: EXPERIMENTAL SET-UP FOR STABILITY TEST AT 75 °C

13.6.2 ***Test 3(c) (ii): SBAT thermal stability test at 75 °C***

13.6.2.1 *Introduction*

This test is used to measure the stability of the substance when subjected to elevated thermal conditions to determine if the substance is too dangerous for transport.

13.6.2.2 *Apparatus and materials*

13.6.2.2.1 The following apparatus is required:

(a) Glass sample tubes of 13 x 100 mm inside a larger tube of 25 x 100 mm. Each 13 x 100 mm tube is surrounded by insulation and placed into the larger tube. Each larger glass tube has insulation surrounding it further isolating it thermally from the metal oven block. The glass sample tube can be sealed to prevent the escape of gases.

(b) A well-insulated multiport metal block that can be heated with resistance heaters to a temperature of at least 260 °C. The heating of the block must be automated or reliably controlled so that the desired temperature can be maintained within ± 0.5 °C. The heated block should have independent protection against excessively heating the block in the event of a primary control system failure. Each port in the metal block should have a diameter of 5 cm and a depth of 10 cm.

(c) The temperature decay time constant, τ, for the configuration outlined in (a) and (b) should be at least 10 minutes. The decay constant, τ, is found by heating 5 grams of an inert material (e.g. dried silica, alumina, or silicone) in the sample tube (13 x 100 mm test tube) to a temperature 50 °C or more higher than the constant temperature of the SBAT. The heated sample tube is placed into the SBAT apparatus (into the larger glass tube with internal and external insulation as previously described). The sample will cool to the constant temperature of the oven. While cooling, the sample temperature is recorded. The decaying temperature will be exponential in shape and is fit to the following equation:

$$(\mathrm{T}-\mathrm{T_a})/(\mathrm{T_i}-\mathrm{T_a}) = \exp(-\mathrm{t}/\tau)$$

where T is the inert reference temperature that varies with time, T_a is the constant oven temperature, T_i is the initial reference temperature, t is time and τ is the temperature decay time constant.

(d) An inert material (e.g. dried silica, alumina or silicone) to be used as a reference which is also placed into insulated glass tubes (13 x 100 mm inside the larger 25 x 100 mm tube) with the same insulation configuration as the sample.

(e) Thermocouples with a data recording system to record the temperature of the reference and sample(s) as well as thermocouple(s) to measure and control the oven temperature.

13.6.2.3 *Procedure*

13.6.2.3.1 Five grams of the sample or an amount that fills the tube to 75 mm height, whichever is less, is placed inside one of the sample tubes. A second sample tube is filled with the same amount of sample. One of the filled sample tubes is not sealed whereas the second filled sample tube is sealed with a screw cap or other method. For the sample tube that is sealed, the thermocouple is attached to the sidewall of the sample tube. For the open sample tube, the thermocouple can be attached to the side of the tube or inserted into the sample.

13.6.2.3.2 Each sample tube is then surrounded with insulation and placed into the larger 25 x 100 mm tube which is also insulated from the side walls of the SBAT oven ports. The approximately 5 gram reference sample must also be present in one of the SBAT ports with the same insulation configuration as the sample. The samples are heated to 75 – 77 °C and maintained at that temperature for 48 hours. Sample and reference temperatures are recorded throughout the test.

13.6.2.3.3 Once the test has been completed, additional test data may be obtained by linearly increasing the temperature of the apparatus to determine the thermal profile of the sample (measuring endotherms and exotherms, as evidenced by departures of the sample from the temperature of the inert reference).

13.6.2.4 *Test criteria and method of assessing results*

13.6.2.4.1 The result from a test is considered "+" if either the sealed or unsealed sample shows more than a 1.5 °C temperature rise during the 48 hour test period indicating self-heating.

13.6.2.4.2 If the test result is "+", the substance should be considered too thermally unstable for transport.

13.6.2.5 *Examples of results*

Substances	Temperature Rise	Result
PETN	Less than 1.5 °C	–
RDX	Less than 1.5 °C	–
TNT	Less than 1.5 °C	–
Composition B, reclaimed	Less than 1.5 °C	–
Double base smokeless powder, 40% NG	Less than 1.5 °C	–
Black powder	Less than 1.5 °C	–
Barium styphnate	Less than 1.5 °C	–
Rocket motor propellant (60-70% AP, 5-16% Al, 12-30% binder)	Less than 1.5 °C	–
Catalyst containing copper acetylide	Greater than 1.5 °C	+

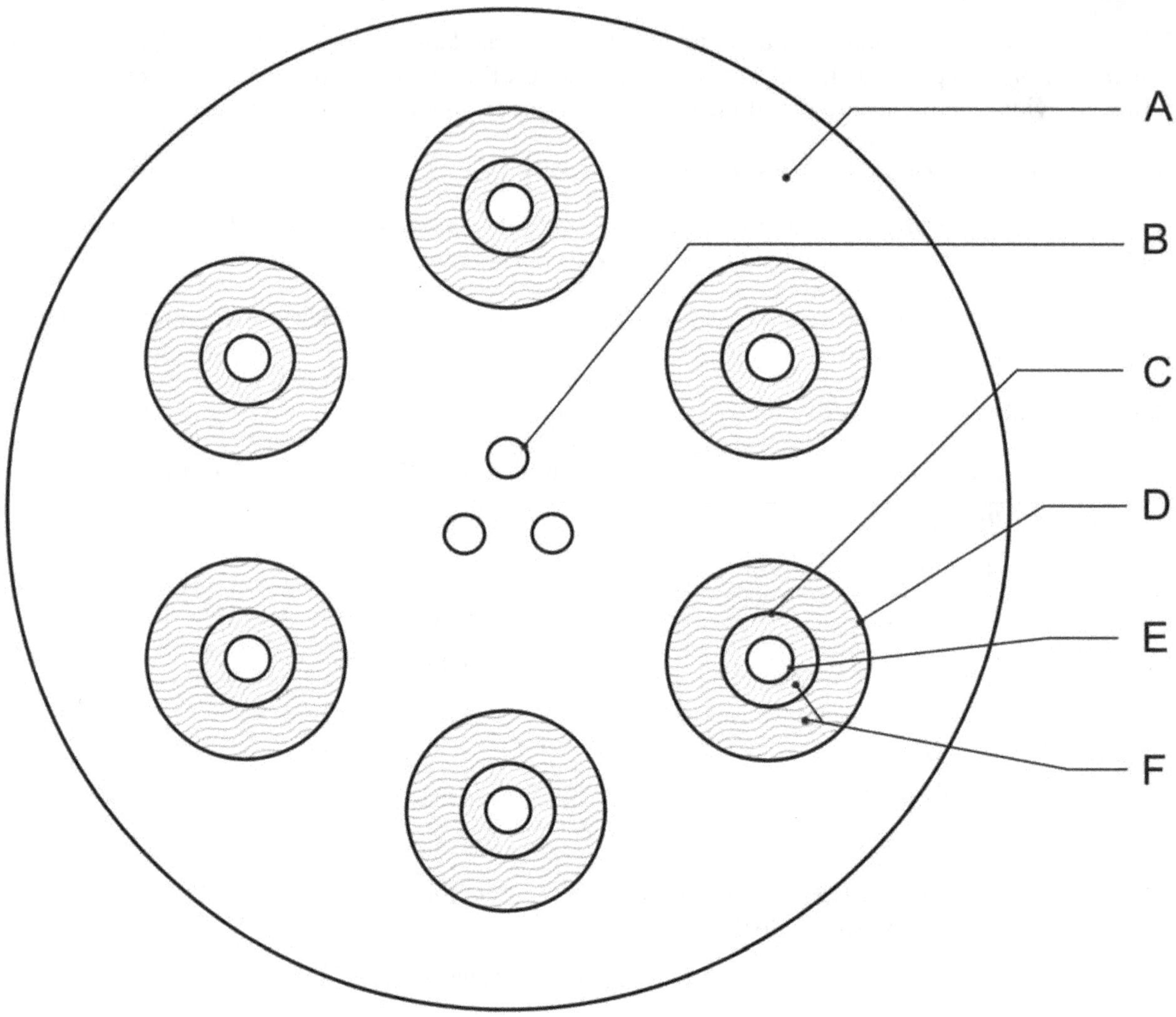

(A)	Metal block	(B)	Cartridge heaters
(C)	Glassware	(D)	Sample port
(E)	Glass sample container	(F)	Insulation

Figure 13.6.2.1: SBAT Heating Block

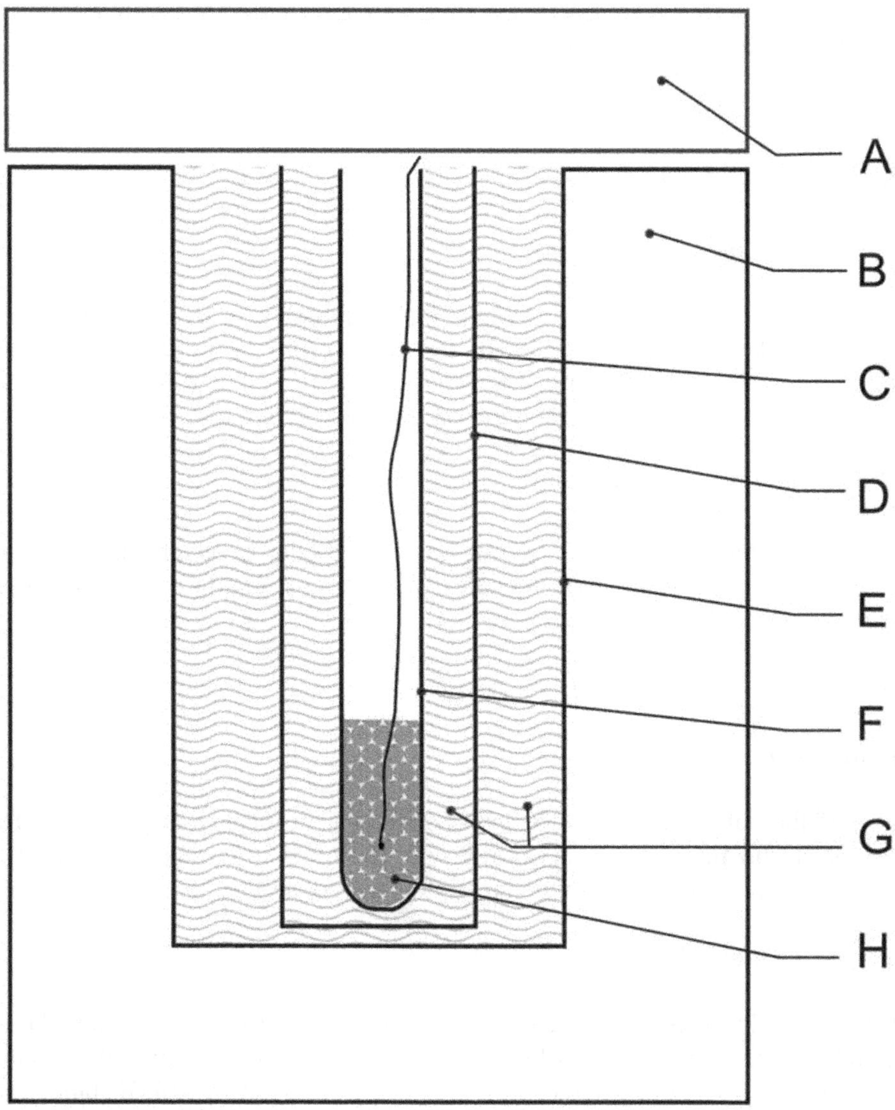

(A)	Insulative cap or blanket	(B)	Metal block
(C)	Thermocouple	(D)	Glassware
(E)	Sample port	(F)	Glass sample container
(G)	Insulation	(H)	Sample

Figure 13.6.2.2: SBAT Port

13.7 Series 3 type (d) test prescription

13.7.1 *Test 3 (d): Small-scale burning test*

This test is used to determine the response of a substance to fire.

13.7.1.1 *Apparatus and materials*

13.7.1.1.1 Solids and liquids

Sufficient Kerosene-soaked sawdust (approximately 100 g of sawdust and 200 cm^3 of kerosene) is required to make a bed 30 cm × 30 cm and 1.3 cm thick. For substances which do not readily ignite, the thickness is increased to 2.5 cm. An electric igniter and a thin-walled plastics beaker, just sufficient in size to contain the test substance and compatible with it, are also required.

13.7.1.1.2 Alternative method (solids only)

A chronometer and a sheet of Kraft paper of 30 cm × 30 cm, placed on a non-flammable surface, are required. A few grammes of fine smokeless powder and a suitable means of ignition are used as described in the procedure and shown in Figure 13.7.1.1.

13.7.1.2 *Procedure*

13.7.1.2.1 Solids and liquids

10 g of the substance is placed in the beaker. The beaker is placed centrally on a bed of kerosene-soaked sawdust and the sawdust is ignited with the electric igniter. The test is performed twice with 10 g of sample and twice with 100 g, unless an explosion is observed.

13.7.1.2.2 Alternative method (solids only)

A conical pile of the substance is placed on the sheet of Kraft paper, the height of the pile being equal to the radius at the base. A trail of smokeless powder surrounds the pile of substance under test and this trail is ignited, by a suitable source of ignition applied from a safe distance, at two diagonally opposite points (see Figure 13.7.1.1). The Kraft paper is ignited by the trail of smokeless powder and transmits the flame to the test substance. The test is performed twice with 10 g and twice with 100 g of sample, unless an explosion is observed.

13.7.1.3 *Test criteria and method of assessing results*

The event is observed visually and the result is reported as one of the following three categories:

(a) Fails to ignite;

(b) Ignites and burns; or

(c) Explodes.

The duration of the combustion, or the time to explosion, may be noted to provide additional information. The test result is considered "+" if any explosion of the test samples occurs and the substance is considered to be too dangerous for transport in the form in which it is tested. Otherwise, the test result is considered "-".

13.7.1.4 *Examples of results*

Substances	Observations	Result
Liquids		
Nitromethane	Burning	-
Solids		
Alternative method		
Blasting gelatin A (NG 92%, cellulose nitrate 8%)	Burning	-
Black powder Pulverin	Burning	-
Lead azide	Explosion	+
Mercury fulminate	Explosion	+

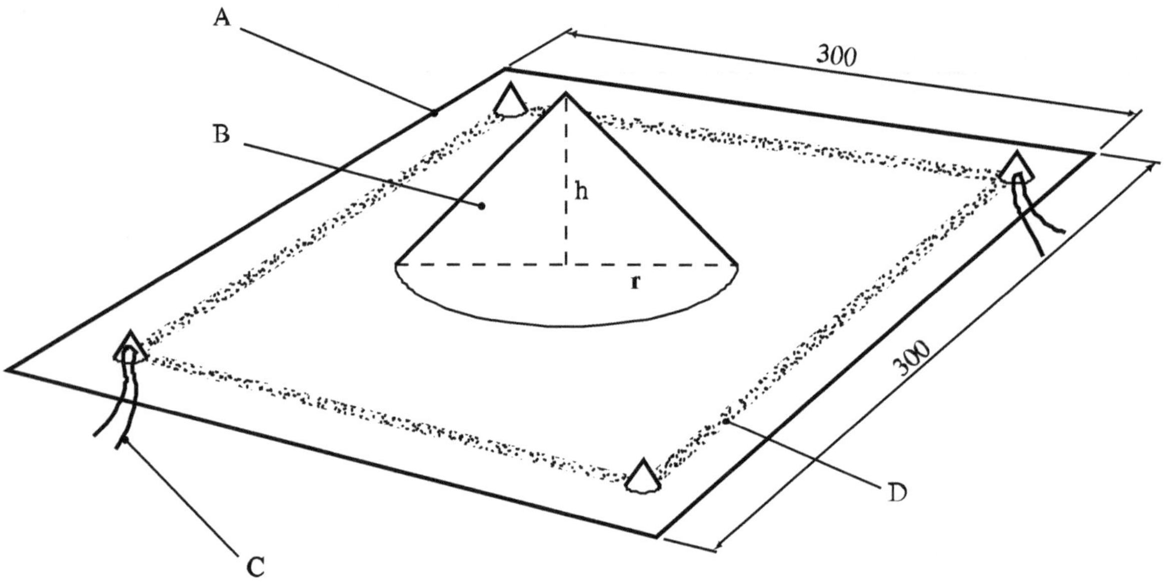

(A) Sheet of kraft paper
(B) Test substance
(C) Ignition by an igniter and a few grams of fine smokeless powder (at two opposite corners)
(D) Trail of fine smokeless powder

Figure 13.7.1.1: SMALL SCALE BURNING TEST (for solids)

SECTION 14

TEST SERIES 4

14.1 Introduction

14.1.1 Series 4 tests are intended to answer the question "Is the article, packaged article or packaged substance too dangerous for transport?" (box 16 of Figure 10.2). Conditions which may occur during transport include high temperature and high relative humidity, low temperature, vibration, bumping and dropping. The two types of test to be carried out are:

Type 4 (a): a test of thermal stability for articles; and

Type 4 (b): a test to determine the danger from dropping.

14.1.2 The question in box 16 is answered "Yes" if a "+" is obtained in either test type 4 (a) or 4 (b).

14.2 Test methods

The test methods currently used are listed in Table 14.1.

Table 14.1: TEST METHODS FOR TEST SERIES 4

Test code	Name of Test	Section
4 (a)	Thermal stability test for unpackaged articles and packaged articles[a]	14.4.1
4 (b) (i)	The steel tube drop test for liquids[a]	14.5.1
4 (b) (ii)	Twelve metre drop test for articles, packaged articles and packaged substances[a]	14.5.2

[a] *Recommended test.*

14.3 Test conditions

14.3.1 The tests are carried out on the packaged substance, packaged article(s) and, if intended to be transported unpackaged, on the article itself. The minimum size unit acceptable for a type 4 (a) test is the smallest packaged unit or an individual article if transported unpackaged. Test 4 (b) (i) should be applied to homogeneous liquids; and test 4 (b) (ii) should be applied to unpackaged and packaged articles and to packaged substances other than homogeneous liquids.

14.4 Series 4 type (a) test prescriptions

14.4.1 *Test 4 (a): Thermal stability test for unpackaged articles and packaged articles*

14.4.1.1 *Introduction*

This test is used to evaluate the thermal stability of articles and packaged articles when subjected to elevated thermal conditions to determine whether the unit being tested is too hazardous for transport. The minimum size unit acceptable for this test is the smallest packaged unit or, if transported unpackaged, the unpackaged article. In general, the package as used for transport should be tested. If this is not possible (e.g. it is too large for the oven), a similar, smaller package should be used filled with the maximum number of articles possible.

14.4.1.2 *Apparatus and materials*

This test requires an oven equipped with a fan and thermostatic control to maintain the temperature at 75 ± 2 °C. It is desirable that the oven should have dual thermostats or similar protection against excessively high temperatures in the event of a thermostat malfunction. A unit should be fitted with a thermocouple connected to a temperature recorder to assess any exothermic temperature increase.

14.4.1.3 *Procedure*

Depending on the unit being tested, a thermocouple is placed either on the outside casing of the unpackaged article, or on the outside casing of an article which is located near the centre of its package. The thermocouple is attached to a temperature recorder. The unit to be tested (together with the thermocouple) is placed in the oven, heated to 75 °C and maintained at that temperature for 48 hours. The oven is then allowed to cool and the unit then removed from the oven and inspected. Temperatures are recorded and signs of reaction, damage or exudation noted.

14.4.1.4 *Test criteria and method of assessing results*

The test result is considered "+" and the article or packaged article(s) considered too dangerous for transport if:

(a) It explodes;

(b) It ignites;

(c) It experiences a temperature rise exceeding 3 °C;

(d) The outer casing of the article or the outer packaging is damaged; or

(e) Dangerous exudation occurs i.e. explosive is visible outside the article(s).

The result is considered "-" if there are no external effects and there is no temperature rise exceeding 3 °C.

14.4.1.5 *Examples of results*

Article tested	**Result**
Cylindrical fountains	-
Delay electric igniter	-
Hand signal device	-
Railway torpedo	-
Roman candle	-
Safety squib	-
Signal flare	-
Small arms ammunition	-
Smoke candle	-
Smoke grenade	-
Smoke pot	-
Smoke signal	-

14.5 Series 4 type (b) test prescriptions

14.5.1 *Test 4 (b) (i): The steel tube drop test for liquids*

14.5.1.1 *Introduction*

This test is used to determine the explosive behaviour of an homogeneous energetic liquid, in a closed steel tube, when it is dropped from various heights onto a steel anvil.

14.5.1.2 *Apparatus and materials*

The steel (A37 type) tube is 33 mm inner diameter, 42 mm outer diameter, and 500 mm long (see figure14.5.1.1). It is filled with the liquid under test and closed at the upper end by a cast-iron threaded cap sealed with polytetrafluoroethylene (PTFE) tape. The cap has an 8 mm hole, for filling, drilled axially and the hole is closed by a plastics plug.

14.5.1.3 *Procedure*

The temperature and density of the liquid are noted. One hour or less before the test, the liquid is agitated for 10 seconds. The height of drop is varied in steps of 0.25 m to a maximum of 5 m. The procedure consists of finding the maximum height at which there is no detonation. The tube falls in a vertical orientation. It is noted whether, and at what height, any of the following occur:

(a) Detonation with fragmentation of the tube;

(b) A reaction causing the tube to burst; or

(c) No reaction with little damage to the tube.

14.5.1.4 *Test criteria and method of assessing results*

If detonation takes place after a drop of 5 m or less, the test result is "+" and the liquid is considered too dangerous to be transported.

Where there is a local reaction without detonation after a 5 m drop, the test result is "-" but metal packaging should not be used unless its suitability for transport has been demonstrated to be safe to the satisfaction of the competent authority.

If there is no reaction after a drop from 5 m, the test result is "-" and it is considered that the liquid can be transported in any form of packaging suitable for liquids.

14.5.1.5 *Examples of results*

Liquid	**Temperature (° C)**	**Drop height for detonation (m)**	**Result**
Nitroglycerine	15	< 0.25	+
Nitroglycerine/Triacetin/2 NDPA (78/21/1)	14	1.00	+
Nitromethane	15	> 5.00	-
TEGDN	13	> 5.00	-

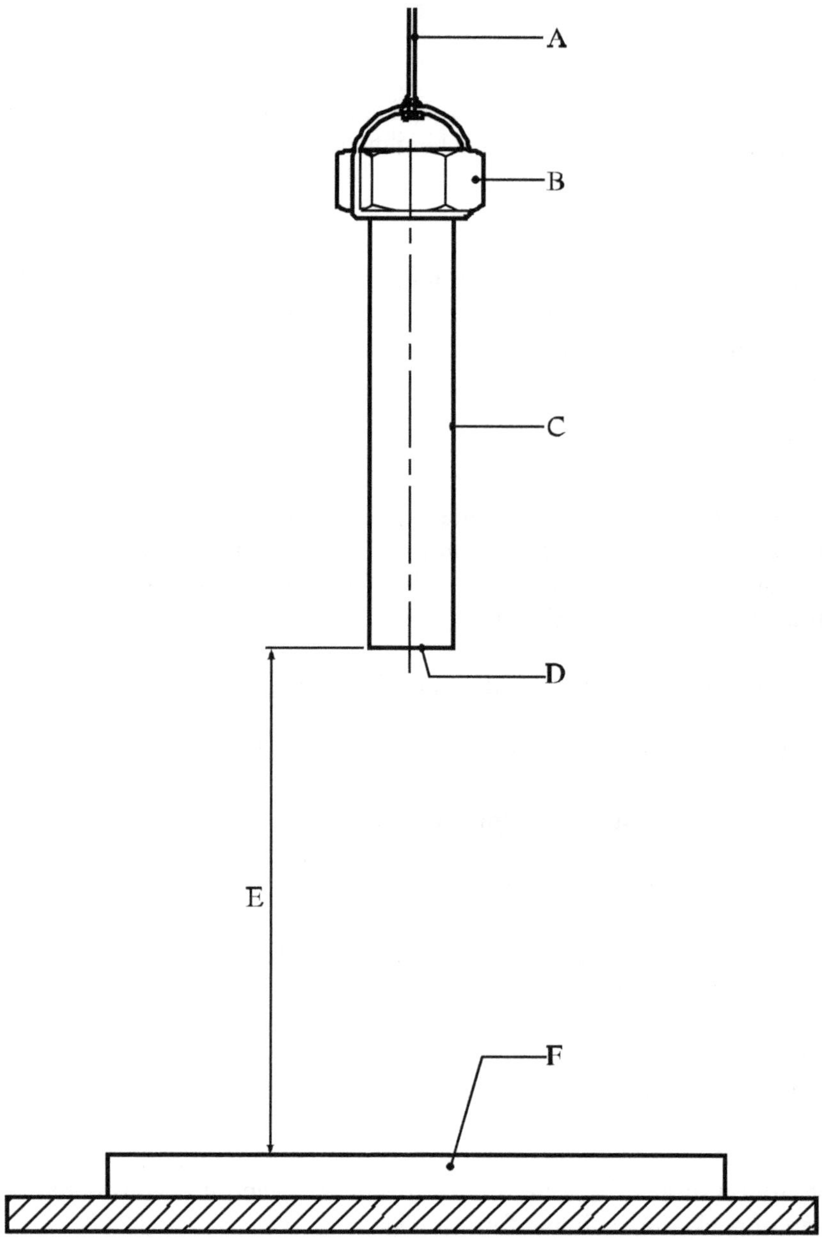

(A)	Release by fusing of wire	(B)	Cast-iron threaded cap
(C)	Seamless steel tube	(D)	Welded-steel base (thickness 4 mm)
(E)	Drop height of 0.25 m to 5.00 m	(F)	Steel anvil (1 m × 0.50 m and thickness 0.15 m)

Figure 14.5.1.1: THE STEEL DROP TEST FOR LIQUIDS

14.5.2 ***Test 4 (b) (ii): Twelve metre drop test for unpackaged articles, packaged articles and packaged substances***

14.5.2.1 *Introduction*

This test determines whether a test unit (article, packaged article(s) or packaged substance (other than homogeneous liquids)) can withstand a free-fall impact without producing any significant fire or explosion hazard. It is not intended as a test to evaluate whether the package will withstand impact.

14.5.2.2 *Apparatus and materials*

14.5.2.2.1 Impact surface

The impact surface is a solid base with a reasonably smooth surface. An example of such a surface is as follows: a steel plate, with a minimum thickness of 75 mm and Brinell hardness of not less than 200, solidly supported by a concrete foundation having a minimum thickness of 600 mm. The length and width of the surface should be not less than one and a half times the dimension of the unit being tested.

14.5.2.2.2 Other apparatus

Photographic or other visual recording devices should be used to verify impact attitude and results. Where impact attitude may be considered to be a significant factor, the test agency may use guidance devices to obtain the desired impact attitude. Such devices should not significantly restrain drop velocity, nor impede rebound after impact.

14.5.2.2.3 Materials

In certain cases, some of the explosive articles in a package of articles under test may be replaced with inert articles. These inert articles should be of the same mass and volume as the explosive articles they replace. The explosive articles should be located in the position where they are most likely to function on impact. If a packaged substance is tested, none of the substance may be replaced by inert material.

14.5.2.3 *Procedure*

The test unit is dropped from a height of 12 m as measured from the lowest point of the test unit to the impact surface. ***A safe waiting period, prescribed by the test agency, should be observed after impact, even if no visible initiation or ignition occurs at impact.*** The test unit should then be further examined to determine if any ignition or initiation has occurred. Three drops are made on the packaged substance or article unless a decisive event (e.g. fire or explosion) occurs earlier. However, each test unit is dropped only once. The data recorded should include the package description and observations. Recorded results should include photographs and recorded visual and audible evidence of ignition, time of occurrence (if any), and indication of severity of the results in terms such as mass detonation or deflagration. The attitude of the test unit at impact should also be recorded. Rupture of the package may be noted but does not affect the conclusion.

14.5.2.4 *Test criteria and method of assessing results*

The test result is considered "+" and the packaged substance or article(s) too dangerous to transport if a fire or explosion resulted from impact. Rupture of the package or article casing alone is not considered a "+" result. The result is considered "-" if no fire or explosion occurred in any of the three drops.

14.5.2.5 *Examples of results*

Substances or article(s)	Number of drops	Observations	Results
Cable cutter cartridge, metal box containing two devices	3	No reaction	-
Cast primers (27.2 kg)	3	No reaction	-
CBI solid propellants 7.11 mm diameter (36.3 kg)	3	No reaction	-
Component (of projectile) containing detonator, primer and fuze assembly	1	Ignition	+
Dynamite, ammonia gelatin (22.7 kg)	3	No reaction	-
Dynamite, 40% strength ammonia (22.7 kg)	3	No reaction	-
Dynamite, 60% strength straight (22.7 kg)	3	No reaction	-
Dynamite, 50% strength straight "ditching" (22.7 kg)	3	No reaction	-
Propellant gas generator 61.7 kg net mass in aluminium container	3	No reaction	-
Demolition firing device, wooden box containing 20 individually packed devices	3	No reaction	-

SECTION 15

TEST SERIES 5

15.1 Introduction

15.1.1 The results from three types of series 5 tests are used to answer the question "Is it a very insensitive explosive substance with a mass explosion hazard?" (box 21 of Figure 10.3). The test types are:

Type 5 (a): a shock test to determine the sensitivity to intense mechanical stimulus;

Type 5 (b): thermal tests to determine the tendency of transition from deflagration to detonation; and

Type 5 (c): a test to determine if a substance, when in large quantities, explodes when subjected to a large fire.

15.1.2 The question in box 21 is answered "No" if a "+" is obtained in any of the three test types i.e. a candidate for Division 1.5 should pass a test of each type.

15.2 Test methods

The test methods currently used are listed in Table 15.1.

Table 15.1: TEST METHODS FOR TEST SERIES 5

Test code	Name of Test	Section
5(a)	Cap sensitivity test[a]	15.4.1
5 (b) (i)	French DDT test	15.5.1
5 (b) (ii)	USA DDT test[a]	15.5.2
5 (b) (iii)	Deflagration to detonation test	15.5.3
5 (c)	External fire test for Division 1.5[a]	15.6.1

[a] *Recommended test.*

A test of each type should be performed.

15.3 Test conditions

15.3.1 As the density of the substance has an important effect on the results from type 5 (a) and 5 (b) tests, the density should be determined. The sample mass and density should always be recorded.

15.3.2 The tests should be performed at ambient temperature unless the substance is to be transported under conditions where it may change its physical state or density.

15.4 Series 5 type (a) test prescription

15.4.1 *Test 5 (a): Cap sensitivity test*

15.4.1.1 *Introduction*

This shock test is used to determine the sensitiveness of a substance to intense mechanical stimulus.

15.4.1.2 *Apparatus and materials*

The experimental set up for the cap sensitivity test is shown in figures 15.4.1.1 and 15.4.1.2 and consists of a cardboard tube of minimum diameter 80 mm and length 160 mm with a maximum wall thickness of 1.5 mm, closed at the base with a membrane just sufficient to retain the sample. The intense mechanical stimulus is provided by a standard detonator (see Appendix 1) inserted centrally in the top of the explosive in the tube to a depth equal to its length. Below the tube is the witness, which consists of a 1.0 mm thick 160 × 160 mm steel plate, placed on a steel ring of 50 mm height, 100 mm inner diameter and 3.5 mm wall thickness (see Figure 15.4.1.1). Alternatively, a 51 mm diameter, 102 mm long cylinder of common (soft) lead (see Figure 15.4.1.2) may be used. The apparatus is placed onto a square shaped steel plate of 25 mm thickness and 152 mm sides.

15.4.1.3 *Procedure*

The substance under test is filled into the tube in three equal increments. For free-flowing granular substances, the sample is consolidated by allowing the tube to fall vertically through a height of 50 mm after filling each increment. Gel-type substances are carefully packed to prevent adding voids. In all cases, the final density of the explosive in the tube should be as close as possible to its shipping density. For high-density cartridged explosives with a diameter greater than 80 mm, the original cartridge is used. Where such original cartridges are inconveniently large for testing, a portion of the cartridge not less than 160 mm long may be cut off and used for testing. In such cases the detonator is inserted into the end in which the substance has not been disturbed by the action of cutting the cartridge. Those explosives whose sensitivity could be temperature dependent should be stored for at least 30 hours at a temperature of 28 - 30 °C prior to testing. Explosives containing prilled ammonium nitrate, which have to be transported in regions of high ambient temperatures should be temperature cycled as follows: 25 °C → 40 °C → 25 °C → 40 °C → 25 °C prior to testing. The tube is placed onto the witness and steel base plate and the standard detonator inserted centrally into the top of the explosive. The detonator is then fired from a safe position and the witness examined. The test is conducted three times unless detonation of the substance occurs.

15.4.1.4 *Test criteria and method of assessing results*

The result is considered "+" and the substance should not be classified in Division 1.5 if in any trial:

(a) The witness plate is torn or otherwise penetrated (i.e. light is visible through the plate) - bulges, cracks or folds in the witness plate do not indicate cap sensitivity; or

(b) The centre of the lead cylinder is compressed from its initial length by an amount of 3.2 mm or greater.

Otherwise, the result is considered "-".

15.4.1.5 *Examples of results*

Substance	Density (kg/m³)	Remarks	Result
Ammonium nitrate prills + fuel oil	840 - 900	Original	-
" " "	750 - 760	2 temperature cycles	+
Ammonium nitrate + TNT + combustible material	1 030 - 1 070	Original	+
Ammonium nitrate prills + DNT (on surface)	820 - 830	Original	-
" " "	800 - 830	30 hours at 40 °C	+
Ammonium nitrate + DNT + combustible material	970 - 1 030	Original	-
" " "	780 - 960	Original	+
Ammonium nitrate + combustible material	840 - 950	Original	-
" " "	620 - 840	Original	+
Ammonium nitrate + alkali nitrate + alkali-earth nitrate + Aluminium + water + combustible material	1 300 - 1 450	Original	-
" " "	1 130 - 1 220	Original	+
Ammonium nitrate + alkali nitrate + nitrate + TNT + Aluminium + water + combustible material	1 500	Original	-
" " "	1 130 - 1 220	Original	+
Ammonium nitrate/methanol (90/10), prilled			-
Ammonium nitrate/nitromethane, 87/13			+
ANFO (94/6), prills			-
ANFO (94/6), 200 μm			+
TNT, granular			+

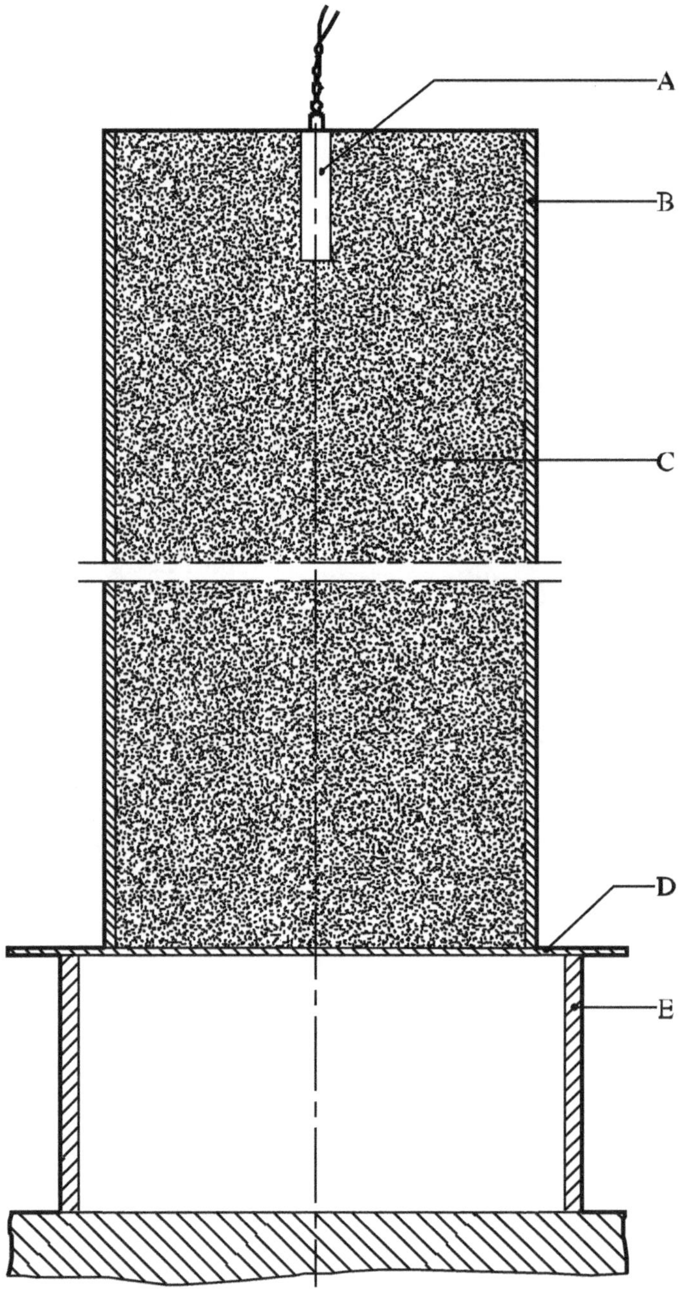

(A) Detonator
(B) Tube fibreboard spirally wound plys
(C) Test substance
(D) Witness plate of normal structural steel
(E) Steel ring

Figure 15.4.1.1: CAP SENSITIVITY TEST (using steel witness plate)

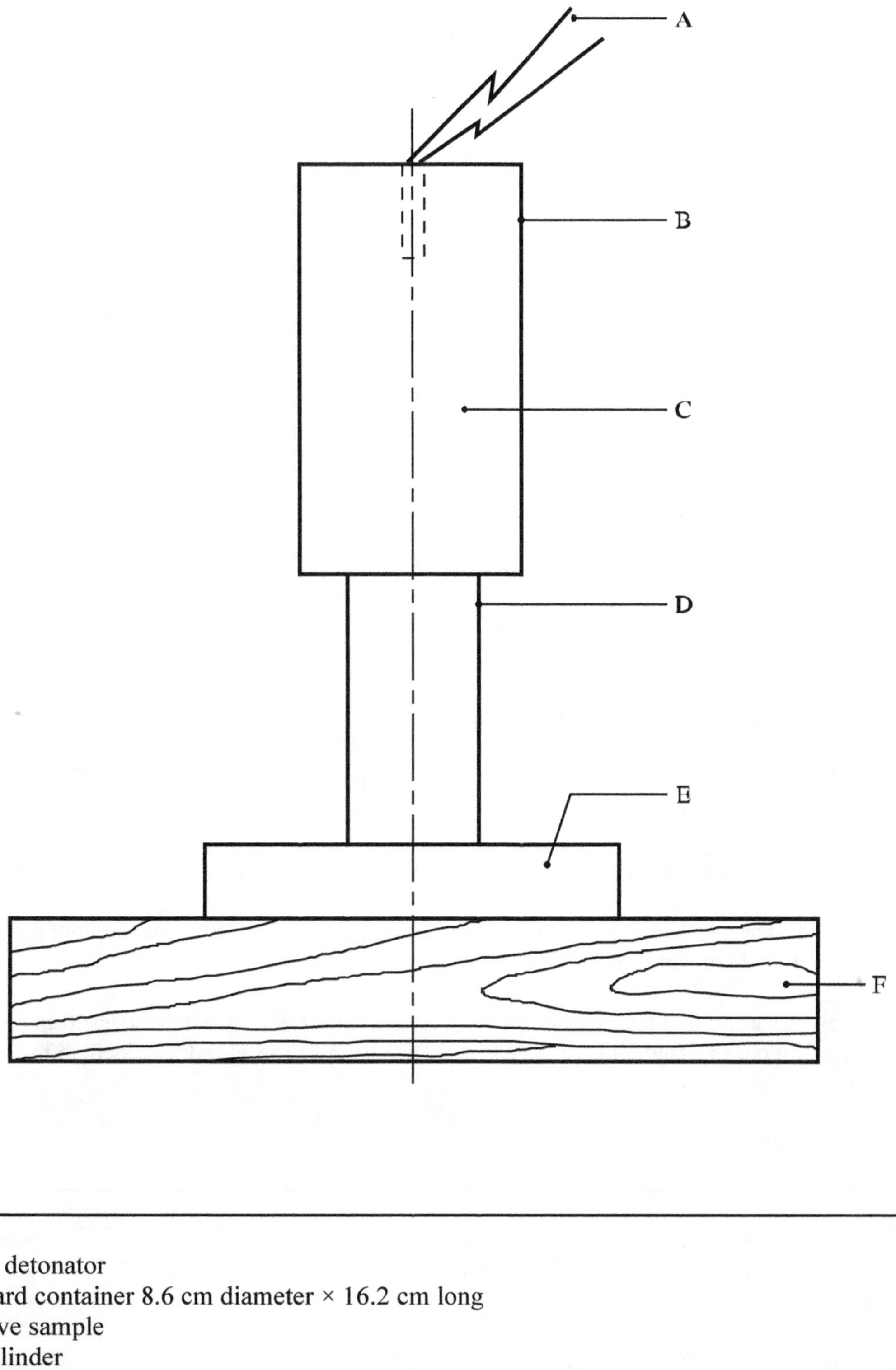

(A) Electric detonator
(B) Cardboard container 8.6 cm diameter × 16.2 cm long
(C) Explosive sample
(D) Lead cylinder
(E) Steel plate 15 × 15 × 2.5 cm
(F) Wood block 30 × 30 × 5 cm

Figure 15.4.1.2: CAP SENSITIVITY TEST (using lead cylinder as witness)

15.5 Series 5 type (b) test prescriptions

15.5.1 *Test 5 (b) (i): French DDT test*

15.5.1.1 *Introduction*

This test is used to determine the tendency for the substance to undergo transition from deflagration to detonation.

15.5.1.2 *Apparatus and materials*

The apparatus consists of a seamless steel tube (A37 type), inner diameter 40.2 mm, wall thickness 4.05 mm, and length 1 200 mm. The static resistance of the tube is 74.5 MPa. As shown in Figure 15.5.1.1, the tube is closed by two screwed caps and a probe for monitoring the shock wave velocity is fitted. The tube is placed horizontally onto a lead witness plate of thickness 30 mm. The substance is ignited by a heated wire, composed of Ni/Cr (80/20) with diameter 0.4 mm and length 15 mm, located at one end of the tube.

15.5.1.3 *Procedure*

The test substance is filled into the tube and compacted by hand compression. The temperature, density and water content of the substance should be recorded. A current of up to 8 A is used for a maximum of three minutes to heat the ignition wire and ignite the substance. The test is performed three times unless deflagration to detonation transition occurs as shown by compression of the lead witness plate or by the measured propagation velocity.

15.5.1.4 *Test criteria and method of assessing results*

The test result is considered "+" and the substance should not be classified in Division 1.5 if in any trial detonation occurs. Evidence of detonation may be assessed by whether:

(a) The lead witness plate is compressed in a manner characteristic of detonation; and

(b) The measured propagation velocity is greater than the speed of sound in the substance and constant in the part of tube furthest from the initiator.

The length before detonation and the detonation velocity should be noted. The test result is considered "-" if the witness plate is not compressed and the speed of propagation, if measured, is less than the speed of sound in the substance.

15.5.1.5 *Examples of results*

Substance	**Density (kg/m^3)**	**Result**
Aluminised gel (62.5% oxidising salts, 15% aluminium, 15% other combustibles)	1 360	-
ANFO (AN size 0.85 mm, oil retention 15%)	860	-
Gelatine-dynamite (NG/EGDN 40%, AN 48%, aluminium 8%, NC)	1 450	+
Guhr-dynamite (NG 60%, Guhr 40%)	820	+
Sensitized slurry explosive	1 570	-

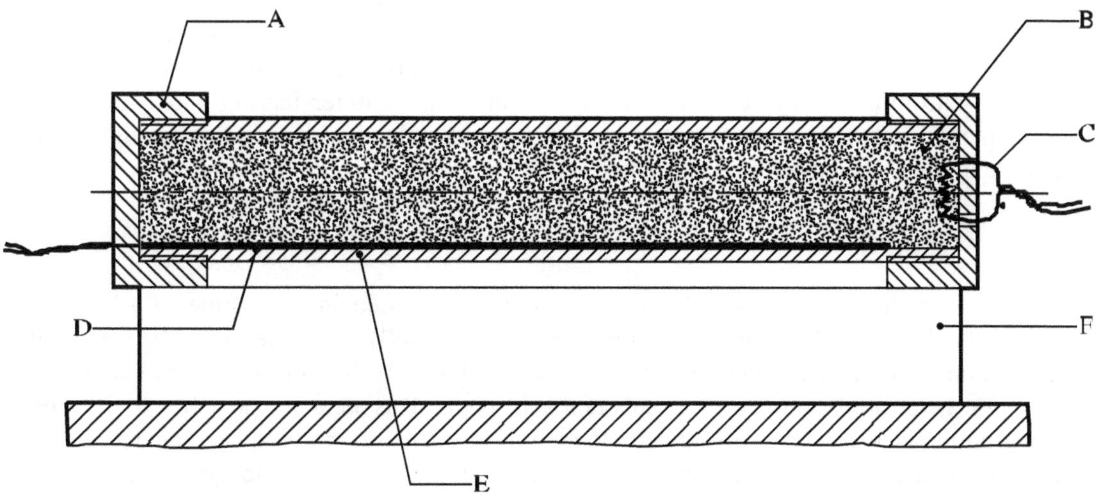

(A) Threaded cast-iron end caps
(B) Test substance
(C) Igniter wire
(D) Velocity probe
(E) Seamless steel tube
(F) Lead witness plate

Figure 15.5.1.1: FRENCH DDT TEST

15.5.2 *Test 5 (b) (ii): USA DDT test*

15.5.2.1 *Introduction*

This test is used to determine the tendency of a substance to undergo transition from deflagration to detonation.

15.5.2.2 *Apparatus and materials*

The experimental arrangement is shown in Figure 15.5.2.1. The sample substance to be tested is contained in a 457 mm length of "3 inch schedule 80" carbon (A53 Grade B) steel pipe with inside diameter 74 mm, wall thickness 7.6 mm, capped at one end with a "3 000 pound" forged steel pipe cap, and at the other with a 13 cm square, 8 mm thick mild steel witness plate which is welded to the pipe. An igniter consisting of 5.0 g of black powder (100% passed through No. 20 sieve, 0.84 mm, and 100% retained by No. 50 sieve, 0.297 mm) is located at the centre of the sample vessel. The igniter assembly consists of a cylindrical container 21 mm in diameter which is made from 0.54 mm thick cellulose acetate which is held together by two layers of nylon filament reinforced cellulose acetate tape. The length of the igniter capsule is approximately 1.6 cm for the 5 g igniter. The igniter capsule contains a small loop formed from a 25 mm length of nickel-chromium alloy resistance wire 0.30 mm in diameter having a resistance of 0.343 ohms. This loop is attached to two insulated copper lead wires. These lead wires are fed through small holes in the wall of the pipe and are sealed with epoxy resin.

15.5.2.3 *Procedure*

After the sample, at ambient temperature, is loaded into the pipe to a height of 23 cm, the igniter (with its leads inserted through small holes in the pipe wall) is inserted into the centre of the pipe and the leads pulled taut and sealed with epoxy resin. The remainder of the sample is then loaded, and the top cap screwed on. For gelatinous samples, the substance is packed as near as possible to its normal shipping density. For granular samples, the substance is loaded to the density obtained by repeated tapping of the pipe against a hard surface. The tube is placed in a vertical position and the igniter is fired by a current of 15 amperes obtained from a 20-volt transformer. Three trials should be performed unless deflagration to detonation transition occurs earlier.

15.5.2.4 *Test criteria and method of assessing results*

The test result is considered "+" and the substance should not be classified in Division 1.5 if a hole is punched through the witness plate. If no hole is punched through the witness plate then the result is considered "-".

15.5.2.5 *Examples of results*

Substance	**Apparent Density (kg/m3)**	**Result**
Ammonium nitrate/fuel oil (94/6)	795	-
Ammonium perchlorate (200 µm)[a]	1 145	-
ANFO blasting agent (with low density combustible additive)	793	+
Emulsion blasting agent (sensitized with microballoons)	1 166	-
Emulsion blasting agent (sensitized with NC)	1 269	-
Emulsion blasting agent (sensitized with oil)	1 339	-
NG dynamite[a]	900	+
PETN (wetted with 25% water)[a]	1 033	+

[a] *Used for calibration purposes - not for classification in Division 1.5.*

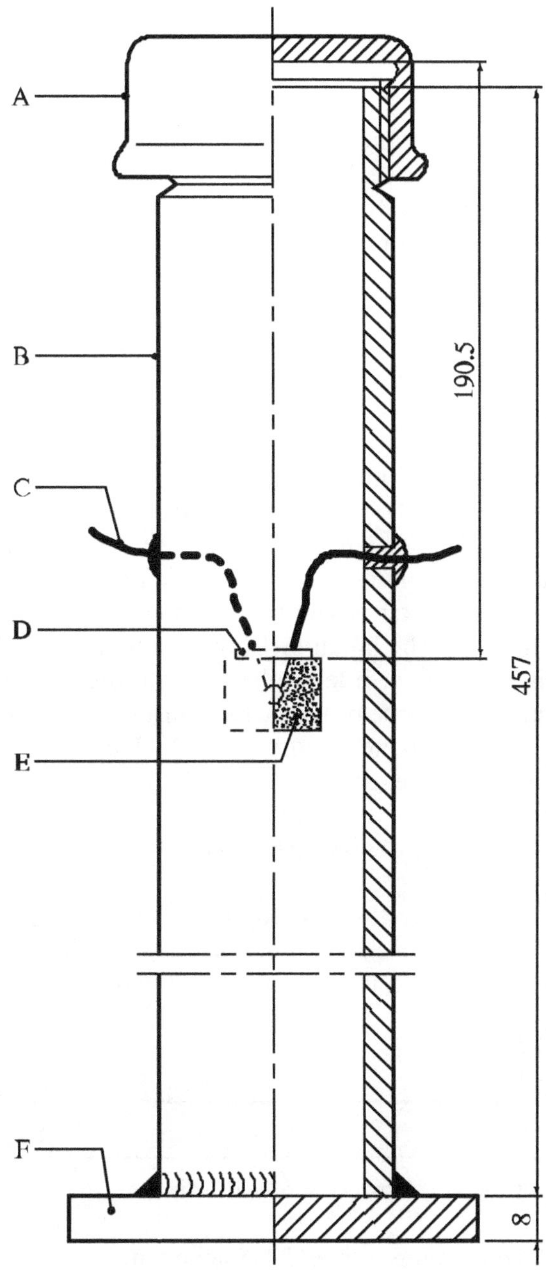

(A) Forged steel cap
(B) Steel pipe
(C) Igniter leads
(D) Seal
(E) Igniter assembly
(F) Witness plate

Figure 15.5.2.1: USA DDT TEST

15.5.3 *Test 5 (b) (iii): Deflagration to detonation transition test*

15.5.3.1 *Introduction*

This test is used to determine the tendency of a substance to undergo transition from deflagration to detonation.

15.5.3.2 *Apparatus and materials*

In comparative tests of DDT susceptibility, steel tubes with inner diameter of 40 mm, wall thickness of 10 mm and length of 1 000 mm are used. The breaking strength of the tube is 130 MPa (see Figure 15.5.3.1). One end of the tube is sealed by a metal threaded plug or other means of fastening such as a stud, screw or welding. The strength of the tube closure should not be less than the breaking strength of the tube. A threaded socket for the igniter is made in the tube wall 100 mm from the plug. The black powder igniter cases are made of mild steel. An electric squib is mounted in the case. Its conductivity is checked by a tester or ohmmeter and then 3 ± 0.01 g of black powder (SGP No 1) is loaded into the case and the case opening sealed with plastic tape.

15.5.3.3 *Procedure*

The test substance is loaded into the tube at normal bulk density. An end of the 10 m length of 12 g/m detonating cord is inserted into the sample at the open end of the tube to a depth of 100 mm and the tube then sealed with plastic tape. The loaded tube is mounted horizontally on a steel plate. The other end of the detonating cord is attached to an aluminium plate of length 200 mm, width 50 mm, and thickness from 2 to 3 mm (Figure 15.5.3.1). The igniter is screwed into the tube wall, making sure that the threads are clean, and is connected to the firing line. The initiation of the test material is then performed. After firing, the tube is inspected. The rupture character is recorded (tube bulging, breaking into large fragments or crushing into small fragments), the presence or absence of unreacted substance and the presence or the absence of the detonating cord traces on the witness plate is determined. Three trials should be performed unless deflagration to detonation transition occurs earlier.

15.5.3.4 *Test criteria and method of assessing results*

Test results are assessed by the tube rupture character or explosion of the detonating cord. The result is considered "+" and the substance should not be classified in Division 1.5 if fragmentation of the tube occurs. The result is considered "-" if no fragmentation of the tube occurs.

15.5.3.5 *Examples of results*

Substance (at 20 °C)	**Density (kg/m^3)**	**Result**
Ammonal (80.5% ammonium nitrate, 15% trotyl, 4.5% aluminium) (powder)	1 000	-
Ammonal No. 1, splitting (66% ammonium nitrate, 24% hexogen, 5% aluminium)	1 100	+
Ammonite 6Zhv (79% ammonium nitrate, 21% trotyl) (powder)	1 000	-
Granulite AS-4 (91.8% ammonium nitrate, 4.2% machine oil, 4% aluminium)	1 000 (1 600)	-
Granulite ASR-8 (70% ammonium nitrate, 4.20% sodium nitrate, 8% aluminium, 2% machine oil)	1 000 (1 600)	-
Ammonium perchlorate	1 100	-
Ammonium perchlorate plus 1.5% of combustible additives	1 100	+

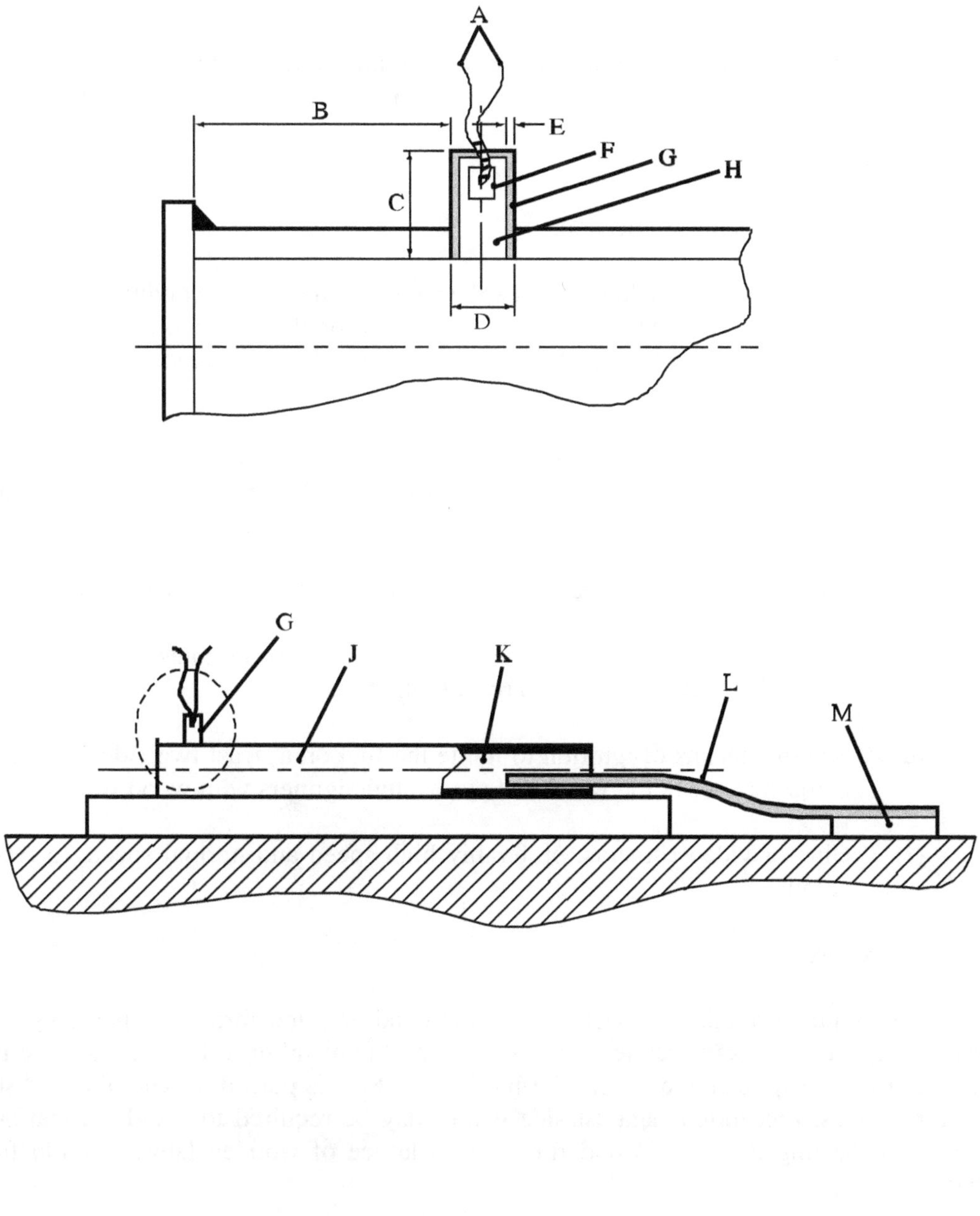

(A)	Igniter leads	(B)	Distance of igniter from end of tube (100 mm)
(C)	Length of igniter (40 mm)	(D)	Outer diameter of igniter (16 mm)
(E)	Thickness of igniter casing (1 mm)	(F)	Squib
(G)	Igniter	(H)	Black powder
(J)	Seamless steel tube with the plug	(K)	Test substance
(L)	Detonating cord	(M)	Aluminium witness plate

Figure 15.5.3.1: DEFLAGRATION TO DETONATION TRANSITION TEST

15.6 Series 5 type (c) test prescription

15.6.1 *Test 5 (c): External fire test for Division 1.5*

15.6.1.1 *Introduction*

This test is used to determine whether a substance, packaged as for transport, can explode if involved in a fire.

15.6.1.2 *Apparatus and materials*

The following items are needed:

(a) A package (or packages) of explosive substance in the condition and form in which it is offered for transport. The total volume of the package (or packages) to be tested should not be less than 0.15 m^3 and need not exceed 200 kg of net explosive substance;

(b) A metal grid to support the products above the fuel and allow adequate heating. If a wooden crib fire is used, the grid should be 1.0 m above the ground and if a liquid hydrocarbon pool fire is used then the grid should be 0.5 m above the ground;

(c) Strapping or wire, if necessary, to hold the packages together on the support grid;

(d) Enough fuel to keep a fire burning for at least 30 minutes or until the substance has clearly had sufficient time to react to the fire;

(e) Suitable means of ignition to ignite the fuel on at least two sides e.g. for a wood fire, kerosene to soak the wood and pyrotechnic igniters with wood wool;

(f) Cine or video cameras, preferably high speed and normal speed, to record events in colour.

15.6.1.3 *Procedure*

15.6.1.3.1 The required number of packages, in the condition and form in which they are offered for transport, are arranged as close as possible to one another on the metal grid. If necessary, the packages may be encircled with a steel strip to support them during the test. Fuel is placed beneath the grid so that the fire will engulf the packages. Precautions against side winds may be required to avoid dissipation of the heat. Suitable methods of heating include a wood fire using a lattice of wooden laths, a liquid fuel fire and a propane burner.

15.6.1.3.2 The recommended method is to use a wood fire which has a balanced air/fuel ratio, thereby avoiding too much smoke which would obscure the events, and which burns with sufficient intensity and duration to bring many kinds of packaged explosives to reaction in 10 to 30 minutes. A suitable method involves using air-dried pieces of wood (approximately 50 mm square section), stacked to form a lattice beneath the grid (1 m off the ground), and up to the base of the grid supporting the packages. The wood should extend beyond the packages to a distance of at least 1.0 m in every direction and the lateral distance between the laths should be about 100 mm. Enough fuel should be used to keep the fire burning for at least 30 minutes or until the substance or article has clearly had sufficient time to react to the fire.

15.6.1.3.3 A receptacle filled with suitable liquid fuel, a combination of both wood and liquid fuel or a gas fire may be used as alternatives to the wood fire providing they are as severe. If a liquid pool fire is used, the receptacle should extend beyond the packages to a distance of at least 1.0 m in every direction. The distance between the grid platform and the receptacle should be approximately 0.5 m. Before using this method, consideration should be given to whether any quenching action or adverse interaction between explosives and liquid fuel can occur such as might bring the results into question. If a gas fire is used, the grid should be at the correct height above the burner to allow for adequate fire engulfment.

15.6.1.3.4 The ignition system should be put into place and the fuel ignited on two sides, one the upwind side, simultaneously. The test should not be performed under conditions where the wind speed exceeds 6 m/s. ***A safe waiting period, prescribed by the test agency, should be observed after the fire has extinguished.***

15.6.1.3.5 Observations are made of evidence of explosion such as a loud noise and the projection of fragments from the fire area.

15.6.1.3.6 The test is normally performed once only but if the wood or other fuel used for the fire is all consumed leaving a significant quantity of unconsumed explosive substance in the remains or in the vicinity of the fire, then the test should be performed again using more fuel, or a different method, to increase the intensity and/or duration of the fire. If the result of the test does not enable the hazard division to be determined, a further test should be performed.

15.6.1.4 *Test criteria and method of assessing results*

A substance which explodes in this test is considered to give a "+" result and should not be classified in Division 1.5.

15.6.1.5 *Examples of results*

Substance	**Result**
ANFO	-
ANFO (with 6% aluminium powder)	-
ANFO (with 6% combustible material)	-
ANFO emulsion (with 1% microspheres)	-
ANFO emulsion (with 3.4% microspheres)	-

SECTION 16

TEST SERIES 6

16.1 Introduction

16.1.1 The results from four types of series 6 tests are used to determine which division, amongst Divisions 1.1, 1.2, 1.3 and 1.4, corresponds most closely to the behaviour of a product if a load is involved in a fire resulting from internal or external sources, or an explosion from internal sources (boxes 26, 28, 30, 32 and 33 of Figure 10.3). The results are also necessary to assess whether a product can be assigned to Compatibility Group S of Division 1.4 and whether or not it should be excluded from Class 1 (boxes 35 and 36 of Figure 10.3). The four types of test are:

Type 6 (a): a test on a single package to determine if there is mass explosion of the contents;

Type 6 (b): a test on packages of an explosive substance or explosive articles, or non-packaged explosive articles, to determine whether an explosion is propagated from one package to another or from a non-packaged article to another;

Type 6 (c): a test on packages of an explosive substance or explosive articles, or non-packaged explosive articles, to determine whether there is a mass explosion or a hazard from dangerous projections, radiant heat and/or violent burning or any other dangerous effect when involved in a fire; and

Type 6 (d): a test on an unconfined package of explosive articles to which special provision 347 of Chapter 3.3 of the Model Regulations applies, to determine if there are hazardous effects outside the package arising from accidental ignition or initiation of the contents.

16.2 Test methods

16.2.1 The test methods currently used are listed in Table 16.1.

Table 16.1: TEST METHODS FOR TEST SERIES 6

Test code	Name of Test	Section
6 (a)	Single package test[a]	16.4.1
6 (b)	Stack test[a]	16.5.1
6 (c)	External fire (bonfire) test[a]	16.6.1
6 (d)	Unconfined package test[a]	16.7.1

[a] *Recommended test.*

16.2.2 Test types 6 (a), 6 (b), 6 (c) and 6 (d) are normally performed in alphabetical order. However, it is not always necessary to follow this order or to conduct tests of all types.

(a) Test type 6 (a) may be waived if explosive articles are carried without packaging or when the package contains only one article, (see also section 16.2.2 (d));

(b) Test type 6 (b) may be waived if in each type 6 (a) test, (see also section 16.2.2 (d));

(i) The exterior of the package is undamaged by internal detonation and/or ignition; or

(ii) The contents of the package fail to explode, or explode so feebly as would exclude propagation of the explosive effect from one package to another in test type 6(b).

(c) Test type 6 (c) may be waived if, in a type 6 (b) test, there is practically instantaneous explosion of virtually the total contents of the stack. In such cases the product is assigned to Division 1.1;

(d) Test type 6(d) is a test used to determine whether a 1.4S classification is appropriate and is only used if special provision 347 on Chapter 3.3 of the Model Regulations applies. When testing articles to which special provision 347 applies, test type 6(d) may be performed first. If the results of test type 6(d) indicate that a 1.4S classification is appropriate, then test types 6(a) and 6(b) may be waived.

16.2.3 If a substance gives a "-" result (no propagation of detonation) in the Series 1 type (a) test, the 6(a) test with a detonator may be waived. If a substance gives a "-" result (no or slow deflagration) in a Series 2 type (c) test, the 6 (a) test with an igniter may be waived.

16.2.4 Explanations of certain terms used in the assignment of divisions and compatibility groups are given in the Glossary in Appendix B of the Model Regulations (e.g. mass explosion, pyrotechnic substance, entire load, total contents, explode, explosion of the total contents).

16.3 Test conditions

16.3.1 Tests from series 6 should be applied to packages of explosive substances and articles in the condition and form in which they are offered for transport. The geometrical arrangement of the products should be realistic in regard to the packing method and the conditions of transport and should be such as to produce the most disadvantageous test results. Where explosive articles are to be carried without packaging, the tests should be applied to the non-packaged articles. All types of packaging containing substances or articles should be subjected to the tests unless:

(a) The product, including any packaging, can be unambiguously assigned to a division by a competent authority on the basis of results from other tests or of available information; or

(b) The product, including any packaging, is assigned to Division 1.1.

16.4 Series 6 type (a) test prescription

16.4.1 *Test 6 (a): Single package test*

16.4.1.1 *Introduction*

This is a test on a single package to determine if there is mass explosion of the contents.

16.4.1.2 *Apparatus and materials*

The following items are required:

(a) A detonator to initiate the substance or article or an igniter just sufficient to ensure ignition of the substance or article (see 16.4.1.3.2 and 16.4.1.3.3);

(b) Suitable confining materials (see 16.4.1.3.4); and

(c) A sheet of 3 mm thick mild steel to act as a witness plate.

Blast measuring equipment may be used.

16.4.1.3 *Procedure*

16.4.1.3.1 The test is applied to packages of explosive substances and articles in the condition and form in which they are offered for transport. Where explosive articles are to be carried without packaging, the tests should be applied to the non-packaged articles. The decision to use either an initiating stimulus or an igniting stimulus is based on the following considerations.

16.4.1.3.2 For packaged substances:

(a) If the substance is intended to function by detonation, it should be tested with a standard detonator (Appendix 1);

(b) If the substance is intended to function by deflagration, it should be tested with an igniter just sufficient (but not more than 30 g of black powder) to ensure ignition of the substance within the package. The igniter should be located in the center of the substance in the package;

(c) Substances not intended for use as explosive, but provisionally accepted into Class 1, should be tested first with a standard detonator (Appendix 1) and, if no explosion occurs, with an igniter as in (b) above. If a substance gives a "-" result (no propagation of detonation) in the Series 1 type (a) test, the test with a detonator may be waived and if a substance gives a "-" result (no or slow deflagration) in a Series 2 type (c) test, the test with an igniter may be waived.

16.4.1.3.3 For packaged articles[1]:

(a) Articles provided with their own means of initiation or ignition:

The functioning of an article near the center of the package is stimulated by the article's own means of initiation or ignition. Where this is impracticable, the article's own means of initiation or ignition is replaced by another form of stimulus having the required effect;

---

[1] *Subject to the proviso that in the case of articles containing a very small quantity of substance(s) of compatibility group A only, a sufficient number of such items should be initiated simultaneously to cause not less than 0.2 g of primary explosive to explode.*

(b) Articles not provided with their own means of initiation or ignition:

(i) An article near the centre of the package is caused to function in the designed mode; or

(ii) An article near the centre of the package is replaced by another article which can be caused to function with the same effect.

16.4.1.3.4 The package is placed on a steel witness plate on the ground. The preferred method of confinement consists of containers, similar in shape and size to the test package, completely filled with earth or sand and placed as closely as possible around the test package to a minimum thickness of confinement, in every direction of 0.5 m for a package not exceeding 0.15 m^3 and 1.0 m for a package greater than 0.15 m^3. Alternative methods of confinement are to use boxes or bags filled with earth or sand placed around and on top of the package or to use loose sand.

16.4.1.3.5 The substance or article should be initiated and observations made on the following: evidence of thermal effects, projection effects, detonation, deflagration or explosion of the total contents of the package. ***A safe waiting period, prescribed by the test agency, should be observed after initiation.*** The test should be performed three times unless a decisive result occurs earlier (e.g. explosion of the total contents). If the results of the recommended number of tests do not enable unambiguous interpretation of the results, the number of tests should be increased.

16.4.1.4 *Test criteria and method of assessing results*

Mass explosion (see definition in Chapter 2.1 of the Model Regulations) indicates a candidate for Division 1.1. Evidence of such an indication includes:

(a) A crater at the test site;

(b) Damage to the witness plate beneath the package;

(c) Measurement of a blast; and

(d) Disruption and scattering of the confining material.

If the product is accepted into Division 1.1, further testing is not necessary; otherwise proceed to a test of type 6 (b).

16.4.1.5 *Examples of results*

Substance	**Packaging**	**Initiation system**	**Events**	**Result**
Ammonium perchlorate (12 µm)	10 kg fibreboard drum	Detonator	Detonation	Candidate Division 1.1
Musk xylene	50 kg fibreboard drum	Detonator	Localised decomposition	Not Division 1.1
Musk xylene	50 kg fibreboard drum	Igniter	Localised decomposition	Not Division 1.1
Single base propellant (non-porous)	60 litre fibreboard drum	Igniter	No explosion	Not Division 1.1
Single base propellant (porous)	60 litre fibreboard drum	Igniter	Explosion	Candidate Division 1.1

16.5 Series 6 type (b) test prescription

16.5.1 *Test 6 (b): Stack test*

16.5.1.1 *Introduction*

This is a test on packages of an explosive substance or explosive articles, or unpackaged explosive articles, to determine whether an explosion is propagated from one package to another or from a unpackaged article to another.

16.5.1.2 *Apparatus and materials*

The following items are required:

(a) A detonator to initiate the substance or article or an igniter just sufficient to ensure ignition of the substance or article (see 16.5.1.4 and 16.5.1.5);

(b) Suitable confining materials (see 16.5.1.3); and

(c) A sheet of 3 mm thick mild steel to act as a witness plate.

Blast measuring equipment may be used.

16.5.1.3 *Procedure*

The test is applied to a stack of packages of an explosive product or a stack of unpackaged articles, in each case, in the condition and form in which they are offered for transport. Where explosive articles are to be carried without packaging, the tests should be applied to the unpackaged articles. Sufficient packages or articles to give a total volume of 0.15 m^3 are stacked on a steel witness plate on the ground. If the volume of an individual package (or unpackaged article) exceeds 0.15 m^3, then the test is performed with at least one acceptor placed in the position most likely to result in communication between the individual products (see 16.3.1). If this position is not known, several acceptors are used. The preferred method of confinement consists of containers, similar in shape and size to the test packages completely filled with earth or sand and placed as closely as possible around the test package to create a minimum thickness of confinement, in every direction, of 1 m. Alternative methods of confinement are to use boxes or bags filled with earth or sand placed around and on top of the stack or to use loose sand. If loose sand is used for confinement, the stack should be covered or protected to ensure that no sand falls into the interstices between adjacent packages or non-packaged articles. Articles carried without packaging are confined in a manner analogous to that used for packaged articles. The decision to use either an initiating stimulus or an igniting stimulus is based on the following considerations.

16.5.1.4 For packaged substances:

(a) If the substance is intended to function by detonation, it should be tested with a standard detonator (Appendix 1);

(b) If the substance is intended to function by deflagration, it should be tested with an igniter just sufficient (but not more than 30 g of black powder) to ensure ignition of the substance within one individual package. The igniter should be located in the centre of the substance in the package;

(c) Substances not intended for use as explosives, but provisionally accepted into Class 1, should be tested using whichever initiation system gave a "+" result in a type 6 (a) test.

16.5.1.5 For packaged articles and unpackaged articles[2]:

(a) Articles provided with their own means of initiation or ignition.

The functioning of an article at the centre of the package near the centre of the stack should be stimulated by the article's own means of initiation or ignition. Where this is impracticable, the article's own means of initiation or ignition may be replaced by another form of stimulus having the required effect;

(b) Articles not provided with their own means of initiation or ignition:

(i) An article at the centre of the package near the centre of the stack is caused to function in the designed mode; or

(ii) An article at the centre of the package near the centre of the stack is replaced by another article which can be caused to function with the same effect.

16.5.1.6 The point of ignition or initiation should be situated in a package near the centre of the stack. Articles carried without packaging are tested in a manner analogous to that used for packaged articles.

16.5.1.7 The substance or article should be initiated and observations made on the following: evidence of thermal effects, projection effects, detonation, deflagration or explosion of the total contents of the package. ***A safe waiting period, prescribed by the test agency, should be observed after initiation.*** The test should be performed three times unless a decisive result occurs earlier (i.e. explosion of the total contents). If the results of the recommended number of tests do not enable unambiguous interpretation of the results, the number of tests should be increased.

16.5.1.8 *Test criteria and method of assessing results*

If in test 6 (b) explosion of the contents of more than one package or unpackaged article occurs practically instantaneously, then the product is assigned to Division 1.1. Evidence of such an occurrence includes:

(a) A crater at the test site appreciably larger than that given by a single package or unpackaged article;

(b) Damage to the witness plate beneath the stack which is appreciably greater than that from a single package or unpackaged article;

(c) Measurement of blast which significantly exceeds that from a single package or unpackaged article; and

(d) Violent disruption and scattering of most of the confining material.

Otherwise proceed to a test of type 6 (c).

16.5.1.9 *Examples of results*

No examples of results are given as these are too specific to the packaging or article tested.

---

2 *Subject to the proviso that in the case of articles containing a very small quantity of substance(s) of compatibility group A only, a sufficient number of such items are initiated simultaneously to cause not less than 0.2 g of primary explosive to explode.*

16.6 Series 6 type (c) test prescription

16.6.1 *Test 6 (c): External fire (bonfire) test*

16.6.1.1 *Introduction*

This is a test performed on packages of an explosive substance or explosive articles, or unpackaged explosive articles, to determine whether there is a mass explosion or a hazard from dangerous projections, radiant heat and/or violent burning when involved in a fire.

16.6.1.2 *Apparatus and materials*

The following items are needed:

(a) If the volume of the package of substance or articles, or unpackaged article, is less than 0.05 m^3, sufficient packages or unpackaged articles to give a total volume of not less than 0.15 m^3;

(b) If the volume of the package of substance or articles, or unpackaged article, is equal to or greater than 0.05 m^3, three packages or unpackaged articles. If the volume of one package or unpackaged article is greater than 0.15 m^3, the competent authority may waive the requirement for three packages or unpackaged articles to be tested;

(c) A metal grid to support the products above the fuel and allow adequate heating. If a wooden crib fire is used, the grid should be 1.0 m above the ground and if a liquid hydrocarbon pool fire is used then the grid should be 0.5 m above the fuel surface at the onset of the test;

(d) Strapping or wire, if necessary, to hold the packages or unpackaged articles together on the support grid;

(e) Enough fuel to keep a fire burning for at least 30 minutes or, if necessary, until the substance or article has clearly had sufficient time to react to the fire (see 16.6.1.3.8);

(f) Suitable means of ignition to ignite the fuel on at least two sides e.g. for a wood fire, kerosene to soak the wood and pyrotechnic igniters with wood wool;

(g) Three 2 000 mm × 2 000 mm × 2 mm 1100-0 aluminium sheets (Brinell Hardness 23, tensile strength 90 MPa), or equivalent, to act as witness screens together with suitable supports to hold them vertically. The witness screens shall be mounted rigidly to their frames. When more than one panel is used to make a witness screen, each panel shall be supported at all joints;

(h) Video equipment capable of recording the events necessary for classification. The type, number and placement of the camera(s) shall be sufficient to record all events to be assessed.

Blast gauges, radiometers and associated recording equipment may also be used. Further equipment may be needed when following the procedure in 16.6.1.3.9.

16.6.1.3 *Procedure*

16.6.1.3.1 The required number of packages or unpackaged articles, in the condition and form in which they are offered for transport, are arranged as close as possible to one another on the metal grid. If directional effects are anticipated, packages or unpackaged articles should be oriented in such a way to maximize probability for projections to hit witness screens and for discrete flame jets to be pointed horizontally. If necessary, the packages or unpackaged articles may be encircled with a steel strip to support them during the test. Fuel is placed beneath the grid so that the fire will engulf the packages or unpackaged articles. Suitable methods of heating include a wood, liquid or gas fuel fire or a combination thereof, which achieves a temperature of 800 °C. Fluctuations of temperature below 800 °C are normal and should not render the test invalid.

16.6.1.3.2 A wood fire should burn the packages or unpackaged articles with sufficient intensity and duration to completely react the explosives (see 16.6.1.2(e)). Dried pallets, boards, laths, or other wood alone or in combination may be stacked to form a lattice beneath the grid 1 m off the ground, and up to the base of the grid supporting the packages or unpackaged articles. The wood should extend at least 1 m beyond the packages or unpackaged articles to ensure that the fire engulfs the product.

16.6.1.3.3 A receptacle filled with suitable liquid fuel, a combination of both wood and liquid fuel fire may be used as an alternative to the wood fire providing it is as severe. If a liquid pool fire is used, the receptacle should extend beyond the packages or unpackaged articles to a distance of at least 1.0 m in every direction. The distance between the grid platform and the receptacle should be approximately 0.5 m. Before using this method, consideration should be given to whether any quenching action or adverse interaction between explosives and liquid fuel can occur such as might bring the results into question.

16.6.1.3.4 If gas is to be used as a fuel, the burning area must extend beyond the packages or unpackaged articles to a distance of at least 1.0 m in every direction. The gas must be supplied in such a manner to ensure that the fire is evenly distributed around the packages. The gas reservoir should be large enough to keep the fire burning for at least 30 minutes. Ignition of the gas may be accomplished either by remotely ignited pyrotechnics or by remote release of the gas adjacent to a pre-existing source of ignition.

16.6.1.3.5 The witness screens are erected vertically in each of three quadrants at a distance of 4 m from the edge of the packages or unpackaged articles. The sheets should be placed so that the centres are approximately level with the centre of the packages or unpackaged articles or, if this is less than 1.0 m above the ground, in contact with the ground. If there are any perforations or indentations in the witness screens before the test, they should be marked so that they can be clearly distinguished from those created during the test.

16.6.1.3.6 The ignition system should be put into place and the fuel ignited on two sides, one the upwind side, simultaneously. The test should not be performed under conditions where the wind speed exceeds 6 m/s. ***A safe waiting period, prescribed by the test agency, should be observed before approaching the test area.***

16.6.1.3.7 Observations are made on the following:

(a) Evidence of explosion;

(b) Potentially hazardous projections; and

(c) Thermal effects.

16.6.1.3.8 The test is normally performed once only but if the wood or other fuel used for the fire is all consumed leaving a significant quantity of unconsumed explosive substance in the remains or in the vicinity of the fire, then the test should be performed again using more fuel, or a different method, to increase the intensity and/or duration of the fire. If the result of the test does not enable the hazard division to be determined, a further test should be performed.

16.6.1.3.9 For candidates to cartridges, small arms (UN No. 0012), this test can be augmented or replaced by the specialised measurement of the energy of projections as described in Appendix 9. This applies to circumstances where the dominant hazard is a projection hazard, for example as known from previous testing of similar explosive articles.

16.6.1.4 *Test criteria and method of assessing results*

16.6.1.4.1 The following criteria are used to answer the questions in Figure 10.3 (boxes 26, 28, 30, 32, 33, 35 and 36) in order to assess the results and classify the product.

16.6.1.4.2 If mass explosion occurs then the product is assigned to Division 1.1. A mass explosion is considered to occur if a substantial proportion explodes so that the practical hazard should be assessed by assuming simultaneous explosion of the whole of the explosive content of the packages or unpackaged articles.

16.6.1.4.3 If a mass explosion does not occur but any one of the following occurs:

(a) A perforation of any of the witness screens (see 16.6.1.3.5);

(b) A metallic projection with a kinetic energy exceeding 20 J as assessed by the distance - mass relation given in Figure 16.6.1.1;

then the product is assigned to Division 1.2.

16.6.1.4.4 If none of the events occur which would require the product to be assigned to Division 1.1 or 1.2 but any one of the following events occurs:

(a) a fireball or jet of flame which extends beyond any of the witness screens;

(b) a fiery projection emanating from the product is thrown more than 15 m from the edge of the packages or unpackaged articles;

(c) a burning time of the product measured to be less than 35 seconds for 100 kg net explosive mass (see 16.6.1.4.8 Notes for scaling time measurements in evaluating thermal flux effects). Alternatively, in the case of articles and low energy substances, the irradiance of the burning product exceeds that of the fire by more than 4 kW/m^2 at a distance of 15 m from the edge of the packages or unpackaged articles. The irradiance is measured over 5 seconds, during the period of maximum output;

then the product is assigned to Division 1.3.

16.6.1.4.5 If none of the events occur which would require the product to be assigned to Division 1.1, 1.2 or 1.3, but any one of the following events occurs:

(a) a fireball or jet of flame which extends more than 1 m from the flames of the fire;

(b) a fiery projection emanating from the product is thrown more than 5 m from the edge of the packages or unpackaged articles;

(c) an indentation in any of the witness screens of more than 4 mm;

(d) a metallic projection with a kinetic energy exceeding 8 J as assessed by the distance - mass relation given in Figure 16.6.1.1;

(e) a burning time of the product measured to be less than 330 seconds for 100 kg net explosive mass (see 16.6.1.4.8: Notes for scaling time measurements in evaluating thermal flux effects);

then the product is assigned to Division 1.4 and to a compatibility group other than Compatibility Group S.

16.6.1.4.6 If none of the events occur which would require the product to be assigned to Division 1.1, 1.2, 1.3 or 1.4 other than Compatibility Group S (see Box 32 of Figure 10.3), then the product is assigned to Division 1.4 Compatibility Group S, unless special provision 347 of Chapter 3.3 of the Model Regulations applies. For candidates to Cartridges, small arms (UN No. 0012), evidence of projections with a kinetic energy not exceeding 8 J as determined by the test procedure in Appendix 9 may be used to assign the product to Compatibility Group S.

16.6.1.4.7 If there are no hazardous effects at all, the product is considered for exclusion from Class 1. The possibilities, as shown by boxes 35 and 36 in Figure 10.3, are:

(a) If the product is an article manufactured with a view to producing a practical explosive or pyrotechnic effect, then:

(i) If there is some effect (projection, fire, smoke, heat or loud noise) external to the device itself, the device is not excluded from Class 1 and the product, as packaged, is assigned to Division 1.4 and to Compatibility Group S. Paragraph 2.1.1.1 (b) of the Model Regulations refers explicitly to the device, rather than the package, so it is usually necessary to make this assessment on the basis of a test involving functioning of the device without packaging or confinement. Sometimes the stated effects are observed in test 6 (c), in which case the product is classified 1.4 S without further tests;

(ii) If there is no effect (projection, fire, smoke, heat or loud noise) external to the device itself, the unpackaged device is excluded from Class 1 in accordance with paragraph 2.1.1.1 (b) of the Model Regulations. Paragraph 2.1.1.1 (b) of the Model Regulations refers explicitly to the device, rather than the package, so it is usually necessary to make this assessment on the basis of a test involving functioning of the device without packaging or confinement;

(b) If the product is not manufactured with a view to producing a practical explosive or pyrotechnic effect, it is excluded from Class 1 in accordance with paragraph 2.1.1.1 of the Model Regulations.

16.6.1.4.8 Notes for scaling time measurements in evaluating thermal flux effects

Notes:

(1) The *value of 35 seconds/100 kg (see 16.6.1.4.4 (c)) correlates to an average thermal flux of 4 kW/m^2 at 15 m and is based on an assumed heat of combustion of 12 500 J/g. If the true heat of combustion is significantly different, the burning time of 35 s may be corrected; for example, a true heat of combustion of 8 372* J/g *burning for (8 372/12 500) × 35 s = 23.4 s would produce the same flux level. Corrections for masses other than 100 kg are made according to the scaling relations and examples in Table 16.2.*

(2) The value of 330 seconds/100 kg (see 16.6.1.4.5(e)) correlates to an average thermal flux of 4 kW/m^2 at 5 m and is based on an assumed heat of combustion of 12 500 J/g . If the true heat of combustion is significantly different, the burning time of 330 s may be corrected; for example, a true heat of combustion of 8 372 J/g burning for (8 372/12 500) × 330 s = 221 s would produce the same flux level. Corrections for masses other than 100 kg are made according to the scaling relations and examples in Table 16.2.

(3) In some burn time trials, individual packages or articles will be observed to burn in separate, identifiable events; in such cases, the burning times and masses of each separate event should be used.

Table 16.2: COMPARATIVE THERMAL FLUX VALUES FOR VARYING MASSES

	1.3/1.4		1.4/1.4S	
Mass (kg)	Flux (15 m)	Burn time (s)	Flux (5 m)	Burn time (s)
20	1.36 kW/m^2	21.7	1.36 kW/m^2	195
50	2.5	29.6	2.5	266
100	4	35	4	330
200	6.3	46.3	6.3	419
500	11.7	63.3	11.7	569

NOTE: Thermal flux is scaled on basis of $(m/m_0)^{2/3}$.
Times are scaled on basis of $(m/m_0)^{1/3}$.

The values of thermal flux can be calculated from the equation:

$$F = \frac{C \times E}{4\pi R^2 t}$$

where:

F = thermal flux in kW/m^2;
C = constant = 0.33;
E = total energy content in joules;
R = distance from fire to exposed position in meters;
T = observed burn time in seconds.

16.6.1.5 *Examples of results*

Substance	**Packaging**	**Events**	**Result**
Musk xylene	3 × 50 kg fibreboard drums	Slow burning only	Not Class 1

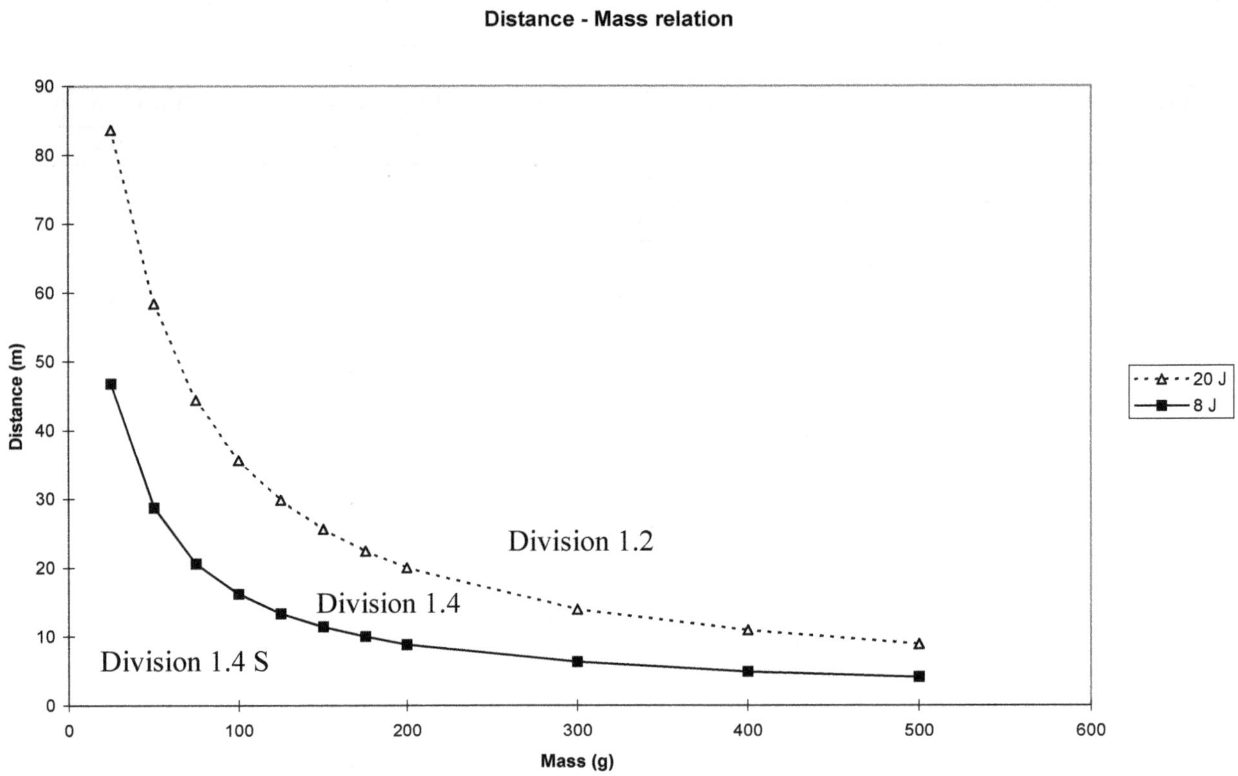

Mass	Projection distance (m)	
(g)	20 J	8 J
25	83.6	46.8
50	58.4	28.7
75	44.4	20.6
100	35.6	16.2
125	29.8	13.3
150	25.6	11.4
175	22.43	10
200	20	8.8
300	13.9	6.3
400	10.9	4.9
500	8.9	4.1

Example data for metallic projections with a kinetic energy of 20 J and 8 J

FIGURE 16.6.1.1: DISTANCE-MASS RELATION FOR METALLIC PROJECTIONS WITH A KINETIC ENERGY OF 20 J AND 8 J [3]

[3] *The data presented in Figure 16.6.1.1 is based upon metallic projections. Non-metallic projections will produce different results and may be hazardous. Hazards from non-metallic projections should also be considered.*

16.7 Series 6 type (d) test prescription

16.7.1 *Test 6 (d): Unconfined package test*

16.7.1.1 *Introduction*

This is a test on a single package to determine if there are hazardous effects outside the package arising from accidental ignition or initiation of the contents.

16.7.1.2 *Apparatus and materials*

The following items are required:

(a) A detonator to initiate the article or an igniter just sufficient to ensure ignition of the substance or article (see 16.7.1.3.2); and

(b) A sheet of 3 mm thick mild steel to act as a witness plate.

Video equipment may be used.

16.7.1.3 *Procedure*

16.7.1.3.1 The test is applied to packages of explosive articles in the condition and form in which they are offered for transport. Where explosive articles are to be carried without packaging, the tests should be applied to the non-packaged articles. The decision to use either an initiating stimulus or an igniting stimulus is based on the following considerations.

16.7.1.3.2 For packaged articles:

(a) Articles provided with their own means of initiation or ignition:

The functioning of an article near the centre of the package is stimulated by the article's own means of initiation of ignition. Where this is impracticable, the article's own means of initiation or ignition is replaced by another form of stimulus having the required effect;

(b) Articles not provided with their own means of initiation or ignition:

(i) an article near the centre of the package is caused to function in the designed mode; or

(ii) an article near the centre of the package is replaced by another article which can be caused to function with the same effect.

16.7.1.3.3 The package is placed on a steel witness plate on the ground without confinement.

16.7.1.3.4 The donor article should be initiated and observations made on the following: denting or perforation of the witness plate beneath the package, a flash or flame capable of igniting an adjacent material, disruption of the package causing projection of the explosives contents; or full perforation of the packaging by a projection. ***A safe waiting period, prescribed by the test agency, should be observed after initiation.*** The test should be performed three times, in different orientations, unless a decisive result is observed earlier. If the results of the recommended number of tests do not enable unambiguous interpretation of the results, the number of tests should be increased.

16.7.1.4 *Test criteria and method of assessing the results*

Inclusion in Compatibility Group S requires that any hazardous effects arising from functioning of the articles in this test are confined within the package. Evidence of a hazardous effect outside the package includes:

(a) Denting or perforation of the witness plate beneath the package;

(b) A flash or flame that ignites an adjacent material such as a sheet of 80 ± 3 g/m² paper at a distance of 25 cm from the package;

(c) Disruption of the package causing projection of the explosives contents; or

(d) A projection which passes completely through the packaging (a projection or fragment retained or stuck in the wall of the packaging is considered as non hazardous).

The competent authority may wish to take into account the expected effect of the initiator when assessing the results of the test, if these are expected to be significant when compared to the articles being tested. If there are hazardous effects outside the package, then the product is excluded from Compatibility Group S.

16.7.1.5 *Examples of results*

Article	Packaging	Initiation system	Events	Result
Cartridges, power device	Fibreboard box containing 20 articles (300 g of propellant each) each in a plastic bag	One of the articles	Articles ignited one by one, producing flames up to 2 m high outside package	Not consistent with Compatibility Group S
Detonator assemblies, non-electric	Fibreboard box containing 60 assemblies each in a plastic bag with its shock tube coiled in a figure 8, with attenuators on the detonators	One of the articles	One out of 60 detonators fired and no visible effects outside the box.	Consistent with Compatibility Group S
Detonators, electric	Fibreboard box containing 84 assemblies, each bundled with its wire so that the blast from a firing detonator would be attenuated	One of the articles	One out of 84 detonators fired. The reaction caused the box to break open and released some of the assemblies but it was judged that there were no hazardous effects outside the package.	Consistent with Compatibility Group S
Charges, shaped (open-face 19 g perforators)	Fibreboard box containing 50 charges in two layers so that pairs of charges were focused toward each other	Detonator with approxim-ately 60 mm of detonating cord	Three trials were conducted. In each of the trials, the witness plate was perforated with three to four charges reacting. The packages were blown apart scattering the remaining charges over a wide area.	Not consistent with Compatibility Group S
Detonators, electric	Fibreboard box containing 50 detonators each with a 450 mm lead wire. Each assembly was contained in its own fibreboard inner box. The boxes were separated by fibreboard panels	One of the articles	One out of 50 detonators fired causing one of the box flaps to open. There were no hazardous effects outside of the package.	Consistent with Compatibility Group S

SECTION 17

TEST SERIES 7

17.1 Introduction

The question "Is it an extremely insensitive explosive article?" (box 40 of Figure 10.3) is answered by series 7 tests and any candidate for Division 1.6 should pass one of each of the eleven types of test comprising the series. The first six types of test (7(a) to 7(f)) are used to establish if a substance is an Extremely Insensitive Substance (EIS) and the remaining five types of test (7 (g), 7 (h), 7 (j), 7(k) and 7(l)) are used to determine if an article containing an EIS(s) may be assigned to Division 1.6. The eleven test types are:

Type 7 (a): a shock test to determine sensitivity to intense mechanical stimulus;

Type 7 (b): a shock test with a defined booster and confinement to determine sensitivity to shock;

Type 7 (c): a test to determine the sensitivity of the explosive substance to deterioration under the effect of an impact;

Type 7 (d): a test to determine the degree of reaction of the explosive substance to impact or penetration resulting from a given energy source;

Type 7 (e): a test to determine the reaction of the explosive substance to an external fire when the material is confined;

Type 7 (f): a test to determine the reaction of the explosive substance in an environment in which the temperature is gradually increased to 365 °C;

Type 7 (g): a test to determine the reaction to an external fire of an article which is in the condition as presented for transport;

Type 7 (h): a test to determine the reaction of an article in an environment in which the temperature is gradually increased to 365 °C;

Type 7 (j): a test to determine the reaction of an article to impact or penetration resulting from a given energy source;

Type 7 (k): a test to determine whether a detonation of an article will initiate a detonation in an adjacent, like article; and

Type 7 (l): a test to determine the sensitivity of the article to shock directed at vulnerable components.

The question in box 40 is answered "no" if a "+" result is obtained in any series 7 test.

17.2 Test methods

The test methods currently used are listed in Table 17.1.

Table 17.1: TEST METHODS FOR TEST SERIES 7

Test code	Name of Test	Section
Tests on substances		
7 (a)	EIS cap test[a]	17.4.1
7 (b)	EIS gap test[a]	17.5.1
7 (c) (i)	Susan test	17.6.1
7 (c) (ii)	Friability test[a]	17.6.2
7 (d) (i)	EIS bullet impact test[a]	17.7.1
7 (d) (ii)	Friability test	17.7.2
7 (e)	EIS external fire test[a]	17.8.1
7 (f)	EIS slow cook-off test[a]	17.9.1
Tests on articles		
7 (g)	1.6 article external fire test[a]	17.10.1
7 (h)	1.6 article slow cook-off test[a]	17.11.1
7 (j)	1.6 article bullet impact test[a]	17.12.1
7 (k)	1.6 article stack test[a]	17.13.2
7(l)	1.6 article fragment impact test	17.14.1

[a] *Recommended test.*

17.3 Test conditions

17.3.1 All explosive components must always be present in articles during Series 7 testing of types 7 (g) to 7 (l). Smaller explosive components containing substances not subjected to tests of type 7 (a) to 7 (f) shall be specifically targeted in tests 7 (j) and 7 (l) when it is assessed that they will cause the most severe reaction from the test article, to ensure the probability of accidental initiation or propagation of a Division 1.6 article remains negligible.

17.3.2 A substance intended for use as a main explosive load in an article of Division 1.6 should be tested in accordance with Test Series 3 and 7. A substance intended for use as a larger (dimensionally) boostering component in an article of Division 1.6, where the volumetric size limit relative to the main explosive load it is boostering is met, should be tested in accordance with Test Series 3 and tests of type 7 (c) (ii) and 7 (e). Test Series 7 should be conducted on the substance in the form (i.e. composition, granulation, density etc.) in which it is to be used in the article.

17.3.3 An article being considered for inclusion in Division 1.6 should not undergo Series 7 testing until after main explosive load and certain boostering component subtances have undergone appropriate tests of type 7 (a) to 7 (f) to determine whether they meet the substance requirements for Division 1.6. Guidance on substance testing determination process is given under section 10.4.3.6.

17.3.4 Tests of types 7 (g), 7 (h), 7 (j), 7 (k) and 7 (l) should be performed to determine if an article with an EIS main load(s) and appropriately insensitive boostering components may be assigned to Division 1.6. These tests are applied to articles in the condition and form in which they are offered for transport, except that non-explosive components may be omitted or simulated if the competent authority is satisfied that this does not invalidate the results of the tests.

17.3.5 Response levels referred to within the following individual Test Series 7 test prescriptions are provided at Appendix 8 (Response descriptors), to aid in the assessment of the results of tests of types 7 (g), 7 (h), 7 (j), 7 (k) and 7 (l) and should be reported to the competent authority to support assignment to Division 1.6.

17.4 Series 7 type (a) test prescription

17.4.1 *Test 7 (a): EIS cap test*

17.4.1.1 *Introduction*

This shock test is designed to determine the sensitivity of an EIS candidate to intense mechanical stimulus.

17.4.1.2 *Apparatus and materials*

The experimental set-up for this test is the same as for test 5 (a) (see 15.4.1).

17.4.1.3 *Procedure*

The experimental procedure is the same as for test 5 (a) (see 15.4.1).

17.4.1.4 *Test criteria and method of assessing results*

The result is considered "+" and the substance should not be classified as an EIS if in any trial:

(a) The witness plate is torn or otherwise penetrated (i.e. light is visible through the plate) - bulges, cracks or folds in the witness plate do not indicate cap sensitivity; or

(b) The centre of the lead cylinder is compressed from its initial length by an amount of 3.2 mm or more.

Otherwise, the result is considered "-".

17.4.1.5 *Examples of results*

Substance	Result
HMX/inert binder (86/14), cast	-
HMX/energetic binder (80/20), cast	+
HMX/aluminium/energetic binder (51/19/14), cast	-
RDX/TNT (60/40), cast	+
TATB/Kel-F (95/5), pressed	-

17.5 Series 7 type (b) test prescription

17.5.1 *Test 7 (b): EIS gap test*

17.5.1.1 *Introduction*

This test is used to measure the sensitivity of an EIS candidate to a specified shock level, i.e. a specified donor charge and gap.

17.5.1.2 *Apparatus and materials*

The set-up for this test consists of an explosive charge (donor), a barrier (gap), a container holding the test charge (acceptor), and a steel witness plate (target).

The following materials are to be used:

(a) United Nations Standard detonator or equivalent;

(b) 95 mm diameter by 95 mm long pellet with a density of 1 600 kg/m^3 ± 50 kg/m^3 of either 50/50 pentolite or 95/5 RDX/WAX;

(c) Tubing, steel, seamless, with an outer diameter of 95 ± 7.0 mm, a wall thickness of 9.75 mm ± 2.75 mm and an inner diameter of 73.0 mm ± 7.0 mm and with a length of 280 mm;

(d) Sample substances, machined to a diameter which is just under the diameter of the steel tubing. The air gap between the sample and tubing wall should be as small as possible;

(e) Polymethyl methacrylate (PMMA) rod, of 95 mm diameter by 70 mm long;

(f) Mild steel plate, 200 mm × 200 mm × 20 mm;

(g) Wood block, 95 mm diameter and 25 mm thick, with a hole drilled through the centre to hold the detonator.

17.5.1.3 *Procedure*

17.5.1.3.1 As shown in Figure 17.5.1.1, the detonator, donor, gap and acceptor charge are coaxially aligned above the centre of the witness plate. A 1.6 mm air gap is maintained between the free end of the acceptor charge and the witness plate with suitable spacers which do not overlap the acceptor charge. Care should be taken to ensure good contact between the detonator and donor, donor and gap and gap and acceptor charge. The test sample and booster should be at ambient temperature for the test.

17.5.1.3.2 To assist in collecting the remains of the witness plate, the whole assembly may be mounted over a container of water with at least a 10 cm air gap between the surface of the water and the bottom surface of the witness plate which should be supported along two edges only.

17.5.1.3.3 Alternative collection methods may be used but it is important to allow sufficient free space below the witness plate so as not to impede plate puncture. The test is performed three times unless a positive result is observed earlier.

17.5.1.4 *Test criteria and method of assessing results*

A clean hole punched through the plate indicates that a detonation was initiated in the sample. A substance which detonates in any trial is not an EIS and the result is noted as "+".

17.5.1.5 *Examples of results*

Substance	Result
HMX/inert binder (86/14), cast	+
HMX/energetic binder (80/20), cast	+
HMX/aluminium/energetic binder (51/19/14), cast	+
RDX/inert binder (85/15), cast	+
RDX/TNT (60/40), cast	+
TATB/Kel-F (95/5), pressed	-
TNT, cast	+

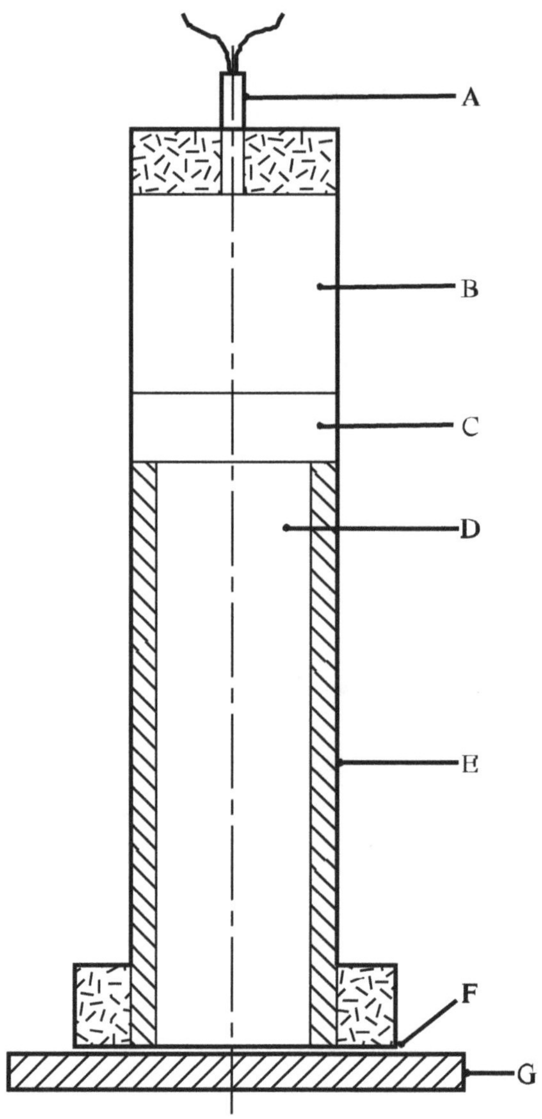

(A)	Detonator	(B)	Booster charge
(C)	PMMA gap	(D)	Substance under test
(E)	Steel tube	(F)	Air gap
(G)	Witness plate		

Figure 17.5.1.1: EIS GAP TEST

17.6 Series 7 type (c) test prescriptions

17.6.1 *Test 7 (c) (i): Susan impact test*

17.6.1.1 *Introduction*

The Susan Impact test is used to assess the degree of explosive reaction under conditions of high velocity impact. The test is conducted by loading the explosives into standardised projectiles and firing the projectiles against a target at a specified velocity.

17.6.1.2 *Apparatus and materials*

17.6.1.2.1 51 mm diameter, 102 mm long explosives billets, fabricated by normal techniques, are employed.

17.6.1.2.2 The Susan test employs the test vehicle shown in Figure 17.6.1.1. The projectile has an assembled weight of 5.4 kg and contains slightly less than 0.45 kg of explosive. The overall dimensions are 81.3 mm in diameter by 220 mm long.

17.6.1.2.3 The projectiles are fired from a 81.3 mm smooth-bore gun. The gun muzzle is positioned about 4.65 m from the 64 mm thick, smooth-surface, armour steel target plate. Projectile impact velocity is obtained by adjusting the propellant charges in the gun.

17.6.1.2.4 A schematic drawing of the firing range showing the target-gun layout and the relative positions of the diagnostic equipment is shown in Figure 17.6.1.2. The flight path is about 1.2 m above ground level.

17.6.1.2.5 The test site is equipped with calibrated blast gauges and recording equipment. The air blast recording system should have a system frequency response of at least 20 kHz. Measurements are made of impact velocities and air shock blast over-pressure. Air blast is measured at a distance of 3.05 m from the impact point (gauges (C) in Figure 17.6.1.2).

17.6.1.3 *Procedure*

17.6.1.3.1 The propellant charge in the gun should be adjusted to produce a projectile velocity of 333 m/s. The projectile is fired and the impact velocity and air blast, produced as a result of its reaction on impact, are recorded. If a velocity of 333 m/s (+ 10%, - 0%) is not obtained, the amount of propellant is adjusted and the test repeated.

17.6.1.3.2 Once an impact velocity of 333 m/s is obtained, the test is repeated until accurate pressure-time records are obtained from at least five separate shots. On each of these accurate shots, the impact velocity should be 333 m/s (+ 10%, - 0%).

17.6.1.4 *Test criteria and method of assessing results*

The maximum air blast overpressure that is determined from each air blast is recorded. The average of the maximum pressures obtained from the five accurate shots is determined. If the average pressure obtained by such a procedure is greater than or equal to 27 kPa, then the substance is not an EIS explosive and the result is noted as "+".

17.6.1.5 *Examples of results*

Substance	Result
HMX/inert binder (86/14), cast	-
HMX/energetic binder (80/20), cast	+
HMX/aluminium/energetic binder (51/19/14), cast	+
RDX/TNT (60/40), cast	+
TATB/Kel-F (95/5), pressed	-

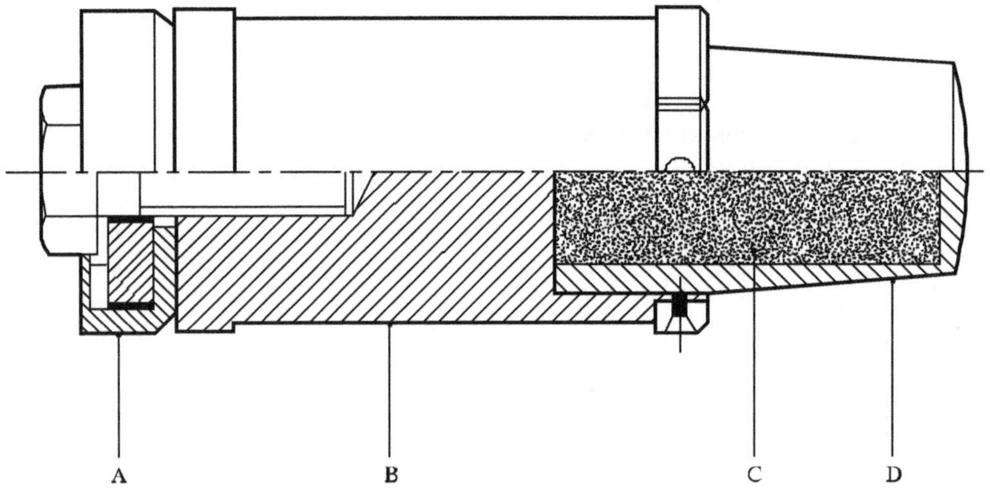

(A) Leather cup seal
(B) Steel body
(C) Explosive under test
(D) Aluminium cup

Figure 17.6.1.1: SUSAN PROJECTILE

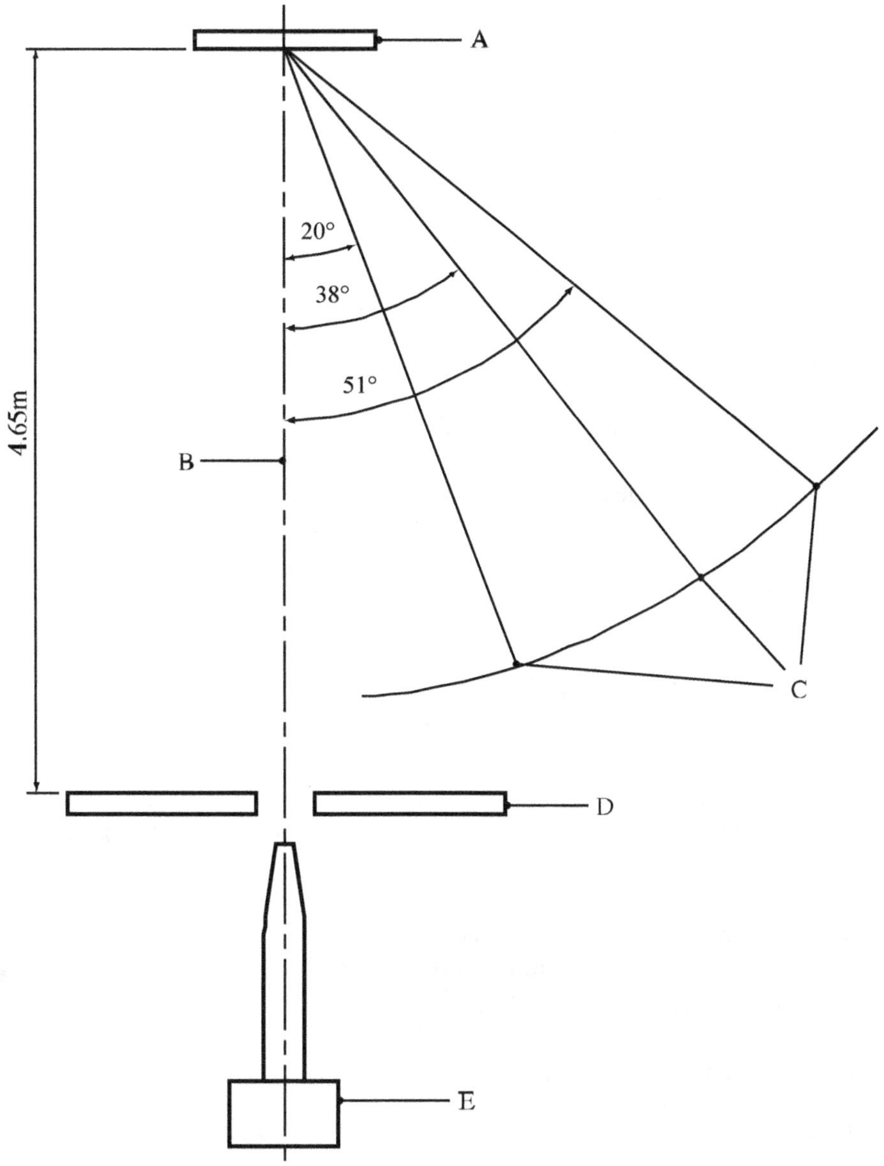

(A) Target plate (6.4 cm thick)
(B) Flight path
(C) Air blast transducers (3.05 m from target point)
(D) Smoke barrier
(E) 81.3 mm gun

Figure 17.6.1.2: SCHEMATIC LAYOUT OF SUSAN TEST (top view)

17.6.2 *Test 7 (c) (ii): Friability test*

17.6.2.1 *Introduction*

The friability test is used to establish the tendency of a compact EIS candidate to deteriorate dangerously under the effect of an impact.

17.6.2.2 *Apparatus and materials*

The following apparatus is required:

(a) A weapon designed to shoot 18 mm diameter cylindrical test pieces at a velocity of 150 m/s;

(b) A Z30C 13 stainless steel plate, 20 mm thick with a front face roughness of 3.2 microns (AFNOR NF E 05-015 and NF E 05-016 standards);

(c) A 108 ± 0.5 cm^3 manometric bomb at 20 °C;

(d) A firing capsule containing a heating wire on 0.5 g of black powder with a mean particle size of 0.75 mm. The composition of the black powder is 74% potassium nitrate, 10.5% sulphur and 15.5% carbon. The moisture content should be less than 1%;

(e) A cylindrical sample of compact substance of diameter 18 ± 0.1 mm. The length is adjusted so as to obtain a mass of 9.0 ± 0.1 g. The sample is brought to and maintained at a temperature of 20 °C;

(f) A fragment recovery box.

17.6.2.3 *Procedure*

17.6.2.3.1 The sample is projected against the steel plate at an initial velocity sufficient to give an impact velocity as close as possible to 150 m/s. The mass of fragments collected after the impact should be at least 8.8 g. These fragments are fired in a manometric bomb. Three tests are carried out.

17.6.2.3.2 The curve of pressure against time p = f (t) is recorded; this enables the curve (dp/dt) = f' (t) to be constructed. From this curve the maximum value $(dp/dt)_{max}$ is read off. This enables the value $(dp/dt)_{max}$, corresponding to an impact speed of 150 m/s, to be estimated.

17.6.2.4 *Test criteria and method of assessing results*

If the average maximum $(dp/dt)_{max}$ value obtained at a speed of 150 m/s is greater than 15 MPa/ms, the substance tested is not an EIS and the result is noted as "+".

17.6.2.5 *Examples of results*

Substance	Result
HMX/inert binder (86/14), cast	-
HMX/energetic binder (80/20), cast	+
HMX/aluminium/energetic binder (51/19/14), cast	-
RDX/TNT (60/40), cast	+
TATB/Kel-F (95/5), pressed	-

17.7 Series 7 type (d) test prescriptions

17.7.1 *Test 7 (d) (i): EIS bullet impact test*

17.7.1.1 *Introduction*

The bullet impact test is used to evaluate the response of an EIS candidate to the kinetic energy transfer associated with impact and penetration of a given energy source, i.e. a 12.7 mm projectile, travelling at a specified velocity.

17.7.1.2 *Apparatus and materials*

17.7.1.2.1 Explosive test samples fabricated by normal techniques are employed. The samples should have a length of 20 cm and a diameter to allow a close fit into a seamless steel pipe having an inside diameter of 45 mm (± 10% variation), a wall thickness of 4 mm (± 10% variation) and a length of 200 mm. The pipes are closed with steel or cast iron end caps, at least as strong as the tube, torqued to 204 Nm.

17.7.1.2.2 The bullet is a standard 12.7 mm armour-piercing bullet with a projectile mass of 0.046 kg, and is fired at the service velocity of about 840 ± 40 m per second from a 12.7 mm gun.

17.7.1.3 *Procedure*

17.7.1.3.1 A minimum of six test articles (explosive substance in a capped steel pipe) should be fabricated for the tests.

17.7.1.3.2 Each test article is positioned on a suitable pedestal at a convenient distance from the muzzle of the gun. Each test article must be secured in a holding device upon its pedestal. This device must be capable of restraining the item against dislodgement by the bullet.

17.7.1.3.3 A test consists of the firing of one projectile into each test item. There should be at least three tests with the test article oriented such that its long axis is perpendicular to the line of flight (i.e. impact through the side of the pipe). There should also be at least three tests with the test article oriented such that its long axis is parallel to the line of flight (i.e. impact through the end cap).

17.7.1.3.4 The remains of the test container are collected. Complete fragmentation of the container is indicative of explosion or detonation.

17.7.1.4 *Test criteria and method of assessing results*

A substance which explodes or detonates in any trial is not an EIS and the result is noted as "+".

17.7.1.5 *Examples of results*

Substance	Result
HMX/inert binder (86/14), cast	-
HMX/energetic binder (80/20), cast	+
HMX/aluminium/energetic binder (51/19/14), cast	-
RDX/TNT (60/40), cast	+
TATB/Kel-F (95/5), pressed	-

17.7.2 *Test 7 (d) (ii): Friability test*

17.7.2.1 *Introduction*

The friability test is used to evaluate the response of an EIS candidate to the kinetic energy transfer associated with impact and penetration of a given energy source travelling at a specified velocity.

17.7.2.2 *Apparatus and materials*

The following apparatus is required:

(a) A weapon designed to shoot 18 mm diameter cylindrical test pieces at a velocity of 150 m/s;

(b) A Z30C 13 stainless steel plate, 20 mm thick with a front face roughness of 3.2 microns (AFNOR NF E 05-015 and NF E 05-016 standards);

(c) A 108 ± 0.5 cm^3 manometric bomb at 20 °C;

(d) A firing capsule containing a heating wire on 0.5 g of black powder with a mean particle size of 0.75 mm. The composition of the black powder is 74% potassium nitrate, 10.5% sulphur and 15.5% carbon. The moisture content should be less than 1%;

(e) A cylindrical sample of compact substance of diameter 18 ± 0.1 mm. The length is adjusted so as to obtain a mass of 9.0 ± 0.1 g. The sample is brought to and maintained at a temperature of 20 °C;

(f) A fragment recovery box.

17.7.2.3 *Procedure*

17.7.2.3.1 The sample is projected against the steel plate at an initial velocity sufficient to give an impact velocity as close as possible to 150 m/s. The mass of fragments collected after the impact should be at least 8.8 g. These fragments are fired in a manometric bomb. Three tests are carried out.

17.7.2.3.2 The curve of pressure against time p = f (t) is recorded; this enables the curve (dp/dt) = f' (t) to be constructed. From this curve the maximum value $(dp/dt)_{max}$ is read off. This enables the value $(dp/dt)_{max}$, corresponding to an impact speed of 150 m/s, to be estimated.

17.7.2.4 *Test criteria and method of assessing results*

If the average maximum $(dp/dt)_{max}$ value obtained at a speed of 150 m/s is greater than 15 MPa/ms, the substance tested is not an EIS and the result is noted as "+".

17.7.2.5 *Examples of results*

Substance	**Result**
HMX/inert binder (86/14), cast	-
HMX/energetic binder (80/20), cast	+
HMX/aluminium/energetic binder (51/19/14), cast	-
RDX/TNT (60/40), cast	+
TATB/Kel-F (95/5), pressed	-

17.8 Series 7 type (e) test prescription

17.8.1 *Test 7 (e): EIS external fire test*

17.8.1.1 *Introduction*

The external fire test is used to determine the reaction of an EIS candidate to external fire when it is confined.

17.8.1.2 *Apparatus and materials*

Explosive test samples fabricated by normal techniques are employed. The samples should have a length of 20 cm and a diameter to allow a close fit into a seamless steel pipe having an inside diameter of 45 mm (± 10% variation), a wall thickness of 4 mm (± 10% variation) and a length of 200 mm. The pipes are closed with steel or cast iron end caps, at least as strong as the tube, torqued to 204 Nm.

17.8.1.3 *Procedure*

17.8.1.3.1 The experimental procedure is the same as for test 6 (c) (see 16.6.1.3) except as noted in paragraph 17.8.1.3.2 below.

17.8.1.3.2 The test is conducted as:

(a) One fire engulfing fifteen confined samples, stacked in three adjacent piles of two samples banded on top of three samples; or

(b) Three fires in which five samples are laid out horizontally and banded together.

Colour photographs are taken to document the condition of the samples after each test. Cratering and the size and location of confining pipe fragments are documented as an indication of the degree of reaction.

17.8.1.4 *Test criteria and method of assessing results*

An explosive substance which detonates or reacts violently with fragments thrown more than 15 m is not an EIS and the result is noted as "+".

17.8.1.5 *Examples of results*

Substance	Result
HMX/inert binder (86/14), cast	-
HMX/inert binder (85/15), cast	-
HMX/energetic binder (80/20), cast	+
HMX/aluminium/energetic binder (51/19/14), cast	-
RDX/inert binder (85/15), cast	+
RDX/TNT (60/40), cast	+
TATB/Kel-F (95/5), pressed	-

17.9 Series 7 type (f) test prescription

17.9.1 *Test 7 (f): EIS slow cook-off test*

17.9.1.1 *Introduction*

This test is used to determine the reaction of an EIS candidate to a gradually increasing thermal environment and to find the temperature at which reaction occurs.

17.9.1.2 *Apparatus and materials*

17.9.1.2.1 Explosive test samples fabricated by normal techniques are employed. The samples should have a length of 200 mm and a diameter to allow a close fit into a seamless steel pipe having an inside diameter of 45 mm (± 10% variation), a wall thickness of 4 mm (± 10% variation) and a length of 200 mm. The pipes are closed with steel or cast iron end caps, at least as strong as the tube, torqued to 204 Nm.

17.9.1.2.2 The sample assembly is placed in an oven which provides a controlled thermal environment over a 40 °C to 365 °C temperature range and can increase the temperature of the surrounding oven atmosphere at the rate of 3.3 °C per hour throughout the temperature operating range and ensure, by circulation or other means, a uniform thermal environment to the item under test.

17.9.1.2.3 Temperature recording devices are used to monitor temperature at 10 minute or less intervals; continuous monitoring is preferred. Instrumentation with an accuracy of ± 2 per cent over the test temperature range is used to measure the temperature of:

(a) The air within the oven; and

(b) The exterior surface of the steel pipe.

17.9.1.3 *Procedure*

17.9.1.3.1 The test item is subjected to a gradually increasing air temperature at a rate of 3.3 °C per hour until reaction occurs. The test may begin with the test item pre-conditioned to 55 °C below the anticipated reaction temperature. The onset temperature at which the sample temperature exceeds the oven temperature should be recorded.

17.9.1.3.2 After the completion of each test, the pipe or any fragments of pipe are recovered in the test area and examined for evidence of violent explosive reaction. Colour photographs may be taken to document the condition of the unit and the test equipment before and after the test. Cratering, and the size and location of any fragments, may also be documented as indications of the degree of reaction.

17.9.1.3.3 Three tests are conducted for each candidate substance unless a positive result is observed earlier.

17.9.1.4 *Test criteria and method of assessing results*

A substance which detonates or reacts violently (fragmentation of one or two end caps and fragmentation of the tube into more than three pieces) is not considered an EIS and the result is noted as "+".

17.9.1.5 *Examples of results*

Substance	Result
HMX/inert binder (86/14), cast	-
HMX/energetic binder (80/20), cast	+
RDX/TNT (60/40), cast	+
TATB/Kel-F (95/5), pressed	-

17.10 Series 7 type (g) test prescription

17.10.1 *Test 7 (g): 1.6 article (or component level) external fire test*

17.10.1.1 *Introduction*

The external fire test is used to determine the reaction of a possible Division 1.6 article to external fire as presented for transport.

17.10.1.2 *Apparatus and materials*

The experimental set-up for this test is the same as for test 6 (c) (see 16.6.1.2).

17.10.1.3 *Procedure*

17.10.1.3.1 The experimental procedure for this test is the same as for test 6 (c) (see 16.6.1.3), except that, if the volume of single item exceeds 0.15 m^3, only one item is required.

17.10.1.3.2 Colour still photographs are taken to document the condition of the test item and the test equipment before and after the test. Explosive substance remains, fragmentation, blast, projections, cratering, witness screen damage, and thrust are documented as an indication of the article's response level.

17.10.1.3.3 Colour video for the duration of each trial can be vital to assessment of response. In sitting the camera(s), it is important to ensure that the field of view will not be obstructed by any of the test facilities or instrumentation and that the field of view will include all necessary information.

17.10.1.3.4 To classify complex articles containing multiple EIS main explosive loads, external fire testing at the individual main load component level should be conducted to fully characterise the article's response level.

17.10.1.4 *Test criteria and method of assessing results*

If there is a response level more severe than burning as outlined in Appendix 8, the result is noted as "+" and the items are not classified as Division 1.6 articles.

17.11 Series 7 type (h) test prescription

17.11.1 *Test 7 (h): 1.6 article or component level slow cook-off test*

17.11.1.1 *Introduction*

This test is used to determine the reaction of a candidate Division 1.6 article to a gradually increasing thermal environment and to find the temperature at which reaction occurs.

17.11.1.2 *Apparatus and materials*

17.11.1.2.1 The test equipment consists of an oven which provides a controlled thermal environment over a 40 °C to 365 °C temperature range and can increase the temperature of the surrounding oven atmosphere at the rate of 3.3 °C per hour throughout the temperature operating range, minimize hot spots, and ensure (by circulation or other means) a uniform thermal environment to the item under test. Secondary reactions (such as those caused by exudate and explosive gases contacting the heating devices) invalidate the test, but these can be avoided by providing a sealed inner container to surround articles transported bare. A means of relief should be provided for the increased air pressure generated during the test due to heating.

17.11.1.2.2 Temperature recording devices (permanent record types) are used to monitor temperature continuously or, at least, every 10 minutes. Instrumentation with an accuracy of ± 2% over the test temperature range is used to measure the temperature at:

(a) The atmosphere air gap adjacent to the unit under test; and

(b) The exterior surface of the unit.

17.11.1.3 *Procedure*

17.11.1.3.1 The test item is subjected to a gradually increasing, at a rate of 3.3 °C per hour, air temperature until unit reaction occurs. The test may begin with the test item pre-conditioned to 55 °C below the predicted reaction temperature. Temperatures and elapsed test time are measured and recorded.

17.11.1.3.2 Colour still photographs are taken to document the condition of the test item and the test equipment before and after the test. Explosive substance remains, fragmentation, blast, projections, cratering, witness plate damage, and thrust are documented as an indication of the article's response level. Colour video for the duration of each trial can be vital to assessment of response. In sitting the camera(s), it is important to ensure that the field of view will not be obstructed by any of the test facilities or instrumentation and that the field of view will include all necessary information.

17.11.1.3.3 The test is conducted twice unless a positive result is obtained earlier. To classify complex articles containing multiple EIS main explosive loads, slow cook-off testing at the individual main load component level should be conducted to fully characterise the article's response level.

17.11.1.4 *Test criteria and method of assessing results*

If there is a response level more severe than burning as outlined in Appendix 8, the result is noted as "+" and the items are not classified as Division 1.6 articles.

17.12 Series 7 type (j) test prescription

17.12.1 *Test 7 (j): 1.6 article or component bullet impact test*

17.12.1.1 *Introduction*

The bullet impact test is used to evaluate the response of a candidate Division 1.6 article to the kinetic energy transfer associated with the impact and penetration by a given energy source.

17.12.1.2 *Apparatus and materials*

Three 12.7 mm gun are used to fire service 12.7 mm armour-piercing ammunition with a projectile mass of 0.046 kg. Standard propellant loads may require adjustment to achieve projectile velocities within tolerance. The guns are fired by remote control and protected from fragment damage by firing through a hole in a heavy steel plate. The firing gun muzzle should be at a maximum range of at least 10 m from the test item to assure stabilization prior to impact, and at a maximum range of 30 m from the test item depending upon the explosive weight of the test item. The test item should be secured in a holding device capable of restraining the item against dislodgement by the projectiles.

17.12.1.3 *Procedure*

17.12.1.3.1 The candidate Division 1.6 article is subjected to a three-round burst fired at 840 ± 40 m/s velocity and 600 rounds/minute rate of fire. The test is repeated in three different orientations striking the test item in the most vulnerable areas as assessed by the competent authority These are areas for which an assessment of the explosive sensitivity (explosiveness and sensitiveness) combined with knowledge of the article design indicate the potential producing the most violent response level.

17.12.1.3.2 Colour still photographs are taken to document the condition of the test item and the test equipment before and after the test. Explosive substance remains, fragmentation, blast, projections, cratering, witness plate damage, and thrust are documented as an indication of the article's response level.

17.12.1.3.3 Colour video for the duration of each trial can be vital to assessment of response. In sitting the camera(s), it is important to ensure that the field of view will not be obstructed by any of the test facilities or instrumentation and that the field of view will include all necessary information.

17.12.1.3.4 To classify complex articles containing multiple EIS main explosive loads, bullet impact testing at the individual main load component level should be conducted to fully characterise the article's response level.

17.12.1.4 *Test criteria and method of assessing results*

If there is a response level more severe than burning as outlined in Appendix 8, the result is noted as "+" and the items are not classified as Division 1.6 articles.

17.13 Series 7 type (k) test prescription

17.13.1 *Test 7 (k): 1.6 article stack test*

17.13.1.1 *Introduction*

This test is used to determine whether a detonation of a candidate Division 1.6 article, as offered for transport, will initiate a detonation in an adjacent, like article.

17.13.1.2 *Apparatus and materials*

The experimental set-up is the same as for test 6 (b) (see 16.5.1.2), with one trial conducted confined, and another unconfined. The test should only be conducted on detonable candidate Division 1.6 articles; the test 7 (k) article stack test is waived for non-detonable candidates for Division 1.6 (evidence is available to demonstrate that the article cannot support a detonation). Where the article is designed to provide a detonation output, the article's own means of initiation or a stimulus of similar power shall be used to initiate the donor. If the article is not designed to detonate but is capable of supporting a detonation, the donor shall be detonated using an initiation system selected to minimise the influence of its explosive effects on the acceptor article(s).

17.13.1.3 *Procedure*

The experimental set-up is the same as for test 6 (b) (see 16.5.1.3). Colour still photographs are taken to document the condition of the test item and the test equipment before and after the test. Explosive substance remains, fragmentation, blast, projections, cratering, witness plate damage, and thrust are documented and used to assess whether or not any acceptor has detonated (including partially). Blast data may be used to supplement this decision. Colour video for the duration of each trial can be vital to assessment of response. In sitting the camera(s), it is important to ensure that the field of view will not be obstructed by any of the test facilities or instrumentation and that the field of view will include all necessary information. Comparing data from the two stack test trials to data from a single donor calibration shot, or to a calculated donor detonation pressure, can be useful in assessing the response level of acceptors.

17.13.1.4 *Test criteria and method of assessing results*

If detonation in the stack is propagated from the donor to an acceptor, the test result is noted as "+" and the article cannot be assigned to Division 1.6. Acceptor article response levels assessed as no reaction, burning, deflagration or explosion as outlined in Appendix 8 are considered as negative results and noted as "-".

17.14 Series 7 type (l) test prescription

17.14.1 *Test 7 (l): 1.6 article (or component level) fragment impact test*

17.14.1.1 *Introduction*

This test is used to determine the response of an article in its transport configuration to a localised shock input representative of a fragment strike typical of that produced from a nearby detonating article.

17.14.1.2 *Apparatus and materials*

To reduce variability due to yaw, a gun system is recommended for firing a standard 18.6 g steel fragment in the shape of a right-circular cylinder with a conical nose, as detailed in Figure 17.14.1, at a candidate Division 1.6 article. The distance between the firing device and the test item should ensure that the fragment is ballistically stable at impact. Barricades should protect the remote control gun system from the potential damaging effects of the test item's reaction.

17.14.1.3 *Procedure*

17.14.1.3.1 The test is repeated in two different orientations, striking the test item in the most vulnerable areas as assessed by the competent authority. These are areas for which an assessment of the explosive sensitivity (explosiveness and sensitiveness) combined with knowledge of the article design indicate the potential for producing the most violent response level. Typically, one test would be conducted targeting a non-EIS boostering component and the second test would target the centre of the main explosive load. The orientation of impact should generally be normal to the outer surface of the article. The fragment impact velocity should be 2530 ± 90 m/s.

17.14.1.3.2 Colour still photographs are taken to document the condition of the test item and the test equipment before and after the test. Explosive substance remains, fragmentation, blast, projections, cratering, witness plate damage, and thrust are documented as an indication of the article's response level.

17.14.1.3.3 Colour video for the duration of each trial can be vital to assessment of response. In sitting the camera(s), it is important to ensure that the field of view will not be obstructed by any of the test facilities or instrumentation and that the field of view will include all necessary information.

17.14.1.3.4 To classify complex articles containing multiple EIS main explosive loads, fragment impact testing at the individual main load component level should be conducted to fully characterise the article's response level.

17.14.1.4 *Test criteria and method of assessing results*

If there is a response level more severe than burning as outlined in Appendix 8, the result is noted as "+" and the items are not classified as Division 1.6 articles.

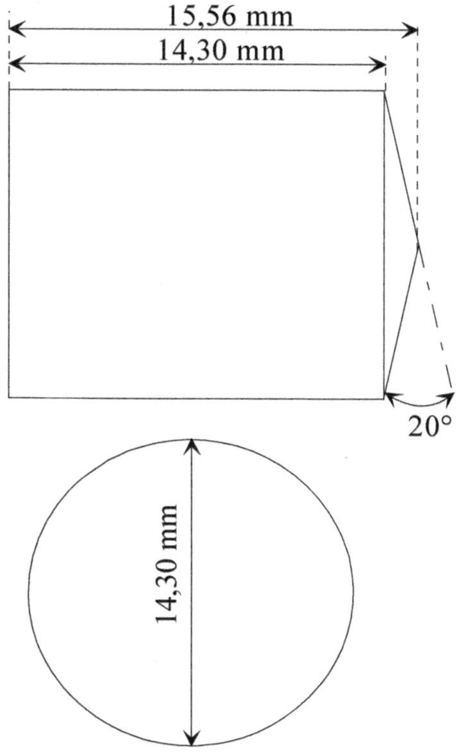

Notes:

Shape: a conical ended cylinder with the ratio $\frac{\text{L (length)}}{\text{D (diameter)}} > 1$ *for stability;*

Tolerances:	*± 0.05 mm and ± 0°30';*
Fragment mass:	*18.6 g;*
Fragment material:	*a mild carbon steel with a Brinell Hardness (HB) less than 270.*

Figure 17.14.1: STANDARD FRAGMENT FOR 1.6 ARTICLE FRAGMENT IMPACT TEST

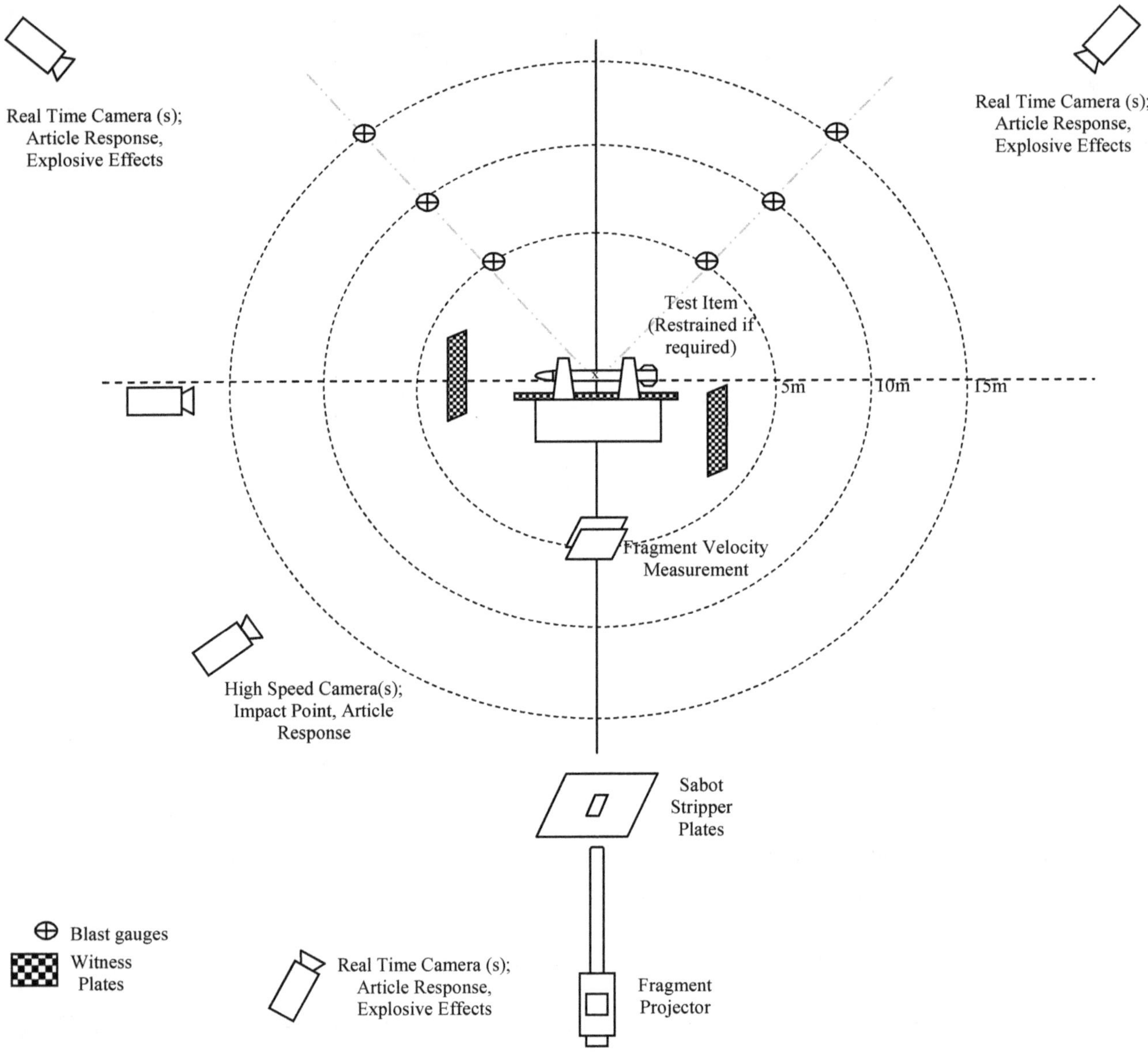

Figure 17.14.2: TYPICAL SETUP FOR 1.6 ARTICLE FRAGMENT IMPACT TEST

SECTION 18

TEST SERIES 8

18.1 Introduction

The assessment whether a candidate for "ammonium nitrate emulsion or suspension or gel, intermediate for blasting explosives (ANE)" is insensitive enough for inclusion in Division 5.1 is answered by series 8 tests and any such candidate for inclusion in Division 5.1 should pass each of the three types of tests comprising the series. The three test types are:

Type 8 (a): a test to determine the thermal stability;

Type 8 (b): a shock test to determine sensitivity to intense shock;

Type 8 (c): a test to determine the effect of heating under confinement;

Test series 8 (d) has been included in this section as one method to evaluate the suitability of ANEs for the transport in portable tanks.

18.2 Test methods

The test methods currently used are listed in Table 18.1.

Table 18.1: TEST METHODS FOR TEST SERIES 8

Test code	Name of Test	Section
8 (a)	Thermal Stability Test for ANE[a]	18.4
8 (b)	ANE Gap Test[a]	18.5
8 (c)	Koenen test[a]	18.6
8 (d)	Vented pipe tests[b]	18.7

[a] *This test is intended for classification.*

[b] *These tests are intended for evaluating the suitability of ANEs for transport in portable tanks.*

18.3 Test conditions

18.3.1 Unless otherwise specified in these tests, the substance should be tested as offered for transport, at the maximum temperature which may occur during transport (see 1.5.4 of this Manual).

18.4 Series 8 Type (a) test prescription

18.4.1 *Test 8 (a): Thermal stability test for ammonium nitrate emulsions, suspension or gels*

18.4.1.1 *Introduction*

18.4.1.1.1 This test is used to determine whether a candidate for "ammonium nitrate emulsion, suspension or gel, intermediate for blasting explosives" is thermally stable at temperatures encountered during transport. In the way this type of test is normally carried out (see 28.4.4), the 500 ml insulated test vessel is only representative for packagings, IBCs and small tanks. For the transport of ammonium nitrate emulsions, suspensions or gels the test is used to measure their thermal stability during tank transport if the test is carried out on candidate products which are at a temperature 20 °C higher than the maximum temperature which may occur during transport, or if higher, at the temperature at the time of loading.

18.4.1.2 *Apparatus and materials*

18.4.1.2.1 The experimental equipment consists of a suitable thermostatically controlled test chamber (which may be fan assisted), appropriate insulated test vessels with closures, temperature probes and recording equipment.

18.4.1.2.2 ***The test should be performed following a risk assessment, taking account of the potential for fire and/or explosion in the test chamber, and the application of appropriate control measures for the protection of persons and property. A number of tests may be run concurrently.*** The recording system should be housed in a separate observation area.

18.4.1.2.3 The test chamber must be large enough to allow air circulation on all sides of the insulated test vessels. The air temperature in the test chamber should be controlled so that the desired temperature for a liquid inert sample in the insulated test vessel can be maintained with a deviation of not more than ± 2 °C for up to 10 days. The air temperature in the test chamber should be measured and recorded.

18.4.1.2.4 Insulated test vessels with a volume of approximately 500 ml with a closure system are used. The closure of the test vessel should be inert.

18.4.1.2.5 The heat loss characteristics of the system used, i.e. insulated test vessel and closure, must be established prior to performance of the test. Since the closure system has a significant effect on the heat loss characteristics, these can be adjusted to some extent by varying the closure system. The heat loss characteristics are determined by measuring the half time of cooling of the vessel filled with a known inert liquid substance e.g. distilled water. The heat loss per unit of mass, L (W/kg.K) is calculated from the half time of cooling, $t_{1/2}$ (s), and the specific heat, Cp (J/kg.K), of the substance using the formula:

$$L = \ln 2 \times \left({C_p}/{t_{1/2}} \right)$$

18.4.1.2.6 Insulated test vessels filled with 400 ml of inert substance, with a heat loss of 100 mW/kg.K or less are suitable.

18.4.1.3 *Procedure*

18.4.1.3.1 Set the test chamber at a temperature which is 20 °C higher than the maximum temperature which may occur during transport or, if higher, at the temperature at the time of loading. Fill the test vessel with the substance under test to about 80% of the capacity of the test vessel, or approximately 400 ml. Insert the temperature probe into the centre of the sample. Seal the lid of the test vessel and place it in the test chamber, connect the temperature recording system and close the test chamber.

18.4.1.3.2 The temperature of the sample and of the test chamber are continuously monitored. The time is noted at which the sample temperature reaches a temperature 2 °C below the test chamber temperature. The test is then continued for a further seven days or until the sample temperature rises to 6 °C or more above the test chamber temperature if this occurs sooner.

18.4.1.3.3 At the end of the test, allow the sample to cool, remove it from the test chamber and carefully dispose of it as soon as possible.

18.4.1.4 *Test criteria and method of assessing results*

18.4.1.4.1 If the sample temperature does not exceed the test chamber temperature by 6 °C or more within the seven day period in any test, the ammonium nitrate emulsion, suspension or gel is considered to be thermally stable and can be further tested as a candidate for "ammonium nitrate emulsion, suspension or gel, intermediate for blasting explosives".

18.4.1.5 *Examples of results*

Substances	Sample mass (g)	Test T (°C)	Result	Comments
Ammonium nitrate	408	102	-	slight discolouration, hardened into lump Mass loss 0.5%
ANE-1 Ammonium nitrate 76%, Water 17%, Fuel/emulsifier 7%	551	102	-	separation of oil and crystallized salts. Mass loss 0.8%
ANE-2 (sensitized) Ammonium nitrate 75%, Water 17%, Fuel/emulsifier 7%	501	102	-	Some discolouration Mass loss 0.8%
ANE-Y Ammonium nitrate 77%, Water 17%, Fuel/emulsifier 7%	500	85	-	Mass loss 0.1%
ANE-Z Ammonium nitrate 75%, Water 20%, Fuel/emulsifier 5%	510	95	-	Mass loss 0.2%
ANE-G1 Ammonium nitrate 74%, Sodium nitrate 1%, Water 16%, Fuel/emulsifier 9%	553	85	-	no rise in temperature
ANE-G2 Ammonium nitrate 74%, Sodium nitrate 3%, Water 16%, Fuel/emulsifier 7%	540	85	-	no rise in temperature
ANE-J1 Ammonium nitrate 80%, Water 13%, Fuel/emulsifier 7%	613	80	-	Mass loss 0.1%
ANE-J2 Ammonium nitrate 76%, Water 17%, Fuel/emulsifier 7%	605	80	-	Mass loss 0.3%
ANE-J4 Ammonium nitrate 71%, Sodium nitrate 11%, Water 12%, Fuel/emulsifier 6%	602	80	-	Mass loss 0.1%

18.5 Series 8 Type (b) Test prescription

18.5.1 *Test 8 (b): ANE Gap Test*

18.5.1.1 *Introduction*

This test is used to measure the sensitivity of a candidate for "ammonium nitrate emulsion or suspension or gel, intermediate for blasting explosives" to a specified shock level, i.e. a specified booster charge and gap.

18.5.1.2 *Apparatus and materials*

18.5.1.2.1 The set-up for this test consists of an explosive charge (booster charge), a barrier (gap), a container holding the sample substance (acceptor charge), and a steel witness plate (target).

The following materials are to be used:

(a) Detonators of sufficient strength to effectively initiate the booster charge;

(b) Booster charges consisting of 95 mm diameter by 95 mm long pellet with a density of 1 600 $kg/m^3 \pm 50$ kg/m^3 of either Pentolite (PETN/TNT with a minimum 50% PETN), Composition B (RDX/TNT with a minimum 50% RDX) or RDX/WAX (with a minimum 95% RDX);

(c) Tubing, steel, with an outer diameter of 95.0 ± 7.0 mm, a wall thickness of 9.75 ± 2.75 mm and an inner diameter of 73.0 ± 7.0 mm, and with a length of 280 mm;

(d) Sample substances (acceptor charges);

(e) Polymethyl methacrylate (PMMA) rod, of 95 mm diameter by 70 mm long;

(f) Mild steel plate, approximately 200 mm × 200 mm × 20 mm;

(g) Wood block, 95 mm diameter and approximately 25 mm thick, with a hole drilled through the centre to hold the detonator in place against the booster charge;

(h) Wood blocks or similar to stand the assembly at least 100 mm off the ground.

18.5.1.3 *Procedure*

18.5.1.3.1 As shown in Figure 18.5.1.1, the detonator, booster charge, PMMA gap and acceptor charge are coaxially aligned above the centre of the witness plate. The bottom end of the tube is sealed with a single layer of cloth adhesive tape, or equivalent, to contain the sample substance which is carefully loaded so as to avoid the formation of voids within the sample or between the sample and the tube walls. The surface of the sample should be level with the rim of the tube. Care should be taken to ensure good contact between the detonator, the booster charge, the PMMA cylinder and the acceptor charge. The sample substance should be at ambient temperature. The wood block holding the detonator, the booster charge, the PMMA cylinder and the steel tube should be held firmly in alignment (e.g. by using a band of adhesive tape at each intersection).

18.5.1.3.2 The whole assembly, including the witness plate, is raised above the ground, with at least a 100 mm air gap between the ground and the bottom surface of the witness plate which is supported along two edges only with wooden blocks, or similar, as shown in Figure 18.5.1.1. The location of the blocks must ensure there is a clear space under where the tube is standing on the witness plate. To assist in collecting the remains of the witness plate, the whole assembly should be vertical (e.g. checked with a spirit level).

18.5.1.3.3 The test is performed three times unless a positive result is observed earlier.

18.5.1.4 *Test criteria and method of assessing results*

A clean hole punched through the plate indicates that a detonation was initiated and propagated in the sample. A substance which detonates and punches a hole in the witness plate in any trial is not to be classified as "ammonium nitrate emulsion or suspension or gel, intermediate for blasting explosives" and the result is noted as "+".

18.5.1.5 *Examples of results*

Substances	**Density g/cm^3**	**Gap mm**	**Result**	**Comments**
Ammonium nitrate (low density)	0.85	35	-	Tube fragmented (large fragments) Plate bent VOD 2.3-2.8 km/s
Ammonium nitrate (low density)	0.85	35	-	Tube fragmented (large fragments) Plate fractured
ANE-FA Ammonium nitrate 69%, Sodium nitrate 12%, Water 10%, Fuel/emulsifier 8%	1.4	50	-	Tube fragmented (large fragments) Plate not perforated
ANE-FA	1.44	70	-	Tube fragmented (large fragments) Plate not perforated
ANE-FB Ammonium nitrate 70%, Sodium nitrate 11%, Water 12%, Fuel/emulsifier 7%	ca 1.40	70	-	Tube fragmented (large fragments) Plate not perforated
ANE-FC (sensitized) Ammonium nitrate 75%, Water 13%, Fuel/emulsifier 10%	1.17	70	+	Tube fragmented (fine fragments) Plate perforated
ANE-FD (sensitized) Ammonium nitrate 76%, Water 17%, Fuel/emulsifier 7%	ca 1.22	70	+	Tube fragmented (fine fragments) Plate perforated
ANE-1 Ammonium nitrate 76%, Water 17%, Fuel/emulsifier 7%	1.4	35	-	Tube fragmented into large pieces. Plate dented VOD: 3.1 km/s
ANE-2 (sensitized) Ammonium nitrate 76%, Water 17%, Fuel/emulsifier 7%	1.3	35	+	Tube fragmented into small pieces Plate perforated VOD: 6.7 km/s
ANE-2 (sensitized) Ammonium nitrate 76%, Water 17%, Fuel/emulsifier 7%	1.3	70	+	Tube fragmented into small pieces Plate perforated VOD: 6.2 km/s
ANE-G1 Ammonium nitrate 74%, Sodium nitrate 1%, Water 16%, Fuel/emulsifier 9%	1.29	70	-	Tube fragmented Plate indented VOD 1 968 m/s
ANE-G2 Ammonium nitrate 74%, Sodium nitrate 3%, Water 16%, Fuel/emulsifier 7%	1.32	70	-	Tube fragmented Plate indented

Substances	Density g/cm^3	Gap mm	Result	Comments
ANE-G3 (sensitized by gassing) Ammonium nitrate 74%, Sodium nitrate 1%, Water 16%, Fuel/emulsifier 9%	1.17	70	+	Tube fragmented Plate punctured
ANE-G4 (sensitized by microballoons) Ammonium nitrate 74%, Sodium nitrate 3%, Water 16%, Fuel/emulsifier 7%	1.23	70	+	Tube fragmented Plate punctured
ANE-G5 Ammonium nitrate 70%, Calcium nitrate 8%, Water 16%, Fuel/emulsifier 7%	1.41	70	-	Tube fragmented Plate indented VOD 2 061 m/s
ANE-J1 Ammonium nitrate 80%, Water 13%, Fuel/emulsifier 7%	1.39	70	-	Tube fragmented Plate indented
ANE-J2 Ammonium nitrate 76%, Water 17%, Fuel/emulsifier 7%	1.42	70	-	Tube fragmented Plate indented
ANE-J4 Ammonium nitrate 71%, Sodium nitrate 11%, Water 12%, Fuel/emulsifier 6%	1.40	70	-	Tube fragmented Plate indented
ANE-J5 (sensitized by microballoons) Ammonium nitrate 71%, Sodium nitrate 5%, Water 18%, Fuel/emulsifier 6%	1.20	70	+	Tube fragmented Plate perforated VOD 5.7 km/s
ANE-J6 (sensitized by microballoons) Ammonium nitrate 80%, Water 13%, Fuel/emulsifier 7%	1.26	70	+	Tube fragmented Plate perforated VOD 6.3 km/s

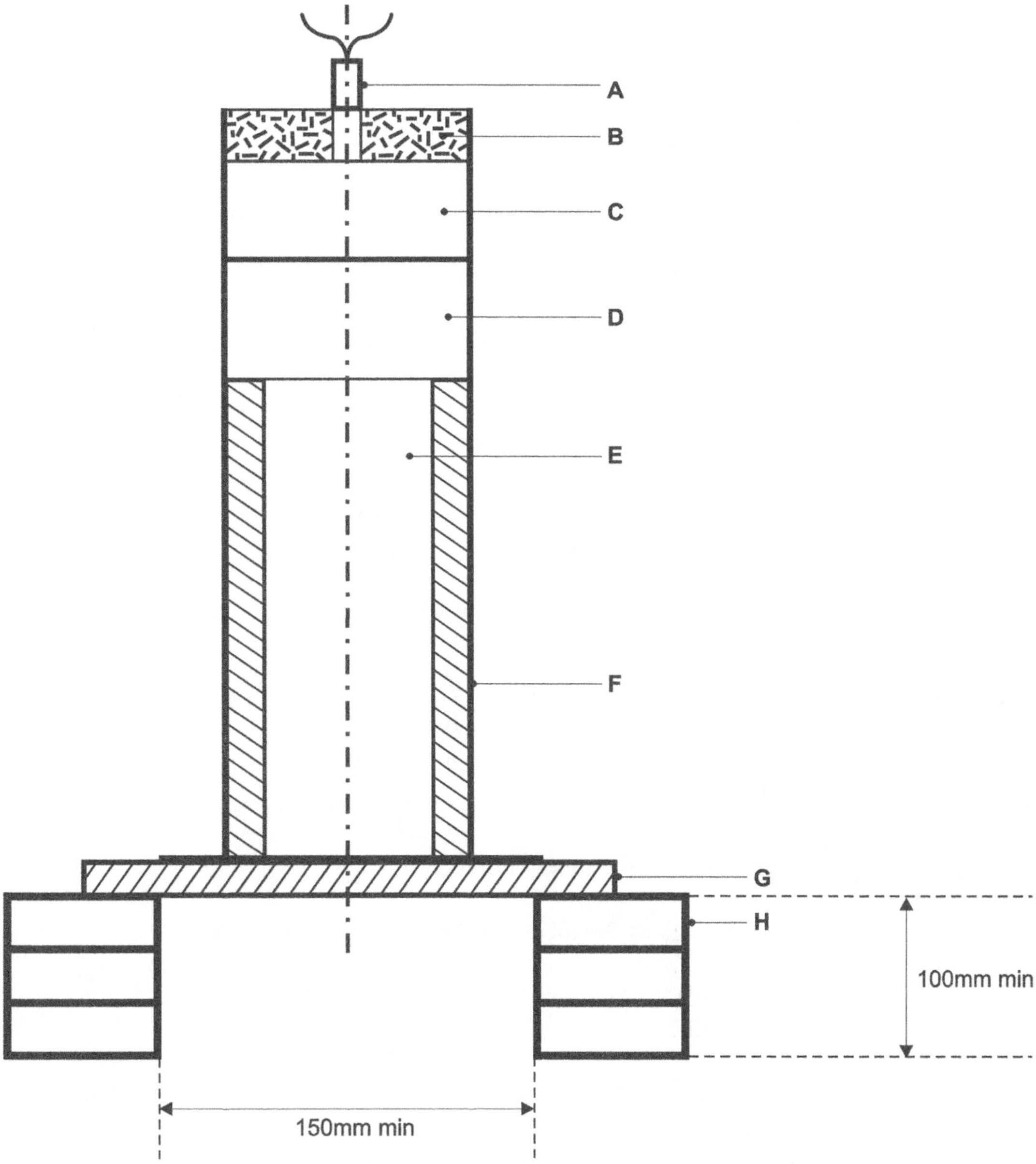

(A) Detonator	(B) Wooden detonator holder	(C) Booster charge
(D) PMMA gap	(E) Substance under test	(F) Steel Tube
(G) Witness plate	(H) Wooden blocks	

Figure 18.5.1.1: ANE GAP TEST

18.6 Series 8 Type (c) Test prescription

18.6.1 *Test 8 (c): Koenen test*

18.6.1.1 *Introduction*

This test is used to determine the sensitiveness of a candidate ammonium nitrate emulsion or suspension or gel, intermediate for blasting explosive, to the effect of intense heat under high confinement.

18.6.1.2 *Apparatus and materials*

18.6.1.2.1 The apparatus consists of a non-reusable steel tube, with its re-usable closing device, installed in a heating and protective device. The tube is deep drawn from sheet steel conforming to specification DC04 (EN 10027-1), or equivalent A620 (AISI/SAE/ASTM), or equivalent SPCEN (JIS G 3141). The dimensions are given in Figure 18.6.1.1. The open end of the tube is flanged. The closing plate with an orifice, through which the gases from the decomposition of the test substance escape, is made from heat-resisting chrome steel and is available with numerous sized orifices. For this test the following diameter holes are used:

- 1.5 mm for the closing plate used in the heating calibration procedure; and
- 2.0 mm for the closing plate used in the test.

The dimensions of the threaded collar and the nut (closing device) are given in Figure 18.6.1.1.

For quality control of the steel tubes, 1% of the tubes from each production lot shall be subjected to quality control and the following data shall be verified:

(a) The mass of the tubes shall be 26.5 ± 1.5 g;

(b) The length of the tubes shall be 75 ± 0.5 mm;

(c) The wall thickness of the tubes measured 20 mm from the bottom of the tube shall be 0.5 ± 0.05 mm; and

(d) The bursting pressure as determined by quasi-static load through an incompressible fluid shall be 30 ± 3 MPa.

18.6.1.2.2 Heating is provided by a gaseous fuel (e.g. propane), from an industrial cylinder fitted with a pressure regulator, via a flow meter and distributed by a manifold to the four burners. The gas pressure is regulated to give a heating rate of 3.3 ± 0.3 K/s when measured by the calibration procedure. Calibration involves heating a tube (fitted with a 1.5 mm orifice plate) filled with 27 cm^3 of dibutyl phthalate or equivalent. The time taken for the temperature of the liquid (measured with a 1 mm diameter thermocouple centrally placed 43 mm below the rim of the tube and inserted through the orifice plate) to rise from 135 °C to 285 °C is recorded and the heating rate calculated.

18.6.1.2.3 Because the tube is likely to be destroyed in the test, heating is undertaken in a protective welded box. A suitable arrangement of the construction and dimensions of the box is given in Figure 18.6.1.2. The tube is suspended between two rods placed through holes drilled in opposite walls of the box. A suitable arrangement of the burners is given in Figure 18.6.1.2. The burners are lit simultaneously by a pilot flame or an electrical ignition device. ***The test apparatus is placed in a protective area.*** Measures should be taken to ensure that any draughts does not affect the burner flames. Provision should be made for extracting any gases or smoke resulting from the test.

18.6.1.2.4 A video camera should be provided to record the test and to ensure all burners are functional during the test. The camera may also provide evidence of blockages of the orifice by solids within the sample.

18.6.1.3 *Procedure*

18.6.1.3.1 The substance is loaded into the tube to a height of 60 mm taking particular care to prevent the formation of voids. The threaded collar is slipped onto the tube from below, the 2 mm orifice plate is inserted and the nut tightened by hand after applying some high temperature anti-seize compound (e.g. molybdenum disulphide based lubricant). It is essential to check that none of the substance is trapped between the flange and the plate, or in the threads.

18.6.1.3.2 Each tube is used for one trial only. The orifice plates, threaded collars and nuts may be used again provided they are undamaged.

18.6.1.3.3 The tube is placed in a rigidly mounted vice and the nut tightened with a spanner. The tube is then suspended between the two rods in the protective box. The test area is vacated, the gas supply turned on and the burners lit. The time to reaction and duration of reaction can provide additional information useful in interpreting the results. If rupture of the tube does not occur, heating is to be continued for at least five minutes before the trial is finished. After each trial the fragments of the tube, if any, should be collected and weighed to ensure all pieces have been recovered.

18.6.1.3.4 The following effects are differentiated:

"O": Tube unchanged;

"A": Bottom of tube bulged out;

"B": Bottom and wall of the tube bulged out;

"C": Bottom of tube split;

"D": Wall of tube split;

"E": Tube split into two[1] fragments;

"F": Tube fragmented into three[1] or more mainly large pieces which in some cases may be connected with each other by a narrow strip;

"G": Tube fragmented into many mainly small pieces, closing device undamaged; and

"H": Tube fragmented into many very small pieces, closing device bulged out or fragmented.

Examples for the effect types "D", "E" and "F" are shown in Figure 18.6.1.3. If a trial results in any of the effects "O" to "E", the result is regarded as "no explosion (negative (-))". If a trial gives the effect "F", "G" or "H", the result is evaluated as "explosion (positive (+))".

18.6.1.3.5 The test is performed to achieve negative (-) results in three tests.

Given the nature of ammonium nitrate emulsions, suspensions or gels and the possibility of varying percentages of solids present, blockages of the orifices may occur during testing potentially leading to a false “+” result. Where this is observed the test may be repeated (maximum twice).

18.6.1.4 *Test criteria and method of assessing results*

The result is considered "+" and the substance should not be classified in Division 5.1 if three negative (-) results cannot be achieved within a maximum of five tests.

[1] *The upper part of the tube remaining in the closing device is counted as one fragment.*

18.6.1.5 *Examples of results*

Substances	Result	Comments
Ammonium nitrate (low density)	-	Limiting diameter: <1 mm
ANE-F1 Ammonium nitrate 71%, Water 21%, Fuel/emulsifier 7%	-	
ANE-F2 Ammonium nitrate 77%, Water 17%, Fuel/emulsifier 7%	-	
ANE-F3 Ammonium nitrate 70%, Sodium nitrate 11%, Water 12%, Fuel/emulsifier 7%	-	
ANE-F4 Ammonium nitrate 42%, Calcium nitrate 35%, Water 16%, Fuel/emulsifier 7%	-	
ANE-F5 Ammonium nitrate 69%, Sodium nitrate 13%, Water 10%, Fuel/emulsifier 8%	-	
ANE-F6 Ammonium nitrate 72%, Sodium nitrate 11%, Water 10%, Fuel/emulsifier 6%	-	
ANE-F7 Ammonium nitrate 76%, Water 13%, Fuel/emulsifier 10%	-	
ANE-F8 Ammonium nitrate 77%, Water 16%, Fuel/emulsifier 6%	-	
ANE-1 Ammonium nitrate 76%, Water 17%, Fuel/emulsifier 7%	-	Limiting diameter: 1.5 mm
ANE-2 (sensitized by microballoons) Ammonium nitrate 75%, Water 17%, Fuel/emulsifier 7%	+	Limiting diameter: 2 mm
ANE-4 (sensitized by microballoons) Ammonium nitrate 70%, Sodium nitrate 11%, Water 9%, Fuel/emulsifier 5.5%	+	Limiting diameter: 2 mm
ANE-G1 Ammonium nitrate 74%, Sodium nitrate 1%, Water 16%, Fuel/emulsifier 9%	-	
ANE-G2 Ammonium nitrate 74%, Sodium nitrate 3%, Water 16%, Fuel/emulsifier 7%	-	
ANE-J1 Ammonium nitrate 80%, Water 13%, Fuel/emulsifier 7%	-	Effect type "O"
ANE-J2 Ammonium nitrate 76%, Water 17%, Fuel/emulsifier 7%	-	Effect type "O"
ANE-J4 Ammonium nitrate 71%, Sodium nitrate 11%, Water 12%, Fuel/emulsifier 6%	-	Effect type "A"

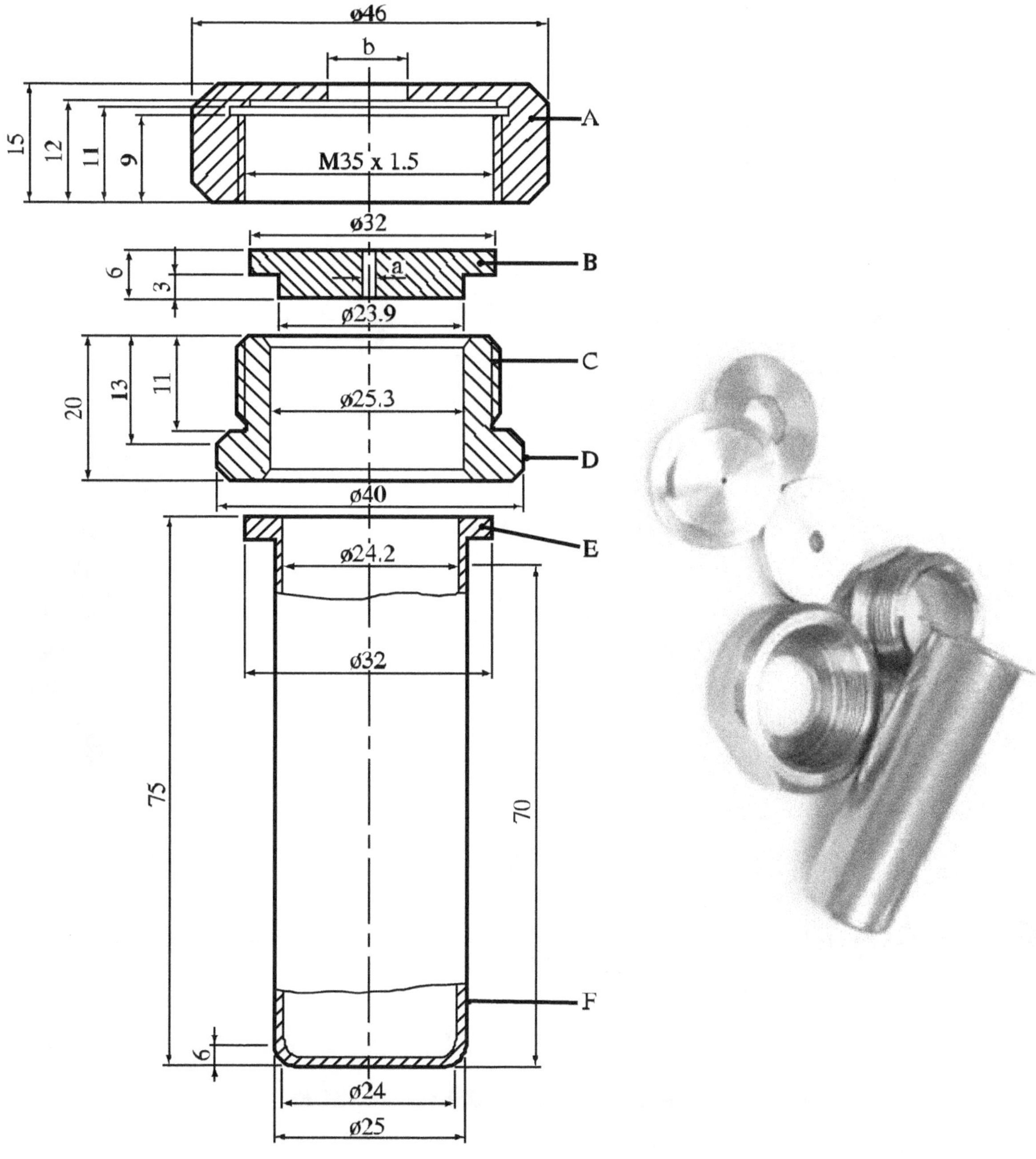

(A)	Nut (b = 10 mm) with flats for size 41 spanner	(B)	Orifice plate (a = 1.5 or 2.0 mm)
(C)	Threaded collar	(D)	Flats for size 36 spanner
(E)	Flange	(F)	Tube

Figure 18.6.1.1: TEST TUBE ASSEMBLY

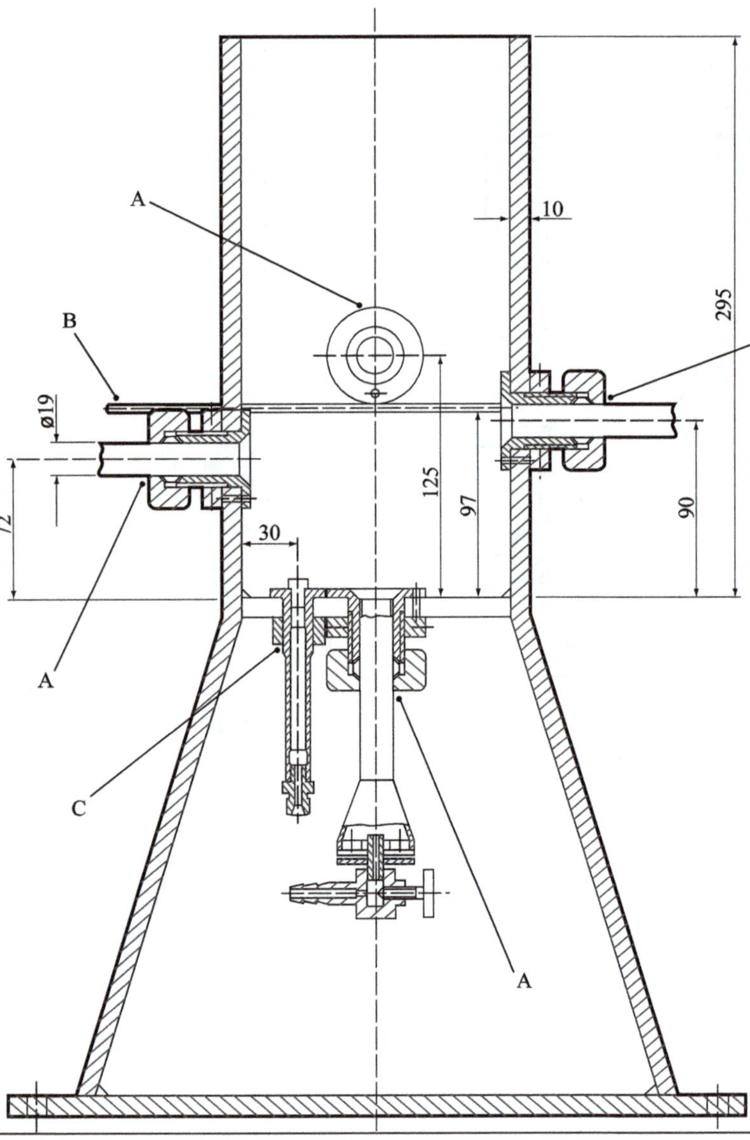

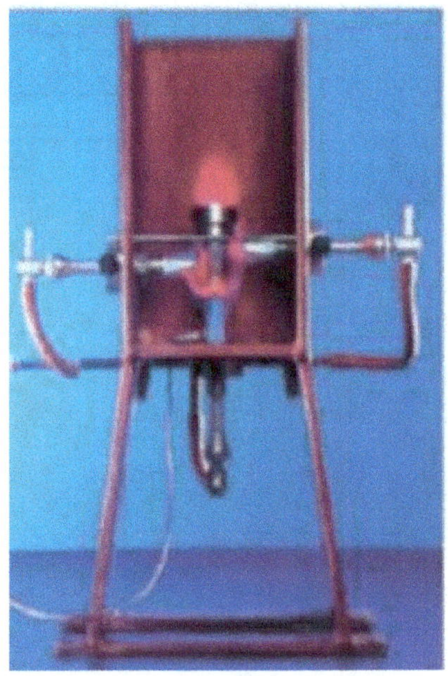

(A)	Burner	(B)	Metal rods to support test tube assembly
(C)	Pilot flame		

Figure 18.6.1.2: HEATING DEVICE

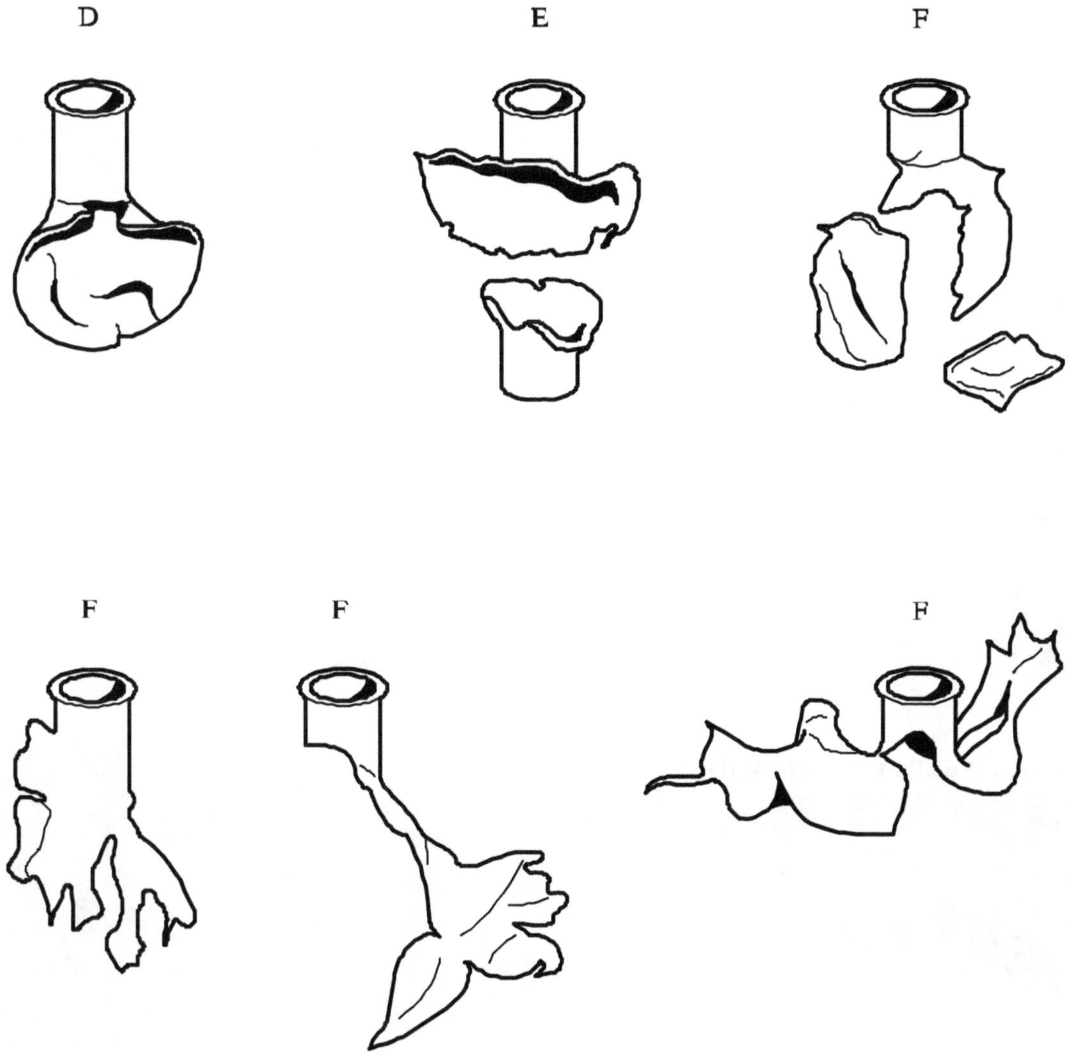

Figure 18.6.1.3: EXAMPLES OF EFFECT TYPES D, E AND F

Examples of Koenen test results

"O": Tube unchanged

"A": Bottom of the tube bulged out

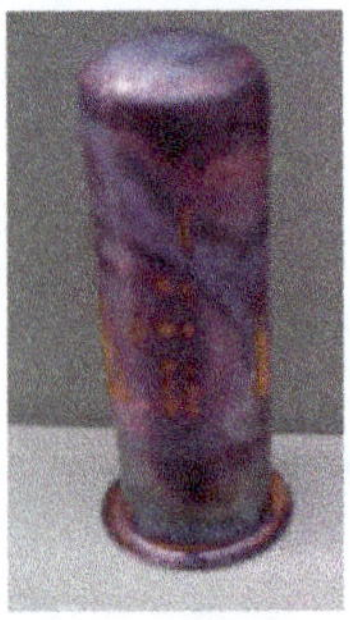

"B": Bottom and wall of the tube bulged out

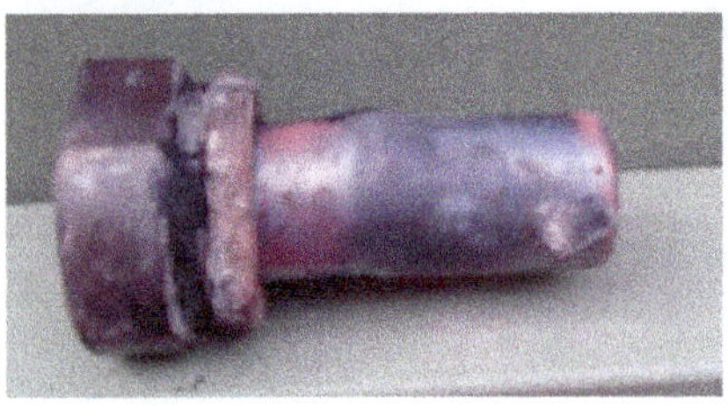

"C": ***Reserved***

"D": Wall of tube split

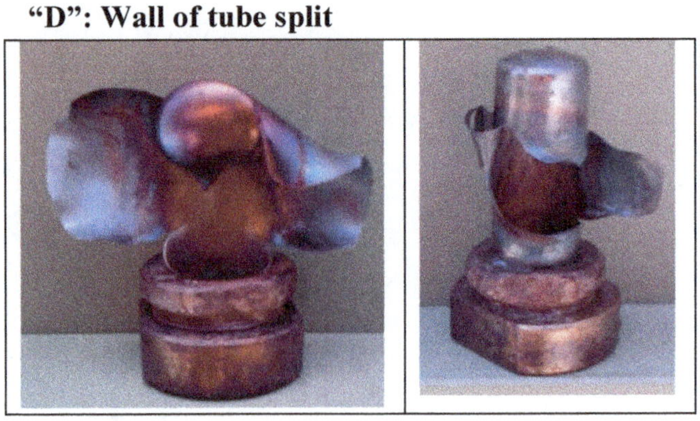

"E": Tube split into two fragments

"F": Tube fragmented into three or more mainly large pieces which in some cases may be connected with each other by a narrow strip;

"G": Tube fragmented into many mainly smaller pieces, closing device undamaged

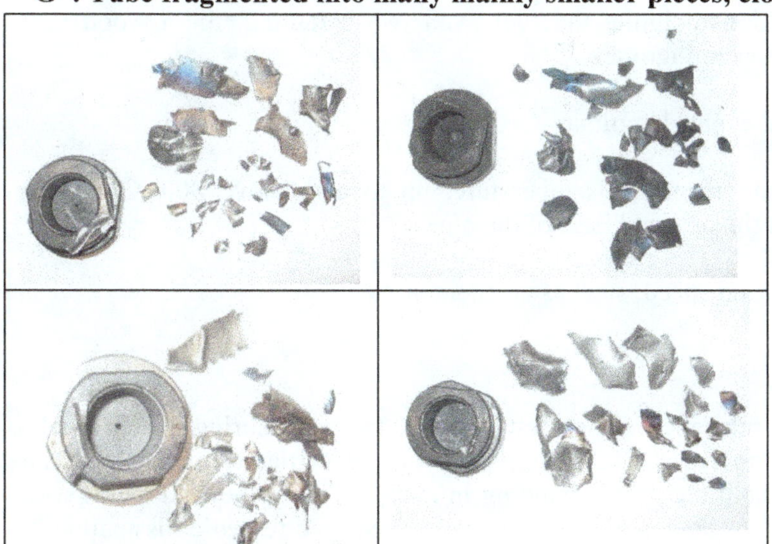

18.7 Series 8 Type (d) Test prescriptions

18.7.1 *Test 8 (d) (i): Vented pipe test*

18.7.1.1 *Introduction*

This test is not intended for classification but is included in this Manual for evaluating the suitability for transport in tanks.

The vented pipe test is used to assess the effect of exposure of a candidate for "ammonium nitrate emulsion or suspension or gel, intermediate for blasting explosives" to a large fire under confined, vented conditions.

18.7.1.2 *Apparatus and materials*

The following items are needed:

(a) A steel pipe 310 ± 10 mm diameter and 610 ± 10 mm long, welded close at the bottom with a 380 mm square, 10 ± 0.5 mm thick mild steel plate. The top of the pipe is welded to a 380 mm square, 10 ± 0.5 mm thick mild steel plate that contains a 78 mm diameter vent hole centrally located in the plate to which a 152 mm long steel pipe nipple of 78 mm internal diameter is welded (see Figure 18.7.1.1). All welding should be to a relevant ISO standard or equivalent. All steel components are to be Schedule 40 carbon steel (A53 Grade B) or equivalent;

(b) A metal grid to support the filled pipe above the fire and allow adequate heating. If a wooden crib fire is used, the grid should be approximately 1.0 m above the ground and if a liquid hydrocarbon pool fire is used then the grid should be approximately 0.5 m above the fuel surface at the onset of the test;

(c) Enough fuel to produce a fire reaching 800 °C (measured at the external base of the pipe) and to keep burning for at least 30 minutes or, if necessary, until the substance has clearly had enough time to react to the fire, evidenced by ejection of material, smoke, fumes, flames, etc., from the top of the pipe. Temporary variation of temperature below 800 °C is normal and should not render the test invalid;

(d) Suitable means of ignition to ignite the fuel from two sides e.g. for a wood fire, kerosene to soak the wood and igniters;

(e) Video cameras to record events in colour;

(f) Means of measuring and recording temperature, up to and above 800 °C, with a thermocouple located at the external base of the pipe;

(g) A means of measuring wind speed, such as an anemometer.

18.7.1.3 *Procedure*

18.7.1.3.1 The pipe is filled with the substance under test without tamping during loading. The substance is carefully packed to prevent adding voids. The steel pipe is placed vertically on the grid and secured from tipping over. Fuel is placed beneath the grid, extending in every direction beyond the pipe so that the fire will fully engulf the pipe. Precautions against side winds may be required to avoid dissipation of the heat. Suitable methods of heating include a wood fire, a liquid or gas fuel fire that produces a flame temperature of at least 800 °C.

18.7.1.3.2 The test should not be performed under conditions where the wind speed consistently exceeds 6 m/s.

18.7.1.3.3 Observations are made on the following:

(a) Wind speed at commencement of the test as per Section 18.7.1.3.2;

(b) Fire duration of at least 30 minutes or until the substance has clearly had enough time to react to the fire, with 800 °C reached at the external base of the pipe;

(c) Temperature at the external base of pipe;

(d) Substance reacting to the fire as described in 18.7.1.2(c);

(e) Evidence of explosion (e.g. fragmentation of the pipe into two or more pieces);

(f) Projection of fragments of the pipe section from the fire area;

(g) Evidence of a rupture (e.g. a split of the pipe or separation of the pipe from the base plate at the weld).

18.7.1.4 *Test criteria and method of assessing results*

A test is considered valid if observation criteria outlined in Section 18.7.1.3.3 (a) to (d) have been met.

The test result is considered "+" and the substance should not be transported in portable tanks as a dangerous good of Division 5.1 if an explosion and/or fragmentation of the pipe, as specified in Section 18.7.1.3.3 (e) and (f) is observed.

The test result is considered "-" if no explosion and/or fragmentation of the pipe is observed. Splitting of the pipe or its separation from the end plates, as specified in Section 18.7.1.3.3 (g) is evidence of a "-" result.

18.7.1.5 *Examples of results*

(*Reserved*)

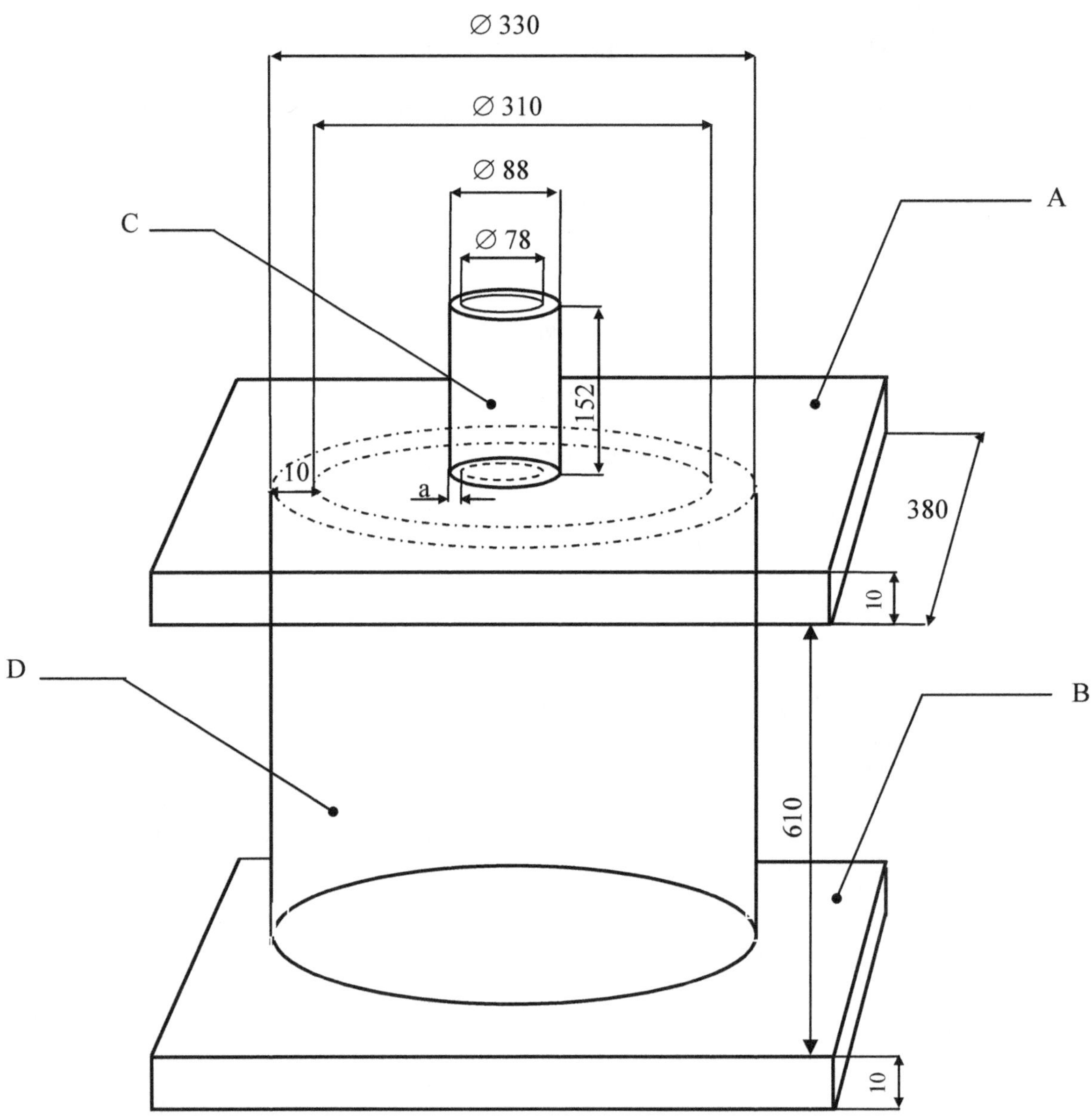

All measurements are in millimetres.

(A)	Top plate (Schedule 40 carbon (A53 grade B))
(B)	Bottom plate (Schedule 40 carbon (A53 grade B))
(C)	Steel pipe nipple (a = 0.5 cm), Schedule 40 carbon (A53 grade B)
(D)	Steel pipe (Schedule 40 carbon (A53 grade B))

Figure 18.7.1.1: VENTED PIPE TEST

18.7.2 ***Test 8 (d) (ii): Modified vented pipe test***

18.7.2.1 *Introduction*

This test is not intended for classification but is included in this Manual for evaluating the suitability of a candidate for "ammonium nitrate emulsion or suspension or gel, intermediate for blasting explosives", to be transported in portable tanks as a dangerous substance of Division 5.1.

The modified vented pipe test is used to assess the effect of exposure of a candidate for "ammonium nitrate emulsion or suspension or gel, intermediate for blasting explosives" to a large fire under confined, vented conditions.

18.7.2.2 *Apparatus and materials*

The following items are needed:

(a) A vented vessel consisting of mild drawn steel pipe with an inner diameter of 265 ± 10 mm, a length of 580 ± 10 mm and a wall thickness of 5.0 ± 0.5 mm. Both the top and the base plates are made from 300 mm square, 6.0 ± 0.5 mm thick mild steel plates. The top and base plates are fixed to the pipe with a fillet weld with a thickness of at least 5 mm. All welding should be to a relevant ISO standard or equivalent. The top plate has a vent diameter of 85 mm ± 1.0 mm. A further two small holes are drilled in the top plate to neatly accommodate thermocouple probes;

(b) A concrete block, or similar solid base, about 400 mm square and 50 to 75 mm thick;

(c) A metal stand for supporting the vessel at a height of approximately 150 mm above the concrete block or similar solid base;

(d) A gas burner capable of accommodating a fuel gas (e.g. propane) flow rate of up to 60 g/min. This rests on the concrete block, or similar solid base, under the stand. A typical example of a suitable burner is a 32-jet Mongolian wok burner;

(e) Enough fuel to produce a fire reaching 800 °C (measured at the external base of the pipe) and to keep burning for at least 60 minutes or, if necessary, until the substance has clearly had enough time to react to the fire, evidenced by ejection of material, smoke, fumes, flames, etc., from the top of the pipe. Temporary variation of temperature below 800 °C is normal and should not render the test invalid;

(f) A sheet metal shield to protect the fuel gas flame from side winds. This can be fabricated from approximately 0.5 mm thick galvanised sheet metal. The diameter of the wind shield is about 600 mm and the height should be about 250 mm. Four adjustable vents approximately 150 mm wide and 100 mm high are spaced equally around the shield to ensure adequate air reaches the gas flame;

(g) Fuel gas bottle(s) connected via a manifold and fed into a pressure regulator. The pressure regulator should reduce the fuel gas bottle pressure from 600 kPa down to about 150 kPa. The gas then flows through a gas rotameter capable of measuring up to 60 g/min of fuel gas and a needle valve. An electrical solenoid valve is used to switch the fuel gas flow on and off remotely. Typically three 9 kg fuel gas bottles will achieve the desired gas flow rate for the duration of up to five tests. The gas pressure and flow are regulated to give a heating rate of 3.3 ± 0.3 K/min when measured by the calibration procedure;

(h) Three thermocouples with approximately 500 (2) and 100 (1) mm long stainless steel probes and fibre-glass coated lead wires;

(i) A data-logger capable of recording the output from the thermocouples;

(j) Cine-cameras or video cameras, preferably high speed and normal speed, to record events in colour;

(k) Pure water for calibration;

(l) The candidate ammonium nitrate emulsion or suspension or gel, intermediate for blasting explosives to be tested;

(m) A means of measuring wind speed at the commencement of the test, such as an anemometer;

(n) Blast gauges, radiometers and other recording equipment may also be used.

18.7.2.3 *Calibration*

18.7.2.3.1 The vessel is filled to the 75% level (i.e. to a depth of about 435 mm) with the pure water, and heated using the procedure specified in 18.7.2.4. Water is heated from ambient temperature up to 90 °C, monitoring temperature by the thermocouple in the water. Temperature-time data must fit a straight line whose slope will be the "calibration heating rate" for the given combination of vessel and heat source.

18.7.2.3.2 The gas pressure and flow must be regulated to give a heating rate of 3.3 ± 0.3 K/min.

18.7.2.3.3 This calibration must be performed prior to the testing of any test substance, though the same calibration can be applied to any test conducted within a day of the calibration provided no change is made to the vessel construction or gas supply. New calibration has to be made every time that the burner is changed.

18.7.2.4 *Procedure*

18.7.2.4.1 The concrete block, or similar solid base, is placed on a sandy base and levelled using a spirit level. The fuel gas burner is positioned in the centre of the solid base and connected to the gas supply line. The metal stand is placed over the burner.

18.7.2.4.2 The vessel is placed vertically on the stand and secured from tipping over. The vessel is filled to 75% of its volume (to a height of approximately 435 mm) with the substance under test without tamping during loading. The initial temperature of the substance must be recorded. The substance is carefully packed to prevent adding voids. The wind shield is positioned around the base of the assembly to protect the gas flame from heat dissipation due to side winds.

18.7.2.4.3 The thermocouple positions are as follows:

(a) The first 500 mm long probe (T1) in the gas flame;

(b) The second 500 mm long probe (T2) extending all the way into the vessel so that the tip is positioned 80 to 90 mm from the bottom of the vessel;

(c) The third 100 mm long probe (T3) in the headspace about 20 mm into the vessel.

The thermocouples are connected to the data-logger and the thermocouple leads and data-logger are adequately protected from the test apparatus in case of explosion.

18.7.2.4.4 Fuel gas pressure and flow is checked and adjusted to the values used during the water calibration described in 18.7.2.3. Video cameras and any other recording equipment are checked and started. Thermocouple functioning is checked and data logging is started, with a time set between thermocouple readings not exceeding 10 seconds, and preferably shorter. ***The test should not be performed under conditions where the wind speed exceeds 6 m/s, unless additional precautions against side winds are taken to avoid dissipation of the heat.***

18.7.2.4.5 The fuel gas burner may be started locally or remotely and all workers immediately retreat to a safe location. Progress of the test is followed by monitoring thermocouple readings and closed circuit television images. The start time of the trial is defined by the time at which the flame thermocouple trace T1 first begins to rise.

18.7.2.4.6 The gas reservoir should be large enough to bring the substance to a possible reaction and provide a fire duration lasting beyond total consumption of the test sample. If the vessel does not rupture, the system should be allowed to cool down before carefully dismantling the test set-up.

18.7.2.4.7 Observations are made on the following:

(a) Wind speed at commencement of the test as per Section 18.7.2.4.4;

(b) Fire duration of at least 60 minutes or until the substance has clearly had enough time to react to the fire, with 800 °C reached at the external base of the pipe;

(c) Temperature at the external base of pipe;

(d) Substance reacting to the fire as described in 18.7.2.2(e);

(e) Evidence of explosion (e.g. fragmentation of the pipe into two or more pieces);

(f) Projection of fragments of the pipe section from the fire area;

(g) Evidence of a rupture (e.g. a split of the pipe or separation of the pipe from the base plate at the weld).

18.7.2.4.8 Test criteria and method of assessing results

A test is considered valid if observation criteria outlined in Section 18.7.2.4.7 (a) to (d) have been met.

The test result is considered "+" and the substance should not be transported in portable tanks as a dangerous good of Division 5.1 if an explosion and/or fragmentation of the pipe, as specified in Section 18.7.2.4.7 (e) and (f) is observed.

The test result is considered "-" if no explosion and/or fragmentation of the pipe is observed. Splitting of the pipe or its separation from the end plates, as specified in Section 18.7.2.4.7 (g) is evidence of a "-" result.

18.7.2.5 *Examples of results*

Substances	**Result**
76.0 ammonium nitrate / 17.0 water / 5.6 paraffin oil / 1.4 PIBSA emulsifier	-
84.0 ammonium nitrate / 9.0 water / 5.6 paraffin oil / 1.4 PIBSA emulsifier	+
67.7 ammonium nitrate / 12.2 sodium nitrate / 14.1 water / 4.8 paraffin oil / 1.2 PIBSA emulsifier	-
67.4 ammonium nitrate / 15.0 methylamine nitrate / 12.0 water / 5.0 glycol / 0.6 thickener	-
71.4 ammonium nitrate / 14.0 hexamine nitrate / 14.0 water / 0.6 thickener	-

PART II

CLASSIFICATION PROCEDURES, TEST METHODS AND CRITERIA RELATING TO SELF-REACTIVE SUBSTANCES OF DIVISION 4.1 AND ORGANIC PEROXIDES OF DIVISION 5.2

CONTENTS OF PART II

NOTE 1: *The country or organisation of origin of each test method is indicated in brackets after each test name.*

NOTE 2: *The test method recommended for use with each test type is indicated in* ***bold*** *and by an asterisk (see sub-section 1.6 of the general Introduction).*

CONTENTS OF PART II (continued)

CONTENTS OF PART II (continued)

SECTION 20

INTRODUCTION TO PART II

20.1 Purpose

20.1.1 Part II of the Manual presents the United Nations schemes for the classification of self-reactive substances of Division 4.1 and organic peroxides of Division 5.2 (see respectively sub-section 2.4.2.3 and section 2.5.3 of the Model Regulations). It includes a description of the procedures, test methods and criteria considered to be the most suitable for providing competent authorities with the necessary information to arrive at a proper classification of these substances for transport. It should be used in conjunction with the classification principles given in paragraphs 20.4.2 (Division 4.1) and 20.4.3 (Division 5.2) and the flow chart given in Figure 20.1 (see also 2.4.2.3.3 and 2.5.3.3 in the Model Regulations).

20.1.2 Self-reactive substances and organic peroxides are classified into seven types according to the hazard. The scheme for classifying these substances is illustrated in Figure 20.1. The tests are performed in two stages. ***In the first stage, for the safety of laboratory workers, preliminary small scale tests should be performed to ascertain the stability and sensitivity of the substance.*** In the second stage, classification tests should be performed.

20.2 Scope

20.2.1 New substances, capable of exothermic decomposition, offered for transport should be subjected to the classification procedures for self-reactive substances as set out in this section unless:

(a) They are explosives according to the criteria of Class 1;

(b) They are oxidizing substances according to the classification procedure for Division 5.1 (see section 34) except that mixtures of oxidizing substances which contain 5.0% or more of combustible organic substances shall be subjected to the classification procedure defined in the NOTE below;

***NOTE**: Mixtures of oxidizing substances meeting the criteria of Division 5.1 which contain 5.0% or more of combustible organic substances, which do not meet the criteria mentioned in (a), (c), (d) or (e) above, shall be subjected to the self-reactive substance classification procedure.*

A mixture showing the properties of a self-reactive substance, type B to F, shall be classified as a self-reactive substance of Division 4.1.

A mixture showing the properties of a self-reactive substance, type G, according to the principle of 20.4.2 (g) shall be considered for classification as a substance of Division 5.1 (see section 34).

(c) They are organic peroxides according to the criteria of Division 5.2;

(d) Their heat of decomposition (see 20.3.3.3) is less than 300 J/g; or

(e) Their self-accelerating decomposition temperature is greater than 75 °C for a 50 kg package.

20.2.2 New organic peroxides offered for transport should be subjected to the classification procedures as set out in this section unless the organic peroxide formulation contains:

(a) Not more than 1.0% available oxygen from the organic peroxides when containing not more than 1.0% hydrogen peroxide; or

(b) Not more than 0.5% available oxygen from the organic peroxides when containing more than 1.0% but not more than 7.0% hydrogen peroxide.

20.2.3 In this context a new substance is one which, in the opinion of the competent authority, involves any of the following:

(a) A new self-reactive substance which meets the definition given in paragraph 2.4.2.3.1.1 of the Model Regulations, or a mixture of already classified self-reactive substances which is considered to be significantly different from those already classified;

(b) An activator is added to an already classified self-reactive substance (see paragraphs 2.4.2.3.2.1 and 2.4.2.3.2.2 of the Model Regulations) which decreases the thermal stability or alters the explosive properties; or

(c) A new organic peroxide which meets the definition given in paragraph 2.5.1 of the Model Regulations, or a mixture of already classified organic peroxides which is considered to be significantly different from those already classified.

20.2.4 Mixtures of organic peroxide formulations currently assigned to a generic entry may be classified as the same type of organic peroxide as that of the most dangerous component and be transported under the conditions of transport given for this type. However, as two components can form a thermally less stable mixture, the self-accelerating decomposition temperature (SADT) of the mixture should be determined and, if necessary, the control and emergency temperature derived from the SADT in accordance with paragraph 2.5.3.4 of the Model Regulations.

20.2.5 The classification procedure should be undertaken before a new product is offered for transport.

20.2.6 Any substance which shows the properties of a self-reactive substance (except type G) should be classified as such, even if this substance gives a positive test result according to the test method for self-heating substances of Division 4.2 (see section 32.5.2 of this Manual and Chapter 2.4 of the Model Regulations).

20.3 Preliminary procedure

20.3.1 *General description*

It is essential for the safety of laboratory workers that small-scale preliminary tests are performed before attempting to handle larger quantities. This involves tests for determining the sensitiveness of the substance to mechanical stimuli (impact and friction), and to heat and flame.

20.3.2 *Test types*

Generally four types of small scale test may be used to make a preliminary safety assessment:

(a) A falling weight test to determine sensitiveness to impact;

(b) A friction or impacted friction test to determine the sensitiveness to friction;

(c) A test to assess thermal stability and the exothermic decomposition energy; and

(d) A test to assess the effect of ignition.

20.3.3 ***Application of the test methods***

20.3.3.1 *For safety, it is usual to perform the tests requiring the smallest quantity of material first.*

20.3.3.2 Impact and friction sensitiveness may be assessed using one of the tests in series 3 of the Class 1 acceptance procedure (see Part I).

20.3.3.3 Thermal stability may be estimated using a suitable calorimetric technique such as differential scanning calorimetry or adiabatic calorimetry. Exothermic decomposition energy may be estimated using a suitable calorimetric technique such as differential scanning calorimetry. In using such techniques, special care should be taken in interpreting the results when:

- Sampling and testing mixtures;
- The material of the sample vessel may influence the result;
- Endotherms immediately precede exotherms;
- Evaporation of constituents will lower the exothermicity (sealed sample vessels should normally be used);
- The presence of air may critically affect the measured decomposition energy;
- There is a large difference between the specific heats of the reactants and products; and
- Using rapid heating rates (when differential scanning calorimetry is used, the heating rates should normally be in the range of 2 to 5 K/min).

If differential scanning calorimetry is used, the extrapolated onset temperature is defined as being the point of intersection of the tangent drawn at the point of greatest slope on the leading edge of the peak with the extrapolated baseline.

20.3.3.4 Any suitable method may be used to assess the effect of ignition provided it adequately identifies those materials which react violently under little or no confinement.

Figure 20.1 (a): FLOW CHART SCHEME FOR SELF-REACTIVE SUBSTANCES AND ORGANIC PEROXIDES

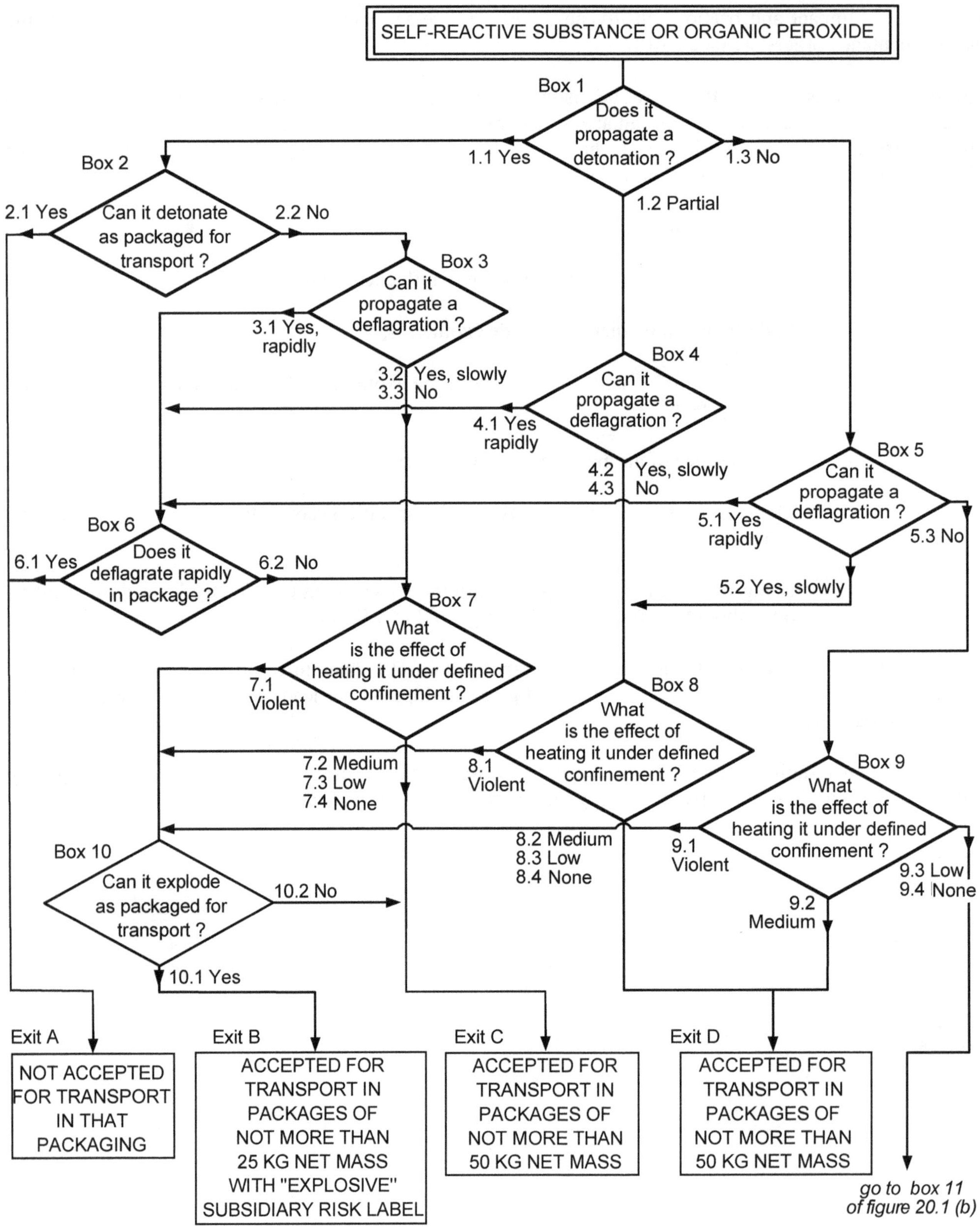

Figure 20.1 (b): FLOW CHART SCHEME FOR SELF-REACTIVE SUBSTANCES AND ORGANIC PEROXIDES (cont'd)

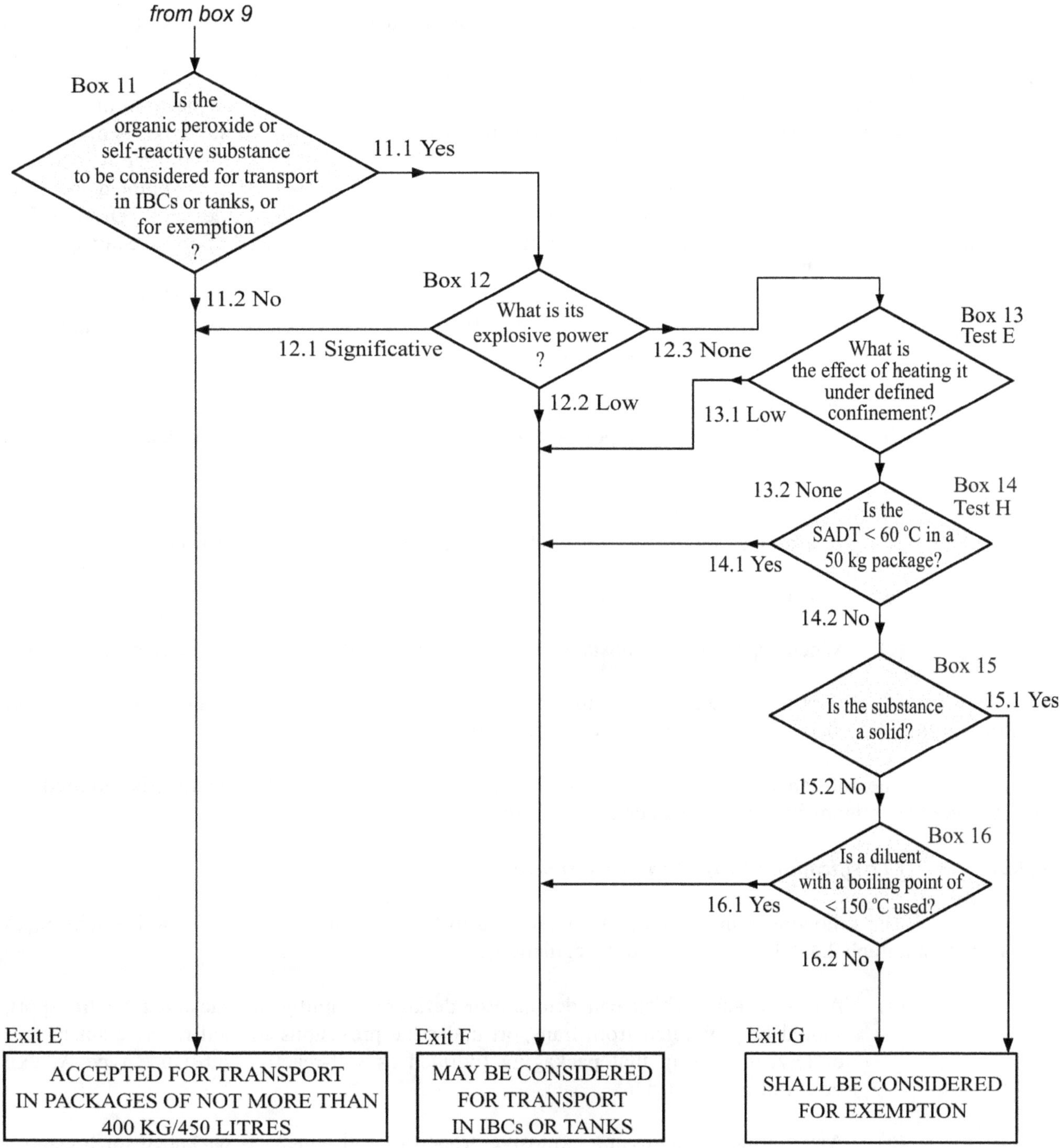

20.4 Classification procedures

20.4.1 *General description*

20.4.1.1 The principles of classification of self-reactive substances and organic peroxides are given in sub-sections 20.4.2 and 20.4.3 respectively (see also paragraphs 2.4.2.3.3 and 2.5.3.3 of the Model Regulations). The general scheme for classification (flow chart) is shown in Figure 20.1. Self-reactive substances and organic peroxides are classified into seven types according to the hazard. In order to determine the substance type, it is necessary to determine the properties as requested in the boxes of the flow chart and the additional requirements in the classification principles. The types range from type A, which is not accepted for transport in the packaging in which it is tested, to type G which is exempted from the provisions for self-reactive substances of Division 4.1 or organic peroxides of Division 5.2. The classification of types B to F is directly related to the maximum quantity allowed in one packaging.

20.4.1.2 A self-reactive substance or organic peroxide formulation should be regarded as possessing explosive properties when in laboratory testing the formulation is liable to detonate, to deflagrate rapidly or to show a violent effect when heated under confinement.

20.4.1.3 The self-accelerating decomposition temperature (SADT) is defined as the lowest ambient temperature at which self-accelerating decomposition may occur in a substance in the packaging as used in transport (see sub-section 2.5.3.4 of the Model Regulations). An SADT should be determined in order to decide if:

(a) A substance should be subjected to temperature control during transport;

(b) When appropriate, a substance conforms to the requirements of type G; or

(c) When appropriate, a substance meets the SADT criterion for self-reactive substances.

20.4.1.4 When appropriate, the physical form of the substance should be determined as set out under definition of liquids in paragraph 1.2.1 of the Model Regulations.

20.4.1.5 The substance type, physical form and whether or not temperature control is required are used to make the assignment of the substance to the appropriate generic entry.

20.4.2 *Classification of self-reactive substances*

The following principles should be applied to the classification of self-reactive substances not listed in paragraph 2.4.2.3.2.3 of the Model Regulations.

(a) Any substance which can detonate or deflagrate rapidly, as packaged for transport, should be prohibited from transport under the provisions for self-reactive substances of Division 4.1 in that packaging (defined as self-reactive substance type A, exit box A of Figure 20.1);

(b) Any substance possessing explosive properties and which, as packaged for transport, neither detonates nor deflagrates rapidly, but is liable to undergo a thermal explosion in that package, should also bear an "EXPLOSIVE" subsidiary risk label. Such a substance may be packaged in amounts of up to 25 kg unless the maximum quantity has to be limited to a lower amount to preclude detonation or rapid deflagration in the package (defined as self-reactive substance type B, exit box B of Figure 20.1);

(c) Any substance possessing explosive properties may be transported without an "EXPLOSIVE" subsidiary risk label when the substance as packaged (maximum 50 kg) for transport cannot detonate or deflagrate rapidly, or undergo a thermal explosion (defined as self-reactive substance type C, exit box C of Figure 20.1);

(d) Any substance which in laboratory testing:

  (i) detonates partially, does not deflagrate rapidly and shows no violent effect when heated under confinement; or

  (ii) does not detonate at all, deflagrates slowly and shows no violent effect when heated under confinement; or

  (iii) does not detonate or deflagrate at all and shows a medium effect when heated under confinement;

  may be accepted for transport in packages of not more than 50 kg net mass (defined as self-reactive substance type D, exit box D of Figure 20.1);

(e) Any substance which, in laboratory testing, neither detonates nor deflagrates at all and shows low or no effect when heated under confinement may be accepted for transport in packages of not more than 400 kg/450 litres (defined as self-reactive substance type E, exit box E of Figure 20.1);

(f) Any substance which, in laboratory testing, neither detonates in the cavitated state nor deflagrates at all and shows only a low or no effect when heated under confinement as well as low or no explosive power may be considered for transport in IBCs or tanks (defined as self-reactive substance type F, exit box F of Figure 20.1); for additional provisions, see Model Regulations, sub-section 4.1.7.2 and packing instruction IBC520, sub-section 4.2.1.13 and portable tank instruction T23;

(g) Any substance which, in laboratory testing, neither detonates in the cavitated state nor deflagrates at all and shows no effect when heated under confinement nor any explosive power should be exempted from classification as a self-reactive substance of Division 4.1 provided that the formulation is thermally stable (self-accelerating decomposition temperature 60 °C to 75 °C for a 50 kg package) and any diluent meets the requirements of paragraph 2.4.2.3.5 of the Model Regulations (defined as self-reactive substance type G, exit box G of Figure 20.1). If the formulation is not thermally stable or a compatible diluent having a boiling point less than 150 °C is used for desensitization, the formulation should be defined as self-reactive liquid/solid type F.

20.4.3 ***Classification of organic peroxides***

The following principles should be applied to the classification of organic peroxide formulations not listed in Table of paragraph 2.5.3.2.4 of the Model Regulations:

(a) Any organic peroxide formulation which can detonate or deflagrate rapidly, as packaged for transport, should be prohibited from transport in that packaging under Division 5.2 (defined as organic peroxide type A, exit box A of Figure 20.1);

(b) Any organic peroxide formulation possessing explosive properties and which, as packaged for transport, neither detonates nor deflagrates rapidly, but is liable to undergo a thermal explosion in that package, should bear an "EXPLOSIVE" subsidiary risk label. Such an organic peroxide may be packaged in amounts of up to 25 kg unless the maximum quantity has to be limited to a lower amount to preclude detonation or rapid deflagration in the package (defined as organic peroxide type B, exit box B of Figure 20.1);

(c) Any organic peroxide formulation possessing explosive properties may be transported without an "EXPLOSIVE" subsidiary risk label when the substance as packaged (maximum 50 kg) for transport cannot detonate or deflagrate rapidly or undergo a thermal explosion (defined as organic peroxide type C, exit box C of Figure 20.1);

(d) Any organic peroxide formulation which in laboratory testing:

(i) detonates partially, does not deflagrate rapidly and shows no violent effect when heated under confinement; or

(ii) does not detonate at all, deflagrates slowly and shows no violent effect when heated under confinement; or

(iii) does not detonate or deflagrate at all and shows a medium effect when heated under confinement

is acceptable for transport in packages of not more than 50 kg net mass (defined as organic peroxide type D, exit box D of Figure 20.1);

(e) Any organic peroxide formulation which, in laboratory testing, neither detonates nor deflagrates at all and shows low or no effect when heated under confinement is acceptable for transport in packages of not more than 400 kg/450 litres (defined as organic peroxide type E, exit box E of Figure 20.1);

(f) Any organic peroxide formulation which, in laboratory testing, neither detonates in the cavitated state nor deflagrates at all and shows only a low or no effect when heated under confinement as well as low or no explosive power may be considered for transport in IBCs or tanks (defined as organic peroxide type F, exit box F of Figure 20.1); for additional provisions, see Model Regulations, sub-section 4.1.7.2 and packing instruction IBC520, sub-section 4.2.1.13 and portable tank instruction T23;

(g) Any organic peroxide formulation which, in laboratory testing, neither detonates in the cavitated state nor deflagrates at all and shows no effect when heated under confinement nor any explosive power should be exempted from Division 5.2, provided that the formulation is thermally stable (self-accelerating decomposition temperature is 60 °C or higher for a 50 kg package) and for liquid formulations diluent type A is used for desensitization (defined as organic peroxide type G, exit box G of Figure 20.1). If the formulation is not thermally stable or a diluent other than type A is used for desensitization, the formulation should be defined as organic peroxide type F.

20.4.4 *Test types*

20.4.4.1 Paragraphs 20.4.2 and 20.4.3 refer only to those properties of self-reactive substances and organic peroxides which are decisive for their classification. These properties should be determined experimentally.

20.4.4.2 The test methods for determining the type of self-reactive substance or organic peroxide are grouped into eight series, lettered A to H, designed to provide the information necessary to answer the questions in the flow chart given in Figure 20.1 and to apply the principles for classification.

20.4.4.3 Test series A comprises laboratory tests and criteria concerning propagation of detonation as requested in box 1 of the flow chart.

20.4.4.4 Test series B comprises a test and criteria concerning the propagation of detonation of the substance as packaged for transport, as requested in box 2 of the flow chart.

20.4.4.5 Test series C comprises laboratory tests and criteria concerning propagation of deflagration as requested in boxes 3, 4 and 5 of the flow chart.

20.4.4.6 Test series D comprises a test and criteria concerning the propagation of a rapid deflagration of the substance as packaged for transport, as requested in box 6 of the flow chart.

20.4.4.7 Test series E comprises laboratory tests and criteria concerning the determination of the effect of heating under defined confinement, as requested in boxes 7, 8, 9 and 13 of the flow chart.

20.4.4.8 Test series F comprises laboratory tests and criteria concerning the explosive power of substances which are considered for transport in Intermediate Bulk Containers (IBCs) or tanks, or for exemption (see box 11 of the flow chart), as requested in box 12 of the flow chart.

20.4.4.9 Test series G comprises tests and criteria concerning the determination of the effect of a thermal explosion of the substance as packaged for transport, as requested in box 10 of the flow chart.

20.4.4.10 Test series H comprises tests and criteria concerning the determination of the self-accelerating decomposition temperature of organic peroxides and self-reactive, or potentially self-reactive substances.

20.4.5 *Application of the test methods*

20.4.5.1 The ordering of test series A to H relates more to the sequence of assessing results than to the order in which tests are conducted. The recommended sequence of laboratory scale testing is test series E, H, F, C and then A. Some tests may not be required - see the introduction to each test series.

20.4.5.2 The package tests of test series B, D and G need be performed only if indicated by the results from the corresponding tests in test series A, C and E.

20.5 Example of a test report

20.5.1 Examples of a test report and a classification are illustrated in figures 20.2 and 20.3 respectively.

20.5.2 To allow for analytical uncertainties, the concentration of the substance to be transported may be up 2% higher than the concentration tested. If an organic peroxide is tested, the available oxygen content should be given in the general Data section of the report. Similarly, if a self-reactive substance is tested, the activator type and concentration, if any, should be given.

20.5.3 If the preliminary test results indicate that the substance is very sensitive to initiation (e.g. by impact, friction or sparks), the results should be recorded in the report.

Figure 20.2: EXAMPLE OF A TEST REPORT

1.	Name of substance	:	tert-BUTYL PEROXYBENZOATE, technically pure
2.	General data		
2.1	Composition	:	98% tert-butyl peroxybenzoate
2.2	Molecular formula	:	$C_{11}H_{14}O_3$
2.3	Available oxygen content	:	8.24%
2.4	Activator content	:	Not applicable
2.5	Physical form	:	Liquid
2.6	Colour	:	Colourless
2.7	Apparent density	:	1 040 kg/m^3
2.8	Particle size	:	Not applicable
3.	Detonation (test series A)		
	Box 1 of the flow chart	:	Does the substance propagate a detonation?
3.1	Method	:	BAM 50/60 steel tube test (test A.1)
3.2	Sample conditions	:	Ambient temperature
3.3	Observations	:	30 cm of tube fragmented, unreacted substance remained in the tube
3.4	Result	:	Partial
3.5	Exit	:	1.2
4.	Deflagration (test series C)		
	Box 4 of the flow chart	:	Does the substance propagate a deflagration?
4.1	Method 1	:	Time/pressure test (test C.1)
4.2	Sample conditions	:	Ambient temperature
4.3	Observations	:	Time 2.5 s
4.4	Result	:	Yes, slowly
4.5	Method 2	:	Deflagration test (test C.2)
4.6	Sample conditions	:	Temperature 50 °C
4.7	Observations	:	Deflagration rate 0.65 mm/s
4.8	Result	:	Yes, slowly
4.9	Overall result	:	Yes, slowly
4.10	Exit	:	4.2
5.	Heating under confinement (test series E)		
	Box 8 of the flow chart	:	What is the effect of heating it under defined confinement?
5.1	Method 1	:	Koenen test (test E.1)
5.2	Sample conditions	:	Mass 26.0 g
5.3	Observations	:	Limiting diameter 3.5 mm (time to reaction 19 s, duration of reaction 22 s)
5.4	Result	:	Violent
5.5	Method 2	:	Dutch pressure vessel test (test E.2)
5.6	Sample conditions	:	10.0 g

5.7	Observations	:	Limiting diameter 10.0 mm (time to reaction 110 s, duration of reaction 4 s)
5.8	Result	:	Violent
5.9	Overall result	:	Violent
5.10	Exit	:	8.1
6.	Explosion in package (test series E)		
	Box 10 of the flow chart	:	Can the substance explode as packaged for transport?
6.1	Method	:	Thermal explosion test in the package (test G.1)
6.2	Sample conditions	:	25 kg of substance in packaging type 6HG2 with a capacity of 30 litres
6.3	Observations	:	Fume off only, no fragmentation of the package
6.4	Result	:	No explosion (packing method OP5)
6.5	Exit	:	10.2
7.	Thermal stability (test series H)		
7.1	Method	:	United States SADT test (test H.1)
7.2	Sample conditions	:	20 litres of substance in packaging type 6HG2 with a capacity of 25 litres
7.3	Observations	:	Auto-accelerating decomposition at 63 °C, no auto-accelerating decomposition at 58 °C, self-accelerating decomposition temperature 63 °C
7.4	Result	:	No temperature control required
8.	Additional data (see 20.5.3)		
8.1	Method	:	BAM fallhammer test (test 3 (a) (ii))
8.2	Sample conditions	:	Ambient temperature
8.3	Observations	:	Limiting impact energy 5 J
8.4	Result	:	Sensitive to impact
9.	Proposed assignment		
9.1	Proper shipping name	:	ORGANIC PEROXIDE TYPE C, LIQUID
9.2	UN number	:	3103
9.3	Division	:	5.2
9.4	Technical name	:	tert-Butyl peroxybenzoate
9.5	Concentration	:	≤ 100%
9.6	Diluent(s)	:	None
9.7	Subsidiary risks	:	None
9.8	Packing group	:	II
9.9	Packing method	:	OP5
9.10	Control temperature	:	Not required
9.11	Emergency temperature	:	Not required

Figure 20.3: CLASSIFICATION OF TERT-BUTYL PEROXYBENZOATE

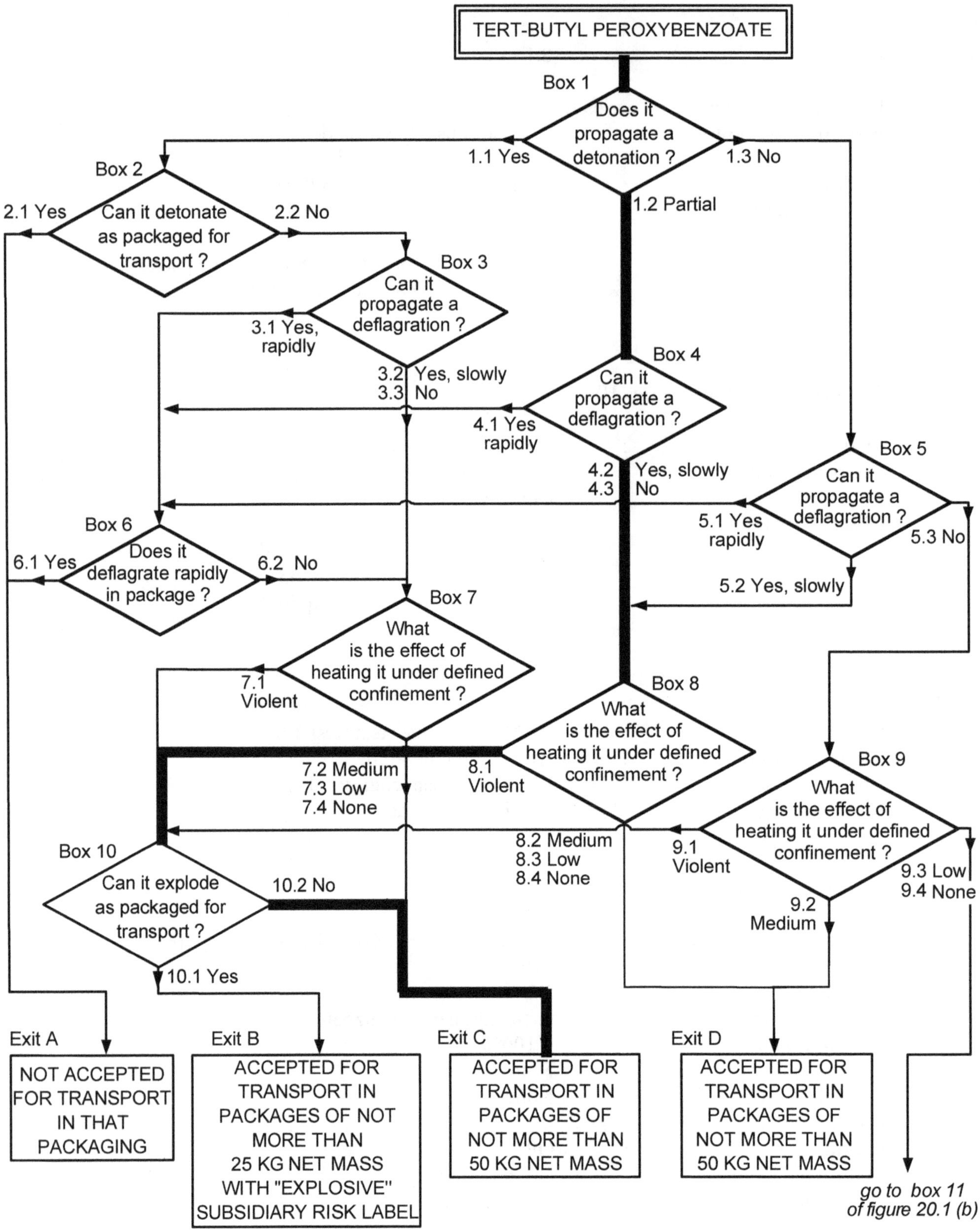

SECTION 21

TEST SERIES A

21.1 Introduction

20.1.1 Test series A comprises laboratory tests and criteria concerning propagation of detonation as requested in box 1 of Figure 20.1.

21.2 Test methods

21.2.1 The question "Does it propagate a detonation?" (box 1 of Figure 20.1) is answered on the basis of the results of one of the test methods in Table 21.1. If a liquid is being considered for transport in tank-containers or IBCs with a capacity exceeding 450 litres, a cavitated version of a Series A test should be performed (see Appendix 3).

Table 21.1: TEST METHODS FOR TEST SERIES A

Test code	Name of test	Section
A.1	BAM 50/60 steel tube test	21.4.1
A.5	UN gap test	21.4.3
A.6	UN detonation test[a]	21.4.4

[a] *Recommended test*

All tests are considered to be equivalent and only one test method has to be used.

21.2.2 For organic peroxides and self-reactive substances, a combination of a test for explosive power (any test of series F except test F.5 for peroxides, and any test of series F except tests F.4 and F.5 for self-reactive substances) with two tests for the effects of heating under confinement may be used as a screening procedure for assessing the ability to propagate a detonation. A test of series A need not be performed if:

(a) A "No" result is obtained from the explosive power test; and

(b) A "No" or "Low" result is obtained from test E.2 and either test E.1 or E.3.

For transport in packages (excluding IBCs), if the screening procedure indicates that a Series A test is not needed, the question in box 1 is answered with a "No". However, if the substance is being considered for transport in tank containers or IBCs, or for exemption, then a series A test is required unless the result of a series A test on a formulation of the substance with a higher concentration and the same physical state is "No".

21.3 Test conditions

21.3.1 As the apparent density of the substance has an important effect on the results from series A tests, it should always be recorded. The apparent density of solids should be determined from measurement of the tube volume and sample mass.

21.3.2 If a mixture can separate out during transport, the test should be performed with the initiator in contact with the potentially most explosive part, if known.

21.3.3 The tests should be performed at ambient temperature unless the substance is to be transported under conditions where it may change its physical state or density. Organic peroxides and self-reactive substances which require temperature control should be tested at the control temperature if below ambient temperature.

21.3.4 ***The preliminary procedure should be performed before performing these tests (see section 20.3).***

21.3.5 When a fresh batch of steel tubes is used, calibration tests using water (for tests on liquids) and an inert, organic solid (for tests on solids) should be performed to determine the average blank fragmentation length. The "No" / "Partial" criteria should be set to 1.5 times the average blank fragmentation length.

21.4 Series A test prescriptions

21.4.1 *Test A.1: BAM 50/60 steel tube test*

21.4.1.1 *Introduction*

This test is used to measure the ability of a substance to propagate a detonation by subjecting it to a detonating booster charge under confinement in a steel tube. It may be used to answer the question in box 1 of Figure 20.1.

21.4.1.2 *Apparatus and materials*

A seamless drawn steel tube of 500 mm length, 60 mm external diameter and 5 mm wall-thickness (e.g. according to DIN 2448) should be used, made from steel St 37.0 with tensile strength of 350 to 480 $N.mm^{-2}$ (e.g. according to DIN 1629). The tube is closed by a malleable cast iron screwing cap or by an appropriate plastic cap, put over the open end of the tube. The booster consists of a cylindrical pellet of 50 g RDX/wax (95/5) compressed to a pressure of 1 500 bar and with dimensions shown in Figure 21.4.1.1. The upper part of the booster has an axial recess of 7 mm diameter and 20 mm depth which accepts a detonator of sufficient strength to initiate the booster reliably. Substances which may react dangerously with steel St. 37.0 are tested in tubes with an internal polythene coating[1].

21.4.1.3 *Procedure*

21.4.1.3.1 Normally, the steel tube is filled with the substance as received, the sample mass determined and, if solid, the apparent density calculated using the measured internal tube volume. However, lumps are crushed and paste-like or gel-type substances are carefully packed to eliminate voids. In all cases, the final density of the substance in the tube should be as close as possible to its shipping density. The booster is placed centrally in the upper end of the tube, so that it is surrounded by the substance. When liquids are tested, the booster is separated from the liquid by wrapping it in a thin foil of aluminium or an appropriate plastic material. The wrapped booster is then attached to the malleable iron cap by means of thin wires passing through four additional drillings in the cap. The cap is carefully screwed onto the tube and the detonator inserted into the booster through the central hole in the screw cap. The detonator is then initiated.

21.4.1.3.2 At least two tests, which may be instrumented (e.g. by a continuous velocity probe), are performed unless detonation of the substance is observed. An instrumented third test may be necessary if no conclusion can be drawn from two uninstrumented tests.

21.4.1.4 *Test criteria and method of assessing results*

21.4.1.4.1 The test results are assessed on the basis of:

(a) The type of fragmentation of the tube;

(b) The completeness of the reaction of the substance; and

(c) If the occasion arises, the measured rate of propagation in the substance.

The test giving the most severe assessment should be used for classification.

---

[1] *In special cases, pure aluminium or steel 1.4571 according to DIN 17440 may be used as tube material.*

21.4.1.4.2 The test criteria are as follows:

"Yes":

- The tube is fragmented completely; or
- The tube is fragmented at both ends; or
- A velocity measurement shows that the rate of propagation in the non-fragmented part of the tube is constant and above the velocity of sound in the substance.

"Partial":

- The tube is fragmented only at the initiator end and the average tube fragmentation length (average over two tests) is greater than 1.5 times the average fragmentation length found with an inert material having the same physical state; and
- A significant portion of unreacted substance remains or a velocity measurement shows that the rate of propagation in the non-fragmented part of the tube is lower than the velocity of sound in the substance.

"No":

- The tube is fragmented only at the initiator end and the average fragmentation length (average of two tests) not more than 1.5 times the average fragmentation length found with an inert material having the same physical state; and
- A significant portion of unreacted substance remains or a velocity measurement shows that the rate of propagation in the non-fragmented part of the tube is lower than the velocity of sound in the substance.

21.4.1.5 *Examples of results*

Substances		Apparent density (km/m³)	Fragmented length (cm)	Result
Azodicarbonamide		627	15	No
2,2'-Azodi(2,4-dimethylvaleronitrile)		793	16	No
Benzene-1,3-disulphohydrazide		640	50	Yes
Benzene sulphohydrazide		630	17	No
tert-Butyl peroxybenzoate		-	30	Partial
tert-Butyl peroxy-2-ethylhexanoate		-	18	No
3-Chloroperoxybenzoic acid, not more than 86% with 3-chlorobenzoic acid		610	24, 6[a]	Yes
Cumyl hydroperoxide, 84% in cumene		-	15	No
Cyclohexanone peroxide(s)		620	50	Yes
2-Diazo-1-naphthol-5-sulphochloride		690	20	No[b]
Dibenzoyl peroxide		730	30, 12[a]	Yes
Dibenzoyl peroxide, 75% with water		740	20	No
Di-tert-butyl peroxide		-	16	No
Dicetyl peroxydicarbonate		590	13	No
Dicumyl peroxide		520	14	No
Diisopropyl peroxydicarbonate		790	50	Yes
Dilauroyl peroxide		580	25	Partial
Dimyristyl peroxydicarbonate		460	20	No
Dimyristyl peroxydicarbonate, 42% stable dispersion in water		-	15	No
N,N'-Dinitrosopentamethylene tetramine, 90% with mineral oil		590	50	Yes[c]
N,N'-Dinitrosopentamethylene tetramine, 80% with 17% inorganic solid and 3% mineral oil		500	50	Yes
N,N'-Dinitrosopentamethylene tetramine, 75% with 15% calcium carbonate and 10% mineral oil		-	26	Partial
Inert substances:	Air		8	
	Dimethyl phthalate		13	
	Icing sugar	682	14	
	Sand		13	
	Water		14	

[a] *Both ends fragmented.*

[b] *Substance completely reacted by deflagration.*

[c] *Detonation velocity 3 040 m/s.*

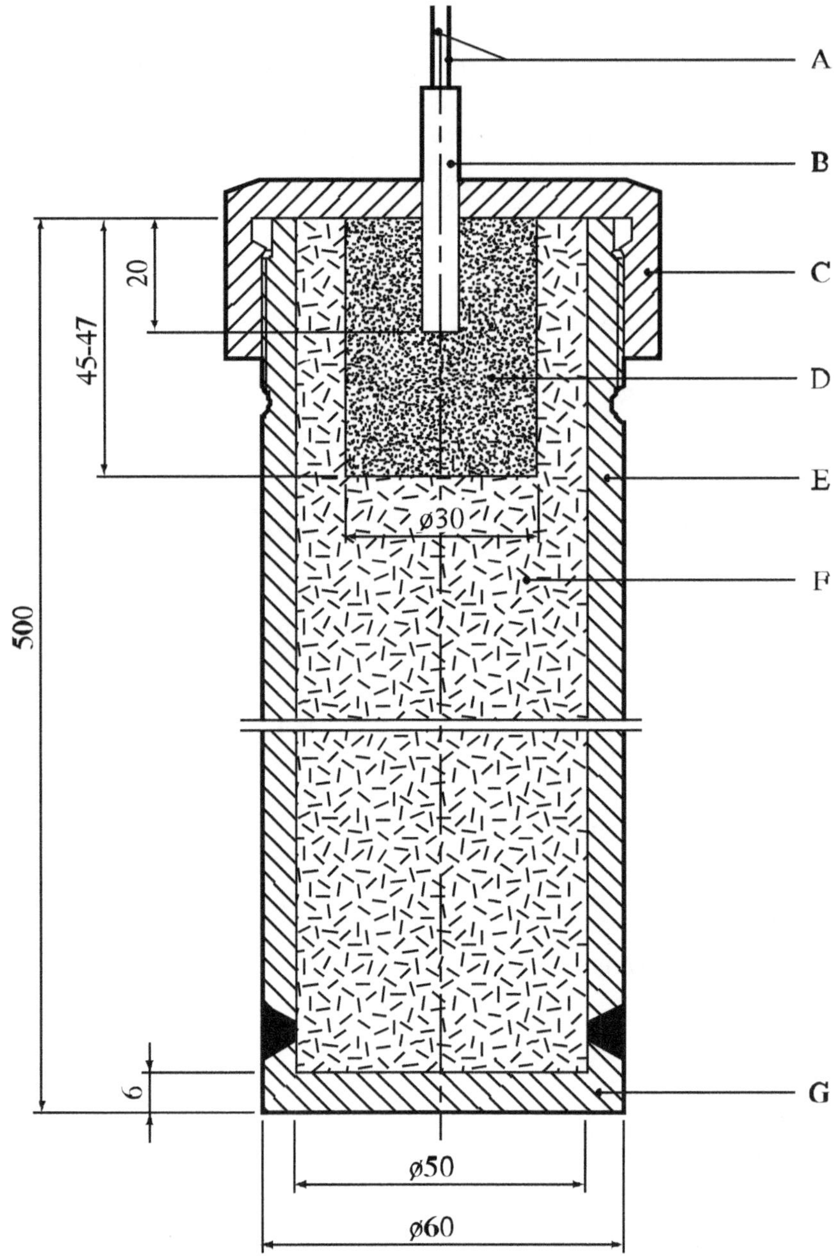

(A)	Detonator wires
(B)	Detonator inserted 20 mm into the booster charge
(C)	Screw cap of malleable cast iron or a plastics cap
(D)	Booster charge of RDX/wax (95/5) with 30 mm diameter and length of approximately 46 mm
(E)	Steel tube 500 mm long with internal diameter 50 mm, external diameter 60 mm
(F)	Substance under test
(G)	Welded steel based 6 mm thick

Figure 21.4.1.1: BAM 50/60 STEEL TUBE TEST

21.4.2 *Test A.2: TNO 50/70 steel tube test*

21.4.2.1 *Introduction*

This test is used to measure the ability of a substance to propagate a detonation by subjecting it to a detonating booster charge under confinement in a steel tube. It may be used to answer the question in box 1 of Figure 20.1.

21.4.2.2 *Apparatus and materials*

21.4.2.2.1 Solids

The apparatus consists of a seamless steel tube (e.g. material St. 35 in accordance with DIN 1629/P3), inner diameter 50 mm, wall thickness 10 mm, and length 1 160 mm (tube type A). At one end (which will be called the bottom), it is closed by welding a 20 mm-thick steel plate to the tube (see Figure 21.4.2.1). The tube is instrumented, e.g. by a continuous wire velocity probe, to measure the velocity of propagation in the substance. The booster charge consists of four RDX/wax (95/5) booster charges of 50 mm diameter, mass 50 g and length 16.4 mm.

21.4.2.2.2 Liquids

The detonability of liquids is determined with the aid of a tube similar to the tube used for solids but with a length of 750 mm. One end of the tube (called the bottom) is closed by a 0.5 mm metal plate under which the four boosters are positioned (tube type B), see Figure 21.4.2.2. The tube is held upright by means of a holder or three supports welded onto the tube. For corrosive liquids and liquids which decompose on contact with St. 35, a tube of stainless steel 316 (passivated if necessary) is used with a length of 750 mm, an internal diameter of 50 mm and an external diameter of 63 mm (tube type C).

21.4.2.3 *Procedure*

21.4.2.3.1 Solids

The substance should be tested at ambient temperature or at the control temperature, if below ambient. After fitting the velocity probe, the solid substance under investigation is introduced via the open end of the tube while the tube is continuously tapped. After filling up to 60 mm below the upper rim of the tube, the sample mass is determined and the apparent density calculated using the measured internal tube volume. The four booster charges are inserted, the final booster charge provided with a detonator and the charge detonated. Two tests are performed unless detonation of the substance is observed.

21.4.2.3.2 Liquids

For tests on liquids the booster charge, which is identical to that used for solids, should be positioned under the metal plate after which the tube is completely filled with the liquid and its mass determined. Subsequently, the procedures used for solids and liquids are the same.

21.4.2.4 *Test criteria and method of assessing results*

21.4.2.4.1 The test results are assessed on the basis of the fragmentation pattern of the tube and, in some cases, on the measured propagation velocity. The test giving the most severe assessment should be used for classification.

21.4.2.4.2 The test criteria are as follows:

"Yes":

- The tube is fragmented completely; or
- A velocity measurement shows that the rate of propagation in the non-fragmented part of the tube is constant and above the velocity of sound in the substance.

"Partial":

- In all tests, the substance failed to detonate but the average fragmentation length (average over two tests) is greater than 1.5 times the average fragmentation length found with an inert material with the same physical state.

"No":

- In all tests, the substance failed to detonate and the average fragmentation length (average over two tests) is not more than 1.5 times the average fragmentation length found with an inert material with the same physical state.

21.4.2.5 *Examples of results*

Substance	Tube type	Apparent density (kg/m^3)	Fragmented length (cm)	Result
tert-Butyl peroxybenzoate	B	-	20	Partial
tert-Butyl peroxy-2-ethylhexanoate	B	-	14	No
tert-Butyl peroxy isopropyl carbonate	B	-	17	Partial
Dibenzoyl peroxide, 75% with water	A	770	30	Partial
1,1-Di-(tert-butylperoxy)-3,3,5-trimethylcyclohexane	C	-	7	No
Dicyclohexyl peroxydicarbonate[a]	A	630	33[b]	Yes
Dicyclohexyl peroxydicarbonate, with 10% water[a]	A	640	33[c]	Yes
2,5-Diethoxy-4-morpholinobenzene-diazonium zinc chloride, 90%	A	-	17	No
2,5-Diethoxy-4-(phenylsulphonyl)-benzenediazonium zinc chloride, 67%	A	-	25	No
Dilauroyl peroxide	A	610	34	Partial
3-Methyl-4-(pyrrolidin-1-yl)benzene-diazonium tetrafluoroborate, 95%	A	-	19	No
Di-n-octanoyl peroxide (liquid)	B	-	10	No

a *Performed at the control temperature.*
b *Propagation velocity, 660 m/s, greater than the sound velocity in the substance.*
c *Propagation velocity, 690 m/s, greater than the sound velocity in the substance.*

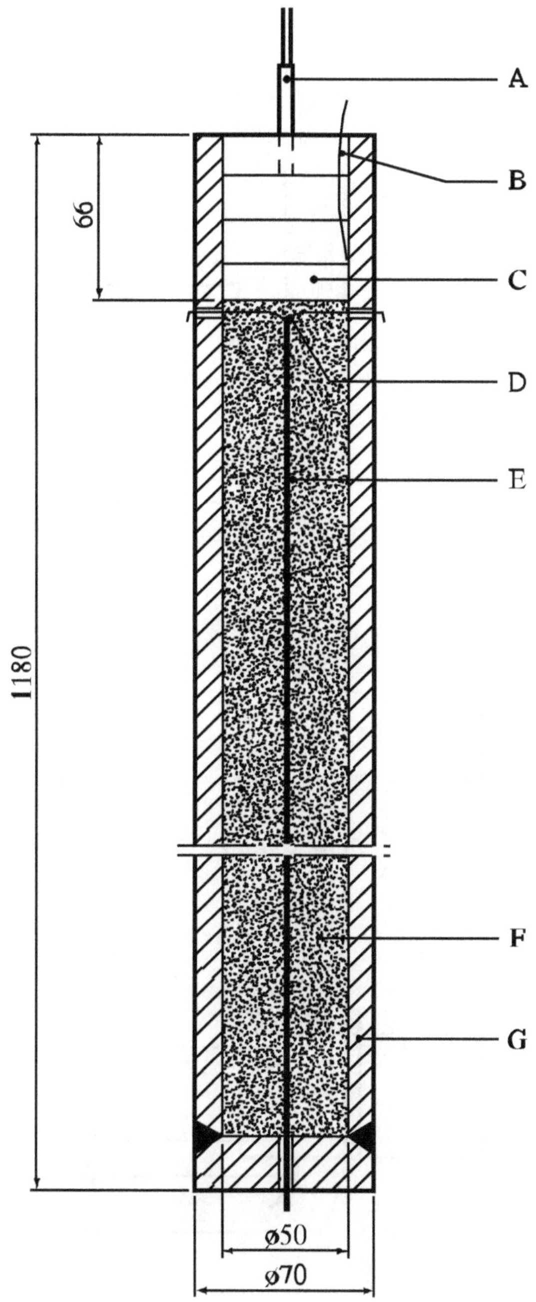

(A)	Detonator	(B)	Ionization probe
(C)	4 RDX/wax cartridges	(D)	Eye
(E)	Velocity probe	(F)	Substance under investigation
(G)	Tube, steel		

Figure 21.4.2.1: TNO 50/70 STEEL TUBE TEST FOR SOLIDS (TUBE TYPE A)

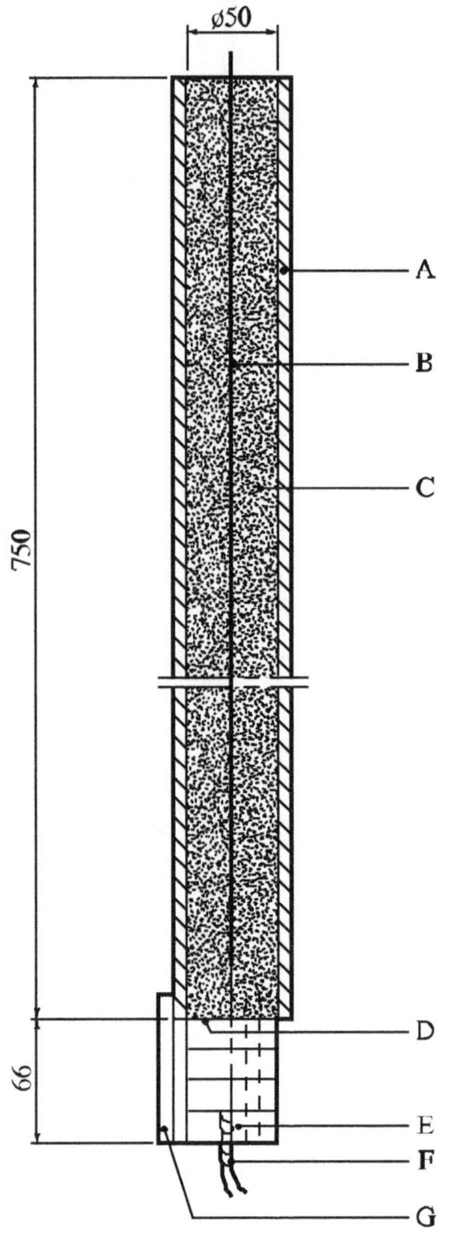

(A)	63.5 (C) or 70 mm (B) outer diameter steel tube	(B)	Velocity probe
(C)	Substance under investigation	(D)	Steel disc
(E)	4 RDX/wax cartridges	(F)	Detonator
(G)	Supports		

Figure 21.4.2.2: TNO 50/70 STEEL TUBE TEST FOR LIQUIDS (TUBE TYPES B AND C)

21.4.3 *Test A.5: UN gap test*

21.4.3.1 *Introduction*

This test is used to measure the ability of a substance to propagate a detonation by subjecting it to a detonating booster charge under confinement in a steel tube. It may be used to answer the question in box 1 of Figure 20.1.

21.4.3.2 *Apparatus and materials*

The apparatus is shown in Figure 21.4.3.1. The test sample is contained in an annealed, seamless, carbon steel tube with an external diameter of 48 ± 2 mm, a wall thickness of 4 mm and a length of 400 ± 5 mm. If the test substance may react with the steel, the inside of the tube may be coated with fluorocarbon resin. The bottom of the tube is closed with a plastics sheet pulled tightly (so that it plastically deforms) over the bottom of the tube and held tightly in place. The plastics sheet shall be compatible with the substance under test. The booster charge consists of 160 g RDX/wax (95/5) or PETN/TNT that has a minimum of 50% PETN in the mixture, 50 ± 1 mm in diameter with a density of 1 600 ± 50 kg/m^3. The charges may be pressed in one or more pieces, as long as the total charge is within the specifications, and the PETN/TNT charge is cast. A mild steel witness plate, 150 ± 10 mm square and 3 mm thick, may be mounted at the upper end of the steel tube and separated from it by spacers 1.6 ± 0.2 mm thick.

21.4.3.3 *Procedure*

21.4.3.3.1 The sample is loaded to the top of the steel tube. Solid samples are loaded to the density attained by tapping the tube until further settling becomes imperceptible. The sample mass is determined and, if solid, the apparent density calculated using the measured internal volume of the tube.

21.4.3.3.2 The tube is placed in a vertical position and the booster charge is placed in direct contact with the sheet which seals the bottom of the tube. The detonator is fixed in place against the booster charge and initiated. Two tests should be performed unless detonation of the substance is observed.

21.4.3.4 *Test criteria and method of assessing results*

21.4.3.4.1 The test results are assessed on the basis of the fragmentation pattern of the tube. The witness plate is used only to provide supplemental information on the violence of the reaction. The test giving the most severe assessment should be used for classification.

21.4.3.4.2 The test criteria are as follows:

"Yes": - The tube is fragmented over its entire length.

"Partial": - The tube is not fragmented over its entire length but the average tube fragmentation (average over the two tests) is greater than 1.5 times the average fragmentation length found with an inert material with the same physical state.

"No": - The tube is not fragmented over its entire length and the average tube fragmentation (average over the two tests) is not more than 1.5 times the average fragmentation length found with an inert material with the same physical state.

21.4.3.5 *Examples of results*

Substance	**Apparent density (kg/m3)**	**Fragmented length (cm)**	**Result**
2,2'-Azodi(isobutyronitrile)	366	40	Yes
tert-Butyl peroxybenzoate		25	Partial
tert-Butyl peroxy-2-ethylhexanoate		25	Partial
Dibenzoyl peroxide, 75% with water	685	40	Yes
2,5-Di-(tert-butylperoxy)-2,5-dimethylhexyne-3		34	Partial
Dilauroyl peroxide	564	28	No

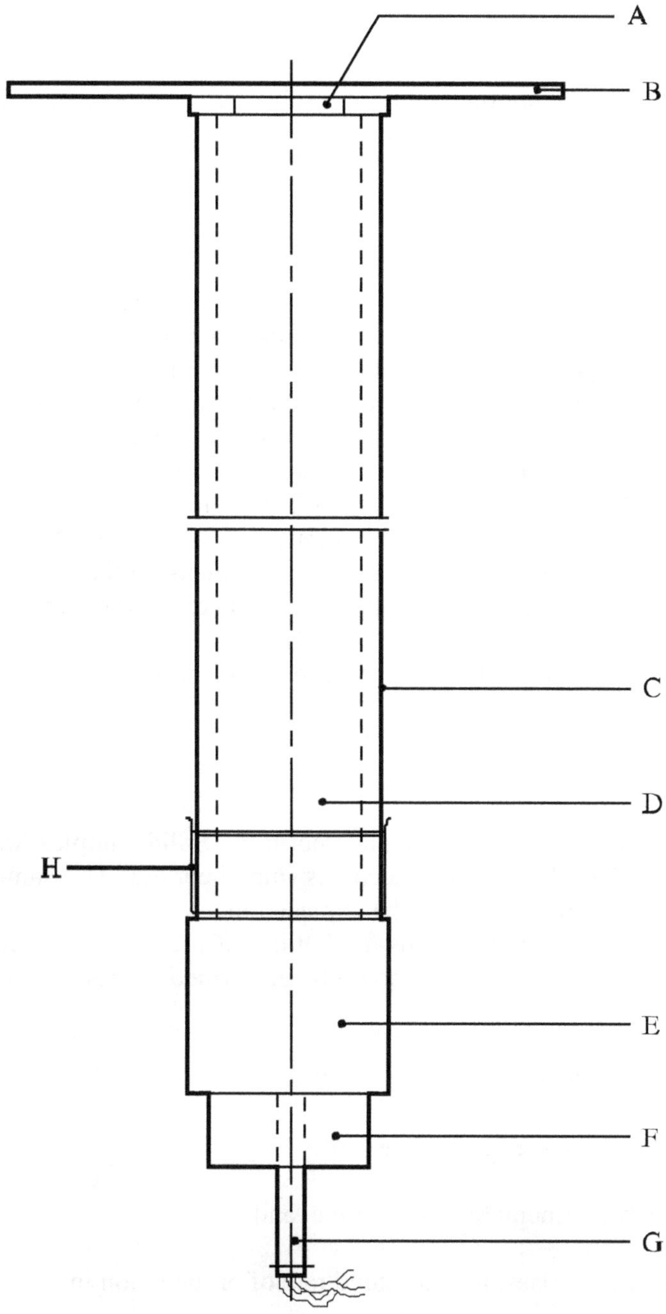

(A)	Spacers	(B)	Witness plate
(C)	Steel tube	(D)	Substance under test
(E)	RDX/wax or PETN/TNT booster charge	(F)	Detonator holder
(G)	Detonator	(H)	Plastics sheet

Figure 21.4.3.1: UN GAP TEST

21.4.4 *Test A.6: UN detonation test*

21.4.4.1 *Introduction*

This test is used to measure the ability of a substance to propagate a detonation by subjecting it to a detonating booster charge under confinement in a steel tube. It may be used to answer the question in box 1 of Figure 20.1.

21.4.4.2 *Apparatus and materials*

The apparatus is shown in Figure 21.4.4.1 and is identical for solids and liquids. The test sample is contained in an annealed, seamless, carbon steel tube with an external diameter of 60 ± 1 mm, a wall thickness of 5 ± 1 mm and a length of 500 ± 5 mm. If the test substance may react with the steel, the inside of the tube may be coated with fluorocarbon resin. The bottom of the tube is closed with a plastics sheet held tightly in place. The plastics sheet shall be compatible with the substance under test. The booster charge is a 200 g RDX/wax (95/5) or PETN/TNT that has a minimum of 50% PETN in the mixture, 60 ± 1 mm in diameter with a density of 1 600 ± 50 kg/m^3. The charges may be pressed in one or more pieces as long as the total charge is within the specifications and the PETN/TNT charge is cast. The tube may be instrumented, e.g. by a continuous wire velocity probe, to measure the velocity of propagation in the substance. Additional information on the explosive behaviour of the test sample can be gained by the use of a witness plate or of a velocity probe, as shown in Figure 21.4.4.1. The mild steel witness plate, 150 mm square and 3 mm thick, may be mounted at the upper end of the tube and separated from it by spacers 1.6 mm thick.

21.4.4.3 *Procedure*

The sample is loaded to the top of the steel tube. Solid samples are loaded to the density attained by tapping the tube until further settling becomes imperceptible. The sample mass is determined and, if solid, the apparent density calculated. The tube is placed in a vertical position and the booster charge is placed in direct contact with the sheet which seals the bottom of the tube. The detonator is fixed in place against the booster charge and initiated. Two tests should be performed unless detonation of the substance is observed.

21.4.4.4 *Test criteria and method of assessing results*

21.4.4.4.1 The test results are assessed on the basis of:

(a) The type of fragmentation of the tube; and

(b) If the occasion arises, the measured rate of propagation in the substance.

The test giving the most severe assessment should be used for classification.

21.4.4.4.2 The test criteria are as follows:

"Yes": - The tube is fragmented completely.

"Partial": - The tube is not fragmented over its entire length but the average tube fragmentation (average over the two tests) is greater than 1.5 times the average fragmentation length found with an inert material with the same physical state.

"No": - The tube is not fragmented over its entire length and the average tube fragmentation (average over the two tests) is not more than 1.5 times the average fragmentation length found with an inert material with the same physical state.

21.4.4.5 *Examples of results*

Substance	**Apparent density (kg/m^3)**	**Fragmented length (cm)**	**Result**
2,2'-Azodi(isobutyronitrile)	346	50	Yes
tert-Butyl peroxybenzoate		28	Partial
tert-Butyl peroxy-2-ethylhexanoate		23	No
Dibenzoyl peroxide, 75% with water	697	22	No
2,5-Di-(tert-butylperoxy)-2,5-dimethylhexyne-3	870	30	Partial
Dilauroyl peroxide	580	32	Partial

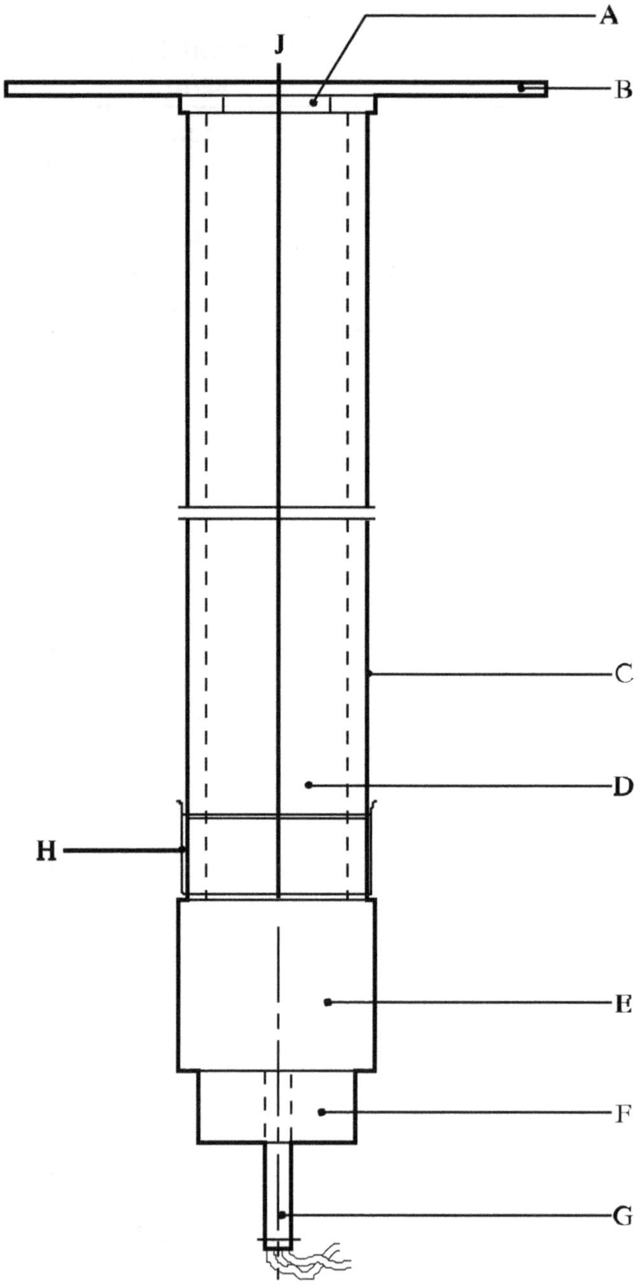

(A)	Spacers	(B)	Witness plate
(C)	Steel tube	(D)	Substance under test
(E)	RDX/wax or PETN/TNT booster charge	(F)	Detonator holder
(G)	Detonator	(H)	Plastics sheet
(J)	Velocity probe		

Figure 21.4.4.1: UN DETONATION TEST

SECTION 22

TEST SERIES B

22.1 Introduction

Test series B comprises a test and criteria concerning the propagation of detonation of a substance as packaged for transport. The test is required only for substances which propagate detonation (box 1 of Figure 20.1).

22.2 Test methods

The question "Can it detonate as packaged for transport?" (box 2 of Figure 20.1) is answered on the basis of the results of the test method in Table 22.1.

Table 22.1: TEST METHOD FOR TEST SERIES B

Test code	Name of test	Section
B.1	Detonation test in package[a]	22.4.1

[a] *Recommended test.*

22.2.2 The test is required only for substances for which the answer to the question in box 1 of Figure 20.1 is "Yes".

22.3 Test conditions

22.3.1 The test from series B should be applied to packages (not larger than 50 kg) of substance in the condition and form in which they are offered for transport.

22.3.2 ***The preliminary procedure should be performed before performing these tests (see section 20.3).***

22.4 Series B test prescription

22.4.1 *Test B.1: Detonation test in package*

22.4.1.1 *Introduction*

This test is used to measure the ability of a substance to propagate a detonation when packaged as for transport. It involves subjecting the substance in a package to the shock from a detonating booster charge. It is used to answer the question in box 2 of Figure 20.1.

22.4.1.2 *Apparatus and materials*

A detonator, detonating cord, plastic explosive and suitable confining material are required. A mild steel sheet of about 1 mm thickness, with a minimum size in each direction of 0.2 m larger than that of the bottom dimensions of the package, serves as a witness plate beneath the package.

22.4.1.3 *Procedure*

The test is applied to packed substances in the condition and form in which they are offered for transport. The package is placed on the steel witness plate from which the edges are supported by bricks or other suitable material so that there is sufficient free air space below the witness plate that puncture is not impeded. Two charges of plastic explosive (each maximum 100 g but in total not more than 1% by mass of the substance in the package) are placed on top of the substance in the package. For liquids, metal-wire support may be needed to ensure that the two explosive charges are located correctly in the middle of each of the two semicircular or triangular parts of the top surface (see Figure 22.4.1.1). Each charge is initiated by the detonator through a detonating cord. Both pieces of detonating cord should be of equal length. The preferred method of confinement is loose sand placed around the test package with a minimum thickness of 0.5 m in every direction. Alternative methods of confinement are to use boxes, bags or drums filled with earth or sand placed around and on top of the package with the same minimum thickness. The test is performed in duplicate unless a detonation is observed. An instrumented third test may be necessary if no conclusion can be drawn from two uninstrumented tests.

22.4.1.4 *Test criteria and method of assessing results*

22.4.1.4.1 The test results are assessed on the basis of evidence of a detonation of the substance under investigation by:

(a) A crater at the test site;

(b) Damage to the witness plate beneath the product;

(c) Disruption and scattering of most of the confining material; and

(d) When appropriate, the measured rate of propagation in the substance.

22.4.1.4.2 The test criteria are as follows:

"Yes": - Formation of a crater at the test site or perforation of the witness plate beneath the product; both in combination with a disruption and scattering of most of the confining material; or the propagation velocity in the lower half of the package is constant and is above the velocity of sound in the substance.

"No": - No formation of a crater at the test site, no perforation of the witness plate beneath the product, a velocity measurement (if made) shows that the rate of propagation is lower than the velocity of sound in the substance and, for solids, retrieval of unreacted substance after the test.

22.4.1.5 *Examples of results*

Substance	**Apparent density (kg/m3)**	**Packaging**	**Result**
Dibenzoyl peroxide	730	1G, 25 kg	Yes[a]
Dicyclohexyl peroxydicarbonate	600	1G, 5 kg	No[b]
Dicyclohexyl peroxydicarbonate, with 10% water	600	1G, 5 kg	No[b]

a *Performed in duplicate. Detection of detonation by formation of crater.*

b *Tests performed in duplicate. Propagation velocity measured instead of using a witness plate.*

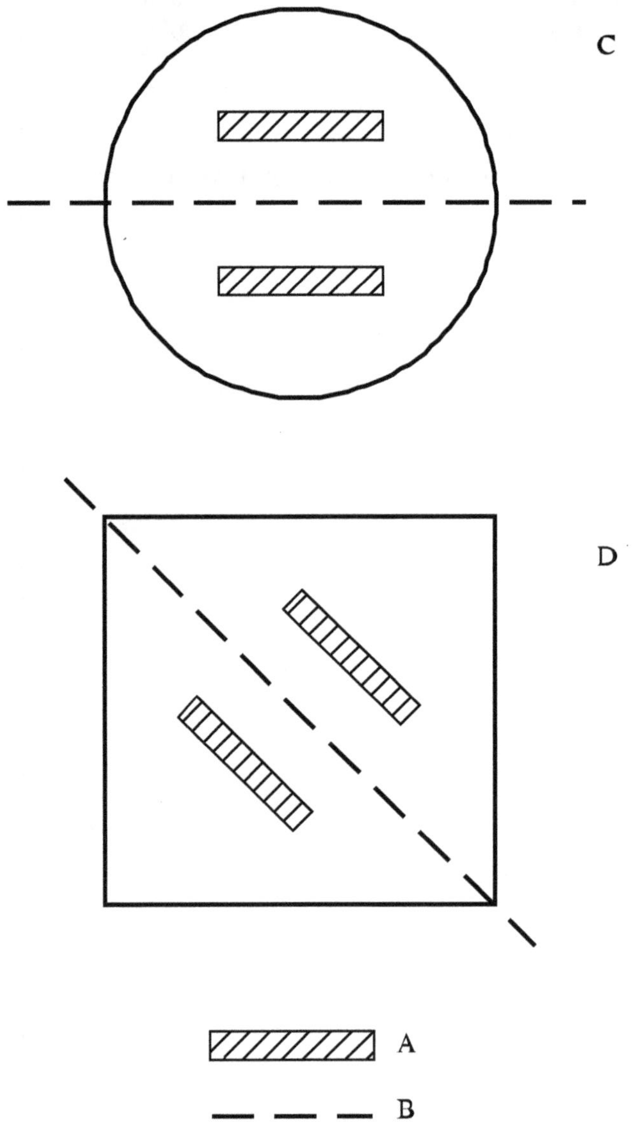

(A) Explosive charges
(B) Line of symmetry
(C) Top view of cylindrical package
(D) Top view of rectangular package

Figure 22.4.1.1: DETONATION TEST IN PACKAGE

SECTION 23

TEST SERIES C

23.1 Introduction

Test series C comprises laboratory tests and criteria concerning propagation of deflagration as requested in boxes 3, 4 and 5 of Figure 20.1.

23.2 Test methods

23.2.1 The question "Can it propagate a deflagration?" (boxes 3, 4 and 5 of Figure 20.1) is answered on the basis of the results of one, or if necessary both, of the test methods in Table 23.1.

Table 23.1: TEST METHODS FOR TEST SERIES C

Test code	Name of test	Section
C.1	Time/pressure test[a]	23.4.1
C.2	Deflagration test[a]	23.4.2

[a] *Recommended test.*

23.2.2 The answer is "Yes, rapidly" if shown to be so by either test. The answer is "Yes, slowly" if the deflagration test result is "Yes, slowly" and the time/pressure test result is not "Yes, rapidly". The answer is "No" if the deflagration test result is "No" and the time/pressure test is not "Yes, rapidly".

23.3 Test conditions

23.3.1 The preliminary procedure (see section 20.3) should be carried out before performing these tests.

23.4 Series C test prescriptions

23.4.1 *Test C.1: Time/pressure test*

23.4.1.1 *Introduction*

This test is used to measure the ability of a substance[1] under confinement to propagate a deflagration. It may be used to answer the question in boxes 3, 4 and 5 of Figure 20.1.

23.4.1.2 *Apparatus and materials*

23.4.1.2.1 The time/pressure apparatus (Figure 23.4.1.1) consists of a cylindrical steel pressure vessel 89 mm in length and 60 mm in external diameter. Two flats are machined on opposite sides (reducing the cross-section of the vessel to 50 mm) to facilitate holding whilst fitting the firing plug and vent plug. The vessel, which has a bore of 20 mm diameter, is internally rebated at either end to a depth of 19 mm and threaded to accept 1" British Standard Pipe (BSP). A pressure take-off, in the form of a side-arm, is screwed into the curved face of the pressure vessel 35 mm from one end and at 90° to the machined flats. The socket for this is bored to a depth of 12 mm and threaded to accept the 1/2" BSP thread on the end of the side-arm. A washer is fitted to ensure a gas-tight seal. The side-arm extends 59 mm beyond the pressure vessel body and has a bore of 6 mm. The end of the side-arm is rebated and threaded to accept a diaphragm type pressure transducer. Any pressure-measuring device may be used provided that it is not affected by the hot gases or decomposition products and is capable of responding to rates of pressure rise of 690 to 2 070 kPa in not more than 5 ms.

23.4.1.2.2 The end of the pressure vessel furthest from the side-arm is closed with a firing plug which is fitted with two electrodes, one insulated from and the other earthed to the plug body. The other end of the pressure vessel is closed by an aluminium bursting disc 0.2 mm thick (bursting pressure approximately 2 200 kPa) held in place with a retaining plug which has a 20 mm bore. A suitable deformable washer or rubber ring is used with both plugs to ensure a good seal. A support stand (Figure 23.4.1.2) holds the assembly in the correct attitude during use. This comprises a mild steel base plate measuring 235 mm × 184 mm × 6 mm and a 185 mm length of square hollow section (S.H.S.) 70 × 70 × 4 mm.

23.4.1.2.3 A section is cut from each of two opposite sides at one end of the length of S.H.S. so that a structure having two flat sided legs surmounted by an 86 mm length of intact box section results. The ends of these flat sides are cut at an angle of 60° to the horizontal and welded to the base plate.

23.4.1.2.4 A slot measuring 22 mm wide × 46 mm deep is machined in one side of the upper end of the base section such that when the pressure vessel assembly is lowered, firing plug end first, into the box section support, the side-arm is accommodated in this slot. A packing piece of steel 30 mm wide and 6 mm thick is welded to the lower internal face of the box section to act as a spacer. Two 7 mm thumb screws, tapped into the opposite face, serve to hold the pressure vessel firmly in place. Two 12 mm wide strips of 6 mm thick steel, welded to the side pieces abutting the base of the box section, support the pressure vessel from beneath.

23.4.1.2.5 The ignition system consists of an electric fusehead of the type commonly used in low tension detonators, together with a 13 mm square piece of primed cambric. Fuseheads with equivalent properties may be used. Primed cambric consists of a linen fabric coated on both sides with a potassium nitrate/silicon/sulphurless gunpowder pyrotechnic composition[2].

23.4.1.2.6 The procedure for the preparation of the ignition assembly for solids starts with separation of the brass foil contacts of an electric fusehead from its insulator, (see Figure 23.4.1.3). The exposed portion of insulation is then cut off. The fusehead is then fixed onto the terminals of the firing plug by means of

---

[1] *When testing liquids, variable results may be obtained because the substance may give two pressure peaks.*

[2] *Obtainable from the national contact for test details in United Kingdom (see Appendix 4).*

the brass contacts such that the tip of the fusehead is 13 mm above the surface of the firing plug. An approximately 13 mm square piece of primed cambric is pierced through the centre and positioned over the attached fusehead around which it is then folded and secured with fine cotton thread.

23.4.1.2.7 For liquids samples, a single piece of thin PVC sheathing, or equivalent, is used to cover the primed cambric in such a way that the primed cambric is not in contact with the liquid sample. The leads of the resistance wire are then fixed onto the terminals of the firing plug such that the tip of the primed cambric is above the surface of the firing plug.

23.4.1.3 *Procedure*

23.4.1.3.1 The apparatus assembled, complete with pressure transducer but without the aluminium bursting disc in position, is supported firing plug end down. 5.0 g[3] of the substance is introduced into the apparatus so as to be in contact with the ignition system. Normally no tamping is carried out when filling the apparatus unless it is necessary to use light tamping in order to get the 5.0 g charge into the vessel. If, even with light tamping, it is impossible to get all the 5.0 g of sample in, then the charge is fired after filling the vessel to capacity. Note should be taken of the charge weight used. The washer or rubber ring and aluminium bursting disc are placed in position and the retaining plug is screwed in tightly. The charged vessel is transferred to the firing support stand, bursting disc uppermost, which should be contained in a suitable, armoured fume cupboard or firing cell. A power source is connected to the external terminals of the firing plug and the charge is fired. The signal produced by the pressure transducer is recorded on a suitable data acquisition system which allows both evaluation and a permanent record of the time/pressure profile to be achieved.

23.4.1.3.2 The test is carried out three times. The time taken for the pressure to rise from 690 kPa to 2 070 kPa above atmospheric is noted. The shortest time interval should be used for classification.

23.4.1.4 *Test criteria and method of assessing results*

23.4.1.4.1 The test results are interpreted in terms of whether a gauge pressure of 2 070 kPa is reached and, if so, the time taken for the pressure to rise from 690 kPa to 2 070 kPa gauge.

23.4.1.4.2 The test criteria are as follows:

"Yes, rapidly": - The time for a pressure rise from 690 kPa to 2 070 kPa is less than 30 ms.

"Yes, slowly": - The time for a pressure rise from 690 kPa to 2 070 kPa is greater than or equal to 30 ms.

"No": - A pressure rise to 2 070 kPa above atmospheric is not achieved.

NOTE: *If necessary, the deflagration test, Test C.2, should be performed to distinguish between "Yes, slowly" and "No".*

[3] ***If preliminary safety-in-handling tests (e.g. heating in a flame) or burning tests (e.g. a series 3 type (d) test) indicate that a rapid reaction is likely to occur, then the sample size should be reduced to 0.5 g until the severity of the confined reaction is known.*** *If it is necessary to use a 0.5 g sample size, the sample size is gradually increased until either a "Yes, rapidly" result is obtained or the test is performed with a 5.0 g sample.*

23.4.1.5 *Examples of results*

Substance	**Maximum pressure (kPa)**	**Time for a pressure rise from 690 to 2 070 kPa (ms)**	**Result**
Azodicarbonamide	> 2 070	63	Yes, slowly
Azodicarbonamide, 67% with zinc oxide	> 2 070	21	Yes, rapidly
2,2'-Azodi(isobutyronitrile)	> 2 070	68	Yes, slowly
2,2'-Azodi(2-methylbutyronitrile)	> 2 070	384	Yes, slowly
tert-Butyl hydroperoxide, 70% with water	1 380	-	No
tert-Butyl peroxybenzoate	> 2 070	2 500	Yes, slowly
tert-Butyl peroxy-2-ethylhexanoate	> 2 070	4 000	Yes, slowly
Cumyl hydroperoxide, 80% with cumene	< 690[a]	-	No
2-Diazo-1-naphthol-5-sulphohydrazide	> 2 070	14	Yes, rapidly
Dibenzoyl peroxide	> 2 070	1	Yes, rapidly
Di-tert-butyl peroxide	> 2 070	100	Yes, slowly
Dicetyl peroxydicarbonate	< 690	-	No
Dicumyl peroxide	< 690[a]	-	No
Dicumyl peroxide, with 60% inert solid	< 690[a]	-	No
2,5-Diethoxy-4-morpholinobenzene-diazonium tetrafluoroborate, 97%	> 2 070	308	Yes, slowly
Dilauroyl peroxide	990	-	No
2,5-Dimethyl-2,5-di-(tert-butylperoxy)-hexyne-3	> 2 070	70	Yes, slowly
Magnesium monoperoxyphthalate hexahydrate, 85% with magnesium phthalate	900	-	No
4-Nitrosophenol	> 2 070	498	Yes, slowly

[a] *No ignition.*

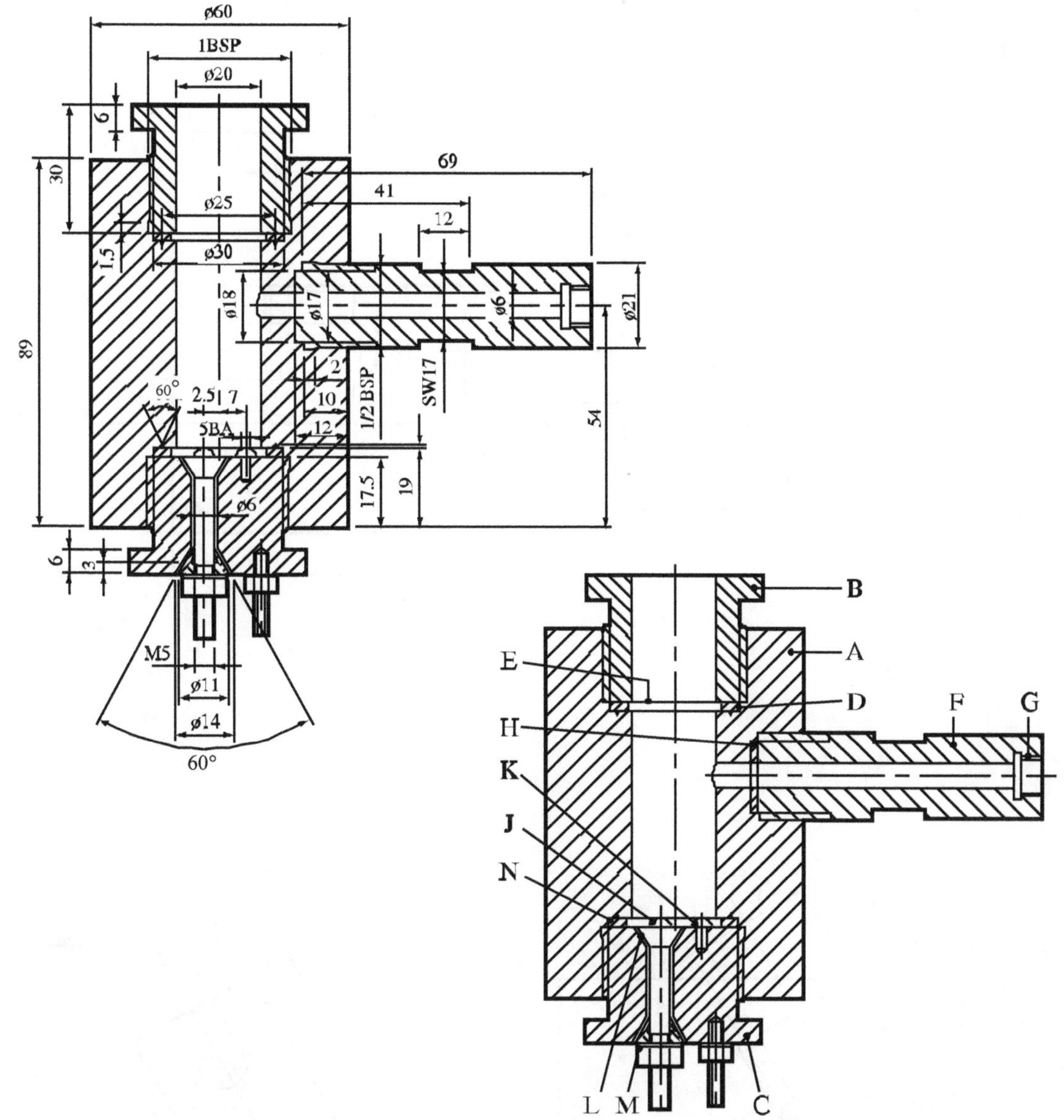

(A) Pressure vessel body
(B) Bursting disc retaining plug
(C) Firing plug
(D) Deformable washer
(E) Bursting disc
(F) Side arm
(G) Pressure transducer thread
(H) Washer
(J) Insulated Electrode
(K) Earthed electrode
(L) Insulation
(M) Steel cone
(N) Washer distorting groove

Figure 23.4.1.1: APPARATUS

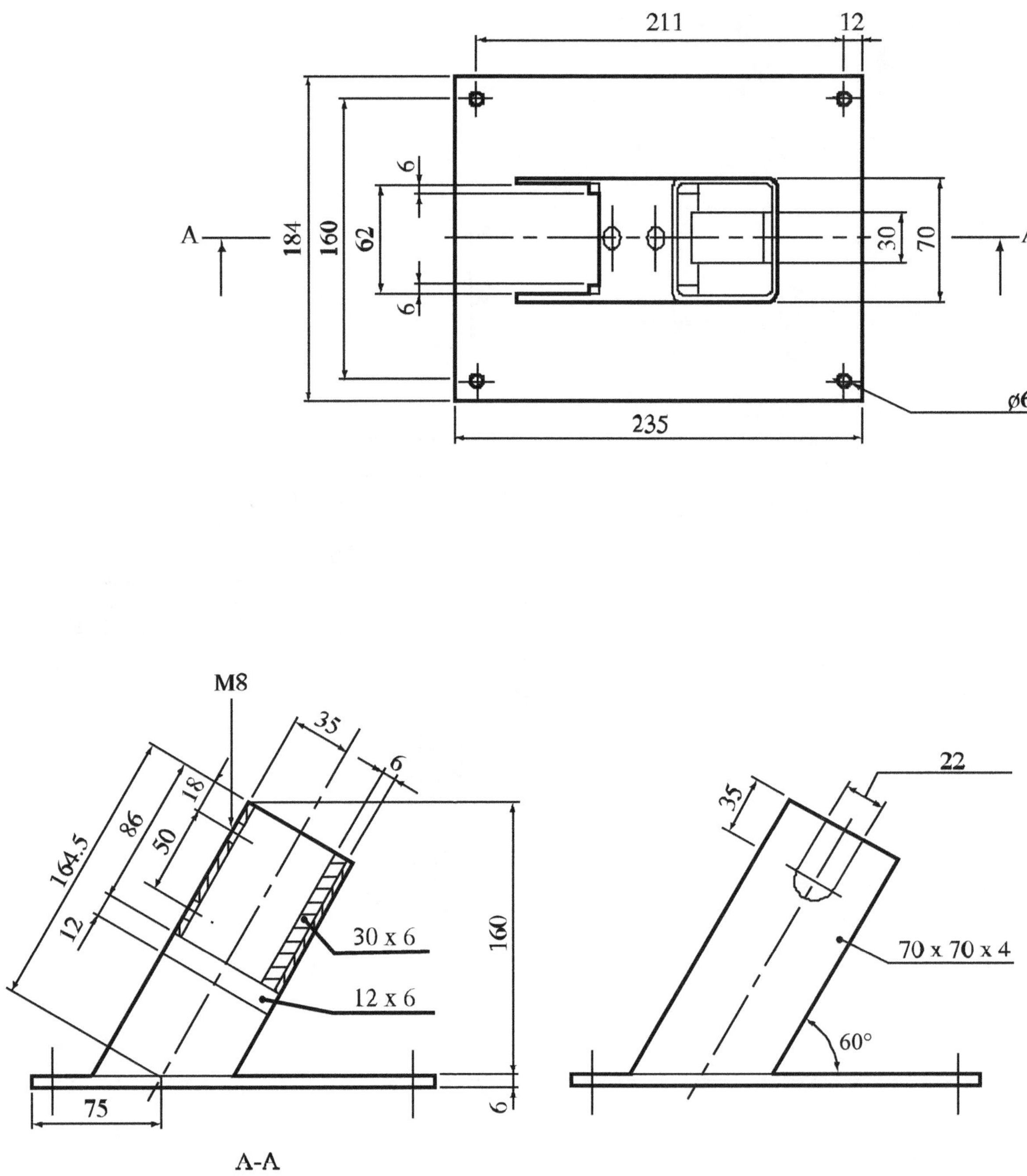

Figure 23.4.1.2: SUPPORT STAND

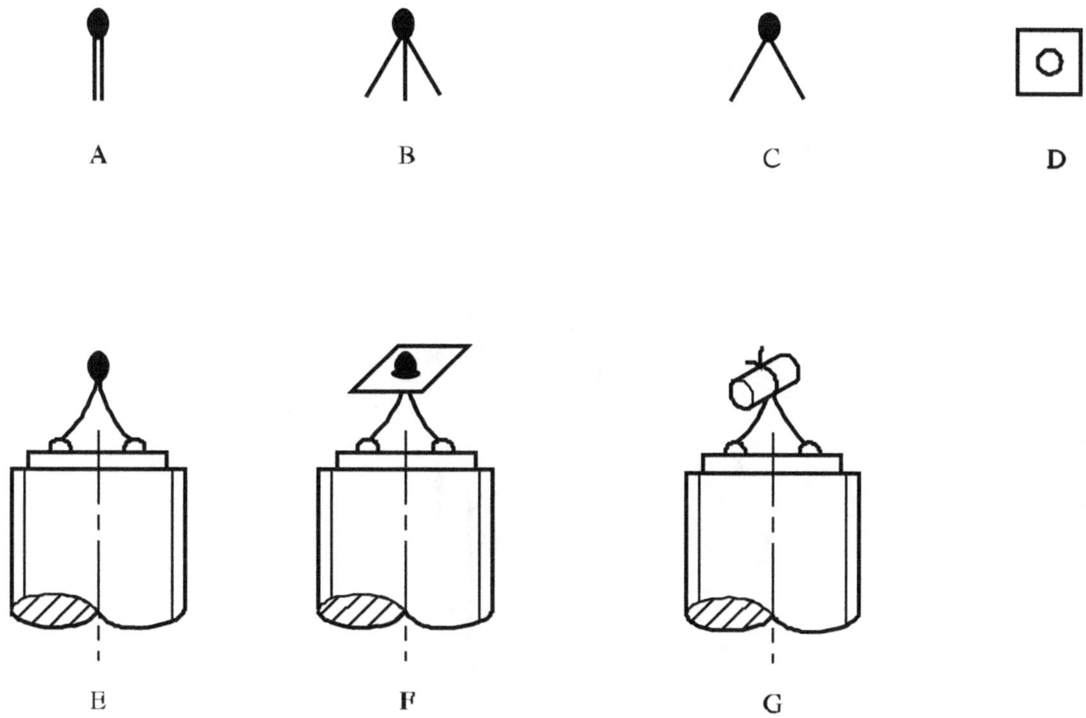

(A)	Electrically ignited fusehead as manufactured
(B)	Brass foil contacts parted from card insulator
(C)	Insulating card cut off
(D)	Primed cambric SR252 13 mm square with centre hole
(E)	Fusehead fixed to pins on firing plug
(F)	Cambric positioned on fusehead
(G)	Cambric folded around and tied with thread

Figure 23.4.1.3: IGNITION SYSTEM FOR SOLIDS

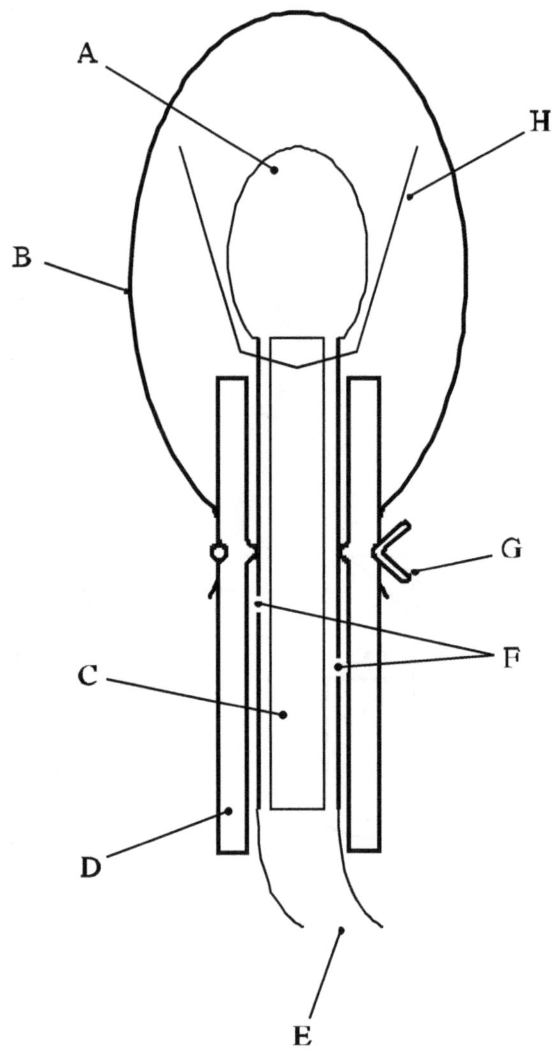

(A)	Fusehead
(B)	PVC sheath
(C)	Insulating card
(D)	Silicone rubber tubing
(E)	Firing leads
(F)	Foil contacts
(G)	Wire to make liquid-tight seal
(H)	Primed cambric

Figure 23.4.1.4: IGNITION SYSTEM FOR LIQUIDS

23.4.2 *Test C.2: Deflagration test*

23.4.2.1 *Introduction*

This test is used to measure the ability of a substance to propagate a deflagration. It may be used to answer the question in boxes 3, 4 and 5 of Figure 20.1.

23.4.2.2 *Apparatus and materials*

23.4.2.2.1 The test is performed with a Dewar vessel (see Figure 23.4.2.1) which is provided with vertical observation windows on opposite sides. The windows are not necessary when using thermocouples to measure the deflagration rate. A timer with an accuracy of 1 second is used to measure the deflagration rate.

23.4.2.2.2 The Dewar vessel has a volume of about 300 cm^3, an internal diameter of 48 ± 1 mm, an external diameter of 60 mm and a length between 180 and 200 mm. The half-time of cooling with water or other suitable material filled to a height of 20 mm below the rim (i.e. 265 cm^3) of Dewar vessel, closed by a tight fitting cork, should be longer than 5 hours. Horizontal graduation marks are drawn at 50 and 100 mm from the top of the Dewar vessel. The time it takes for the decomposition front to propagate from the 50 mm mark to the 100 mm mark yields the deflagration rate. A glass thermometer with an accuracy of 0.1 °C is used to measure the temperature of the test substance prior to ignition. Alternatively, the deflagration rate and sample temperature may be determined by using two thermocouples at distances of 50 mm and 100 mm from the top of the Dewar vessel.

23.4.2.2.3 Any gas flame with a flame length of at least 20 mm can be used for igniting the substance.

23.4.2.2.4 For personal protection, the test is performed in an explosion-proof fume-chamber or in a well-ventilated test-cell. The capacity of the exhaust fan shall be large enough to dilute the decomposition products to the extent that no explosive mixtures with air can be obtained. A shield is placed between the observer and the Dewar vessel.

23.4.2.3 *Procedure*

23.4.2.3.1 ***If preliminary safety-in-handling tests (e.g. heating in a flame) or a small scale burning test (e.g. the series 3 type (d) test) indicate that a rapid reaction is likely to occur, exploratory tests in borosilicate glass tubes should be performed, with suitable safety precautions, prior to the Dewar vessel test.*** In this case, it is recommended that the test be performed in a 14 mm diameter tube first and subsequently in a 28 mm diameter tube. If the deflagration rate in any of these exploratory tests exceeds 5 mm/s, the substance can forthwith be classified as a rapidly deflagrating substance and the main test, using a Dewar vessel, can be omitted.

23.4.2.3.2 The Dewar vessel and the substance are brought to the emergency temperature as defined in the Model Regulations. If the substance is sufficiently stable as to require no emergency temperature, a test temperature of 50 °C is used. The Dewar vessel is filled to a height of 20 mm below the rim with the substance. Granular substances are filled into the Dewar vessel in such a way that the bulk density of the substance will be comparable with that in transport and there are no lumps.

23.4.2.3.3 Pasty materials are introduced into the Dewar vessel in such a way that no air pockets will be present in the sample to be tested. The height of filling shall be about 20 mm below the rim of the Dewar vessel. The mass and the temperature of the substance are recorded. The Dewar vessel is placed in the test-cell or fume-chamber behind a shield, after which the substance is heated at the top by means of a gas burner. At the moment when ignition is observed or, alternatively, if no ignition occurs within five minutes, the gas burner is removed and extinguished. The period of time that is required for the reaction zone to pass the distance between the two marks is measured with the timer. If the reaction stops before reaching the lower mark, the substance is considered to be non-deflagrative. The test is performed in duplicate and the shortest time interval is used for the calculation of the deflagration rate. Alternatively, the rate may be determined by locating thermocouples down the centre of the Dewar at distances 50 mm and 100 mm from the top of the Dewar vessel. The thermocouple outputs are monitored continuously. The passage of the reaction front causes a steep increase in output. The time between the increases in output is determined.

23.4.2.4 *Test criteria and method of assessing results*

23.4.2.4.1 The test results are interpreted in terms of whether a reaction zone will propagate downwards through the substance and, if so, the rate of propagation. The participation of oxygen from the air in the reaction at the sample surface is negligible after the reaction zone has propagated over a distance of 30 mm. The reaction zone will extinguish if the substance does not deflagrate under the test conditions. The propagation velocity of the reaction zone (deflagration rate) is a measure of the susceptibility of the substance to deflagration under atmospheric pressure.

23.4.2.4.2 The test criteria are as follows:

"Yes, rapidly": - the deflagration rate is greater than 5.0 mm/s.

"Yes, slowly": - the deflagration rate is less than or equal to 5.0 mm/s and greater than or equal to 0.35 mm/s.

"No": - the deflagration rate is less than 0.35 mm/s or the reaction stops before reaching the lower mark.

NOTE: *The time/pressure test, test C.1, is carried out if "Yes, rapidly" is not obtained.*

23.4.2.5 *Examples of results*

Substance	Sample mass (g)	Temperature (°C)	Propagation rate (mm/s)	Result
Azodicarbonamide	174	50	0.35	Yes, slowly
2,2'-Azodi(isobutyronitrile)	101	45	[a]	No
tert-Butyl peroxybenzoate	276	50	0.65	Yes, slowly
tert-Butyl peroxy-2-ethylhexanoate	237	25	0.74	Yes, slowly
tert-Butyl peroxy-3,5,5-trimethylhexanoate, 75% in solvent	238	50	0.27	No
Cumyl hydroperoxide, 80% with cumene	273	50	0.12	No
Dibenzoyl peroxide	158	20	100[b]	Yes, rapidly
Di-tert-butyl peroxide	212	50	0.27	No
Di-(4-tert-butylcyclohexyl) peroxydicarbonate	123	35	4.3	Yes, slowly
Dicetyl peroxydicarbonate	159	35	No ignition	No
Dicumyl peroxide	292	50	No ignition	No
Dicyclohexyl peroxydicarbonate	-	26	26	Yes, rapidly
Dicyclohexyl peroxydicarbonate, 90% with water	-	15	13	Yes, rapidly
Dilauroyl peroxide	130	45	No ignition	No
Dilauroyl peroxide, 42% stable dispersion in water	265	45	No ignition	No
2,5-Dimethyl-2,5-di-(tert-butylperoxy)-hexyne-3	235	50	2.9	Yes, slowly
2,5-Dimethyl-2,5-di-(benzoylperoxy)-hexane	231	50	6.9	Yes, rapidly
4-Nitrosophenol	130	35	0.90	Yes, slowly

[a] *Pulsating flame followed by extinguishing of flame; no stable propagation under test conditions.*

[b] *Performed with an exploratory test using a glass tube with a diameter of 14 mm at 20 °C instead of 50 °C.*

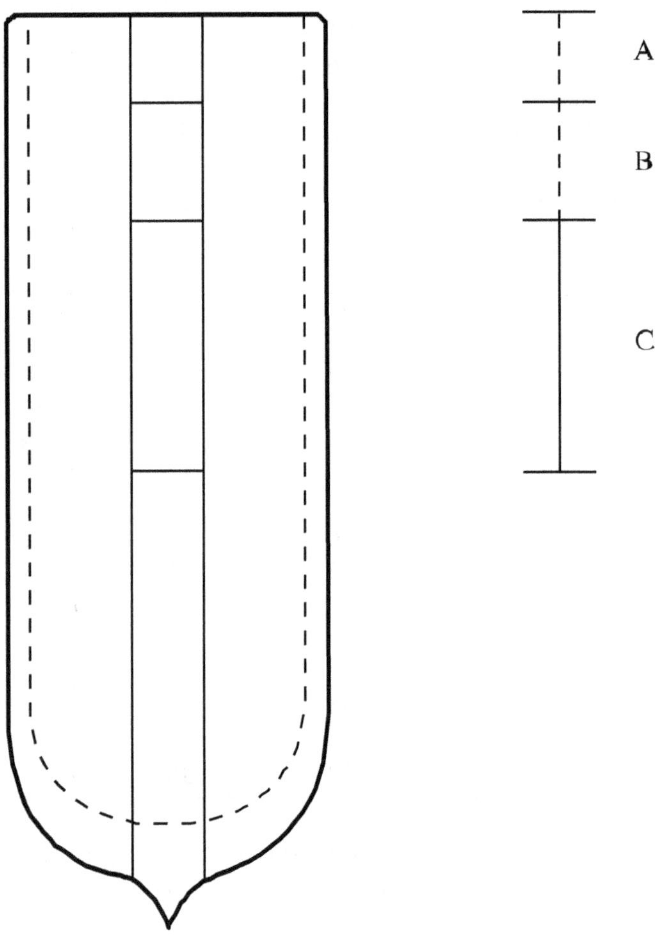

(A)	Height of filling 20 mm below rim
(B)	30 mm region for deflagration to be established
(C)	50 mm region for measuring the deflagration rate

Figure 23.4.2.1: DEWAR VESSEL WITH OBSERVATION WINDOWS

SECTION 24

TEST SERIES D

24.1 Introduction

Test series D comprises a test and criteria concerning the propagation of a rapid deflagration of a substance as packaged for transport. The test is required for substances which deflagrate rapidly in a series C test.

24.2 Test method

24.2.1 The question "Does it deflagrate rapidly in package?" (box 6 of Figure 20.1) is answered on the basis of the results of the test method in Table 24.1.

Table 24.1: TEST METHOD FOR TEST SERIES D

Test code	Name of test	Section
D.1	Deflagration test in package[a]	24.4.1

[a] *Recommended test.*

24.2.2 The test is only required for substances for which the answer from the series C test is "Yes, rapidly".

24.3 Test conditions

24.3.1 The test of series D should be applied to packages (not larger than 50 kg) of substance in the condition and form in which they are offered for transport.

24.3.2 ***The preliminary procedure (see section 20.3) should be carried out before performing these tests.***

24.4 Series D test prescription

24.4.1 *Test D.1: Deflagration test in the package*

24.4.1.1 *Introduction*

This test is used to measure the ability of a substance to rapidly propagate a deflagration when packaged as for transport. It may be used to answer the question in box 6 of Figure 20.1.

24.4.1.2 *Apparatus and materials*

An igniter just sufficient to ensure ignition of the substance (e.g. a squib with maximum 2 g of a slow burning pyrotechnic composition wrapped in thin plastics foil) and suitable confining material are required.

24.4.1.3 *Procedure*

The test is applied to packaged substances in the condition and form in which they are offered for transport. The package is placed on the ground and the igniter is placed in the centre of the substance. For liquids, a support of metal wires may be needed to fix the igniter in the desired place. The igniter should be protected against the liquid. The test is performed under confinement. The preferred method of confinement is loose sand placed around the test package with a minimum of 0.5 m in every direction. Alternative methods of confinement are to use boxes, bags or drums filled with earth or sand placed around and on top of the package with the same minimum thickness. The test is performed in triplicate, unless an explosion is observed. ***If no deflagration is observed after ignition, the package should not be approached for at least 30 minutes. It is recommended that a thermocouple be inserted near the igniter to allow monitoring of its operation and to check when it is safe to approach the package.***

24.4.1.4 *Test criteria and method of assessing results*

24.4.1.4.1 The test results are assessed on the basis of the evidence of a rapid deflagration of the substance under investigation by:

(a) Fragmentation of the packaging; and

(b) Disruption and scattering of most of the confining material.

24.4.1.4.2 The test criteria are as follows:

"Yes" : - Fragmentation of inner or outer packaging into more than three pieces (excluding the bottom and top parts of the packaging) shows that the substance under investigation deflagrated rapidly in that package.

"No" : - No fragmentation or a fragmentation of the inner or outer packaging into not more than three pieces (excluding the bottom and top parts of the packaging) shows that the substance under investigation did not deflagrate rapidly in that package.

24.4.1.5 *Examples of results*

Substance	**Packaging**	**Number of fragments**	**Result**
Dibenzoyl peroxide	1A2, 25 kg	>40	Yes
Dibenzoyl peroxide	4G, 25 kg	>40	Yes
Dibenzoyl peroxide, 94% with water	1A2, 25 kg	>40	Yes
Dibenzoyl peroxide, 75% with water	4G, 25 kg	No fragmentation	No

SECTION 25

TEST SERIES E

25.1 Introduction

25.1.1 Test series E comprises laboratory tests and criteria concerning the determination of the effect of heating under defined confinement as requested in boxes 7, 8, 9 and 13 of Figure 20.1.

25.2 Test methods

25.2.1 The question "What is the effect of heating it under defined confinement?" (boxes 7, 8, 9 and 13 of Figure 20.1) is answered on the basis of combinations of the results from the test methods in Table 25.1.

Table 25.1: TEST METHODS FOR TEST SERIES E

Test code	Name of test	Section
E.1	Koenen test[a]	25.4.1
E.2	Dutch pressure vessel test[b]	25.4.2
E.3	USA pressure vessel test	25.4.3

[a] *Recommended test for self-reactive substances in combination with one of the other tests.*
[b] *Recommended test for organic peroxides in combination with one of the other tests.*

25.2.2 The combination of the Koenen test and either the Dutch pressure vessel or USA pressure vessel test should be used for self-reactive substances. The combination of the Dutch pressure vessel test and either the Koenen test or the USA pressure vessel test should be used for organic peroxides. For classification, the highest hazard rating should be applied.

25.3 Test conditions

25.3.1 ***The preliminary procedure (see section 20.3) should be carried out before performing these tests.***

25.4 Series E test prescriptions

25.4.1 *Test E.1: Koenen test*

25.4.1.1 *Introduction*

This test is used to determine the sensitiveness of substances to the effect of intense heat under high confinement. It may be used, in conjunction with an additional heating under confinement test, to answer the question in boxes 7, 8, 9 and 13 of Figure 20.1.

25.4.1.2 *Apparatus and materials*

25.4.1.2.1 The apparatus consists of a non-reusable steel tube, with its re-usable closing device, installed in a heating and protective device. The tube is deep drawn from sheet steel conforming to specification DC04 (EN 10027-1), or equivalent A620 (AISI/SAE/ASTM), or equivalent SPCEN (JIS g 3141). The dimensions are given in Figure 25.4.1.1. The open end of the tube is flanged. The closing plate with an orifice, through which the gases from the decomposition of the test substance escape, is made from heat-resisting chrome steel. For classification the following diameter holes shall be used: 1.0 - 1.5 - 2.0 - 2.5 - 3.0 - 5.0 - 8.0 - 12.0 - 20.0 mm. In addition, other diameters can be used for hazard assessment. The dimensions of the threaded collar and the nut (closing device) are given in Figure 25.4.1.1.

For quality control of the steel tubes, 1% of the tubes from each production lot shall be subjected to quality control and the following data shall be verified:

(a) The mass of the tubes shall be 26.5 ± 1.5 g, tubes to be used in one test sequence shall not differ in mass by more than 1 g;

(b) The length of the tubes shall be 75 ± 0.5 mm;

(c) The wall thickness of the tubes measured 20 mm from the bottom of the tube shall be 0.5 ± 0.05 mm; and

(d) The bursting pressure as determined by quasi-static load through an incompressible fluid shall be 30 ± 3 MPa.

25.4.1.2.2 Heating is provided by propane, from an industrial cylinder fitted with pressure regulator, via a flow meter and distributed by a manifold to the four burners. Other fuel gases may be used provided the specified heating rate is obtained. The gas pressure is regulated to give a heating rate of 3.3 ± 0.3 K/s when measured by the calibration procedure. Calibration involves heating a tube (fitted with a 1.5 mm orifice plate) filled with 27 cm^3 of dibutyl phthalate or equivalent. The time taken for the temperature of the liquid (measured with a 1 mm diameter thermocouple centrally placed 43 mm below the rim of the tube) to rise from 135 °C to 285 °C is recorded and the heating rate calculated.

25.4.1.2.3 Because the tube is likely to be destroyed in the test, heating is undertaken in a protective welded box, the construction and dimensions of which are given in Figure 25.4.1.2. The tube is suspended between two rods placed through holes drilled in opposite walls of the box. The arrangement of the burners is given in Figure 25.4.1.2. The burners are lit simultaneously by a pilot flame or an electrical ignition device. ***The test apparatus is placed in a protected area.*** Measures should be taken to ensure that the burner flames are not affected by any draughts. Provision should be made for extracting any gases or smoke resulting from the test.

25.4.1.3 *Procedure*

25.4.1.3.1 The tube is filled to a height of 60 mm from the bottom of the tube. Cast solids should be cast to the internal dimensions of the steel tube with a height of 60 mm and then placed inside the tube. Powders are filled in approximately three equal increments with tamping[1] to 80 N force between each increment. Liquids and gels are loaded into the tube to a height of 60 mm taking particular care with gels to prevent the formation of voids. Determine the total mass used to fill the tube to this level and use this amount of solid for each trial filling being performed. The threaded collar is slipped onto the tube from below, the appropriate orifice plate is inserted and the nut tightened by hand after applying some molybdenum disulphide based lubricant. It is essential to check that none of the substance is trapped between the flange and the plate, or in the threads.

25.4.1.3.2 With orifice plates from 1.0 mm to 8.0 mm diameter, nuts with an orifice of 10.0 mm diameter should be used; if the diameter of the orifice is above 8.0 mm, that of the nut should be 20.0 mm. Each tube is used for one trial only. The orifice plates, threaded collars and nuts may be used again provided they are undamaged.

25.4.1.3.3 The tube is placed in a rigidly mounted vice and the nut tightened with a spanner. The tube is then suspended between the two rods in the protective box. The test area is vacated, the gas supply turned on and the burners lit. The time to reaction and duration of reaction can provide additional information useful in interpreting the results. If rupture of the tube does not occur, heating is continued for at least five minutes before the trial is finished. After each trial the fragments of the tube, if any, should be collected and weighed.

25.4.1.3.4 The following effects are differentiated:

"O": Tube unchanged;
"A": Bottom of tube bulged out;
"B": Bottom and wall of the tube bulged out;
"C": Bottom of tube split;
"D": Wall of tube split;
"E": Tube split into two[2] fragments;
"F": Tube fragmented into three[2] or more mainly large pieces which in some cases may be connected with each other by a narrow strip;
"G": Tube fragmented into many mainly small pieces, closing device undamaged; and
"H": Tube fragmented into many very small pieces, closing device bulged out or fragmented.

Examples for the effect types "D", "E" and "F" are shown in Figure 25.4.1.3. If a trial results in any of the effects "O" to "E", the result is regarded as "no explosion". If a trial gives the effect "F", "G" or "H", the result is evaluated as "explosion".

25.4.1.3.5 The series of trials is started with a single trial using an orifice plate with a certain diameter mm. If, in this trial, the result "explosion" is observed, the series is continued with single trials at increasing diameters until only negative results in three tests are obtained at the same level. If in the first trial "no explosion" occurs, the series is continued with single trials using plates with the decreasing diameters and finally 1.0 mm until, at one of these diameters, the result "explosion" is obtained. Subsequently, trials are carried out at increasing diameters, until only negative results in three tests are obtained at the same diameter. The limiting diameter of a substance is the largest diameter of the orifice at which the result "explosion" is obtained. If no "explosion" is obtained with a diameter of 1.0 mm, the limiting diameter is recorded as being less than 1.0 mm.

---

1 ***For reasons of safety, e.g. the substance is friction sensitive, the substance need not be tamped.*** *In cases where the physical form of the sample can be changed by compression or compression of the sample is not related to the transport conditions, e.g. for fibrous materials, more representative filling procedures may be used.*

2 *The upper part of the tube remained in the closing device is counted as one fragment.*

25.4.1.4 *Test criteria and method of assessing results*

25.4.1.4.1 The test criteria are as follows:

"Violent": - The limiting diameter is greater than or equal to 2.0 mm.

"Medium": - The limiting diameter is equal to 1.5 mm.

"Low": - The limiting diameter is equal to or less than 1.0 mm and the effect in any test is different from type "O".

"No": - The limiting diameter is less than 1.0 mm and the effect in all tests is of type "O".

25.4.1.5 *Examples of results*

Substance	**Sample mass (g)**	**Limiting diameter (mm)**	**Type of fragmentation** [a]	**Result**
Azodicarbonamide	20.0	1.5	"F"	Medium
Azodicarbonamide, 67% with zinc oxide	24.0	1.5	"F"	Medium
2,2'-Azodi(2,4-dimethylvaleronitrile)	17.5	< 1.0	"O"	No
2,2'-Azodi(isobutyronitrile)	15.0	3.0	"F"	Violent
Benzene-1,3-disulphohydrazide		12.0	"F"	Violent
Benzene-1,3-disulphohydrazide, 70% with mineral oil		2.0	"F"	Violent
Benzene sulphohydrazide	18.5	1.0	"F"	Low
tert-Butyl peroxybenzoate	26.0	3.5	"F"	Violent
tert-Butyl peroxy-2-ethylhexanoate	24.2	2.0	"F"	Violent
Cumyl hydroperoxide, 84.1% in cumene	27.5	1.0	"F"	Low
2-Diazo-1-naphthol-5-sulphochloride	19.0	2.5	"F"	Violent
Dibenzoyl peroxide	17.5	10.0	"F"	Violent
Dibenzoyl peroxide, 75% with water	20.0	2.5	"F"	Violent
Di-tert-butyl peroxide,	21.5	<1.0	"O"	No
Dicetyl peroxydicarbonate	16.0	<1.0	"O"	No
2,4-Dichlorobenzoyl peroxide	21.0	6.0[b]	"F"	Violent
Dicumyl peroxide	18.0	<1.0	"O"	No
Diisopropyl peroxydicarbonate	21.0	8.0	"F"	Violent
Dilauroyl peroxide	14.0	<1.0	"O"	No
2,5-Dimethyl-2,5-di(tert-butylperoxy)-hexane	23.0	1.5	"F"	Medium
Dimyristyl peroxydicarbonate	16.0	<1.0	"O"	No
N,N'-Dinitroso-N,N'-dimethyl-terephthalamide 70%, with mineral oil	18.0	4.0	“F”	Violent
Diperoxy isophthalic acid	18.0	24.0	“H”	Violent
Disuccinic acid peroxide	18.0	6.0	“F”	Violent
4-Nitrosophenol	17.0	< 1.0	“A”	Low

[a] *At the limiting diameter.*

[b] *With a sample mass of 13 g the limiting diameter is < 1.0 mm*

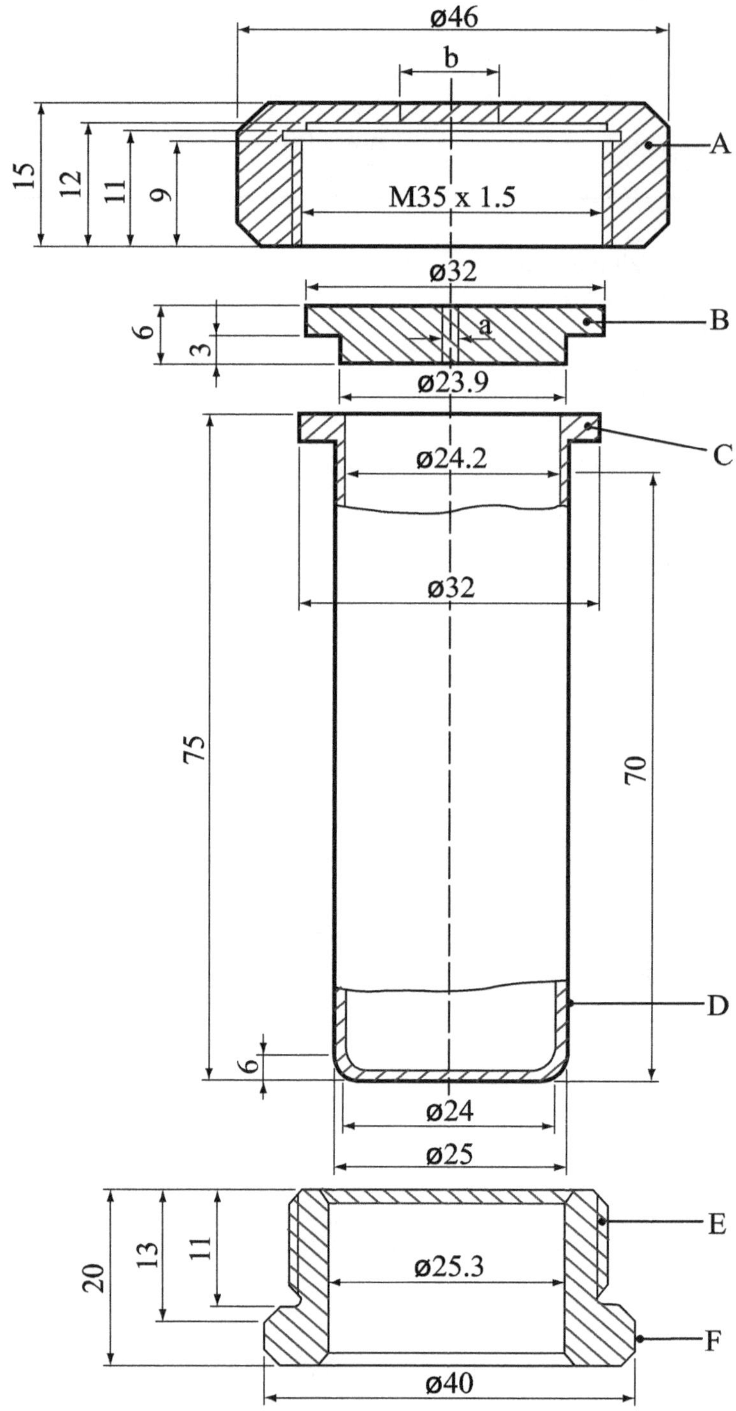

(A) Nut (b = 10.0 or 20.0 mm) with flats for size 41 spanner
(B) Orifice plate (a = 1.0 to 20.0 mm diameter)
(C) Flange
(D) Tube
(E) Threaded collar
(F) Flats for size 36 spanner

Figure 25.4.1.1: TEST TUBE ASSEMBLY

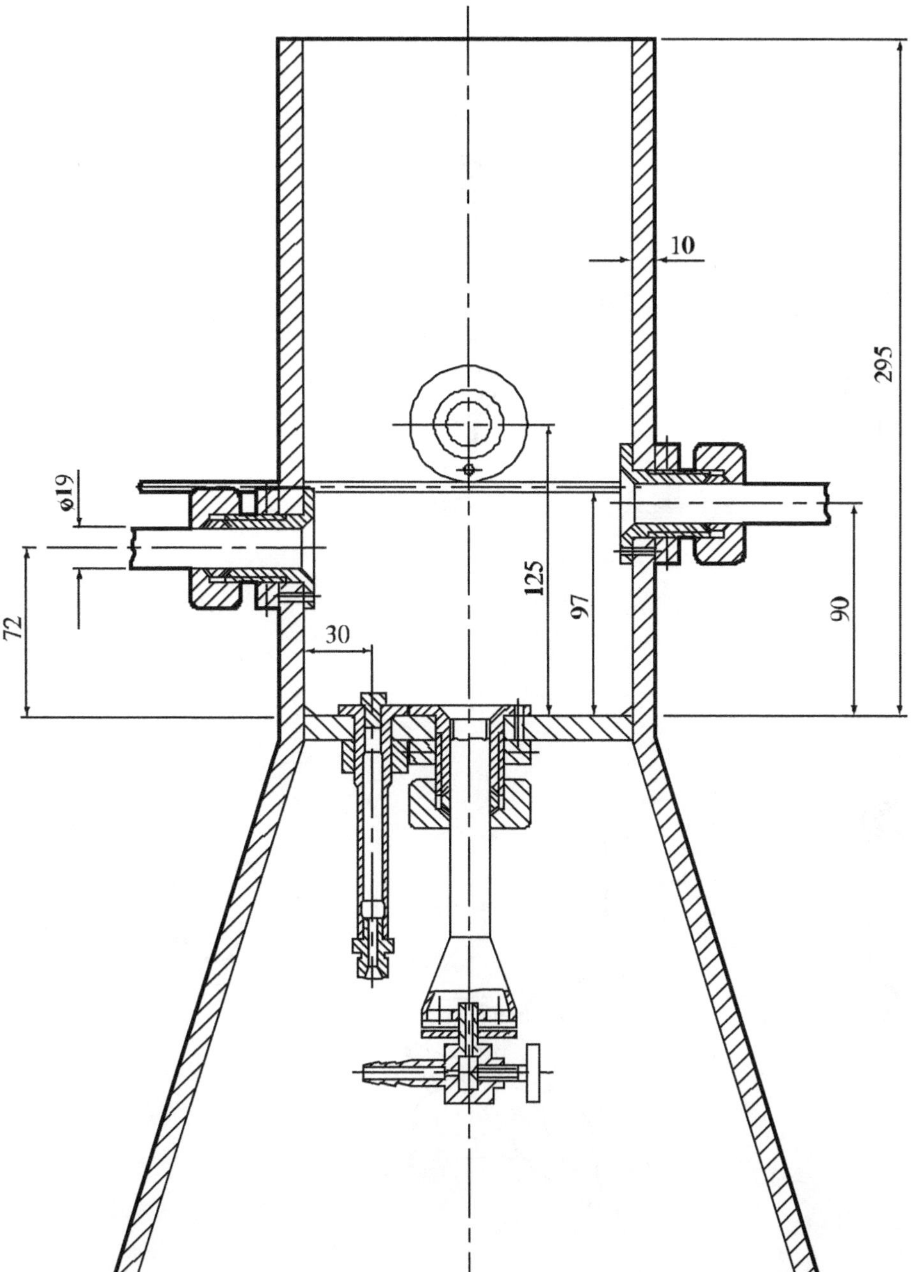

Figure 25.4.1.2: HEATING AND PROTECTIVE DEVICE

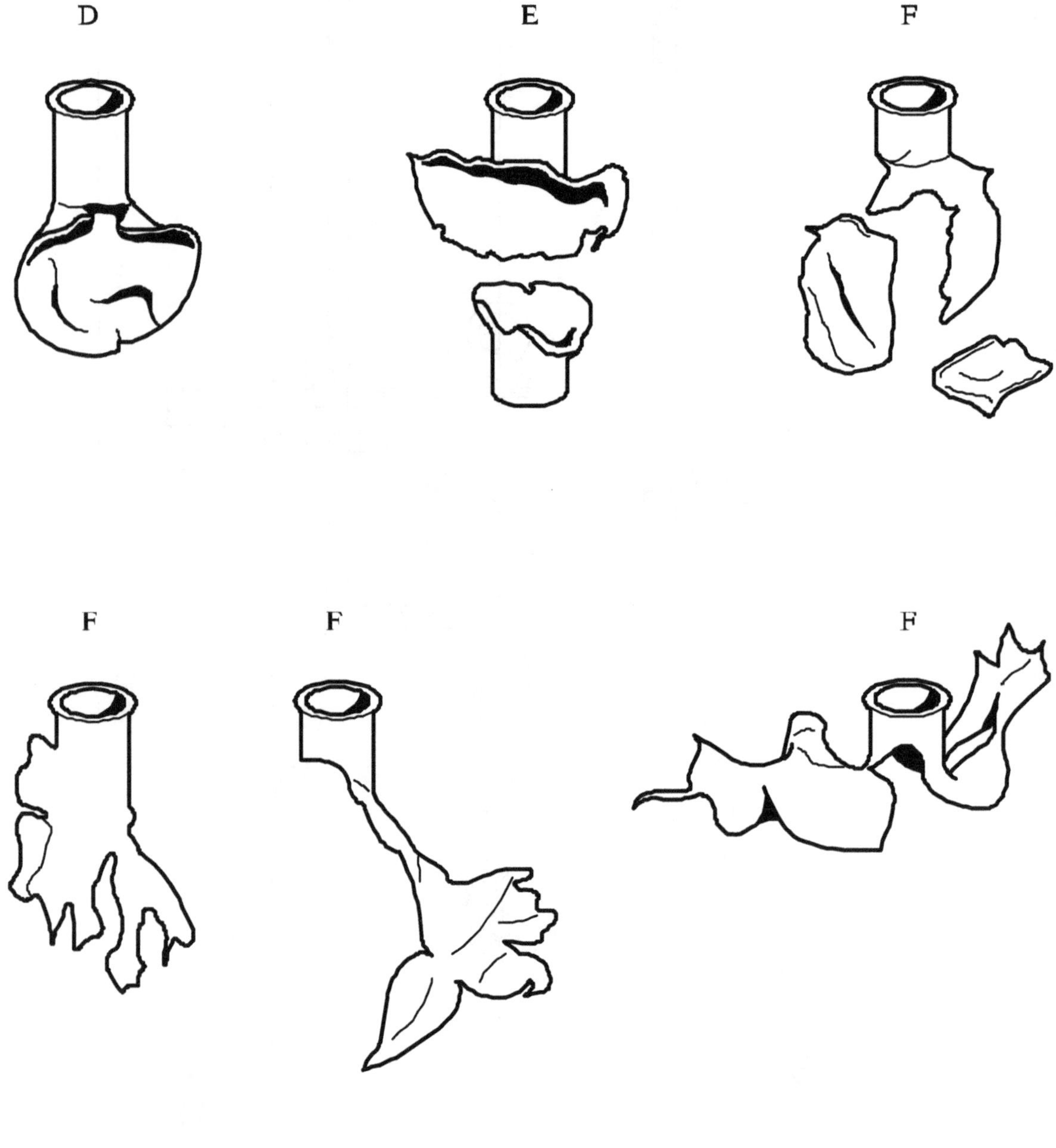

Figure 25.4.1.3: EXAMPLES OF EFFECT TYPES D, E AND F

25.4.2 *Test E.2: Dutch pressure vessel test*

25.4.2.1 *Introduction*

This test is used to determine the sensitiveness of substances to the effect of intense heat under defined confinement. It may be used, in conjunction with an additional heating under confinement test, to answer the question in boxes 7, 8, 9 and 13 of Figure 20.1.

25.4.2.2 *Apparatus and materials*

25.4.2.2.1 Description of the pressure vessel

Figure 25.4.2.1 shows the apparatus used. The vessel is made of stainless steel, type AISI 316. Eight aperture discs are used, the diameters of the orifices being: 1.0 - 2.0 - 3.5 - 6.0 - 9.0 - 12.0 - 16.0 and 24.0 mm. In addition other diameters may be used for hazard assessment. These discs have a thickness of 2.0 mm ± 0.2 mm. The bursting discs are 38 mm diameter aluminium discs rated to burst at 620 ± 60 kPa at 22 °C (see Figure 25.4.2.2).

25.4.2.2.2 Heating device

The pressure vessel is heated by technical-grade butane taken from a cylinder fitted with a pressure regulator. A Teclu burner is used. Other gases may be used, with a suitable burner, provided that a heating rate of 3.5 ± 0.3 K/s is obtained. The heating rate should be checked by heating 10 g of dibutyl phthalate or equivalent in the pressure vessel and measuring its temperature. The time taken for the temperature of the oil to rise from 50 °C to 200 °C is recorded and the heating rate calculated.

25.4.2.3 *Procedure*

25.4.2.3.1 For a normal test, 10.0 g of the substance should be placed in the vessel. The bottom of the vessel should be evenly covered with the substance. The bursting disc, the selected orifice plate and retaining ring are then put in place. The wing nuts are tightened by hand and the box nut with a spanner. The bursting disc is covered by enough water to keep it at a low temperature. The pressure vessel is placed on a tripod (with an inside ring diameter of 67 mm) which may be placed inside a protective cylinder. The ring at the middle of the vessel rests on the tripod.

25.4.2.3.2 The burner is lit, the flow of gas is set to the required rate and the flow of air adjusted so that the colour of the flame is blue and the inner cone of the flame light blue. The tripod should be of such a height that the inner cone just touches the bottom of the vessel. The burner is then placed under the vessel through an opening in the protective casing. ***The test area should be very well ventilated and entry prohibited during the test.*** The vessel is observed from outside the test area either by mirrors or by an aperture in the wall fitted with armoured glass. The time to reaction and duration of reaction can provide additional information useful in interpreting the results. Finally the receptacle is cooled in water and cleaned.

25.4.2.3.3 The series of trials is started with a single trial using an orifice plate with a certain diameter. If there is no rupture of the disc with this orifice, experiments are performed with single trial using plates with decreasing diameters until rupture of the disc occurs. In cases where no disc rupture is observed with an orifice of 1.0 mm, the next test with an orifice of 1.0 mm is carried out with 50.0 g of the substance instead of 10.0 g. If still no rupture of the disc is observed the experiment is repeated until three successive experiments without rupture are obtained. In the event of rupture of the disc, the experiments are repeated at the next higher level (10 g instead of 50 g or next higher diameter of the orifice) until the level is found at which there are no ruptures in three successive experiments.

25.4.2.4 *Test criteria and method of assessing results*

25.4.2.4.1 The relative degree of sensitivity of a substance to heating in a pressure vessel is expressed by the limiting diameter. This being the orifice with the largest diameter in millimetres with which, in three tests, the bursting disc is broken at least once, while having remained unbroken during three tests with the next larger diameter.

25.4.2.4.2 The test criteria are as follows:

"Violent": - Rupture of the disc with an orifice of 9.0 mm or greater and a sample mass of 10.0 g.

"Medium": - No rupture of the disc with an orifice of 9.0 mm but rupture of the disc with an orifice of 3.5 mm or larger but smaller than 9.0 mm and a sample mass of 10.0 g.

"Low": - No rupture of the disc with an orifice of 3.5 mm and a sample mass of 10.0 g but rupture of the disc with an orifice of 1.0 mm or larger but smaller than 3.5 mm and a sample mass of 10.0 g or rupture of the disc with an orifice of 1.0 mm and a sample mass of 50.0 g.

"No": - No rupture of the disc with an orifice of 1.0 mm and a sample mass of 50.0 g.

25.4.2.5 *Examples of results*

Substance	**Limiting diameter (mm)**	**Result**
Azodicarbonamide	1.5	Low
2,2'-Azodi-(2,4-dimethylvaleronitrile)	6.0	Medium
2,2'-Azodi(isobutyronitrile)	5.5	Medium
2,2'-Azodi(2-methylbutyronitrile)	6.0	Medium
tert-Butyl peroxybenzoate	9.0	Violent
tert-Butyl peroxy-2-ethylhexanoate	6.0	Medium
Cumyl hydroperoxide, 80% with cumene	1.0	Low
Dibenzoyl peroxide, 75% with water	6.0	Medium
Di-tert-butyl peroxide	3.5	Medium
Dicetyl peroxydicarbonate	1.0	Low
Dicumyl peroxide	3.5	Medium
2,5-Diethoxy-4-morpholinobenzenediazonium zinc chloride, 90%	< 1.0[a]	No
2,5-Diethoxy-4-morpholinobenzenediazonium tetrafluoroborate, 97%	< 1.0	No
2,5-Diethoxy-4-(phenylsulphonyl)-benzenediazonium zinc chloride, 67%	< 1.0[a]	No
Dilauroyl peroxide	2.0	Low
Dilauroyl peroxide, 42%, stable dispersion in water	< 1.0[a]	No
3-Methyl-4-(pyrrolidin-1-yl)benzene-diazonium tetrafluoroborate, 95%	< 1.0[a]	No
4-Nitrosophenol	1.0[a]	Low

[a] *Test carried out with a 50 g sample.*

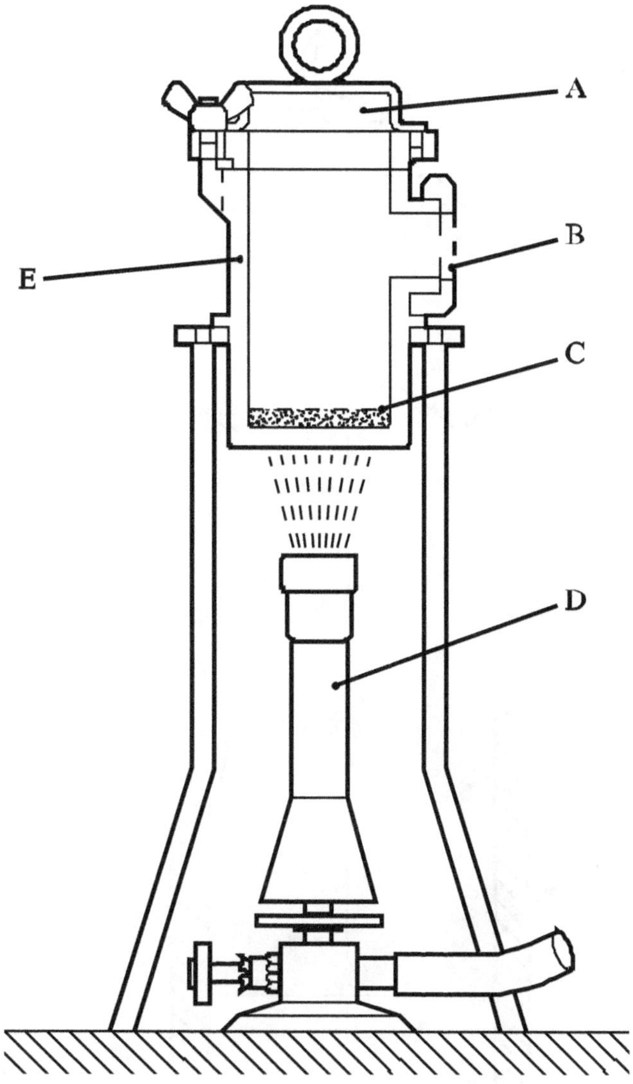

(A) Bursting disc
(B) Orifice plate
(C) Test sample (10 g or 50 g)
(D) Teclu burner
(E) Pressure vessel with an internal diameter of 50 mm and internal height of 94.5 mm

Figure 25.4.2.1: DUTCH PRESSURE VESSEL TEST

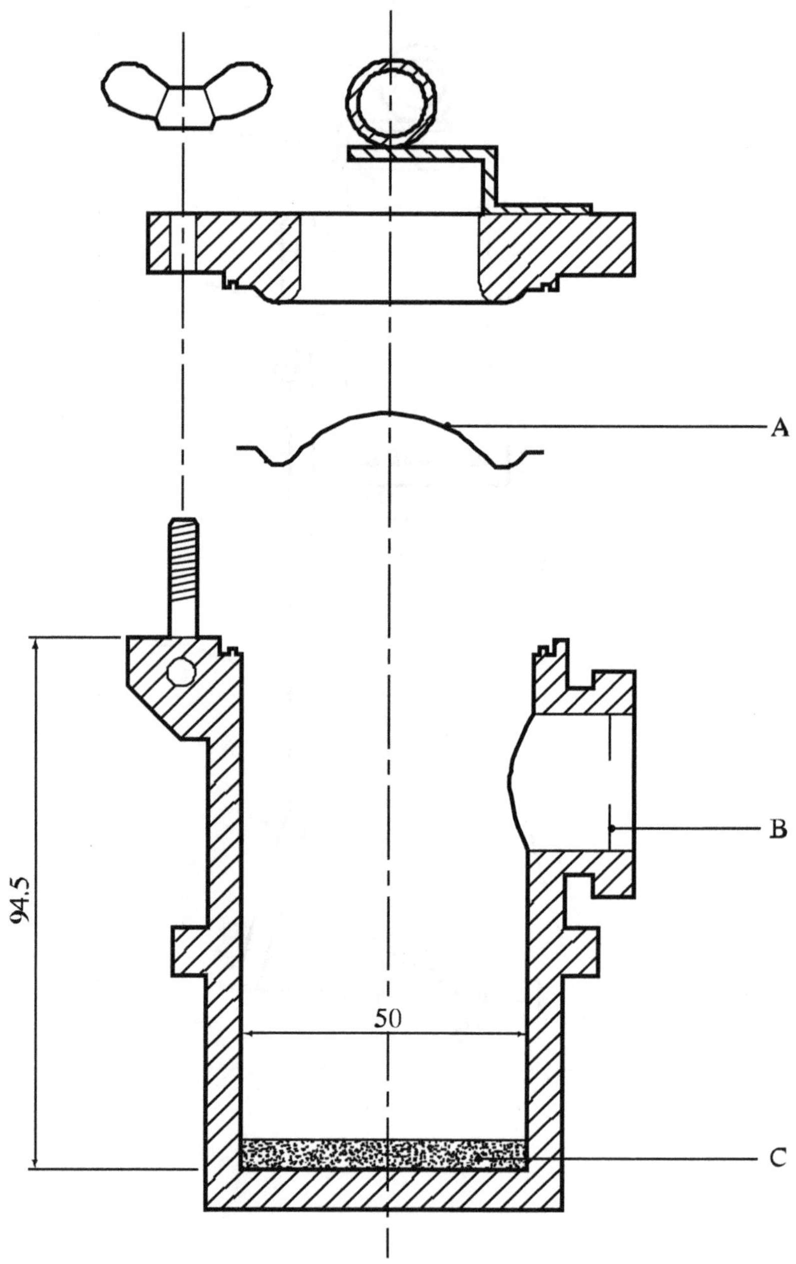

(A) Bursting disc
(B) Orifice plate
(C) Test sample

Figure 25.4.2.2: BURSTING DISC ASSEMBLY

25.4.3 ***Test E.3: United States pressure vessel test***

25.4.3.1 *Introduction*

This test is used to determine the sensitiveness of substances to the effect of intense heat under defined confinement. It may be used, in conjunction with an additional heating under confinement test, to answer the question in boxes 7, 8, 9 and 13 of Figure 20.1.

25.4.3.2 *Apparatus and materials*

The following apparatus and materials are used:

(a) Test-vessel: 316 stainless steel cylindrical pressure vessel (see Figure 25.4.3.1);

(b) Pressure vessel holder (see Figure 25.4.3.2);

(c) Electrical heater (e.g. 700 W);

(d) Sample holder: aluminium cup 28 mm × 30 mm;

(e) Rupture discs: 38 mm aluminium rupture discs rated 620 ± 50 kPa at 22 °C;

(f) 2 mm thick orifice plates with the following orifice diameters (mm): 1.0, 1.2, 2.0, 3.0, 3.5, 5.0, 6.0, 8.0, 9.0, 12.0, 16.0 and 24.0.

25.4.3.3 *Procedure*

25.4.3.3.1 The general assembly of the apparatus is shown in Figure 25.4.3.1. The heating rate should be checked by heating 5.0 g of dibutyl phthalate in a sample cup in the pressure vessel and measuring the temperature of the dibutyl phthalate. The time taken for the temperature of the dibutyl phthalate to rise from 50 °C to 200 °C is recorded and the heating rate calculated. The heating rate should be 0.5 ± 0.1 K/s. An orifice plate, with an orifice diameter larger than the expected vent needed to cause rupture, is selected and inserted into the side port.

25.4.3.3.2 A 5.0 g sample of the substance to be tested is weighed accurately into an aluminium cup. The cup is than lowered and positioned with the aid of forceps in the centre of the pressure vessel. The rupture disc is positioned and secured tightly by the flange bolts. Water is poured over the rupture disc to keep the disc relatively cool. The heating is switched on to the correct setting at least 30 minutes prior to the start of the test. The test-vessel is inserted into the pressure vessel holder onto the heater. This plate holder prevents the test-vessel from falling over. It also prevents escaping vent vapours from reaching the hot plate. The time to decomposition is noted.

25.4.3.3.3 If the disc does not rupture, the experiment is repeated using a smaller orifice diameter until rupture of the disc occurs. In the event of a rupture of the disc, the experiment is repeated at the next higher diameter of the orifice until the level is found at which there are no ruptures in three successive experiments.

25.4.3.4 *Test criteria and method of assessing results*

25.4.3.4.1 The smallest orifice diameter which does not cause a rupture disc to burst during decomposition is designated as the USA-PVT number. This value is used as a measure of the effects of heating a substance under defined confinement. USA-PVT ratings for all substances are based on the same test conditions and heating rate.

25.4.3.4.2 The effect of heating under confinement of the substance is defined by the following criteria:

"Violent": - Substances having USA-PVT numbers 9.0 to 24.0.

"Medium": - Substances having USA-PVT numbers 3.5 to 8.0.

"Low": - Substances having USA-PVT numbers 1.2 to 3.0.

"No": - Substances having USA-PVT number 1.0.

25.4.3.5 *Examples of results*

Substance	**USA-PVT No**	**Result**
tert-Butyl hydroperoxide, 70% with water	1.0	No
tert-Butyl peroxy acetate, 75% in solution	8.0	Medium
tert-Butyl peroxybenzoate	8.0	Medium
tert-Butyl peroxy isopropyl carbonate, 75% in solution	2.0	Low
tert-Butyl peroxypivalate, 75% in solution	4.5[a]	Medium
Cumyl hydroperoxide, 85% with cumene	1.0	No
Dibenzoyl peroxide	18.0[a]	Violent
Di-tert-butyl peroxide	1.0	No
Dicumyl peroxide	2.0	Low
Dicumyl peroxide, with 60% inert solid	1.0	No
Dilauroyl peroxide	6.0	Medium
2,5-Dimethyl-2,5-di-(tert-butylperoxy) hexyne-3	9.0	Violent

[a] *Intermediate diameters, no longer used.*

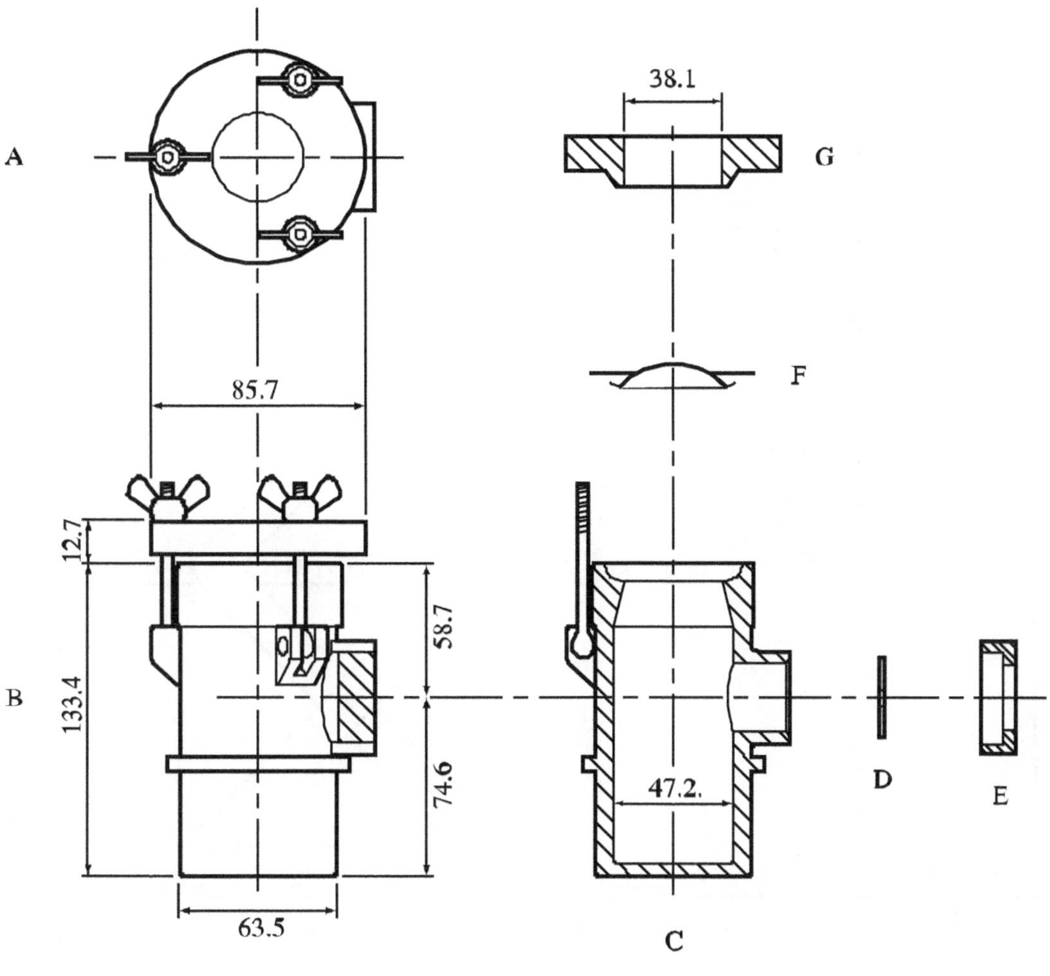

(A) Top view of lid
(B) Side view of assembly
(C) Body of pressure vessel
(D) Orifice plate
(E) Orifice plate retaining nut
(F) Bursting disc
(G) Cap

Figure 25.4.3.1: UNITED STATES PRESSURE VESSEL TEST

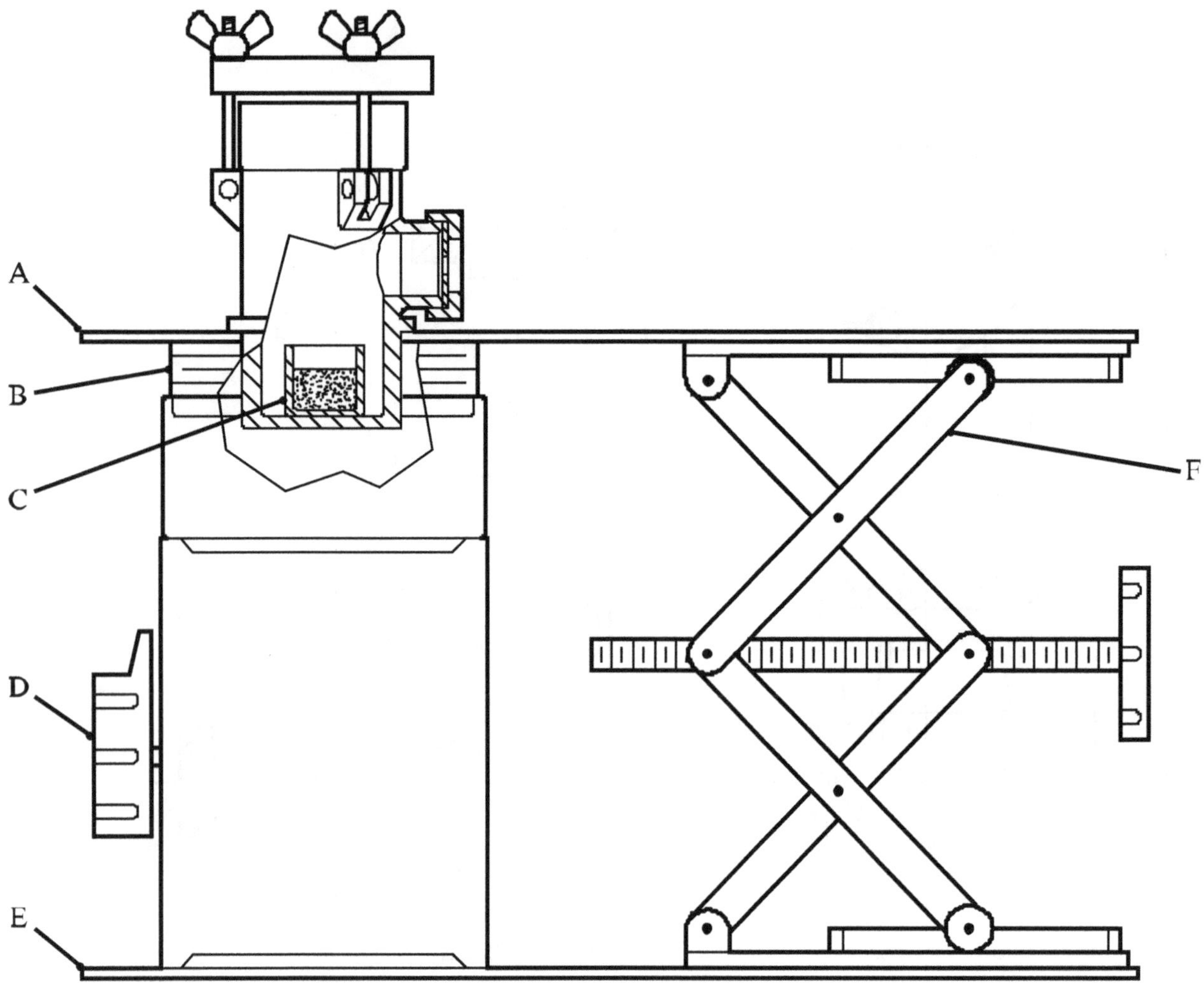

(A) Shield
(B) Thermal insulation
(C) Sample cup
(D) Electrical heater
(E) Base
(F) Laboratory jack

Figure 25.4.3.2: TEST APPARATUS AND STAND (side view)

SECTION 26

TEST SERIES F

26.1 Introduction

26.1.1 Test series F comprises laboratory tests and criteria concerning the explosive power of substances as requested in box 12 of Figure 20.1. It is applicable to those substances which are being considered for transport in Intermediate Bulk Containers (IBCs) or tank-containers or for exemption from the provisions for self-reactive substances of Division 4.1 or organic peroxides of Division 5.2 (see box 11 of Figure 20.1). Substances not being considered for these may be assigned to type E without further testing.

26.1.2 All the tests, except test F.5, may also be used for those substances which are being screened for their ability to propagate detonation (see 21.2.2).

26.2 Test methods

The question "What is its explosive power?" (box 12 of Figure 20.1) is answered on the basis of the results of one of the test methods in Table 26.1.

Table 26.1: TEST METHODS FOR TEST SERIES F

Test code	Name of test	Section
F.1	Ballistic mortar Mk.IIId test	26.4.1
F.2	Ballistic mortar test	26.4.2
F.3	BAM Trauzl test	26.4.3
F.4	Modified Trauzl test[a]	26.4.4
F.5	High-pressure autoclave	26.4.5

[a] *Recommended test.*

26.3 Test conditions

26.3.1 ***The preliminary procedure (see section 20.3) should be performed before carrying out these tests.***

26.3.2 The results obtained in tests F.1 to F.4 depend on the sensitivity to detonative shock of the substance and the strength of the detonator used. If the results are much lower than those obtained with similar substances, a more powerful initiation system may be used with a suitable adjustment (e.g. by firing with inert samples) to the test criteria.

26.3.3 The tests should be performed at ambient temperature. However, if the control temperature is below ambient temperature, adequate safety precautions should be taken. If the substance is transported under temperature control as a solid but is liquid at ambient temperature then the test should be performed just below the melting point.

26.4 Series F test prescriptions

26.4.1 *Test F.1: Ballistic mortar Mk.IIId test*

26.4.1.1 *Introduction*

This test is used to measure the explosive power of a substance. A detonator is initiated in the substance whilst the substance is confined in the bore of a mortar. The recoil (swing) of the mortar is measured and, after allowing for the effect of the detonator, the power is calculated as a percentage equivalence of picric acid - the explosive standard. The test may be used to answer the question in box 12 of Figure 20.1.

26.4.1.2 *Apparatus and materials*

26.4.1.2.1 The mortar consists of a steel tube 457 mm long by 203 mm external diameter, closed at one end and having a bore 229 mm long with an internal diameter of 25 mm, flaring out at the open end. The mortar is suspended by 4 stainless steel cables attached at either end by bottle screws and pivots, thus allowing it to swing freely. The total weight is approximately 113.2 kg and the suspension length is 2,080 mm. A pen, secured at the extremity of a hinged trailing arm attached to the mortar, scribes on a horizontal table a line equal to the horizontal swing of the mortar. The mortar body is made in two pieces, an outer casing of mild steel and an inner cylinder of Vibrac V30 steel heat-treated to a proof stress of 772 MPa. This cylinder is held in place by a retaining annular plate (Figure 26.4.1.1).

26.4.1.2.2 The sand used for stemming is clean, dry, quartz sand sieved to pass a BS 600 micron sieve and to be retained on a BS 250 micron sieve. The picric acid is in the form of pure, dry crystals sieved in the same manner. Boric Acid (analytical grade) is sieved to pass a BS 500 micron sieve. Charge bags are cylindrical, 25 mm in diameter and made of thin paper. The inner charge bag is 90 mm long and the outer charge bag 200 mm long.

26.4.1.2.3 The detonator used is a flat-based aluminium sheathed detonator containing 0.6 g PETN.

26.4.1.3 *Procedure*

26.4.1.3.1 Impact, friction and electric spark tests are normally carried out on the substance before it is tested in the mortar. 10.00 ± 0.01 g of the substance is tamped into the inner charge bag. The detonator is inserted into a 6 mm cavity made in the substance with a phosphor-bronze rod and the neck of the bag is twisted around the detonator. The charge is then inserted in the outer bag and pressed down to the bottom with a special tool. 57 g of sieved sand is poured into the outer bag and gently compacted by tapping. The neck of the outer bag is twisted around the detonator wires and the whole charge is inserted into the bore of the mortar and rammed with the special tool. The detonator is fired and the total horizontal swing (S) is measured. The test is performed three times and the average swing (S_m) for the substance calculated.

26.4.1.3.2 Liquids are tested by using a cylindrical glass vessel[1] of approximately 16 ml volume instead of the paper charge bag. The open end of the vessel is reduced to a narrow tube 8 mm in diameter and 8 mm in length. The standard detonator sheathed in a suitable length of polythene tubing makes a push fit seal in the neck of the vessel. The vessel is then inserted into the outer charge bag, as for solids.

26.4.1.3.3 The picric acid standard value and the detonator standard value established with paper charge bags may be employed in the power equivalence calculation for liquids fired in glass vessels.

26.4.1.3.4 When a new inner cylinder is inserted into the outer casing of the mortar, the mean swing (average of 10 shots) given by boric acid (B_m) and picric acid (P_m) are found.

---

[1] *The use of glass vessels with powerful explosives can damage the mortar liner. Test firings with picric acid in glass vessels caused severe wear but produced the same results as picric acid fired in paper charge bags.*

26.4.1.4 *Test criteria and method of assessing results*

26.4.1.4.1 The explosive power (P) is calculated from $P = 100 \times \frac{(S_m^2 - B_m^2)}{(P_m^2 - B_m^2)}$ as a percentage (rounded to the nearest whole number) of the value given by picric acid.

26.4.1.4.2 The test criteria are as follows:

"Not Low": - Explosive power values of 7% or greater of that given by picric acid.

"Low": - Explosive power values of less than 7% of that given by picric acid but more than 1% of that given by picric acid.

"No": - Explosive power values of 1% or less of that given by picric acid.

26.4.1.5 *Examples of results*

Substance	**Mean % of picric acid**	**Result**
tert-Butyl hydroperoxide, 70% with water	2	Low
tert-Butyl peroxybenzoate	13	Not low
tert-Butyl peroxy-2-ethylhexanoate	8	Not low
Cumyl hydroperoxide, 80% in cumene	4	Low
Dibenzoyl peroxide	8	Not low
Dibenzoyl peroxide, 75% with water	6	Low
Di-tert-butyl peroxide	8	Not low
Dicetyl peroxydicarbonate	1	No
Dicumyl peroxide, with 60% inert solid	1	No
Dilauroyl peroxide	1	No
2,5-Dimethyl-2,5-di-(tert-butylperoxy)hexyne-3	17	Not low
Magnesium monoperoxyphthalate hexahydrate, 85% with magnesium sulphate	1	No

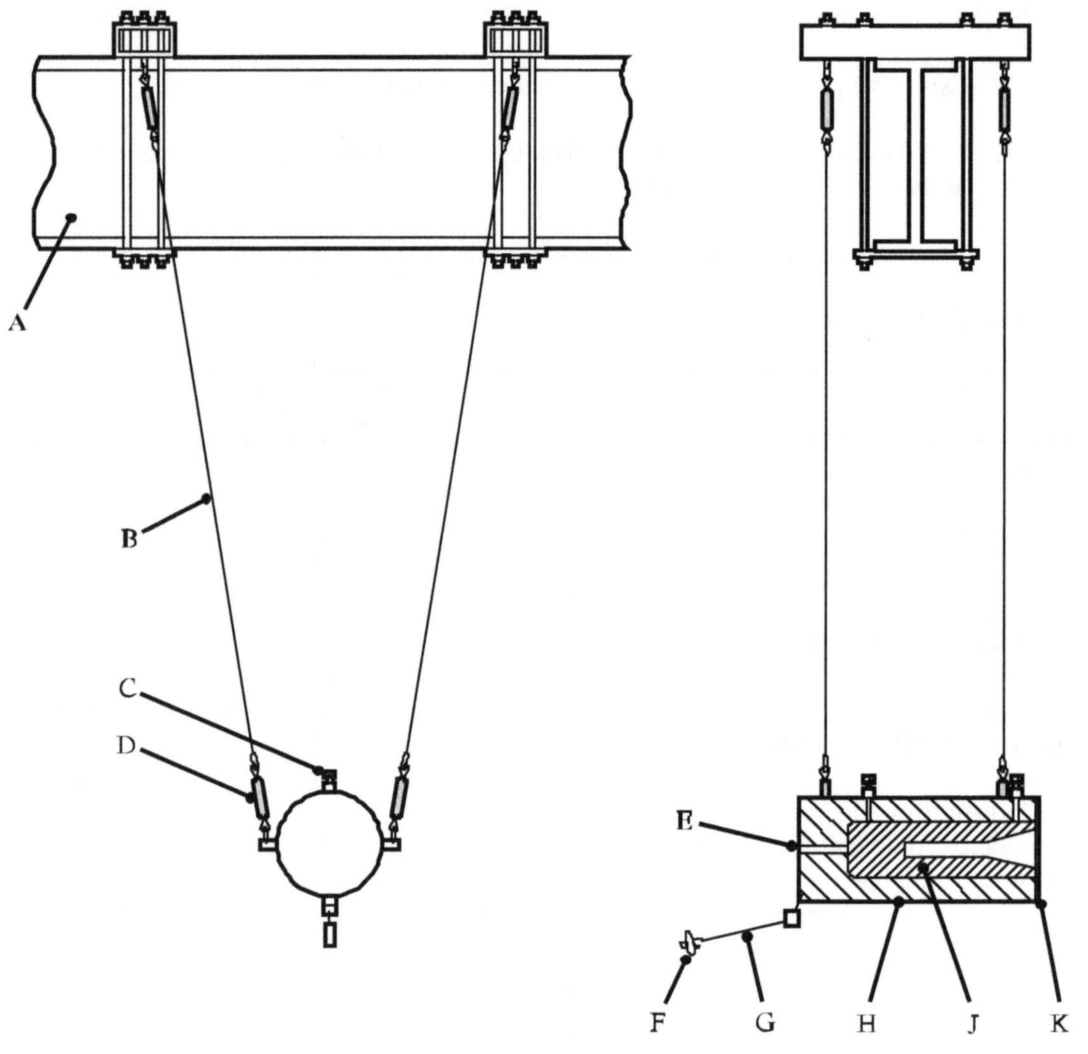

(A)	Suspension beam	(B)	Suspension wire
(C)	Liner set-screws	(D)	Bottle screws
(E)	Hole to aid liner removal	(F)	Stylus holder
(G)	Hinged trailing stylus arm	(H)	Mortar body outer casing
(J)	Inner liner	(K)	Annular retaining plate

Figure 26.4.1.1: BALLISTIC MORTAR MK.IIID

26.4.2 *Test F.2: Ballistic mortar test*

26.4.2.1 *Introduction*

This test is used to measure the explosive power of a substance. A detonator is initiated in the substance whilst the substance is confined in the bore of a mortar by a steel projectile. The recoil of the mortar is measured and the power is calculated as a percentage equivalence of picric acid - the explosive standard. The test may be used to answer the question in box 12 of Figure 20.1.

26.4.2.2 *Apparatus and materials*

26.4.2.2.1 Mortar

The mortar is made of chrome-nickel steel with a mass of 248.50 ± 0.25 kg. It is cylindrical and the axial cavity consists, from front to rear, of the projectile holder, the firing chamber and access for the detonator lead wires. The dimensions of the mortar will change during use (in particular, the firing chamber is enlarged) resulting in diminished recoil with the firing of a given charge. The reference charge used is 10.00 g picric acid (see 26.4.2.2.7). The used mortar is replaced by a new one when the average energy released in the last 10 firings is less than 90 per cent of the average energy produced in the first 10 firings (for similar temperatures and using a projectile in good condition).

26.4.2.2.2 Ballistic Pendulum

The mortar is suspended by steel arms from a roller-bearing mounted horizontal axis. The mortar pendulum is rendered ballistic by attaching a steel weight underneath it. Figure 26.4.2.1 is a representation of the ballistic mortar-pendulum, which has the following principal characteristics:

oscillation period	3.47 s
oscillating mass	479 kg
distance between axis of rotation and axis of mortar	2.99 m

A cursor on the graduated sector displaced by a lateral arm attached to the mortar is used to measure the recoil. The scale marked on the sector is proportional to (1 - cos A), A being the angle of recoil of the pendulum, i.e. proportional to the work performed.

26.4.2.2.3 Projectile

The projectile is a steel cylinder. The diameters (diameter 127 mm, length 162 mm) are in practice adjusted to meet the following conditions:

(a) The gap between the projectile, when new, and its housing in the mortar should be less than 0.1 mm;

(b) The mass of the projectile, when new, should be 16.00 ± 0.01 kg.

A worn projectile should be replaced when the gap between it and its housing in the mortar exceeds 0.25 mm. On firing, the projectile is normally expelled at a speed of between 100 and 200 km/h. The use of a receptacle lined with shock absorbing material enables it to be stopped without being damaged.

26.4.2.2.4 Sample vessels

When liquids are tested, glass ampoules of 16 g mass with a filling orifice and a recess to take the detonator (see Figure 26.4.2.3) are used to contain the test samples. Other substances (solid, granular, pastes, etc.) are loaded in cylindrical sample vessels, 20 mm in diameter, made of tinfoil of thickness 0.03 mm and mass about 2 g. This applies, for instance, to picric acid charges.

26.4.2.2.5 Supports

A steel-wire ring support with three feet, shown in Figure 26.4.2.2, is used to centre the charge in the firing chamber (to reduce wear on the mortar).

26.4.2.2.6 Detonators

These are European standard detonators charged with 0.6 g PETN, as shown in Appendix 1.

26.4.2.2.7 Picric acid (reference substance)

This is pure crushed picric acid, grain size less than 0.5 mm, dried at 100 °C and kept in a tightly-stoppered flask.

26.4.2.3 *Procedure*

26.4.2.3.1 Preparation of charge

26.4.2.3.1.1 Compact solids have to be tested in the form of cylindrical blocks, diameter 20 mm ± 1 mm, with at one end an axial cavity (diameter 7.3 ± 0.2 mm, depth 12 mm) to accommodate the detonator. The mass of the block should be 10.0 ± 0.1 g. To prepare the charge, the block is wrapped in tinfoil, thickness 0.03 mm, mass approximately 2 g. The detonator is placed in its housing and the end of the foil wrapping pressed around the head of the detonator (see Figure 26.4.2.3).

26.4.2.3.1.2 Substances other than liquids are packed at normal density in tinfoil wrappings, each test charge weighing 10.0 ± 0.1 g. The detonator is inserted to a depth of approximately 12 mm in the substance. The end of the wrapping is twisted around the head of the detonator (see Figure 26.4.2.3).

26.4.2.3.1.3 For liquids, a charge of 10.0 ± 0.1 g of the test substance is placed in a glass ampoule. The detonator is placed in its housing. If necessary, metal wire can be used to hold it in place (but under no circumstances combustible material).

26.4.2.3.1.4 Charges of 10.0 ± 0.1 g of picric acid, at normal density, are also prepared in foil wrappings. The detonator is inserted to a depth of approximately 12 mm in the picric acid. The end of the wrapping is twisted around the head of the detonator.

26.4.2.3.2 Trial run

26.4.2.3.2.1 The charge, placed in a charge holder (see 26.4.2.2.5) is pressed home in the firing chamber, the detonator head being in contact with the rear surface of the chamber (see Figure 26.4.2.3).

26.4.2.3.2.2 The projectile is greased with a consistent quality lubricant, inserted and pushed home in its housing in the mortar. To avoid any scatter in the results, due to possible distortion of the mortar or projectile, the position of the projectile in relation to the mortar housing is checked and recorded.

26.4.2.3.2.3 The cursor (see 26.4.2.2.2) is placed in contact with the moving arm so that the recoil of the pendulum can be measured. After firing, the deviation (D) of the pendulum is noted, i.e. the point at which the cursor remains on the graduated sector at the end of the swing.

26.4.2.3.2.4 The projectile and bore should then be carefully cleaned.

26.4.2.4 *Procedure*

26.4.2.4.1 Four firings are first carried out with picric acid. The average of the four deviations obtained is calculated. This average should have a value around 100, in the arbitrary unit of power adopted for the scale of the sector on which the pendulum recoil is measured. The results of the four firings should not vary by more than one unit, and if that is the case, the value D_o will be the average of the four deviations measured. If one of the results varies by more than one unit from the average, it is discarded and the value D_o is the average of the other three results.

26.4.2.4.2 The ambient temperature is recorded.

26.4.2.4.3 The procedure is then repeated with the test substance, for not less than three firings, the deviations obtained being D_1, D_2, D_3 etc. The corresponding powers, expressed as a percentage of the finding for picric acid, are calculated by the equation:

$$T_k = 100 \times D_k/D_o, \text{ where } k = 1, 2, 3, ...,$$

or, for liquids fired in glass ampoules:

$$T_k = 200 \times D_k/D_o, \text{ where } k = 1, 2, 3, ...$$

26.4.2.4.4 The average value of T_k is then calculated. The result, rounded to the nearest whole number, is called the "ballistic mortar power" (BMP) of the substance.

NOTE: *When several substances are to be tested in succession during the same half-day, only one series of four firings with picric acid is made in the half-day.*

26.4.2.5 *Test criteria and method of assessing results*

26.4.2.5.1 The result of the test is presented in detail by giving the results T_1, T_2, T_3 ... and the value of BMP and the ambient temperature. In summary form, only the BMP value is given.

26.4.2.5.2 The test criteria are as follows:

"Not Low": - BMP values of 7 or more.

"Low": - BMP values less than 7 but more than 1.

"No": - BMP values of 1 or less.

26.4.2.6 *Examples of results*

Substance	BMP value	Result
2,2'-Azodiisobutyronitrile	13	Not low
tert-Butyl peroxybenzoate	16	Not low
tert-Butyl peroxy-2-ethylhexanoate	7	Not low
Dibenzoyl peroxide	16	Not low
Dibenzoyl peroxide, 75% with water	8	Not low
Di-tert-butyl peroxide	7	Not low
Dilauroyl peroxide	5	Low
p-Menthyl hydroperoxide, 55% in p-menthane	3	Low
Peroxyacetic acid, 36% in a mixture with 19% water, 36% acetic acid and 6% hydrogen peroxide, with stabiliser	27	Not low
Pinanyl hydroperoxide, 54% in pinane	2	Low
Picric acid	100	
Trinitrotoluene	95	
Sand	1	

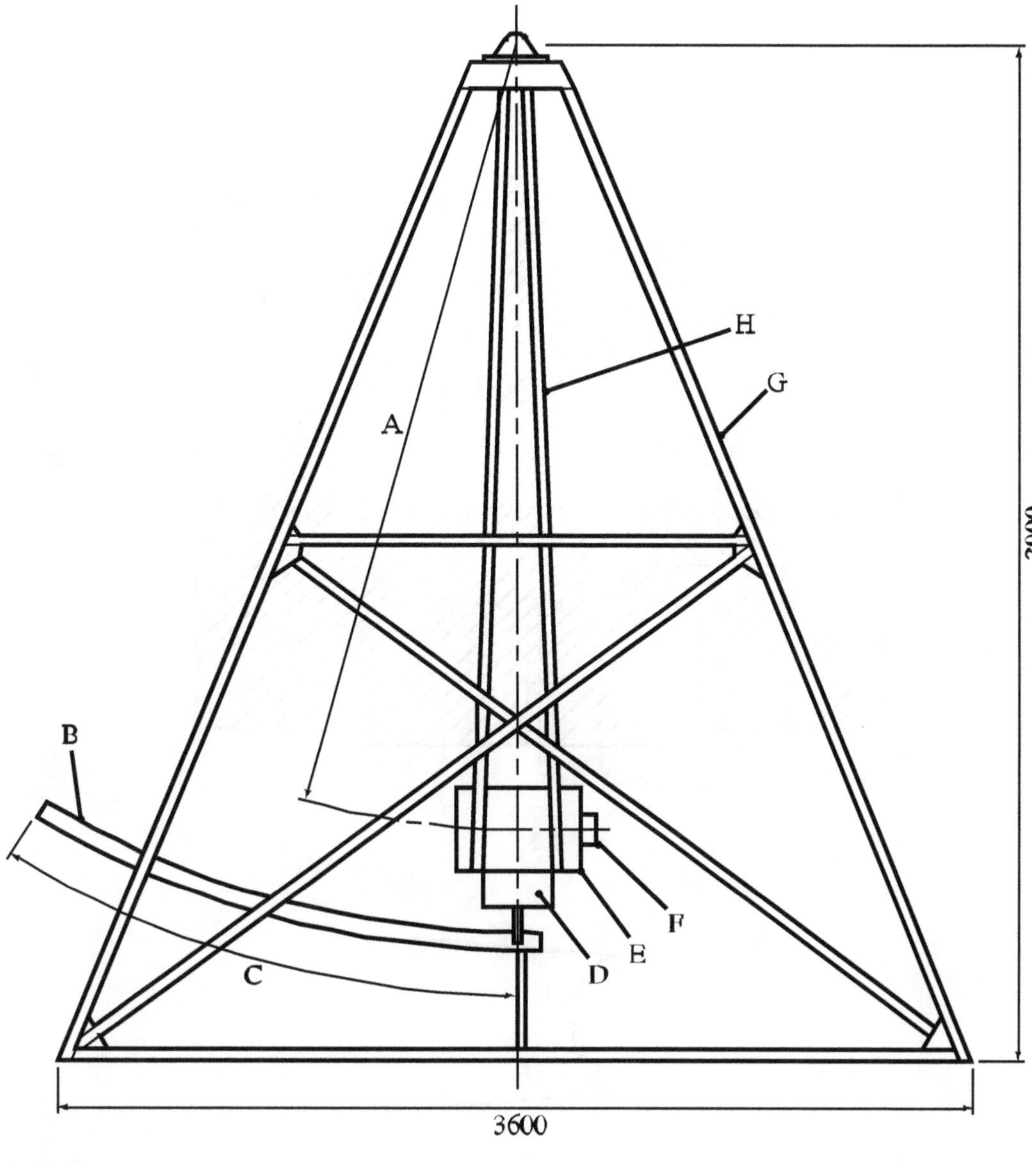

(A) 2 905 mm distance between axes
(B) Graduated scale
(C) 30° swing angle
(D) Weight
(E) Mortar
(F) Projectile
(G) Frame
(H) Swinging arm

Figure 26.4.2.1: BALLISTIC MORTAR TEST

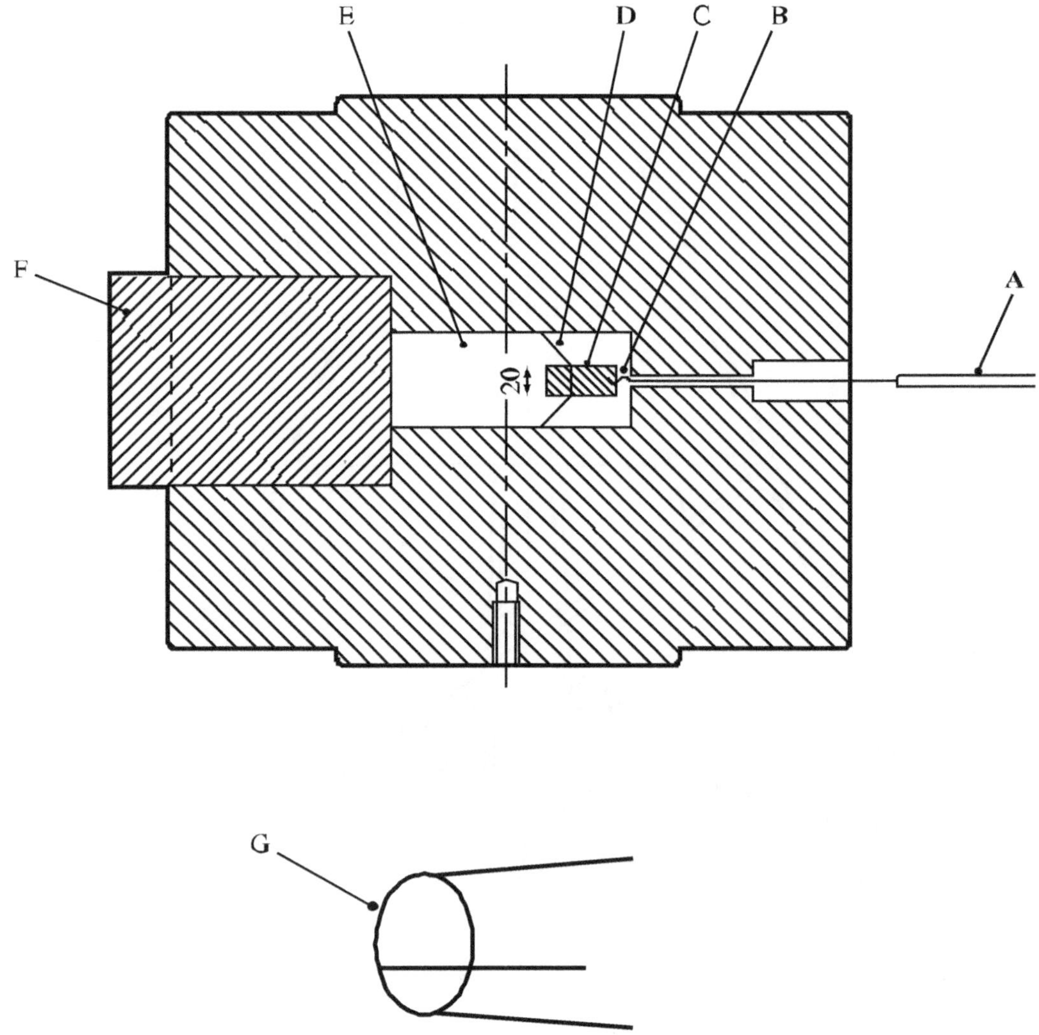

(A) To igniter
(B) Detonator
(C) 20 mm diameter charge
(D) Charge holder
(E) Firing chamber
(F) Projectile
(G) Expanded view of charge holder

Figure 26.4.2.2: MORTAR (TOP) AND CHARGE HOLDER (BOTTOM)

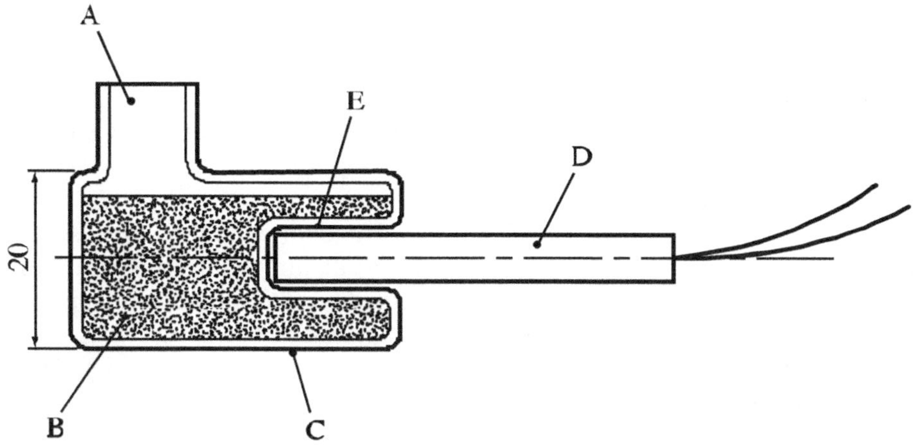

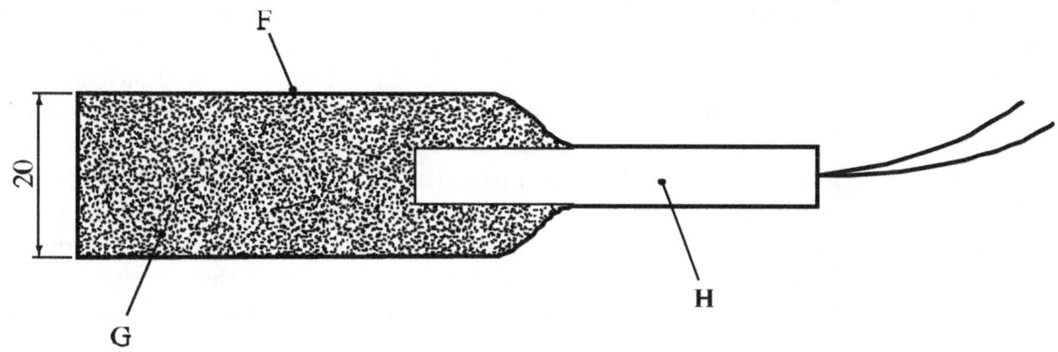

(A) Filling orifice
(B) 20 mm diameter charge containing 10 g of substance
(C) Glass ampoule (16 g)
(D) 0.6 g PETN detonator
(E) Detonator housing
(F) 2 g foil wrapping
(G) 20 mm diameter charge containing 10 g of substance
(H) 0.6 g PETN detonator

Figure 26.4.2.3: CHARGE FOR LIQUIDS (TOP) AND SUBSTANCES OTHER THAN LIQUIDS (BOTTOM)

26.4.3 *Test F3: BAM Trauzl test*

26.4.3.1 *Introduction*

This test is used to measure the explosive power of a substance. A detonator is initiated in the substance whilst the substance is confined in a hole in a lead block. The explosive power is expressed in the form of the increase in volume of the cavity in the lead block per 10 g of substance. The test may be used to answer the question in box 12 of Figure 20.1.

26.4.3.2 *Apparatus and materials*

26.4.3.2.1 Initiation of the substance is achieved by a European standard detonator with 0.6 g of PETN (see Appendix 1).

26.4.3.2.2 For the test, standardized Trauzl lead blocks of cylindrical shape are used with a height of 200 mm and a diameter of 200 mm. They have an axial recess of 25 mm diameter and 125 mm depth with a volume of 61 cm^3, see Figure 26.4.3.1. The lead blocks are made by casting pure soft lead into moulds at a casting temperature of 390 °C to 400 °C. The quality of a casting series of lead blocks is checked by three test explosions each using 10 cm^3 of crystallized picric acid (loading density 1.0 g/cm^3). The mean of the three net expansion values obtained should be between 287 cm^3 and 300 cm^3.

26.4.3.3 *Procedure*

26.4.3.3.1 Solid substances are shaped to cylindrical test charges with a volume of 10 cm^3 by wrapping in a weighed piece of tinfoil and the mass of the charge determined. The test charges have an external diameter of 24.5 mm, a height of 22.2 mm, and a coaxial recess of 7 mm diameter and 12 mm depth to take the detonator. The test charge is prepared in a device consisting of a piston, a two-part matrix, a clamping frame and a base piece (Figure 26.4.3.2). For this, a trapezoidal piece of tinfoil (thickness approx. 0.01 mm) 55 mm wide is wound round the piston. The die with its tinfoil sheath is then introduced into the matrix until it reaches the backstop of the base. The matrix is clamped with the aid of the frame and the piston is slowly drawn out of the tinfoil tube after pressing it firmly against the base piece. The bottom of the tinfoil tube is carefully pierced in its centre using a thin wooden rod. From the side of the clamping frame, the standard detonator PETN 0.6 g is inserted into the base piece so that the detonator wires are drawn through the hole in the adjusting screw until the detonator touches the screw. The screw is so adjusted that the detonator projects 12.0 mm from the base piece. The substance to be tested is put into the tinfoil tube, compressing it lightly by means of a wooden rod. The projecting tinfoil edge is folded inwards and the piston is pressed into the matrix up to the collar. After withdrawing the piston, the prepared test charge, in the tinfoil tube, with the inserted detonator, is carefully taken out of the matrix.

26.4.3.3.2 Liquid substances are tested in thin-walled glass cylinders of similar shape and of such a capacity as to take the sample volume of 10 cm^3 and that of the detonator, when it is introduced to a depth of 12 mm into the liquid. The neck of the vessel is of such a length that it keeps the detonator in a central position. After having determined the sample mass, the test charge is carefully introduced into the recess of the lead block right to the bottom. The lead blocks are stored in a temperature-controlled room, so that the temperature, measured in the depth of the recess, immediately before introducing the charge, is in the range 10 °C and 20 °C. For the test, the lead block should rest on a flat, massive support of steel laid on the ground. For stemming, the remaining space in the recess is filled with dry quartz sand, passed through a sieve of 144 meshes per cm^2, with a gravimetric density of 1.35 g/cm^3. The lead block is then hit at the side three times with a hammer of 2 kg mass and the excess of sand is removed from the upper surface.

26.4.3.3.3 The detonator is fired and the block emptied of any residues. The volume of the enlarged cavity is measured using water and the expansion which would be caused by a 10 g sample calculated using the expression:

$$10 \times \frac{\text{Volume of enlarged caity (cm}^3) - 61}{\text{Sample mass (g)}}$$

26.4.3.3.4 Normally, the test is conducted in duplicate and the highest expansion value obtained used for the assessment.

26.4.3.4 *Test criteria and method of assessing results*

26.4.3.4.1 The explosive power is expressed in the form of the increase in volume of the cavity in the lead block per 10 g of substance. At a given strength of initiation, the explosive power increases with the volume of expansion.

26.4.3.4.2 The test criteria are as follows:

"Not Low": - The expansion of the lead block is 25 cm^3 or more per 10 g of sample.

"Low": - The expansion of the lead block is less than 25 cm^3 but more than or equal to 10 cm^3 per 10 g of sample.

"No": - The expansion of the lead block is less than 10 cm^3 per 10 g of sample.

26.4.3.5 *Examples of results*

Substance	Sample mass (g)	Expansion (cm3/10g)	Result
Azodicarbonamide		9	No
2,2'-Azodi(isobutyronitrile)		26[a]	Not low
Benzene-1,3-disulphohydrazide		50[a]	Not low
Benzene-1,3-disulphohydrazide, 70% with mineral oil		11[a]	Low
Benzene sulphohydrazide	8.4	8	No
tert-Butyl peroxybenzoate	9.1	32[a]	Not low
tert-Butylperoxy-2-ethylhexanoate	7.2	28[a]	Not low
3-Chloroperoxybenzoic acid, not more than 86% with 3-chlorobenzoic acid	7.1	42[a]	Not low
Cumyl hydroperoxide, 84.1% with cumene	9.3	10	Low
Cyclohexanone peroxide(s)	6.4	50[a]	Not low
Dibenzoyl peroxide	8.0	31[a]	Not low
Dibenzoyl peroxide, 75% with water	8.0	21[a]	Low
Di-tert-butyl peroxide	7.2	28[a]	Not low
Dicetyl peroxydicarbonate	7.3	5[a]	No
Dicumyl peroxide	6.9	12[a]	Low
Diisopropyl peroxydicarbonate	7.8	78[a]	Not low
Dilauroyl peroxide	8.0	11[a]	Low
Dimyristyl peroxydicarbonate	7.4	11	Low
Dimyristyl peroxydicarbonate, 42%, stable dispersion in water	7.8	5[a]	No
N,N'-Dinitrosopentamethylene tetramine		147[a]	Not low
N,N'-Dinitrosopentamethylene tetramine 80%, with 17% inorganic solid and 3% mineral oil	10.2	7[b]	No
Diperoxy isophthalic acid	8.7	144	Not low
4-Nitrosophenol	7.3	11	Low
Boric acid		0	
Dimethyl phthalate		5	
Water		6	

[a] *Determined by using test charges of 11 cm^3 volume and a slightly stronger test blasting cap with a copper tube of 6.85 mm external diameter and 45 mm length, with flat bottom containing 0.5 g PETN with 10% wax, 0.2 g tetryl, and 0.5 g mercury fulminate as explosive charge, all pressed at 400 bar.*

[b] *Initiation with 3 detonators gives 123 cm^3/10 g, Not low*

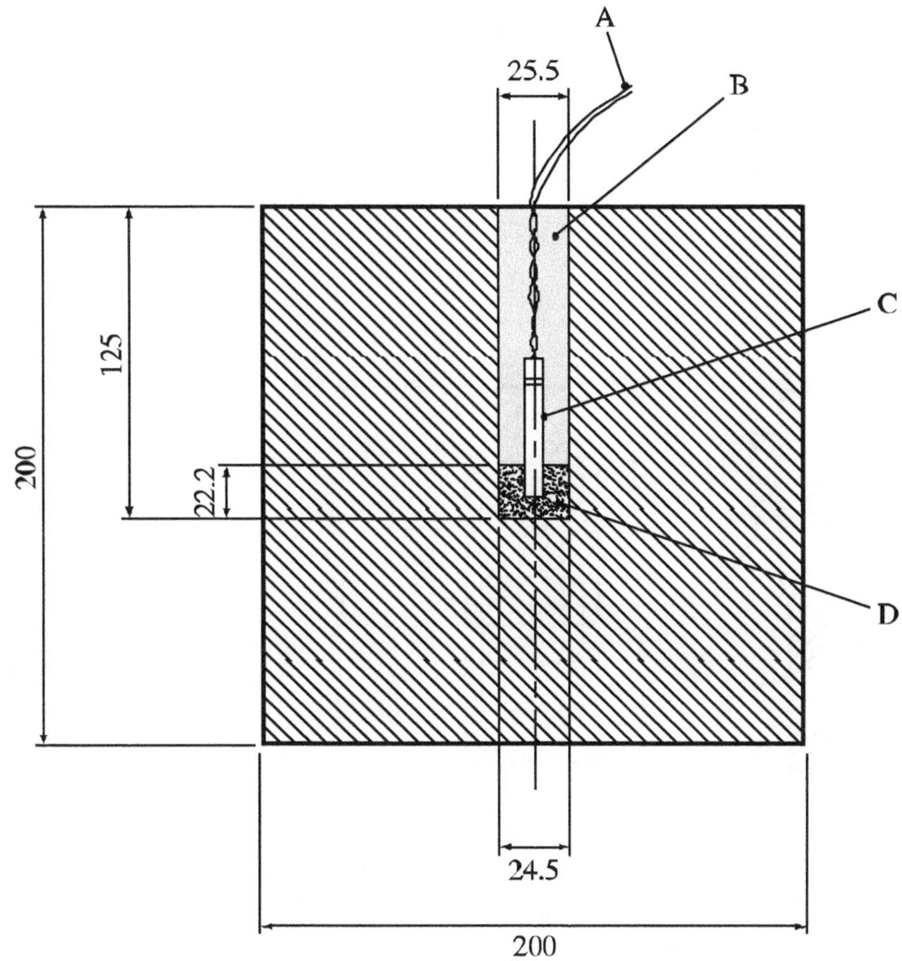

(A)	Detonator wires
(B)	Dry sand stemming
(C)	European standard detonator
(D)	Test sample

Figure 26.4.3.1: BAM TRAUZL TEST

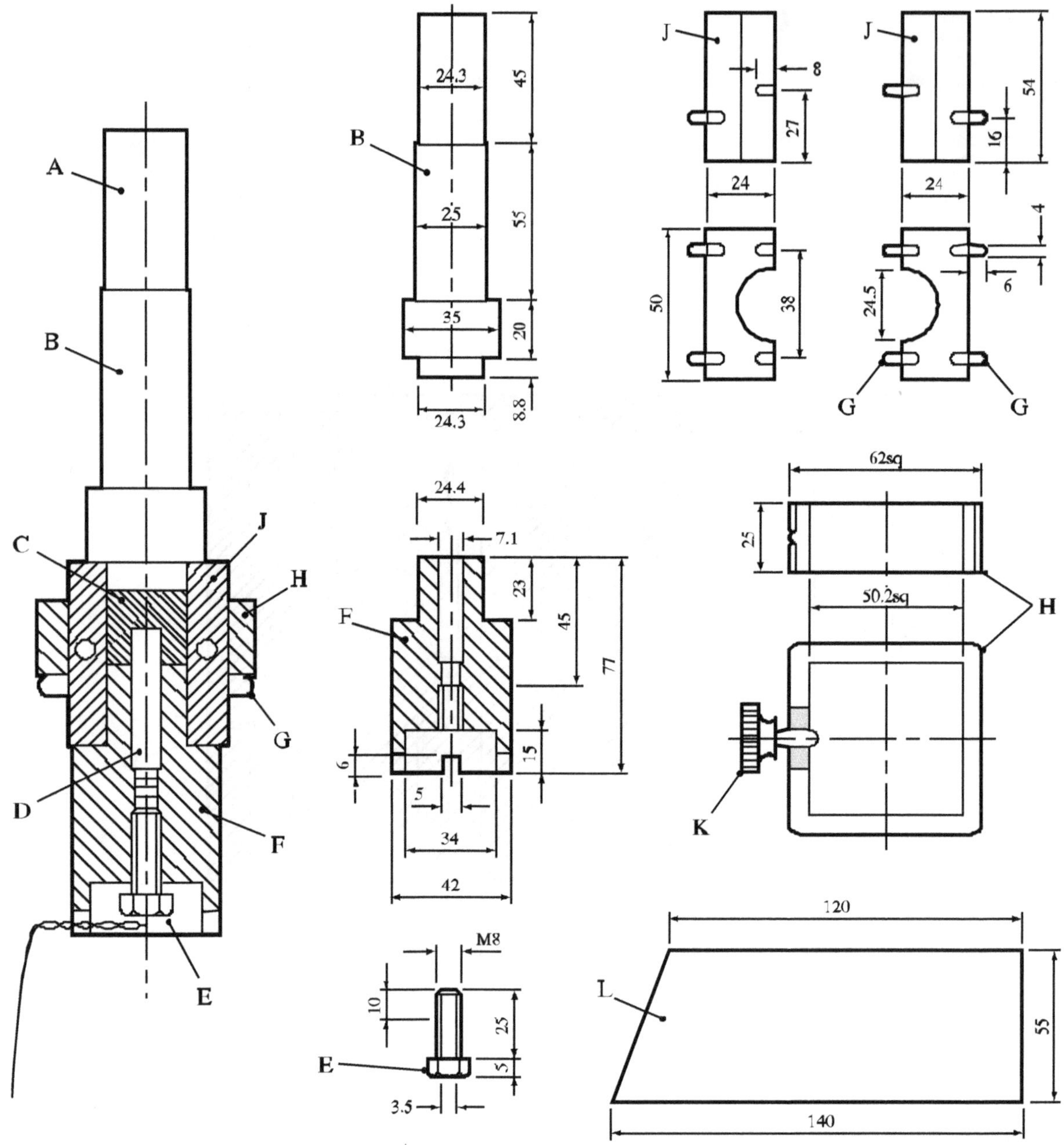

(A)	End part of piston for preparing the tinfoil tube	(B)	Piston
(C)	Test sample	(D)	Detonator
(E)	Adjusting screw (axial bore 3.5 mm, slot 1 mm, 100 mm long)	(F)	Base piece
(G)	Pins	(H)	Clamping frame
(J)	Matrix	(K)	Knurled screw
(L)	Tin foil		

Figure 26.4.3.2: DEVICE FOR MAKING THE CHARGES (10 cm^3 volume, 24.6 mm diameter and 22.2 mm height) FOR THE BAM TRAUZL TEST

26.4.4 ***Test F.4: Modified Trauzl test***

26.4.4.1 *Introduction*

This test is used to measure the explosive power of a substance. A detonator is initiated in the substance whilst the substance is confined in a hole in a lead block. The explosive power is expressed in the form of the average increase in volume of the cavity in the lead block over the value obtained from an inert substance with similar physical properties. The test may be used to answer the question in box 12 of Figure 20.1.

26.4.4.2 *Apparatus and materials*

The lead block is machined from cast or extruded lead rod 50 ± 1 mm diameter by 70 mm long with a cavity 25.4 mm in diameter and 57.2 mm long as shown in Figure 26.4.4.1. The cavity is cut with a flat-tipped drill to prevent puncture of the lead block. The blasting cap used is a standard No. 8 (USA) detonator (see Appendix 1). Assembly A is used for liquids and pastes and assembly B for solids (see Figure 26.4.4.2). The sample vial used for assembly A is a commercial 12 ml, 21 mm outer diameter vial. The sample vial used for assembly B is a commercial 16 ml, 24.9 mm outer diameter vial. The polyethylene stoppers are the standard ones supplied with the vials. The glass tube used to locate the blasting cap in assemblies A and B is a borosilicate glass culture tube. It is 10 mm outer diameter × 75 mm long. A 10 mm diameter hole drilled in the centre of the polyethylene stopper gives a tight fit which holds the tube secure. Two rubber "0" rings (16.5 mm inner diameter, 2.5 mm diameter cross section) are used to centre the vial in the cavity in the lead block in assembly A.

26.4.4.3 *Procedure*

A 6.0 g sample is placed in the sample vial which is assembled as required and placed in the lead block. The lead block is placed on a solid surface in a protected area, the blasting cap is fully inserted and, when the area is vacated, the cap is fired. The volume of the cavity in the lead block is measured accurately, to the nearest 0.2 ml, using water before and after the test. Three tests are conducted on the substance and on the inert reference substance using the same type of assembly.

26.4.4.4 *Test criteria and method of assessing results*

26.4.4.4.1 The explosive power of the test sample is expressed in terms of the average increase of the volume of the cavity in the lead block over that given by the inert reference substance.

26.4.4.4.2 The test criteria are as follows:

"Not low": - The average net block expansion is equal to or greater than 12 cm^3.

"Low": - The average net block expansion is less than 12 cm^3 but greater than 3 cm^3.

"No": - The average net block expansion is 3 cm^3 or less.

26.4.4.5 *Examples of results*

Substance	Average net expansion (cm^3)	Result
2,2'-Azodi(isobutyronitrile)	18	Not low
2,2'-Azodi(2-methylbutyronitrile)	14	Not low
tert-Butyl hydroperoxide, 72% with water	7	Low
tert-Butyl peroxyacetate, 75% in solution	25	Not low
tert-Butyl peroxybenzoate	19	Not low
tert-Butyl peroxy-2-ethylhexanoate	10	Low
Cumyl hydroperoxide, 85% with cumene	5	Low
Dibenzoyl peroxide	16	Not low
Di-tert-butyl peroxide	12	Not low
Di-sec-butyl peroxydicarbonate	23[a]	Not low
Diisopropyl peroxydicarbonate	45[a]	Not low
2,5-Dimethyl-2,5-di-(tert-butylperoxy)hexyne-3	31	Not low
2,5-Dimethyl-2,5-di-(benzoylperoxy) hexane	9	Low
Di-n-propyl peroxydicarbonate	32[a]	Not low

Expansion with inert reference substances in assembly A

Air	6
Dimethyl phthalate	10
Mineral spirits	10.5
Paste of 60% $CaCO_3$ + 40% dimethyl phthalate	8
Water	10

Expansion with inert reference substances in assembly B

Air	5.5
Benzoic acid	7
Calcium carbonate (powder)	5
Kaolin clay	6

[a] *Performed at ambient temperature.*

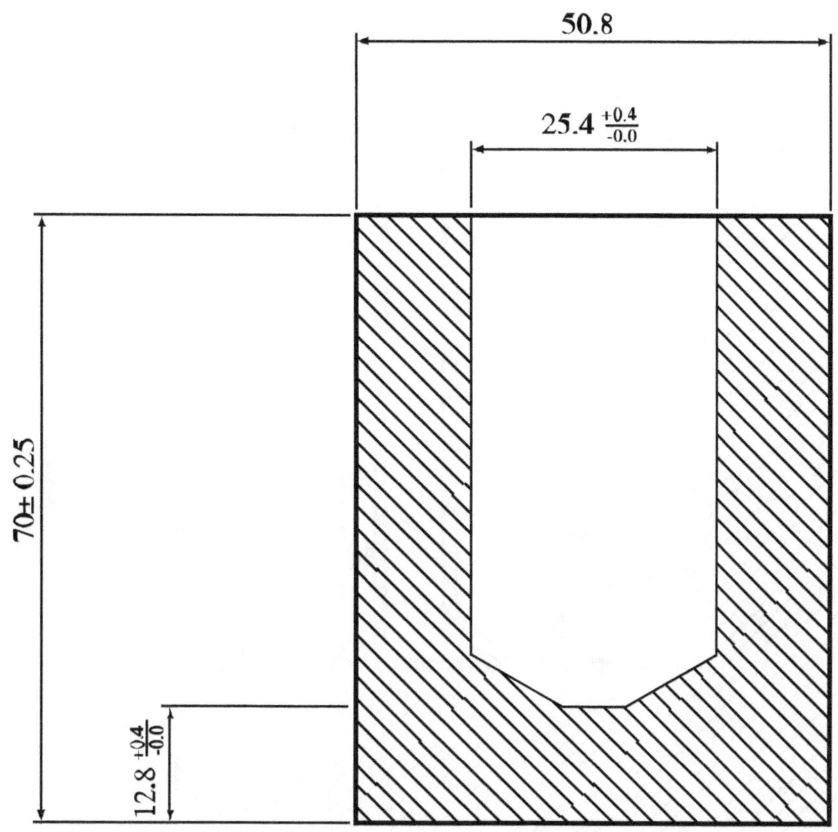

Figure 26.4.4.1: MODIFIED TRAUZL TEST

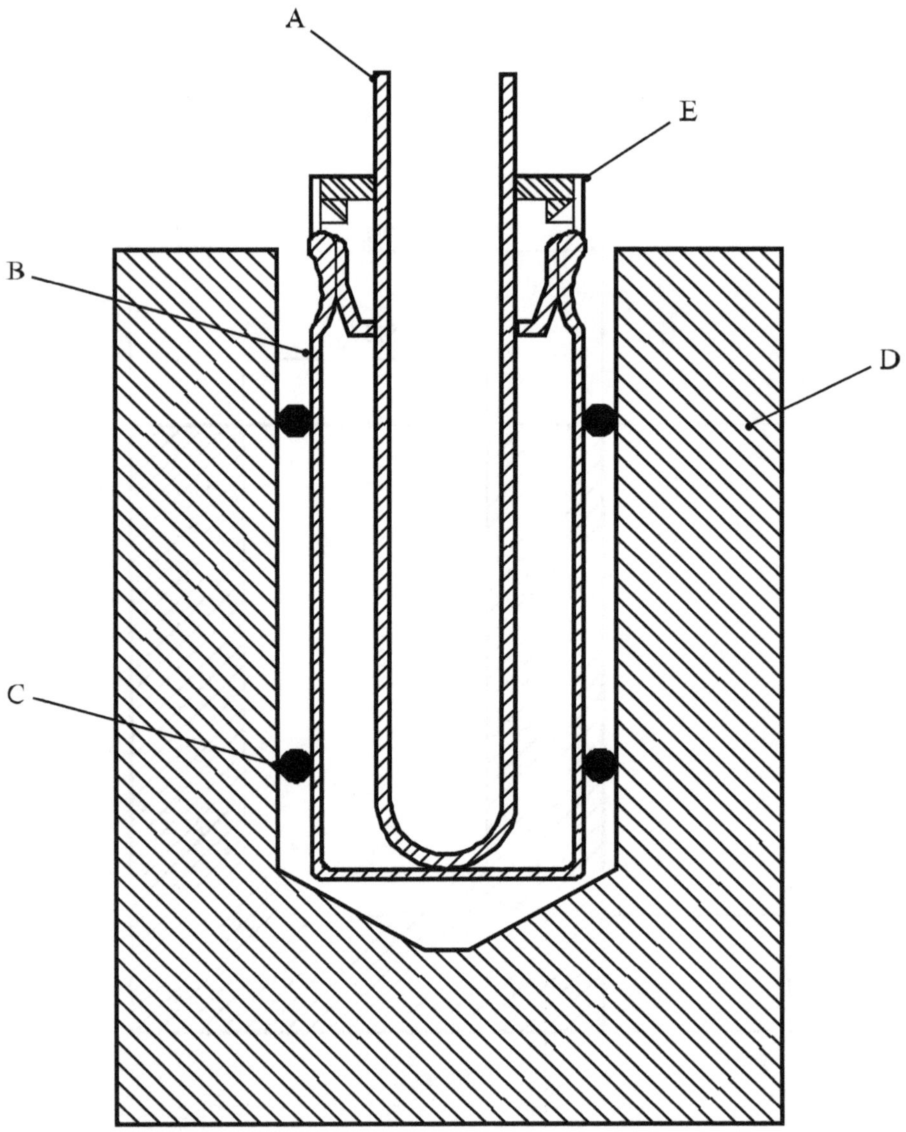

(A) Glass tube
(B) Glass vial (12 ml assembly A, 16 ml assembly B)
(C) "O"-ring (assembly A only)
(D) Lead block
(E) Stopper

Figure 26.4.4.2: ASSEMBLIES A AND B

26.4.5 *Test F.5: High pressure autoclave*

26.4.5.1 *Introduction*

This test is used to measure the specific energy of a substance. Varying quantities of substance are heated in a sealed vessel and the maximum pressure rise obtained for each sample size measured. The specific energy is a function of the maximum pressure rise. The test may be used to answer the question in box 12 of Figure 20.1.

26.4.5.2 *Apparatus*

The high pressure autoclave (Figure 26.4.5.1) is a cylindrical stainless steel vessel with a capacity of 96 ml and a working pressure of 150 MPa at a temperature of 473 K. The vessel is made from stainless steel type AISI 431, has an internal diameter of 38 mm and an internal height of 84 mm. An inner sample vessel (stainless steel type AISI 316, internal diameter 32 mm and internal height 77 mm) is used inside the autoclave. A coated Ni/Cr resistance wire (specific resistance of approximately 10 ohms/m), skip-wound around a piece of glass tube, is used to heat the sample by means of a constant current supply; a heating power of 50-150 W is supplied. Owing to the use of the inner vessel, the heat transfer from the substance to the autoclave is relatively small compared to the heat transfer that would exist in the situation without inner vessel. Therefore rapid heating will take place which will lead to an exothermic reaction involving self-heating and explosion. The pressure-time history of the explosion is determined. The pressure is registered by means of a piezo-electric transducer.

26.4.5.3 *Procedure*

The required amount of substance is weighed into the sample vessel after which the sample vessel is placed in the autoclave. The heating coil is connected to the lid of the autoclave which is then closed. Provisions are made to ensure that the whole heating coil is completely immersed in the substance. The ends of the heating wire are then connected to the power supply electrodes by low-resistance wire. The sample is then heated until explosion occurs. Experiments are normally performed using 5, 10, 15, 20 and 25 g of substance and the maximum pressure recorded. However, these quantities may need to be varied according to the bulk density and detonability of the substance.

26.4.5.4 *Test criteria and method of assessing results*

26.4.5.4.1 The results are assessed on the basis of the specific energy (F) which is a function of the maximum pressure rise (P_m). The initial sample mass (M_o) and the volume (V) of the reaction vessel are required to calculate the specific energy using the following equation:

$$V/M_o = F/Pm + C$$

Where V = internal volume of pressure vessel - volume of steel of inner sample vessel;
C = constant under the test conditions; and
F = determination from the slope of the plot of V/M_o versus $1/P_m$.

The explosive power of a substance is based on the value of the specific energy F only.

26.4.5.4.2 The test criteria are as follows:

"Not Low": - Specific energy greater than 100 J/g.

"Low": - Specific energy $\geq$ 5 J/g and $\leq$ 100 J/g.

"No": - Specific energy less than 5 J/g.

26.4.5.5 *Examples of results*

Substance	**F-value (J/g)**	**Result**
2,2'-Azodi(isobutyronitrile)	101	Not low
tert-Butyl peroxybenzoate	110	Not low
tert-Butyl peroxy-2-ethylhexanoate	56	Low
Cumyl hydroperoxide, 80% in cumene	60	Low
Dibenzoyl peroxide, 75% with water	41	Low
Di-tert-butyl peroxide	140	Not low
Dicetyl peroxydicarbonate	No reaction	No
Dicumyl peroxide, 40% with inert solid	No reaction	No
Dilauroyl peroxide	8	Low
Dilauroyl peroxide, 42%, stable dispersion, in water	1.3	No

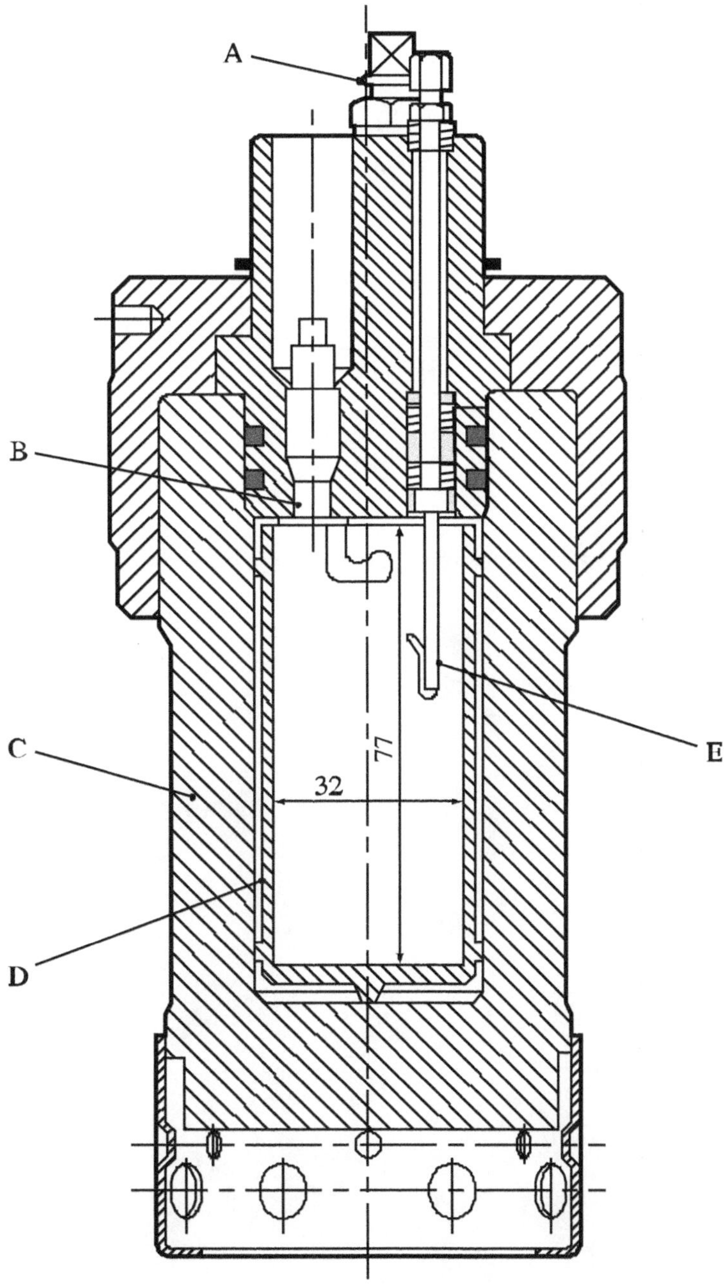

(A) Valve
(B) Pressure transducer
(C) Autoclave
(D) Inner sample vessel
(E) Electrodes

Figure 26.4.5.1: HIGH-PRESSURE AUTOCLAVE

SECTION 27

TEST SERIES G

27.1 Introduction

27.1.1 Test series G comprises tests and criteria concerning the determination of the effect of a thermal explosion of a substance as packaged for transport, requested in box 10 of Figure 20.1. The test is only needed for substances which show a violent effect in tests involving heating under defined confinement (test series E).

27.2 Test methods

27.2.1 The question "Can it explode as packaged for transport?" (box 10 of Figure 20.1) is answered on the basis of the results of one of the test methods in Table 27.1.

Table 27.1: TEST METHODS FOR TEST SERIES G

Test code	Name of test	Section
G.1	Thermal explosion test in package[a]	27.4.1
G.2	Accelerating decomposition test in package	27.4.2

[a] *Recommended test.*

27.2.2 Both tests are considered equivalent for all substances excluding water-wetted solids. For water-wetted solids, in the event that the results of the two tests are not in agreement, the results of test G.1 have precedence over the results of test G.2.

27.3 Test conditions

27.3.1 The test of Series G should be applied to packages of substance (not more than 50 kg) in the condition and form in which they are offered for transport.

27.3.2 ***The preliminary procedure (see section 20.3) should be performed before carrying out these tests.***

27.4 Series G test prescriptions

27.4.1 *Test G.1: Thermal explosion test in package*

27.4.1.1 *Introduction*

This test is used to determine the potential for thermal explosion in a package. It may be used to answer the question in box 10 of Figure 20.1.

27.4.1.2 *Apparatus and materials*

27.4.1.2.1 Packaging (not larger than that needed for 50 kg of substance), substance, a suitable heating device (e.g. 2 kW for 25 kg of substance) and equipment for measuring temperatures.

27.4.1.3 *Procedure*

The test is applied to packed substances in the condition and form in which they are offered for transport. The method of obtaining the thermal explosion is by heating the substance as homogeneously as possible with an electric heating coil within the package. The surface temperature of the heating coil should not be so high that premature ignition of the substance can take place. It may be necessary to use more than one heating coil. The package should be mounted on a stand to keep it upright. The heating system is switched on and the temperature of the substance continuously recorded. The heating rate should be about 60 °C/hour. The difference in temperature between the substance at the top and bottom of the package should be as small as possible. ***It is advisable to make provision in advance for the remote destruction of the package in the event of heater failure.*** The test is performed in duplicate unless an explosion is observed.

27.4.1.4 *Test criteria and method of assessing results*

27.4.1.4.1 Observations are made on the evidence of an explosion of the package under investigation by fragmentation of the package. The results obtained are only valid for the package tested.

27.4.1.4.2 The test criteria are as follows:

"Yes": - A fragmentation of inner and/or outer packaging(s) into more than three pieces (excluding bottom and top parts of the packaging(s) show(s) that the substance under investigation can give an explosion of that package.

"No": - No fragmentation or a fragmentation into not more than three pieces shows that the substance under investigation does not explode in the package.

27.4.1.5 *Examples of results*

Substance	Packagings	Number of fragments [a]	Result
2,2'-Azodi-(isobutyronitrile)	4G, 30 kg	N.F.	No
tert-Butyl peroxybenzoate	1B1, 25 litres	> 30	Yes
tert-Butyl peroxybenzoate	6HG2, 30 litres	N.F.	No
tert-Butyl peroxy-2-ethylhexanoate	1B1, 25 litres	> 5	Yes
tert-Butyl peroxy-2-ethylhexanoate	6HG2, 30 litres	N.F.	No
tert-Butyl peroxy isopropyl carbonate	1B1, 25 litres	> 80	Yes
tert-Butyl peroxy isopropyl carbonate	6HG2, 30 litres	> 20	Yes
tert-Butyl peroxypivalate, 75% in solution	6HG2, 30 litres	N.F.	No
Dibenzoyl peroxide, 75% with water	4G, 25 kg	N.F.	No
2,2-Di-(tert-butylperoxy)butane, 50% in solution	3H1, 25 litres	N.F.	No
2,2-Di-(tert-butylperoxy)butane, 50% in solution	6HG2, 30 litres	N.F.	No

[a] *N.F. means no fragmentation.*

27.4.2 *Test g.2: Accelerating decomposition in the package test*

27.4.2.1 *Introduction*

This test is used to determine the potential for thermal explosion in a package. It may be used to answer the question in box 10 of Figure 20.1.

27.4.2.2 *Apparatus and materials*

The equipment for this test is a typical SADT test cell as represented in Figure 27.4.2.1 and as described for test H.1 in section 28.

27.4.2.3 *Procedure*

The test is carried out in accordance with the procedure described for test H.1 in section 28. The temperature of the test cell may be set approximately 10 °C above the SADT, if known. The SADT test cell is designed to readily vent any significant pressure developed during the decomposition of a test sample. The cell covers are simply set in place and held by gravity alone.

27.4.2.4 *Test criteria and method of assessing results*

27.4.2.4.1 The condition of the sample, the package, the test cell, and the immediate surroundings provide a measure of the violence of the decomposition reaction of the product/package configuration tested.

27.4.2.4.2 The test criteria are defined in terms of the condition of the test cell, the package and the occurrence of an explosion and are as follows:

"Yes": - Significant disruption of the inside of the test cell would be noted. The outer cover may be lifted off and carried at least two metres indicating significant internal cell pressure. The test package would be severely damaged, breaking into at least three fragments.

"No": - Slight to no disruption of the test cell. The outer cover may be lifted off but not carried further than two meters from the cell. The test package may be disrupted and damaged e.g. split inner packaging, torn carton.

27.4.2.5 *Examples of results*

Substance	**Package**	**Result**
tert-Butyl peroxyacetate, 75% in solution	6HG2, 20 litres	Yes
tert-Butyl peroxybenzoate	6HG2, 20 litres	No
tert-Butyl peroxy-2-ethylhexanoate	6HG2, 20 litres	No
tert-Butyl peroxy isopropyl carbonate, 75% in solution	6HG2, 20 litres	No
tert-Butyl peroxypivalate, 75% in solution	6HG2, 20 litres	No
Dibenzoyl peroxide	4G, 0.454 kg	Yes
2,5-Dimethyl-2,5-di-(tert-butylperoxy)-hexyne-3	6HG2, 20 litres	Yes

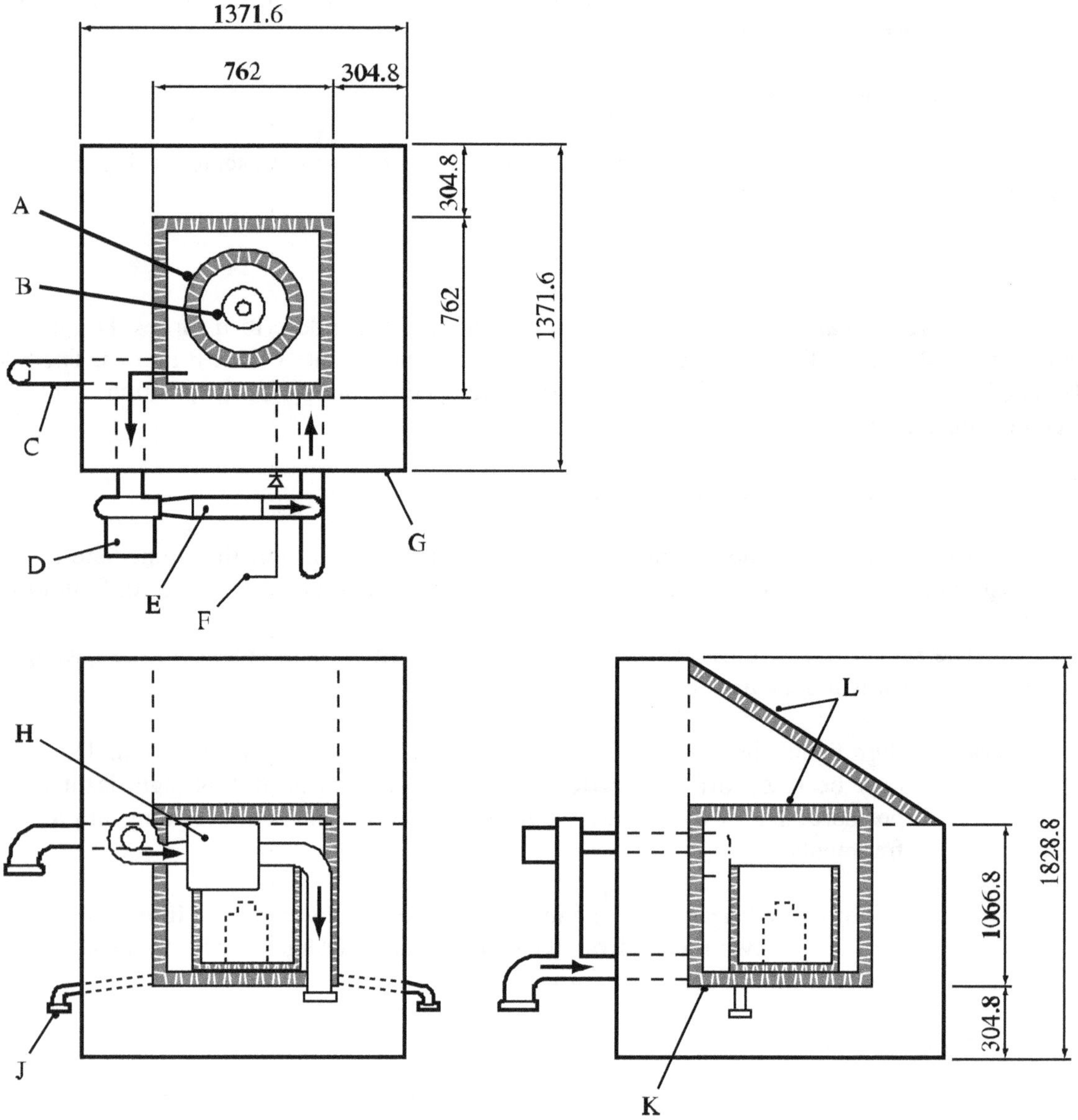

(A)	Test container	(B)	Test package
(C)	Air duct	(D)	Fan
(E)	Heater	(F)	Carbon dioxide
(G)	Frame	(H)	Recirculator
(J)	Pipe drains with caps	(K)	Insulation
(L)	Insulated covers		

Figure 27.4.2.1: ACCELERATING DECOMPOSITION TEST IN THE PACKAGE

SECTION 28

TEST SERIES H

28.1 Introduction

This test series contains test methods for the determination of the self-accelerating decomposition temperature (SADT). The SADT is defined as the lowest temperature at which self-accelerating decomposition may occur with a substance in the packaging as used in transport. The SADT is a measure of the combined effect of the ambient temperature, decomposition kinetics, package size and the heat transfer properties of the substance and its packaging. To assist in interpreting the results, models can be used in which the main resistance to heat flow is:

(a) At the boundary i.e. packaging (Semenov model);

(b) Within the substance (Frank-Kamentskii model); or

(c) A combination of these (Thomas model).

The text should be used in conjunction with the temperature control requirements given in sub-section 2.5.3.4 of the Model Regulations.

References: *N.N. Semenov, Z. Physik, 48, 1928, 571.*
D.A. Frank-Kamentskii, Zhur. Fiz. Khim., 13, 1939, 738.
P.H. Thomas, Trans. Faraday Soc., 54, 1958, 60.

28.2 Test methods

28.2.1 Test series H comprises tests and criteria concerning the thermal stability of substances at transport temperatures or concerning the determination of whether a substance meets the definition of a self-reactive substance.

28.2.2 Each test involves either storage at a fixed external temperature and observation of any reaction initiated or storage under near adiabatic conditions and measurement of the rate of heat generation versus temperature. The test methods included in test series H are given in Table 28.1. Each of the methods listed is applicable to solids, liquids, pastes and dispersions.

Table 28.1: TEST METHODS FOR TEST SERIES H

Test code	Name of test	Section
H.1	United States SADT test[a]	28.4.1
H.2	Adiabatic storage test (AST)[b]	28.4.2
H.3	Isothermal storage test (IST)	28.4.3
H.4	Heat accumulation storage test[c]	28.4.4

a *Recommended test for substances transported in packagings.*
b *Recommended test for substances transported in packagings, IBCs or tanks.*
c *Recommended test for substances transported in packagings, IBCs or small tanks.*

The list of tests is not exhaustive; other tests may be used provided that they give the correct SADT of the substance as packaged for transport.

28.2.3 When necessary (SADT ≤ 50 °C for organic peroxides, SADT ≤ 55 °C for self-reactive substances), control and emergency temperatures may be derived from the SADT using Table 28.2.

Table 28.2: DERIVATION OF CONTROL AND EMERGENCY TEMPERATURES

Type of receptacle	SADT [a]	Control temperature	Emergency temperature
Single packagings and IBCs	≤ 20 °C	20 °C below SADT	10 °C below SADT
	over 20 °C to 35 °C	15 °C below SADT	10 °C below SADT
	over 35 °C	10 °C below SADT	5 °C below SADT
Portable tanks	< 50 °C	10 °C below SADT	5 °C below SADT

[a] *i.e. the SADT of the substance as packaged for transport.*

28.2.4 If a substance is being tested to determine whether it is a self-reactive substance of Division 4.1, a test of series H, or a suitable alternative test, should be performed to determine if its SADT would be less than or equal to 75 °C when transported in a 50 kg package.

28.2.5 The results obtained for the largest commercial package are applicable to smaller packages of similar construction and material provided that the heat transfer per unit mass is no smaller than from the larger package.

28.3 Test conditions

28.3.1 The preliminary procedure (see section 20.3) should be performed and the effect of heating under confinement (test series E) determined before performing SADT tests. ***Safety precautions should be taken which allow for the possibility of catastrophic failure of the test vessel and for the hazards arising from the ignition of secondary fuel-air mixtures and the evolution of toxic decomposition products. Substances liable to detonate should only be tested with special precautions.***

28.3.2 The test selected should be conducted in a manner which is representative, both in size and material, of the package to be transported. For transport in metal packagings, IBCs or tanks, it may be necessary to include a representative amount of the metal in the sample tested i.e. representative of both the metal(s) and the area of contact.

28.3.3 ***Extra care should be taken when handling samples which have been tested since changes may have occurred rendering the substance more unstable and more sensitive. Tested samples should be destroyed as soon as possible after the test.***

28.3.4 Samples which have been tested at a particular temperature and are apparently unreacted may be used again, for screening purposes only, provided extra care is taken. Fresh samples should be used for the actual determination of the SADT.

28.3.5 If the complete package is not tested, the heat loss data used for the determination of the SADT should be representative of the package, IBC or tank offered for transport. The heat loss per unit of mass of the package, IBC or tank may be determined by calculation (taking account of the quantity of substance, dimensions of the package, heat transfer in the substance and the heat transfer through the packaging to the environment) or by measuring the half-time of cooling of the package filled with the substance or another substance having similar physical properties. The heat loss per unit mass, L (W/kg.K), can be calculated from the half-time of cooling, $t_{1/2}$ (s), and the specific heat, C_p (J/kg.K), of the substance using the formula:

$$L = \ln 2 \times \left(C_p / t_{1/2} \right)$$

28.3.6 The half-time of cooling can be determined by measuring the period of time in which the temperature difference between the sample and surroundings is decreased by a factor of 2. For example, for liquids, the packaging may be filled with dibutyl phthalate or dimethyl phthalate and this heated to about 80 °C. Water should not be used as erratic results may be obtained through evaporation / condensation. The temperature drop is measured at the centre of the package over the temperature range which includes the expected SADT. For scaling, it may be necessary continuously to monitor the temperature of the substance and surroundings and then use linear regression to obtain the coefficients of the equation:

$$\ln \{T - T_a\} = c_o + c \times t$$

where:

	T	=	substance temperature (°C);
	T_a	=	ambient temperature (°C);
	c_o	=	ln{Initial substance temperature - initial ambient temperature}; and
	c	=	L/C_p;
	t	=	time (s).

28.3.7 Examples of the heat loss characteristics of some typical packages are given in Table 28.3. The actual value obtained will depend on the shape, wall thickness, surface coating etc. of the packaging.

Table 28.3: HEAT LOSS PER UNIT MASS FROM PACKAGES, IBCs AND TANKS

Type of receptacle	Normal capacity (litres)	Filling	Heat loss per unit mass (L) (mW/K.kg)
For liquids:			
1A1	50	47.5 kg DMP[a]	63
1H1	50	47.5 kg DMP[a]	94
1H1	200	200 kg water	56
3H1 (black)	60	47.5 kg DMP[a]	105
6HG2	30	35.0 kg DMP[a]	69
IBC 31 HA1	500	500 kg water	51
Tank	3 400	3 400 kg water	18[b]
Tank container (insulated)	20 000	14 150 kg isododecane	1.7
For solids:			
1G	38	28.0 kg DCHP[c]	35
1G	50	37.0 kg DCHP[c]	29
1G	110	85.0 kg DCHP[c]	22
4G	50	32.0 kg DCHP[c]	27

[a] *DMP = dimethyl phthalate.*

[b] *Calculated using a heat transfer coefficient of 5 $W/m^2.K$.*

[c] *Dicyclohexyl phthalate (solid).*

28.4 Series H test prescriptions

28.4.1 *Test H.1: United States SADT test*

28.4.1.1 *Introduction*

This method determines the minimum constant temperature air environment at which auto-accelerative decomposition occurs for a substance in a specific package. Packages of up to 220 litres may be tested by this method. An indication of the explosion hazard from the decomposition reaction can also be obtained.

28.4.1.2 *Apparatus and materials*

28.4.1.2.1 The test substance and packaging should be representative of that intended for commercial use. The packaging forms an essential part of the test.

28.4.1.2.2 The apparatus consists of a test chamber in which the air surrounding the package under test can be maintained at a constant temperature for a period of at least ten days.

28.4.1.2.3 The test chamber should be constructed such that:

(a) It is well insulated;

(b) Thermostatically controlled air circulation is provided so as to maintain a uniform air temperature within ± 2 °C of the desired temperature; and

(c) The minimum separation distance from the package to the wall is 100 mm.

Any type of oven may be used provided that it is capable of meeting the temperature control requirements and will not ignite any decomposition products. Examples of a suitable small-package and large-package oven are given in 28.4.1.2.4 and 28.4.1.2.5 respectively.

28.4.1.2.4 A small-package oven may be constructed from an open head 220 litre steel drum. This will readily accommodate packages of up to 25 litres capacity. Construction details are shown in Figure 28.4.1.1. Larger packages may be tested as long as a 100 mm clearance between the package and the oven wall is maintained.

28.4.1.2.5 An expendable large-package oven may be made from 50 mm × 100 mm lumber formed into a cubic frame 1.2 m on each side, which is lined inside and out with 6 mm thick waterproof plywood and insulated all around with 100 mm fibreglass insulation. Construction details are shown in Figure 28.4.1.2. The frame should be hinged on one side to permit loading and unloading of test drums. The floor should be provided with 50 mm × 100 mm lumber on edge, spaced 200 mm on centres to keep the test container off the floor and allow free air circulation around the package. Cleats should run normal to the door to permit a fork-lift to move the drums. A circulating fan should be located on the side opposite the door. The air flow should be from the upper corner of oven to the fan discharge at the diagonally opposite lower corner. An electric heater of 2.5 kW is suitable for heating the air. Thermocouples should be placed in the air intake and outflow ducts as well as the top, middle and bottom of the oven. For substances with an SADT below ambient temperature, the test should be performed in a cooling chamber or solid carbon dioxide should be used for cooling the oven.

28.4.1.2.6 The package should be equipped with a thermowell which positions the thermocouple at the package mid-point. The thermowell may be constructed of glass, stainless steel or other suitable material but should be introduced in a manner which does not reduce package strength or venting capability.

28.4.1.2.7 Continuous temperature-measuring and recording equipment, which is protected from fire and explosion hazards, is required.

28.4.1.2.8 ***The tests should be performed in an area which provides adequate protection from fire and explosion hazards, and from toxic fumes. A safety distance, e.g. 90 m, from public roads and occupied buildings is recommended. If there may be toxic fumes, longer safety distances may be required.***

28.4.1.3 *Procedure*

28.4.1.3.1 The package is weighed. A thermocouple is inserted in the package to be tested such that the temperature in the centre of the sample can be monitored. If the required oven temperature is below ambient, the oven is switched on and the inside of the oven cooled to the desired temperature before placing the package in the oven. If the required oven temperature is at or above ambient temperature, the package is inserted in the oven at ambient temperature and the oven then switched on. There should be a minimum clearance of 100 mm between the package and the sides of the oven.

28.4.1.3.2 The sample is heated and the temperature of the sample and test chamber continuously monitored. The time is noted at which the sample temperature reaches a temperature 2 °C below the test chamber temperature. The test is then continued for a further seven days or until the sample temperature rises to 6 °C or more above the test chamber temperature if this occurs sooner. Note the time taken for the sample to rise from the 2 °C below the test chamber temperature to its maximum temperature.

28.4.1.3.3 When the test is completed, cool the sample and remove it from the test chamber. Note the variation of temperature with time. If the package remains intact, note the percentage mass loss and determine any changes in composition. Dispose of the sample as soon as possible.

28.4.1.3.4 If the temperature of the sample does not exceed the oven temperature by 6 °C or more then repeat the test with a new sample in an oven at a 5 °C higher temperature. The self-accelerating decomposition temperature (SADT) is defined as the lowest oven temperature at which the sample temperature exceeds the oven temperature by 6 °C or more. If the substance is being tested to determine if temperature control is necessary, perform sufficient tests to determine the SADT to the nearest 5 °C or to determine if the SADT is greater than or equal to 60 °C. If the substance is being tested to determine if it meets the SADT criterion for a self-reactive substance, perform sufficient tests to determine if the SADT for a 50 kg package is 75 °C or less.

28.4.1.4 *Test criteria and method of assessing results*

28.4.1.4.1 The SADT is reported as the lowest temperature at which the sample exceeds the oven temperature by 6 °C or more. If the sample temperature does not exceed the oven temperature by 6 °C or more in any test, the SADT is recorded as being greater than the highest oven temperature used.

28.4.1.5 *Examples of results*

Substance	**Sample mass (kg)**	**Packaging**	**SADT (°C)**
tert-Amyl peroxybenzoate	18.2	6HG2, 22.8 litres	65
tert-Butyl peroxyacetate (60%)	7.2	6HG2, 22.8 litres	75
Dibenzoyl peroxide	0.45	1G	70
Di-(4-tert-butylcyclohexyl) peroxydicarbonate	43	1G	40
2,5-Diethoxy-4-morpholinobenzene-diazonium zinc chloride (66%)	30	1G, 50 litres	50
2-(N-ethoxycarbonyl-N-phenylamino)-3-methoxy-4-(N-methyl-N-cyclohexyl-amino)benzenediazonium zinc chloride (62 %)	10	6HG1, 25 litres	50

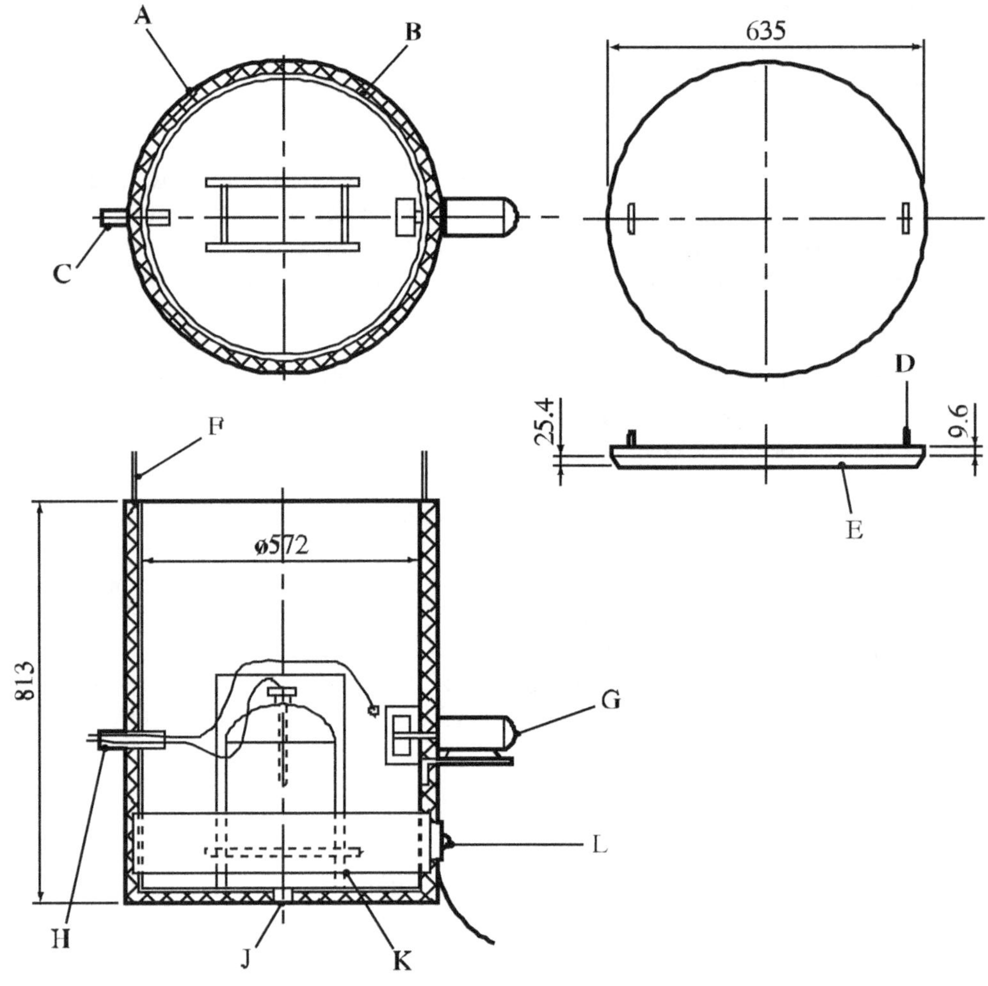

(A)	Insulation 25 mm thick	(B)	220 litres open top drum
(C)	19 mm pipe	(D)	9.6 mm eye bolt in steel cover
(E)	Insulation on steel cover	(F)	3 mm control cable
(G)	Fan	(H)	Thermocouples and controls
(J)	Drain	(K)	25 mm angle stand
(L)	2 kW drum heater		

Figure 28.4.1.1: SMALL PACKAGE OVEN

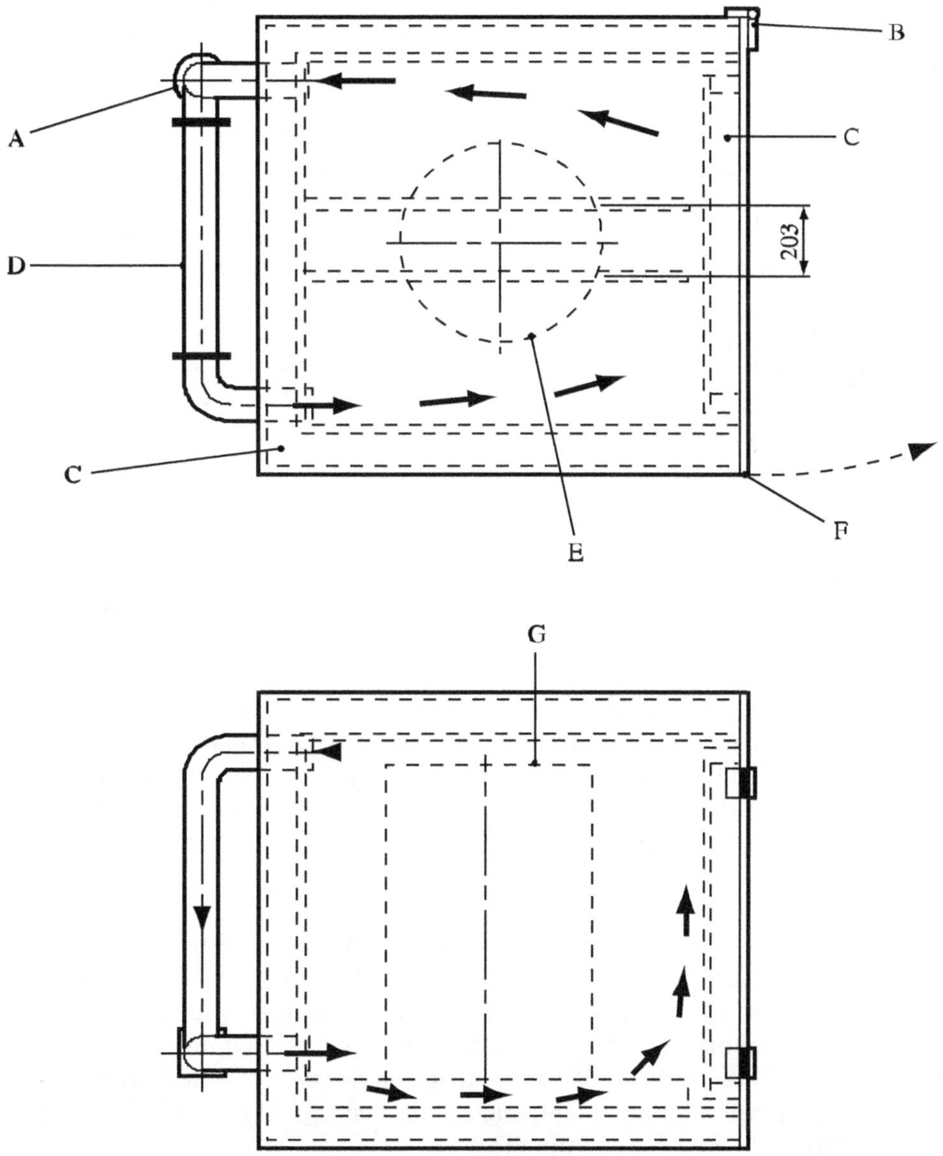

(A)	Fan	(B)	Hinges (2)
(C)	Insulation	(D)	Heater
(E)	Drum	(F)	Latch
(G)	Drum (e.g. 0.58 m x 0.89 m)		

Figure 28.4.1.2: LARGE PACKAGE OVEN (top view and side view)

28.4.2 ***Test H.2: Adiabatic storage test***

28.4.2.1 *Introduction*

28.4.2.1.1 This test method determines the rate of heat generation produced by a reacting substance as a function of temperature. The heat generation parameters obtained are used with the heat loss data relating to the package to determine the SADT of a substance in its packaging. The method is appropriate for every type of packaging, including IBCs and tanks.

28.4.2.1.2 Measurements can be performed in the temperature range from -20 °C to 220 °C. The smallest temperature rise that can be detected corresponds to a rate of heat generation of 15 mW/kg. The upper limit is determined by the capacity of the cooling system to safely cool the substance (up to 500 W/kg if water is used as the coolant). Although the test is not perfectly adiabatic, the heat losses are less than 10 mW. The maximum error is 30% at 15 mW/kg and 10% from 100 mW/kg to 10 W/kg.

28.4.2.1.3 If the cooling system is activated at a stage where the rate of heat generation exceeds the cooling capacity, an explosion may occur. ***The test site should therefore be carefully selected to reduce to a minimum the possible hazards from an explosion and of a possible subsequent gas explosion of the decomposition products (secondary explosion).***

28.4.2.2 *Apparatus and materials*

28.4.2.2.1 The apparatus consists of a glass Dewar vessel (1.0 or 1.5 litre) to contain the sample, an insulated oven with a differential control system to keep the temperature in the oven to within 0.1 °C of the sample temperature and an inert lid for the Dewar vessel. In special cases, sample holders of other construction materials may have to be used. An inert heating coil and cooling tube pass through the lid into the sample. Pressure build-up in the Dewar vessel is prevented by a 2 m long PTFE capillary tube which passes through the insulated lid. A constant power heating unit is used for internal heating of the substance to a pre-set temperature or for calibration. Internal heating and cooling may be stopped or started automatically at pre-set temperatures. In addition to the cooling system, a secondary safety device is fitted which disconnects the power supply to the oven at a pre-set temperature. A schematic drawing of the AST is given in Figure 28.4.2.1.

28.4.2.2.2 The temperature of the substance is measured at its centre by means of thermocouples or platinum resistance sensors placed in a steel or glass tube. The temperature of the surrounding air is measured at the same height as the sample temperature, also with thermocouples or platinum resistance sensors. Continuous temperature measuring and recording equipment is required to monitor the temperature of substance and the air in the oven. This equipment should be protected from fire and explosion. For substances with an SADT below ambient temperature, the test should be performed in a cooling chamber or solid carbon dioxide should be used for cooling the oven.

28.4.2.3 *Procedure*

28.4.2.3.1 Calibration procedure

The calibration procedure is as follows:

(a) Fill the Dewar vessel with sodium chloride, dibutyl phthalate or with a suitable oil and place it in the vessel holder of the AST oven;

(b) Heat the sample in 20 °C steps using the internal heating system at a known power rating, e.g. 0.333 or 1.000 W, and determine the heat losses at 40 °C, 60 °C, 80 °C and 100 °C;

(c) Use the data to determine the heat capacity of the Dewar vessel using the method given in 28.4.2.4.

28.4.2.3.2 Test procedure

The test procedure is as follows:

(a) Fill the Dewar vessel with the weighed sample, including a representative amount of packaging material (if metal), and place it in the vessel holder of the AST oven;

(b) Start the temperature monitoring and then increase the sample temperature using the internal heater to a pre-set temperature at which detectable self-heating may occur. The specific heat of the substance can be calculated from the temperature rise, heating time and heating power;

(c) Stop the internal heating and monitor the temperature. If no temperature rise due to self-heating is observed within 24 hours, increase the temperature by 5 °C. Repeat this procedure until self-heating is detected;

(d) When self-heating is detected, the sample is allowed to heat up under adiabatic conditions to a pre-set temperature where the rate of heat generation is less than the cooling capacity, upon which the cooling system is activated;

(e) On cooling, determine the loss in mass, if any, and determine the change in composition (if desired).

28.4.2.4 *Test criteria and method of assessing results*

28.4.2.4.1 Calculate the rate of temperature drop, A (°C/h), of the Dewar vessel at the different temperatures used in the calibration procedure. Plot a graph through these values in order to allow determination of the rate of temperature drop at any temperature.

28.4.2.4.2 Calculate the heat capacity, H (J/°C), of the Dewar vessel using the formula:

$$H = \frac{3600 \times E_1}{A + B} - (M_1 \times Cp_1)$$

where:

E_1	=	power applied to the internal heater (W)
A	=	rate of temperature drop at the temperature of calculation (°C/h)
B	=	slope of the curve on internal heating (calibration substance) at the temperature of calculation (°C/h)
M_1	=	mass of calibration substance (kg)
Cp_1	=	specific heat of calibration substance (J/kg.°C)

28.4.2.4.3 Determine the heat loss, K (W), using:

$$K = \frac{A \times (H + M_1 \times Cp_1)}{3600}$$

at each desired temperature and draw a graph through these values.

28.4.2.4.4 Calculate the specific heat, Cp_2 (J/kg.°C) of the substance using:

$$Cp_2 = \frac{3600 \times (E_2 + K)}{C \times M_2} - \frac{H}{M_2}$$

where: E_2 = power applied to the internal heater (W)
C = slope of the curve on internal heating (sample) at the temperature of calculation (°C/h)
M_2 = mass of sample (kg)

28.4.2.4.5 Calculate the heat generation, Q_T (W/kg), of the substance at intervals of 5 °C using the following formula for each temperature:

$$Q_T = \frac{(M_2 \times Cp_2 + H) \times \frac{D}{3600} - K}{M_2}$$

where: D = slope of the curve during the self-heating at the temperature of calculation (°C/h)

28.4.2.4.6 Plot the calculated rates of heat generation per unit of mass as a function of the temperature in a graph with linear scales and draw the best-fit curve through the plotted points. Determine the heat losses per unit of mass, L (W/kg.°C) of the specific package, IBC or tank (see 28.3.5). Draw a straight line of gradient L tangential to the heat generation curve. The intersection of the straight line and the abscissa is the critical ambient temperature i.e. the highest temperature at which the substance as packaged does not show self-accelerating decomposition. The SADT is the critical ambient temperature (°C) rounded to the next higher multiple of 5 °C. An example is given in Figure 28.4.2.2.

28.4.2.5 *Examples of results*

Substance	**Mass (kg)**	**Packaging**	**Heat loss per unit mass (mW/kg.K)**	**SADT (°C)**
Azodicarbonamide	30	1G	100	> 75
tert-Butyl peroxybenzoate	25	6HG2	70	55
tert-Butyl peroxy-2-ethylhexanoate	25	6HG2	70	40
tert-Butyl peroxypivalate	25	6HG2	70	25

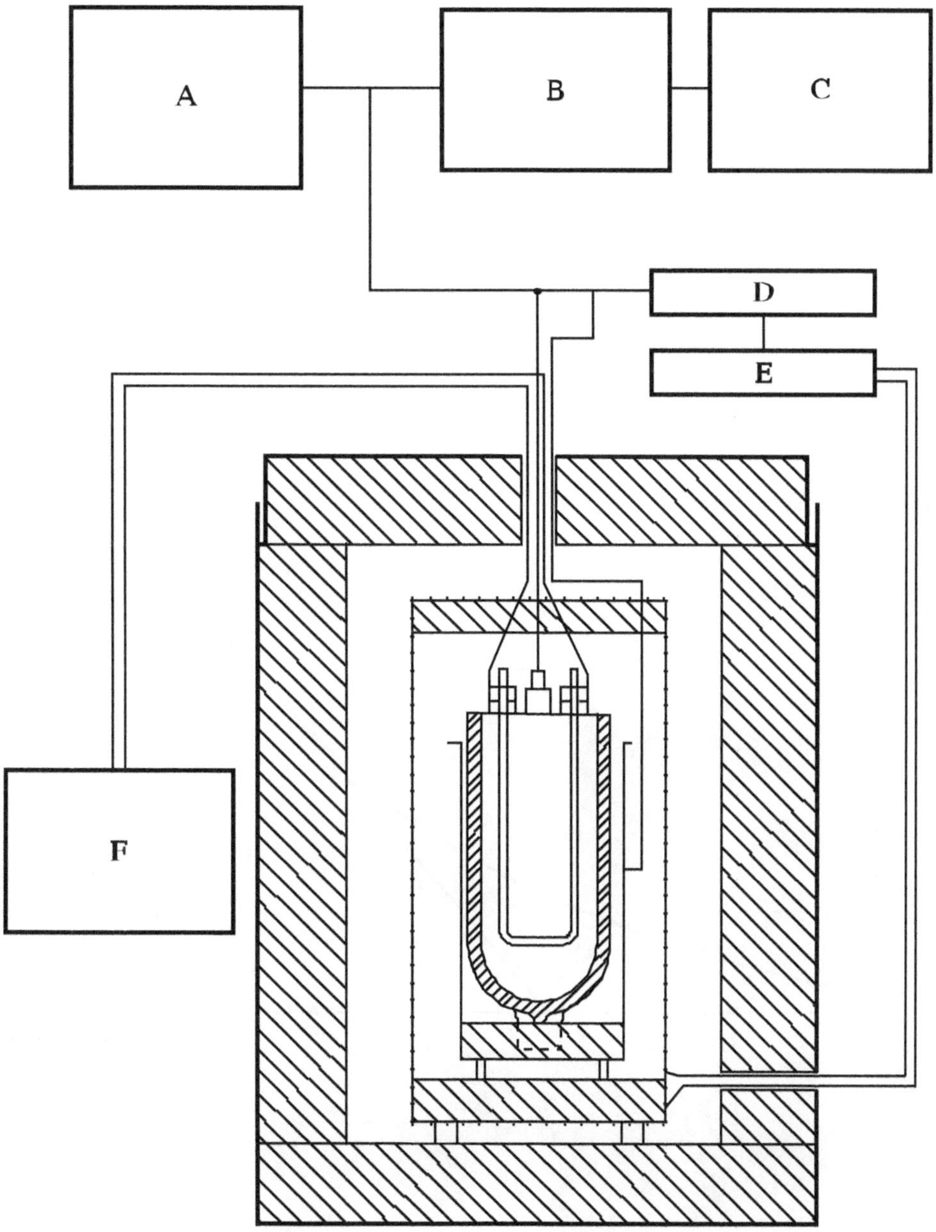

(A) Multi-point recorder and temperature controller (10 mV)
(B) External zero set
(C) Recorder set for greatest accuracy
(D) Control
(E) Relay
(F) Internal pre-heater

Figure 28.4.2.1: ADIABATIC STORAGE TEST

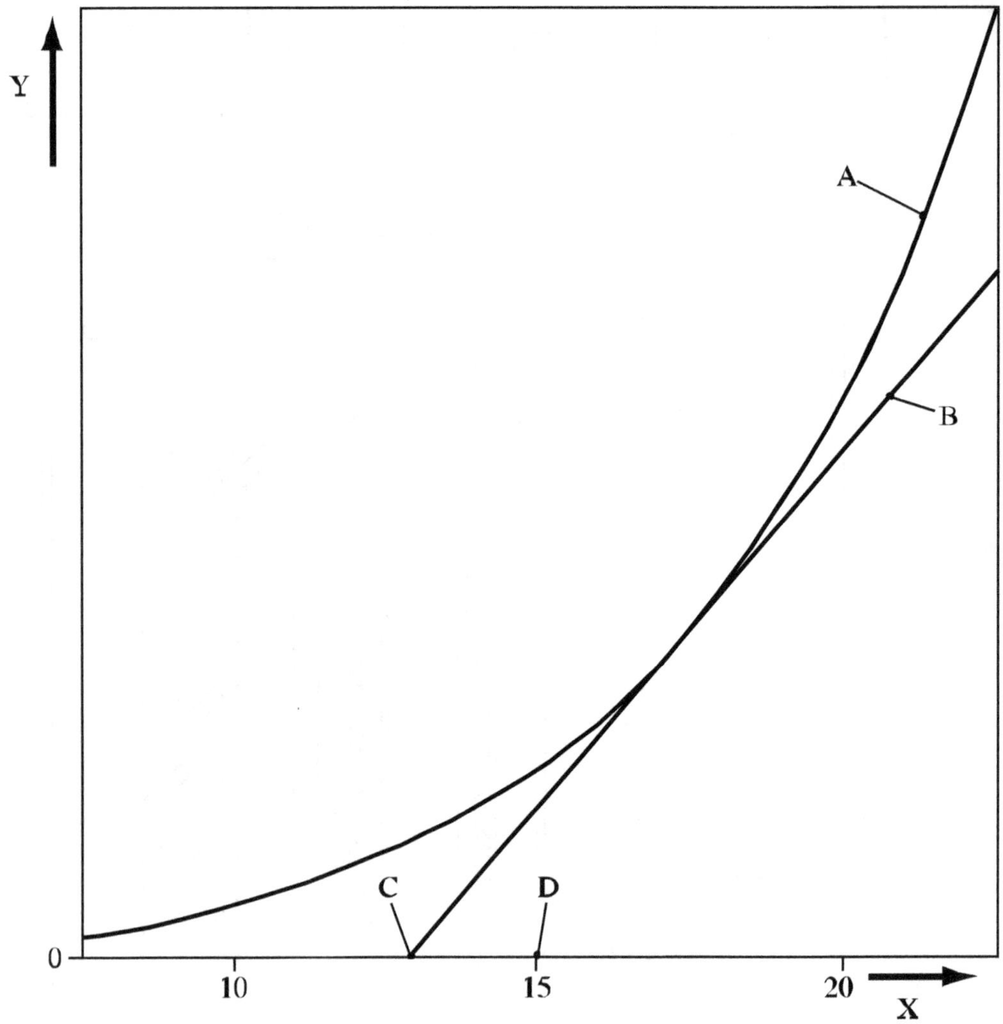

(A) Heat generation curve
(B) Line with gradient equal to the rate of heat loss and tangential to the heat generation curve
(C) Critical ambient temperature (intercept of heat loss line with the abscissa)
(D) Self-accelerating decomposition temperature (SADT) - critical ambient temperature rounded up to next higher multiple of 5 °C
(X) Temperature
(Y) Heat flow (generation or loss) per unit mass

Figure 28.4.2.2: EXAMPLE OF DETERMINATION OF SADT

28.4.3 ***Test H.3: Isothermal storage test (IST)***

28.4.3.1 *Introduction*

28.4.3.1.1 This method determines the rate of heat generation produced by reacting or decomposing substances as a function of time at constant temperature. The heat generation parameters obtained are used with the heat loss data relating to the package to determine the SADT of a substance in its packaging. The method is appropriate for every type of packaging including IBCs and tanks. Some substances may show an increase in the rate of heat generation with increasing decomposition (e.g. due to autocatalysis or induced decomposition). This property is also taken into account by this test method.

28.4.3.1.2 Measurements can be performed in the temperature range of -20 °C to 200 °C. Heat generation values can be measured from 5 mW/kg to 5 W/kg. The thermal resistance between the sample holder and aluminium block via the heat flow meters is about 0.1 W/°C. The apparatus is able to measure rates of heat generation from 15 mW/kg to 1 500 mW/kg with a maximum error of 30% at 15 mW/kg and 5% from 100 to 1 500 mW/kg.

28.4.3.1.3 The robust construction of the apparatus, the relatively small sample size and the well-defined operating conditions enable the test to be performed in an ordinary laboratory. The effects of a thermal explosion, e.g. fragmentation of the sample holder and generation of pressure, should be contained within the apparatus.

28.4.3.2 *Apparatus and materials*

28.4.3.2.1 The apparatus consists of an air-insulated heat sink (an aluminium block) which is kept at a constant temperature by means of controlled heating. A cryostat is used to maintain temperatures below 40 °C. The heat controller is capable of maintaining the temperature to within 0.2 °C of the set temperature. The temperature of the heat sink is measured with a platinum-resistance sensor. The two holes drilled in the block contain heat flow meters (e.g. Peltier elements). A schematic drawing of the IST is given in Figure 28.4.3.1. For substances with an SADT below ambient temperature, the test should be performed in a cooling chamber or solid carbon dioxide should be used for cooling the oven.

28.4.3.2.2 Holders are placed on both heat flow meters: one with the sample and one with an inert substance. Both holders are identical and have a volume of 70 cm^3. The amount of substance in each holder is about 20 g. The holders are made of glass or stainless steel. The steel must be compatible with the test substance. When using a glass holder, it is provided with a long capillary tube which prevents pressure build-up in the holder and evaporation of the sample.

28.4.3.2.3 The voltage difference resulting from the difference in heat flow from sample holder to the heat sink and inert substance holder to the heat sink is continuously recorded as a function of time (differential measurement) by a recorder or computer.

28.4.3.3 *Procedure*

28.4.3.3.1 Calibration procedure

Before a measurement can be performed, the blank signal and the sensitivity of the heat flow meter need to be determined by the following calibration procedure.

(a) Set the IST at the selected test temperature;

(b) Insert a heating coil in the sample holder. Fill the sample and reference holders with inert material (e.g. sodium chloride or milled glass beads) ensuring that the heating coil is completely covered with material. Place both holders in the IST;

(c) Determine the blank signal (the output of the recorder when no electric power is applied to the heating coil);

(d) Determine the sensitivity of the heat flow meter using two or three different electrical heating powers which lie within the expected range of heat generation of the sample to be tested.

28.4.3.3.2 Test procedure

The test procedure is as follows:

(a) Set the IST for the desired test temperature;

(b) Fill the sample holder with the weighed sample and with a representative quantity of packaging material (if metal) and insert the holder into the apparatus. The amount of sample should be sufficient to give a rate of heat generation between 5 mW to 1 500 mW per kg of substance;

(c) Start monitoring the rate of heat production. The results from the first 12 hours of the test should not be used because this period is required for temperature equilibration. The duration of each test depends on the test temperature and on the rate of heat production. The test should last for at least 24 hours after the 12 hour equilibration period but can then be stopped if the rate of heat generation is falling from the maximum or if the rate of heat generation is greater than 1.5 W/kg;

(d) At the end of the test the change in sample mass should be determined;

(e) The test is repeated with new samples at temperature intervals of 5 °C so that there are seven results with a maximum heat generation rate between 15 and 1 500 mW/kg.

28.4.3.4 *Test criteria and method of assessing results*

28.4.3.4.1 Calculate the sensitivity, S (mW/mV), of the instrument at the various electrical powers used in the calibration procedure using the following formula:

$$S = \frac{P}{U_d - U_b}$$

where: P = electric power (mW)
U_d = dummy signal (mV)
U_b = blank signal (mV)

28.4.3.4.2 Use these values and the test data to calculate the maximum rate of heat generation, Q (mW/kg), at the different test temperatures using the formula:

$$Q = \frac{(U_s - U_b) \times S}{M}$$

where: U_s = sample signal (mV)
M = mass (kg)

28.4.3.4.3 Plot the calculated maximum rate of heat generation per unit of mass as a function of the test temperature on a graph with linear scales and draw the best-fit curve through the plotted points. Determine the heat losses per unit of mass, L (W/kg.°C) of the specific package, IBC or tank (see 28.3.5). Draw a straight line of gradient L tangential to the heat generation curve. The intersection of the straight line and the abscissa is the critical ambient temperature i.e. the highest temperature at which the substance as packaged does not show self-accelerating decomposition. The SADT is the critical ambient temperature (°C) rounded to the next higher multiple of 5 °C. An example is given in Figure 28.4.3.2.

28.4.3.5 *Examples of results*

Substance	**Mass (kg)**	**Packaging**	**Heat loss per unit mass (mW/kg.K)**	**SADT (°C)**
Azodicarbonamide	30	1G	100	> 75
tert-Butyl peroxybenzoate	25	6HG2	70	55
tert-Butyl peroxy-2-ethylhexanoate	25	6HG2	70	40
tert-Butyl peroxypivalate	25	6HG2	70	25
2,5-Diethoxy-4-morpholinobenzenediazonium zinc chloride (90%)	25	1G	150	45
2,5-Diethoxy-4-morpholinobenzenediazonium tetrafluoroborate (97%)	25	1G	15	55
2,5-Diethoxy-4-(phenylsulphonyl)-benzenediazonium zinc chloride (67%)	25	1G	15	50
2-(N-ethoxycarbonyl-N-phenylamino)-3-methoxy-4-(N-methyl-N-cyclohexylamino)-benzenediazonium zinc chloride (62%)	25	1G	15	45
3-Methyl-4-(pyrrolidin-1-yl) benzenediazonium tetrafluoroborate (95%)	25	1G	15	55

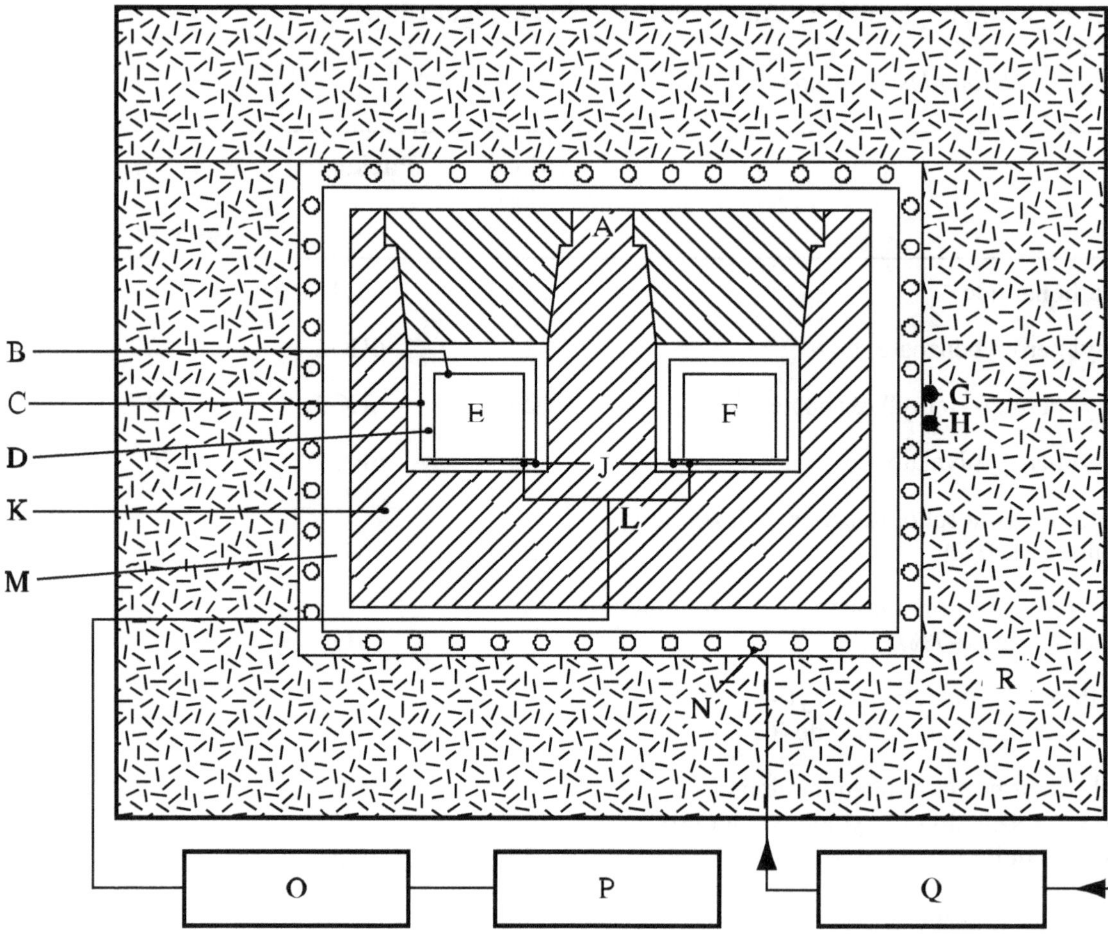

(A)	Platinum resistance thermometer	(B)	Sample vessel
(C)	Cylindrical holder	(D)	Air spaces
(E)	Sample	(F)	Inert material
(G)	Platinum resistance sensor for temperature control	(H)	Platinum resistance sensor for safety control
(J)	Peltier elements	(K)	Aluminium block
(L)	Electric circuit	(M)	Air space
(N)	Heating wires	(O)	Amplifier
(P)	Recorder	(Q)	Temperature controller
(R)	Glass wool		

Figure 28.4.3.1: ISOTHERMAL STORAGE TEST

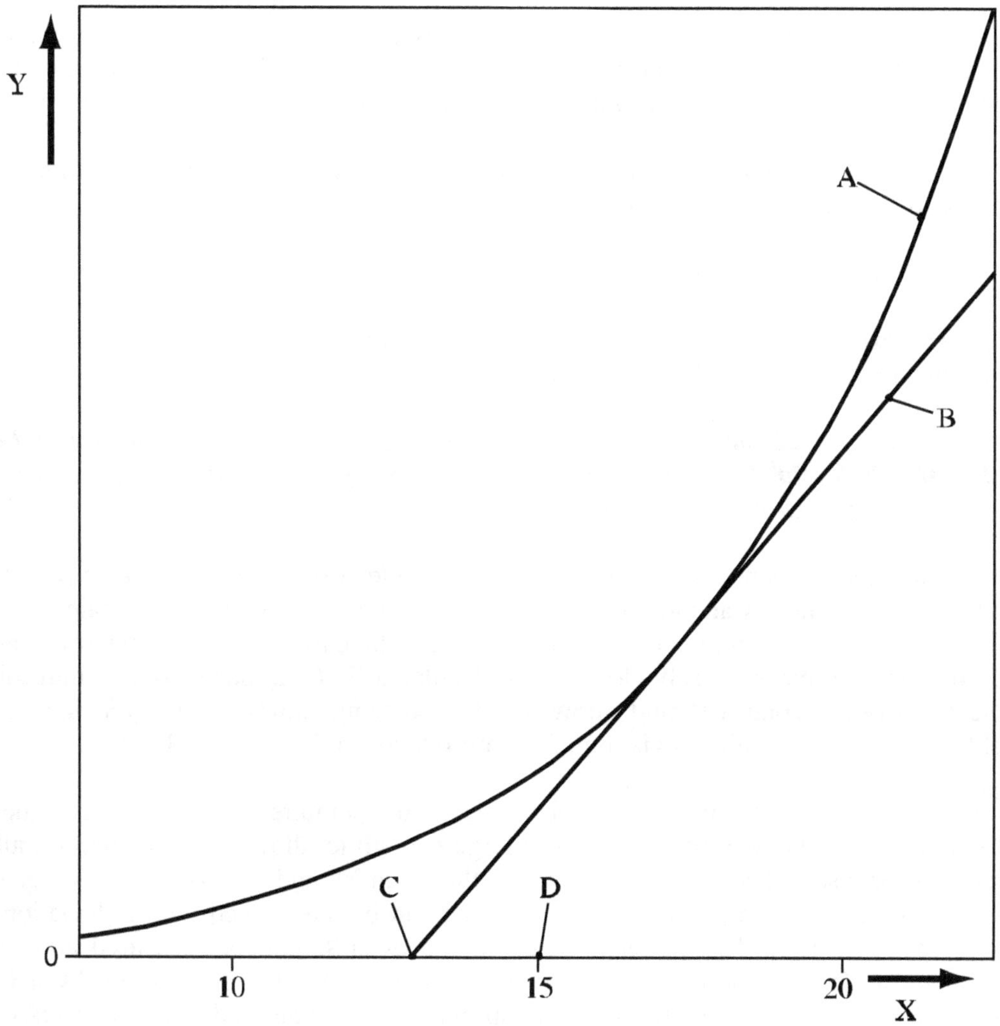

(A) Heat generation curve
(B) Line with gradient equal to the rate of heat loss and tangential to the heat generation curve
(C) Critical ambient temperature (intercept of heat loss line with the abscissa)
(D) Self-accelerating decomposition temperature (SADT) - critical ambient temperature rounded up to next higher multiple of 5 °C
(X) Temperature
(Y) Heat flow (generation or loss) per unit mass

Figure 28.4.3.2: EXAMPLE OF DETERMINATION OF SADT

28.4.4 ***Test H.4: Heat accumulation storage test***

28.4.4.1 *Introduction*

28.4.4.1.1 This method determines the minimum constant air environment temperature at which thermally unstable substances undergo exothermic decomposition at conditions representative of the substance when packaged as for transport. The method is based on the Semenov theory of thermal explosion i.e. the main resistance to heat flow is considered to be at the vessel walls. The method can be used for the determination of the SADT of a substance in its packaging, including IBCs and small tanks (up to 2 m^3).

28.4.4.1.2 The effectiveness of the method depends on selecting a Dewar vessel with heat loss per unit mass characteristics similar to the package offered for transport.

28.4.4.2 *Apparatus and materials*

28.4.4.2.1 The experimental equipment consists of a suitable test chamber, appropriate Dewar vessels with closures, temperature probes and measuring equipment.

28.4.4.2.2 ***The test should be performed in a test cell capable of withstanding fire and overpressure and, preferably, should be fitted with a pressure relief system e.g. a blow out panel.*** The recording system should be housed in a separate observation area.

28.4.4.2.3 For tests at temperatures up to 75 °C, a double walled metal chamber (ca. 250 mm inner diameter, 320 mm outer diameter and 480 mm high made from 1.5 to 2.0 mm thick stainless steel sheet) is used with fluid from a temperature-controlled circulating bath passed between the walls at the desired temperature. The test chamber is loosely closed by an insulated lid (e.g. made from 10 mm thick polyvinyl chloride). The temperature control should allow the desired temperature for a liquid inert sample in the Dewar vessel to be maintained with a deviation of not more than ± 1 °C for up to 10 days.

28.4.4.2.4 Alternatively, and particularly for tests at temperatures above 75 °C, a thermostatically controlled drying oven (which may be fan-assisted) large enough to allow air circulation on all sides of the Dewar vessel may be used. The air temperature in the oven should be controlled so that the desired temperature for a liquid inert sample in the Dewar vessel can be maintained with a deviation of not more than ± 1 °C for up to 10 days. The air temperature in the oven should be measured and recorded. It is recommended that the door of the oven be fitted with a magnetic catch or replaced by a loosely fitting insulated cover. The oven may be protected by an appropriate steel liner and the Dewar vessel housed in a wire mesh cage.

28.4.4.2.5 For tests at sub-ambient temperatures, a double walled chamber (e.g. a freezer) of suitable size provided with a loose door or lid (e.g. with a magnetic closure) may be used. The temperature of the air in the chamber should be controlled to ± 1 °C of the set temperature.

28.4.4.2.6 Dewar vessels, with their closure system, are used with heat loss characteristics which are representative of the maximum size of package offered for transport. The closure of the Dewar vessel should be inert. Particularly for solids, cork or rubber bungs may be used. A closure system for use with liquids with low or medium volatility is illustrated in Figure 28.4.4.1. Samples which are highly volatile at the test temperature should be tested in a pressure-tight metal vessel fitted with a pressure relief valve. The pressure vessel is placed in the Dewar vessel and the effect of the heat capacity of the metal vessel taken into account by calculation.

28.4.4.2.7 The heat loss characteristics of the system used, i.e. Dewar vessel and closure, should be established (see 28.3.6) prior to performance of the test. Since the closure system has a significant effect on the heat loss characteristics, these can be adjusted to some extent by varying the closure system. In order to achieve the required level of sensitivity, Dewar vessels with sample capacities below 0.5 litres should not be used.

28.4.4.2.8 Dewar vessels, filled with 400 ml of substance, with a heat loss of 80 to 100 mW/kg.K are normally suitable for representing a 50 kg package. For larger packages, IBCs or small tanks larger Dewar vessels with lower heat losses per unit mass should be used. For example, spherical 1 litre Dewar vessels having heat loss characteristics in the range 16 to 34 mW/kg.K may be suitable for IBCs and small tanks.

28.4.4.3 *Procedure*

28.4.4.3.1 Set the test chamber at the selected storage temperature. Fill the Dewar vessel, to 80% of its capacity, with the substance under test and note the mass of the sample. Solids should be moderately compressed. Insert the temperature probe into the centre of the sample. Seal the lid of the Dewar in place and insert the Dewar vessel in the test chamber, connect the temperature recording system and close the test chamber.

28.4.4.3.2 The sample is heated and the temperature of the sample and test chamber continuously monitored. The time is noted at which the sample temperature reaches a temperature 2 °C below the test chamber temperature. The test is then continued for a further seven days or until the sample temperature rises to 6 °C or more above the test chamber temperature if this occurs sooner. Note the time taken for the sample to rise from 2 °C below the test chamber temperature to its maximum temperature.

28.4.4.3.3 If the sample survives, cool and remove it from the test chamber and carefully dispose of it as soon as possible. The percentage mass loss and change in composition may be determined.

28.4.4.3.4 Repeat the test with fresh samples, varying the storage temperature in 5 °C steps. If the substance is being tested to determine if temperature control is necessary, perform sufficient tests to determine the SADT to the nearest 5 °C or to determine if the SADT is greater than or equal to 60 °C. If the substance is being tested to determine if it meets the SADT criterion for a self-reactive substance, perform sufficient tests to determine if the SADT for a 50 kg package is 75 °C or less.

28.4.4.4 *Test criteria and method of assessing results*

28.4.4.4.1 The SADT is reported as the lowest temperature at which the sample undergoes exceeds the test chamber temperature by 6 °C or more. If the sample temperature does not exceed the test chamber temperature by 6 °C or more in any test, the SADT is recorded as being greater than the highest storage temperature used.

28.4.4.5 *Examples of results*

Substance	**Sample mass (kg)**	**Dewar heat loss (mW/kg.K)**	**SADT (°C)**
Azodicarbonamide	0.28	74	> 75
Azodicarbonamide, 90% with 10% activator	0.21	70	55
2,2'-Azodi(isobutyronitrile)	0.18	62	50
Benzene-1,3-disulphohydrazide, 50%	0.52	81	70
tert-Butyl hydroperoxide, 80% with 12% di-tert-butyl peroxide	0.30	72	100[a]
tert-Butyl peroxyneodecanoate, 40%	0.42	65	25
tert-Butyl peroxy-3,5,5-trimethylhexanoate	0.38	79	60
Dibenzoyl peroxide, 50%	0.25	91	60
Di-(4-tert-butylcyclohexyl) peroxydicarbonate	0.19	79	45
2,2-Di-(tert-butylperoxy)butane, 50%	0.31	88	80
Di-(2-ethylhexyl) peroxydicarbonate	0.39	64	0
2,5-Diethoxy-4-morpholinobenzenediazonium zinc chloride (66%)	0.25	58	45
Diisotridecyl peroxydicarbonate	0.38	80	10
Peroxyacetic acid, 15%, with 14% hydrogen peroxide (type F)	1.00	33	> 50[b]

[a] *In a pressure vessel contained with a 2 litres Dewar vessel.*
[b] *In a spherical, 1 litre Dewar vessel.*

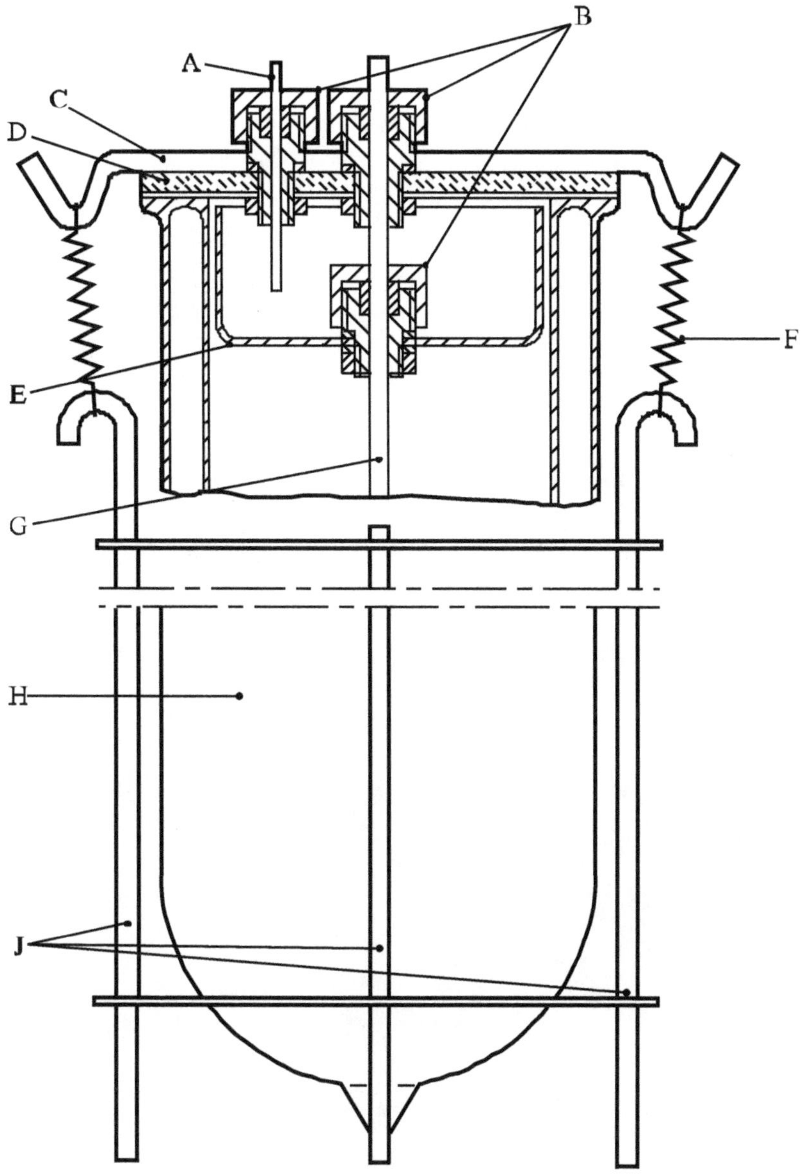

(A)	PTFE capillary tube	(B)	Special screw fittings (PTFE or Al) with O-ring seal
(C)	Metal strip	(D)	Glass lid
(E)	Glass beaker base	(F)	Spring
(G)	Glass protective tube	(H)	Dewar vessel
(J)	Steel retaining device		

Figure 28.4.4.1: DEWAR VESSEL WITH CLOSURE FOR TESTING LIQUIDS AND WETTED SOLID

PART III

CLASSIFICATION PROCEDURES, TEST METHODS AND CRITERIA RELATING TO CLASS 2, CLASS 3, CLASS 4, DIVISION 5.1, CLASS 8 AND CLASS 9

CONTENTS OF PART III

NOTE 1: *The country or organisation of origin of each test method is indicated in brackets after each test name.*

NOTE 2: *Unless otherwise specified, all test methods given in Part III of the Manual are recommended tests as only one test is given for each property.*

CONTENTS OF PART III (continued)

CONTENTS OF PART III (continued)

CONTENTS OF PART III (continued)

SECTION 30

INTRODUCTION TO PART III

30.1 Purpose

30.1.1 Part III of the Manual presents the United Nations schemes for the classification of:

(a) Flammable aerosols (see section 31 of this Manual and special provision 63 of Chapter 3.3 of the Model Regulations);

(b) Flammable liquids and liquid desensitized explosives of Class 3 (see section 32 of this Manual and Chapter 2.3 of the Model Regulations);

(c) Readily combustible solids and solid desensitized explosives of Division 4.1 (see sub-section 33.2 of this Manual and Chapter 2.4 of the Model Regulations);

(d) Pyrophoric and self-heating substances of Division 4.2 (see sub-section 33.3 of this Manual and Chapter 2.4 of the Model Regulations);

(e) Substances which in contact with water emit flammable gases of Division 4.3 (see sub-section 33.4 of this Manual and Chapter 2.4 of the Model Regulations);

(f) Oxidizing substances of Division 5.1 (see section 34 of this Manual and Chapter 2.5 of the Model Regulations);

(g) Corrosive properties of substances of class 8 (see section 37 of this Manual and Chapter 2.8 of the Model Regulations);

(h) Ammonium nitrate fertilizers capable of self-sustaining decomposition of Class 9 (see sub-section 38.2 of this Manual); and

(i) Lithium cells and batteries of Class 9 (see sub-section 38.3 of this Manual).

30.1.2 Part III contains some classification procedures, test methods and criteria which are also given in the Model Regulations. Sections 35 and 36 are reserved, to allow for possible future developments, for Classes 6 and 7 respectively.

30.2 Scope

The appropriate classification procedure should be undertaken before a new product is offered for transport. The producer or other applicant for classification of a new product should provide:

(a) Adequate information concerning the names and characteristics of the substance or article;

(b) The results of all relevant tests which have been done; and

(c) The proposed classification with any subsidiary risk requirements.

SECTION 31

CLASSIFICATION PROCEDURES, TEST METHODS AND CRITERIA RELATING TO FLAMMABLE AEROSOLS OF CLASS 2

31.1 Purpose

31.1.1 This section presents the United Nations scheme for the classification of flammable aerosols. The text should be used in conjunction with the classification principles given in Chapters 2.2 and 3.3 (special provision 63) of the Model Regulations, the flow charts given here in figures 31.1, 31.2 and 31.3 and the tests prescriptions given in sub-sections 31.4, 31.5 and 31.6 of this Manual.

31.1.2 The test procedures outlined here adequately assess the relative hazard of flammable aerosols so that an appropriate classification can be made.

31.1.3 For the purposes of this section the following definitions apply:

Aerosols or aerosol dispensers are non-refillable receptacles meeting the requirements of section 6.2.4 of the Model Regulations, made of metal glass or plastics and containing a gas, compressed, liquefied or dissolved under pressure, with or without a liquid, paste or powder, and fitted with a release device allowing the contents to be ejected as solid or liquid particles in suspension in a gas, as a foam, paste or powder or in a liquid state or in a gaseous state;

Flammable components are flammable liquids, flammable solids or flammable gases and gas mixtures. This designation does not cover pyrophoric, self-heating or water-reactive substances.

NOTE 1: *A flammable liquid means a liquid having a flash point of not more than 93 °C. Test methods for determining the flash point are given in sub-section 32.4 of this Manual;*

NOTE 2: *For the definition of flammable solids, see paragraph 2.4.2.2 of the Model Regulations. Classification procedures, test methods and criteria relating to flammable solids of Division 4.1 are given in sub-section 33.2 of this Manual;*

NOTE 3: *A flammable gas is a gas having a flammable range with air at 20 °C and a standard pressure of 101.3 kPa.*

31.2 Scope

31.2.1 Aerosols offered for transport shall be subjected to the classification procedures as set out in special provision 63 of Chapter 3.3 of the Model Regulations and, for flammability, shall be subjected to the classification procedures as set out in this section. The classification procedure shall be undertaken before a new product is offered for transport.

NOTE: *Aerosol dispensers not submitted to flammability classification procedures in this sub-section shall be classified as extremely flammable.*

31.3 Classification procedure for flammable aerosols

31.3.1 Aerosols shall be classified as flammable or extremely flammable according to their heat of combustion and to their flammable component contents, as follows:

(a) The aerosol product is classified as extremely flammable if the product contains 85% or more flammable components and the chemical heat of combustion exceeds or is equal to 30 kJ/g;

(b) The aerosol product is classified as non-flammable if the product contains 1% or less flammable components and the chemical heat of combustion is less than 20 kJ/g.

31.3.2 In the case of spray aerosols the classification shall be made taking into account the chemical heat of combustion and on the basis of the results of the ignition distance test, as follows:

(a) If the chemical heat of combustion is less than 20 kJ/g:

(i) The aerosol is classified as flammable if ignition occurs at a distance equal or greater than 15 cm but less than 75 cm;

(ii) The aerosol is classified as extremely flammable if ignition occurs at a distance of 75 cm or more;

(iii) If no ignition occurs in the ignition distance test, the enclosed space test shall be performed and in this case, the aerosol is classified as flammable if the time equivalent is less than or equal to 300 s/m^3 or the deflagration density is less than or equal to 300 g/m^3; otherwise the aerosol is classified as non-flammable;

(b) If the chemical heat of combustion is equal to or more than 20 kJ/g, the aerosol is classified as extremely flammable if ignition occurs at a distance of 75 cm or more; otherwise the aerosol is classified as flammable.

31.3.3 The chemical heat of combustion shall be determined following one of the methods described in the following standards: ASTM D 240, ISO/FDIS 13943:1999 (E/F) 86.1 to 86.3 and NFPA 30B.

31.3.4 In the case of foam aerosols, the classification shall be made on the basis of the results of the foam flammability test (see sub-section 31.6 of this Manual).

(a) The aerosol product shall be classified as extremely flammable if:

(i) the flame height is 20 cm or more and the flame duration is 2 s or more; or

(ii) the flame height is 4 cm or more and the flame duration is 7 s or more;

(b) The aerosol product which does not meet the criteria in (a) is classified as flammable if the flame height is 4 cm or more and the flame duration is 2 s or more.

31.3.5 The classification criteria for aerosols, spray aerosols and foam aerosols is summarized in figures 31.1, 31.2 and 31.3 respectively.

FIGURE 31.1: OVERALL PROCEDURE FOR CLASSIFICATION OF FLAMMABLE AEROSOLS

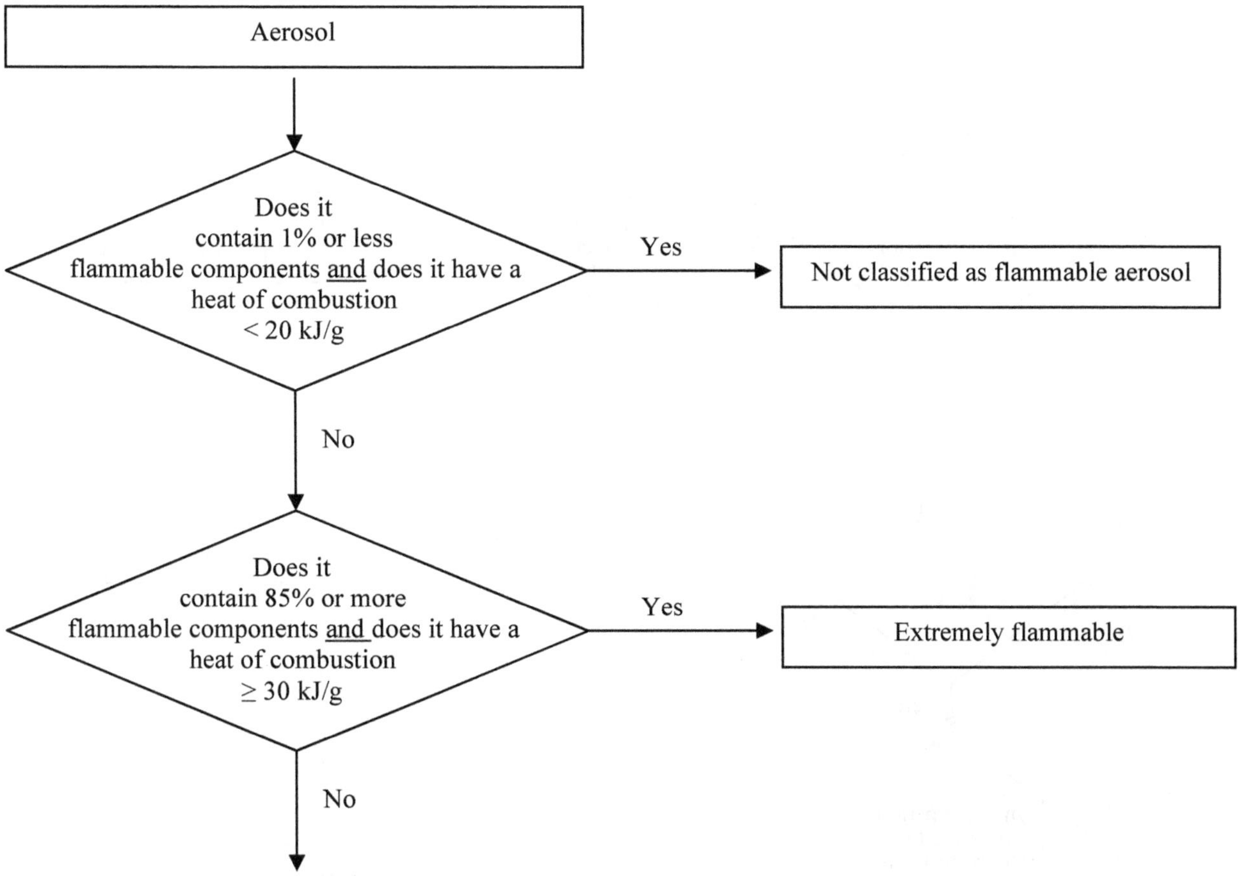

For spray aerosols, go to Figure 31.2.
For foam aerosols, go to Figure 31.3.

FIGURE 31.2: PROCEDURE FOR CLASSIFICATION OF SPRAY AEROSOLS

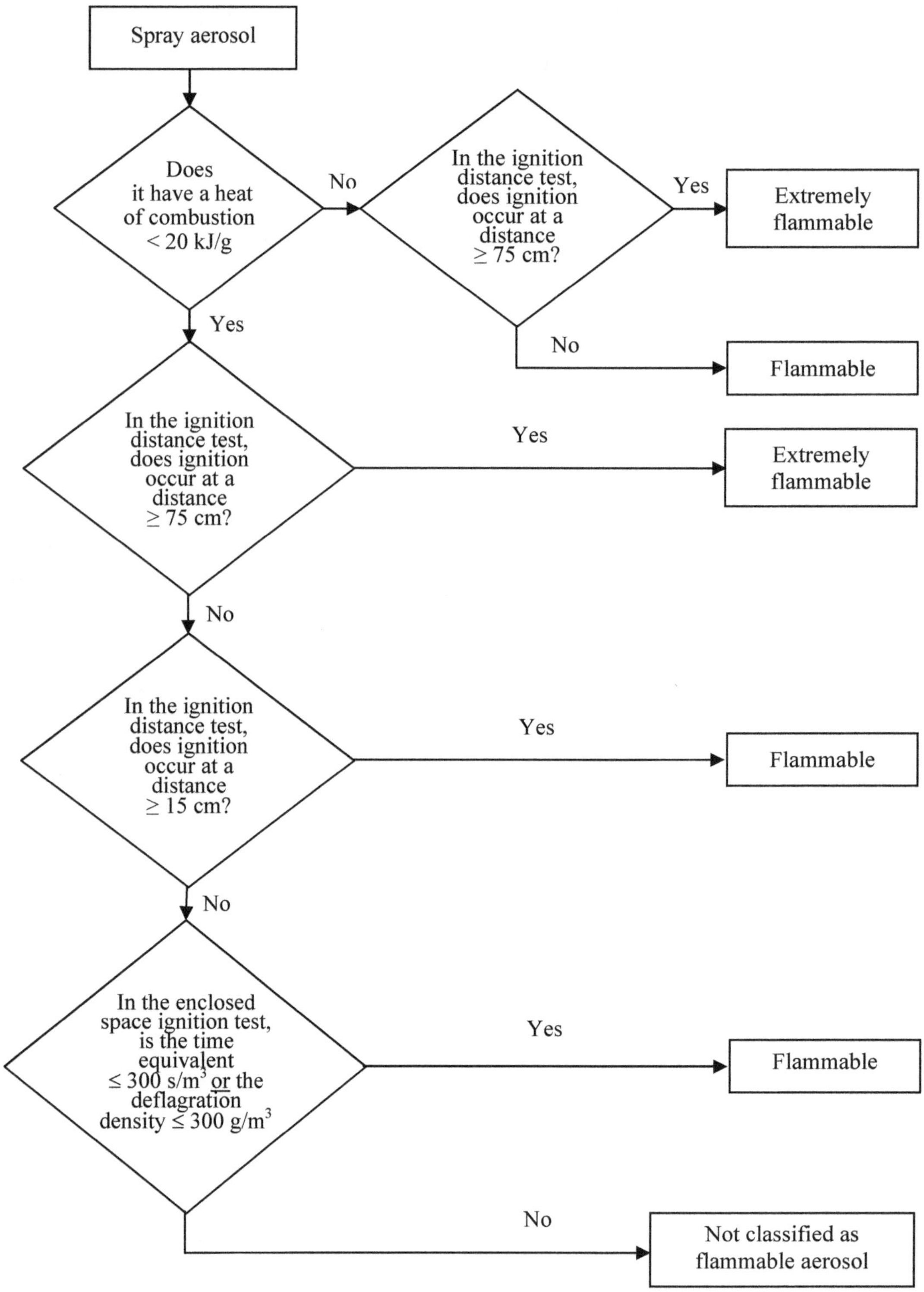

FIGURE 31.3: PROCEDURE FOR CLASSIFICATION OF FOAM AEROSOLS

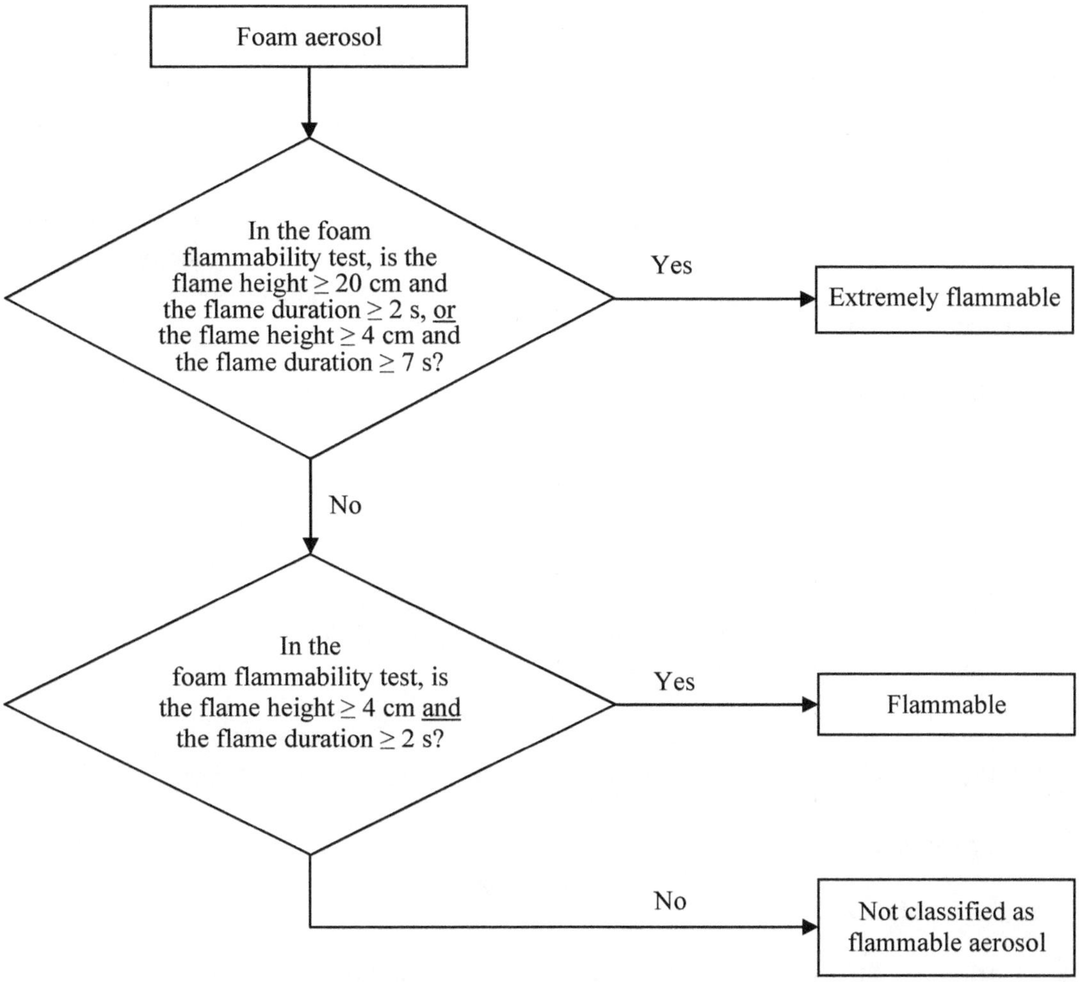

31.4 Ignition distance test for spray aerosols

31.4.1 *Introduction*

31.4.1.1 This test standard describes the method to determine the ignition distance of an aerosol spray in order to assess the associated flame risk. The aerosol is sprayed in the direction of an ignition source at intervals of 15 cm to observe if ignition and sustained combustion of the spray takes place. Ignition and sustained combustion is defined as when a stable flame is maintained for at least 5 s. The ignition source is defined as a gas burner with a blue, non-luminous flame 4-5 cm in height.

31.4.1.2 This test is applicable to aerosol products with a spray distance of 15 cm or more. Aerosol products with a spray distance of less than 15 cm such as dispensing foams, mousses, gels and pastes or fitted with a metering valve, are excluded from this test. Aerosol products that dispense foams, mousses, gels or pastes are subject to testing under the aerosol foam flammability test.

31.4.2 *Apparatus and material*

31.4.2.1 The following apparatus is required:

Water bath maintained at 20 °C	accurate to ± 1 °C
Calibrated laboratory scales (balance)	accurate to ± 0.1 g
Chronometer (stopwatch)	accurate to ± 0.2 s
Graduated scale, support and clamp	graduations in cm
Gas burner with support and clamp	
Thermometer	accurate to ± 1 °C
Hygrometer	accurate to ± 5%
Pressure gauge	accurate to ± 0.1 bar

31.4.3 *Procedure*

31.4.3.1 *General requirements*

31.4.3.1.1 Before testing, each aerosol dispenser shall be conditioned and then primed by discharging for approximately 1 s. The purpose of this action is to remove non-homogeneous material from the diptube.

31.4.3.1.2 The instructions of use shall be strictly followed, including whether the dispenser is intended to be used in the upright or inverted position. When shaking is required, shake immediately before testing.

31.4.3.1.3 The test shall be carried out in a draught-free environment capable of ventilation, with the temperature controlled at 20 °C ± 5 °C and relative humidity in the range 30 - 80%.

31.4.3.1.4 Each aerosol dispenser is to be tested:

(a) When full according to the complete procedure, with the gas burner in the range of 15 - 90 cm distance from the actuator of the aerosol can;

(b) When 10 - 12% full nominal (% by mass) only one test, either at 15 cm distance from the actuator when the spray from a full can did not ignite at all, or at the flame ignition distance of the spray of a full can plus 15 cm.

31.4.3.1.5 During the test, the can shall be positioned as indicated by label instructions. The ignition source shall be positioned accordingly.

31.4.3.1.6 The following procedure requires testing the spray at intervals of 15 cm between the burner flame and the aerosol actuator, in the range of 15 - 90 cm. It is efficient to start at 60 cm distance between burner flame and aerosol actuator. The distance between burner flame and aerosol actuator shall be increased by 15 cm in the case of an ignition of the spray at 60 cm distance. The distance shall be decreased by 15 cm in the case of no ignition at 60 cm distance between burner flame and aerosol actuator. The aim of the procedure is to determine the maximum distance between aerosol actuator and burner flame that leads to sustained combustion of the spray or to determine that ignition could not be obtained at 15 cm distance between the burner flame and the aerosol's actuator.

31.4.3.2 *Test procedure*

(a) A minimum of 3 full aerosol dispensers per product shall be conditioned to 20 °C ± 1 °C with at least 95% of the dispenser immersed in the water for at least 30 min before each test (if the aerosol is fully immersed, 30 min conditioning is sufficient);

(b) Comply with general requirements. Record the temperature and relative humidity of the environment;

(c) Weigh an aerosol dispenser and note its mass;

(d) Determine the internal pressure and initial discharge rate at 20 °C ± 1 °C (to eliminate faulty or partly filled aerosol dispensers);

(e) Support the gas burner on a flat horizontal surface or fix the burner to a support by means of a clamp;

(f) Ignite the gas burner; the flame shall be non-luminous and approximately 4-5 cm high;

(g) Place the actuator's exit orifice at the required distance from the flame. The aerosol shall be tested in the position it is designed to be used, e.g. upright or inverted;

(h) Level the actuator's orifice and burner flame, ensuring that the orifice is properly directed towards and aligned with the flame (see Figure 31.4.1). The spray shall be expelled through the top half of the flame;

(i) Comply with the general requirements regarding shaking of the dispenser;

(j) Actuate the valve of the aerosol dispenser, to discharge its contents for 5 s, unless ignition occurs. If ignition occurs, continue discharging and time the duration of the flame for 5 s, from the start of ignition;

(k) Note the ignition results for the distance between the gas burner and the aerosol dispenser in the table provided;

(l) If no ignition occurs during step (j), the aerosol shall be tested in alternative orientations, e.g. inverted for upright use products, to check if ignition is obtained;

(m) Repeat steps (g) to (l) twice more (a total of 3) for the same can at the same distance between the gas burner and the aerosol actuator;

(n) Repeat the test procedure for another two aerosol cans of the same product at the same distance between gas burner and aerosol actuator;

(o) Repeat steps (g) to (n) of the test procedure at a distance between 15 and 90 cm between the actuator of the aerosol can and the burner flame depending on the outcome of each test (see also 31.4.3.1.4 and 31.4.3.1.5);

(p) If no ignition occurs at 15 cm, the procedure is finished for initially full cans. The procedure is also finished when ignition and sustained combustion is obtained at a distance of 90 cm. If ignition could not be obtained at 15 cm distance, record that ignition did not occur. The maximum distance between burner flame and the aerosol's actuator for which an ignition and sustained combustion was observed is noted as the "ignition distance", in all other circumstances;

(q) One test shall also be conducted on 3 cans of 10 - 12% nominal fill level. These cans shall be tested at a distance between the aerosol's actuator and the burner flame of "the flame ignition distance of full cans + 15 cm";

(r) Discharge an aerosol can to a 10 - 12% nominal fill level (by mass) in bursts of 30 s maximum. Observe a 300 s minimum time period between bursts. During this interim period dispensers shall be placed in the water bath for conditioning;

(s) Repeat steps (g) to (n) for 10 - 12% nominal fill aerosol cans, omitting steps (l) and (m). This test shall only be performed with the aerosol in one position, e.g. upright or inverted, corresponding with that which produced the ignition (if any) for filled cans;

(t) Record all results in the Table 31.4 as shown below.

31.4.3.2.1 All experiments shall be performed in a fume hood in a room that may be well ventilated. Ventilation of the fume hood and room can be applied for at least 3 min after each test. Take all necessary safety precautions to prevent the inhalation of combustion products.

31.4.3.2.2 The cans with a 10 - 12% nominal fill level shall be tested only once. The result tables need only one result per can indicated.

31.4.3.2.3 When the test in the position in which the dispenser is designed to be used gives a negative result, the test shall be repeated in the position of the dispenser most likely to result in a positive result.

31.4.4 ***Test criteria and method of assessing results***

31.4.4.1 All the results shall be recorded. Table 31.4 below shows the model of "result table" to be used.

Table 31.4: MODEL OF "RESULT TABLE"

Date		Temperature °C Relative humidity %		
Name of product				
Net volume		Can 1	Can 2	Can 3
Initial level of filling		%	%	%
Dispenser distance	Test	1 2 3	1 2 3	1 2 3
15 cm	Ignition? Y or N			
30 cm	Ignition? Y or N			
45 cm	Ignition? Y or N			
60 cm	Ignition? Y or N			
75 cm	Ignition? Y or N			
90 cm	Ignition? Y or N			
Observations – including can position				

31.4.4.2 Spray aerosols shall be classified as flammable, extremely flammable or non flammable according to the following criteria:

(a) An aerosol with a chemical heat of combustion less than 20 kJ/g is classified as flammable if ignition occurs at a distance equal or greater than 15 cm but less than 75 cm;

(b) An aerosol with a chemical heat of combustion less than 20 kJ/g is classified as extremely flammable if ignition occurs at a distance of 75 cm or more;

(c) If for an aerosol with a chemical heat of combustion less than 20 kJ/g, no ignition occurs in the ignition distance test, the enclosed space ignition test described in sub-section 31.5 of this Manual shall be performed;

(d) An aerosol with a chemical heat of combustion equal to or more than 20 kJ/g is classified as extremely flammable if ignition occurs at a distance of 75 cm or more. Otherwise the aerosol is classified as flammable.

Figure 31.4.1: SYSTEM FOR IGNITION DISTANCE TEST

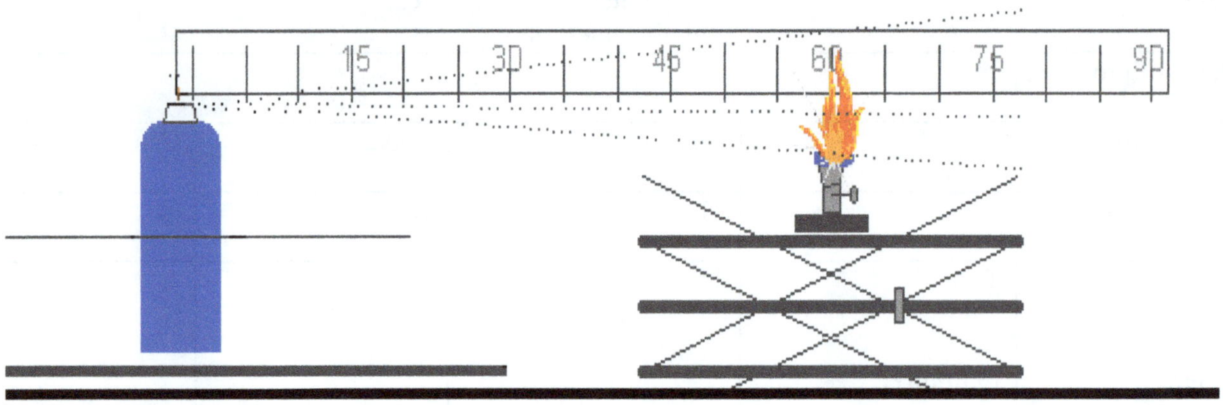

31.5 Enclosed space ignition test

31.5.1 *Introduction*

31.5.1.1 This test standard describes the method to assess the flammability of products emerging from aerosol dispensers due to their propensity to ignite in an enclosed or confined space. The contents of an aerosol dispenser are sprayed into a cylindrical test vessel containing a burning candle. If an observable ignition occurs, the elapsed time and amount discharged is noted.

31.5.2 *Apparatus and material*

31.5.2.1 The following apparatus is required:

Chronometer (stopwatch)	accurate to ± 0.2 s
Water bath maintained at 20 °C	accurate to ± 1 °C
Calibrated laboratory scales (balance)	accurate to ± 0.1 g
Thermometer	accurate to ± 1 °C
Hygrometer	accurate to ± 5%
Pressure gauge	accurate to ± 0.1 bar
Cylindrical test vessel	as detailed below

31.5.2.2 *Preparation of test apparatus*

31.5.2.2.1 A cylindrical vessel approximately 200 dm^3 (55 gallons) volume, approximately 600 mm in diameter and approximately 720 mm long and open at one end shall be modified as follows:

(a) A closure system consisting of a hinged cover should be matched to the open end of the receptacle; or

(b) A plastic film 0.01 to 0.02 mm thick may be used as a closure system. If the test is carried out with a plastic film this must be used as described below:

Stretch the film over the open end of the drum and hold it in place with an elastic band. The strength of the band shall be such that when placed around the drum resting on its side, it stretches by only 25 mm when a mass of 0.45 kg is attached to its lowest point. Cut a 25 mm slit in the film, starting 50 mm from the edge of the drum. Ensure that the film is taut;

(c) At the other end of the drum drill a 50 mm diameter hole 100 mm from the edge in such a way that the orifice is uppermost when the receptacle is laid down and ready for the test (Figure 31.5.1);

(d) On a 200 × 200 mm metal support place a paraffin wax candle 20 to 40 mm in diameter and 100 mm high. The candle shall be replaced when having a height of less than 80 mm. The candle's flame is protected from the action of the spray by a 150 mm wide, 200 mm high deflector. This includes the plane inclined at 45° produced 150 mm from the base of the deflector (Figure 31.5.2);

(e) The candle placed on the metal support shall be positioned midway between the two ends of the drum (Figure 31.5.3);

(f) The drum is laid on the ground or on a support at a spot where the temperature is between 15 °C and 25 °C. The product to be tested will be sprayed within the drum of roughly 200 dm^3 in which there will be a source of ignition.

31.5.2.2.2 Usually, the product leaves the aerosol can at an angle of 90° relevant to the vertical axis of the can. The layout and procedure described refers to this kind of aerosol product. In the case of unusually operating aerosols (e.g. vertical-spray aerosol dispensers) it will be necessary to record changes to equipment and procedures in accordance with good laboratory practice, such as ISO/IEC 17025:1999 General requirements for the competence of testing and calibration laboratories.

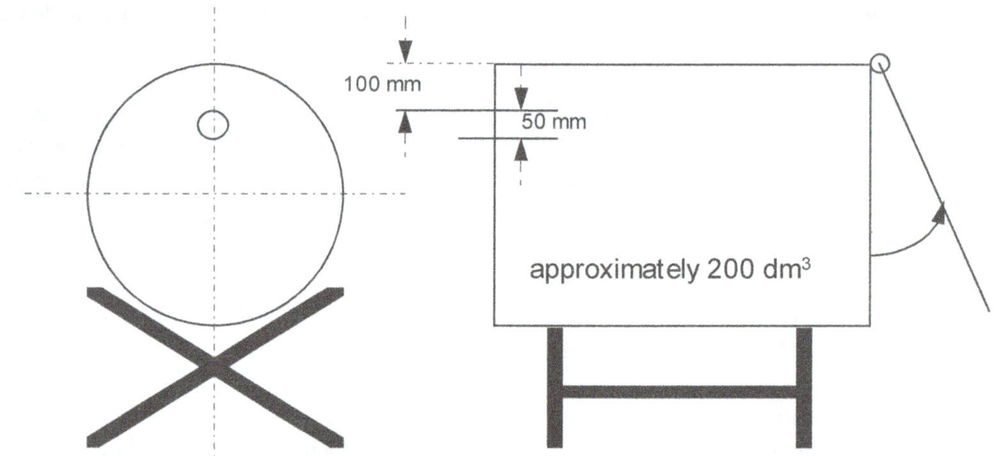

Figure 31.5.1: DRUM FOR ENCLOSED SPACE IGNITION TEST

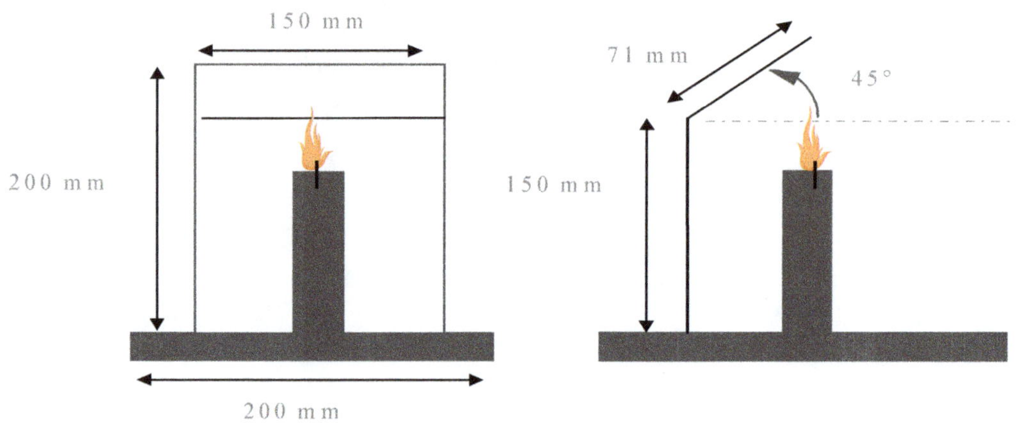

Figure 31.5.2: CANDLE FOR ENCLOSED SPACE IGNITION TEST

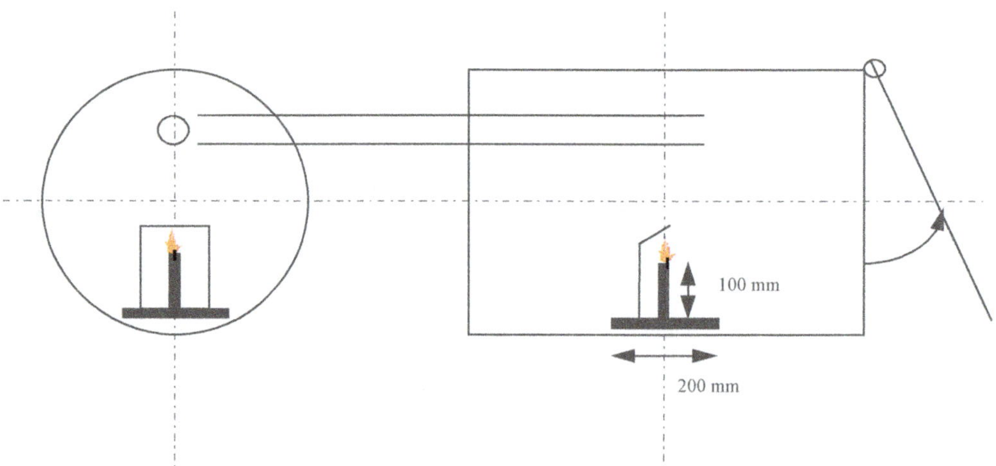

Figure 31.5.3: CANDLE ON ITS METAL SUPPORT

31.5.3 ***Procedure***

31.5.3.1 *General requirements*

31.5.3.1.1 Before testing, each aerosol dispenser shall be conditioned and then primed by discharging for approximately 1 second. The purpose of this action is to remove non-homogeneous material from the diptube.

31.5.3.1.2 The instructions of use shall be strictly followed, including whether the dispenser is intended to be used in the upright or inverted position. When shaking is required, shake immediately before testing.

31.5.3.1.3 The tests shall be carried out in a draught-free environment capable of ventilation, with the temperature controlled at 20 °C ± 5 °C and relative humidity in the range 30 - 80%.

31.5.3.2 *Test procedure*

(a) A minimum of 3 full aerosol dispensers per product shall be conditioned to 20 °C ± 1 °C in a water bath with at least 95% of the dispenser immersed in the water for at least 30 min (if the aerosol is fully immersed, 30 min conditioning is sufficient);

(b) Measure or calculate the actual volume of the drum in dm^3;

(c) Comply with general requirements. Record the temperature and relative humidity of the environment;

(d) Determine the internal pressure and initial discharge rate at 20 °C ± 1 °C (to eliminate faulty or partly filled aerosol dispensers);

(e) Weigh one of the aerosol dispensers and note its mass;

(f) Light the candle and apply the closure system (cover or plastic film);

(g) Place the aerosol dispenser actuator orifice 35 mm or closer for a wide spray product, from the centre of the entrance hole in the drum. Start the chronometer (stopwatch) and following the instructions for use of the product; direct the spray towards the centre of the opposite extremity (cover or plastic film). The aerosol shall be tested in the position it is designed to be used, e.g. upright or inverted;

(h) Spray until ignition occurs. Stop the chronometer and note the time elapsed. Re-weigh the aerosol dispenser and note its mass;

(i) Ventilate and clean the drum removing any residue likely to affect subsequent tests. Allow the drum to cool if necessary;

(j) Repeat the test procedure steps (d) to (i) for another two aerosol dispensers of the same product (3 in total, note: each dispenser is only tested once);

31.5.4 ***Test criteria and method of assessing results***

31.5.4.1 A test report containing the following information shall be drawn up:

(a) The product tested and its references;

(b) The internal pressure and discharge rate of the aerosol dispenser;

(c) The temperature and relative air humidity of the room;

(d) For each test, the discharge time (s) needed to achieve ignition (if the product does not ignite, state this);

(e) The mass of the product sprayed during each test (in g);

(f) The actual volume of the drum (in dm^3).

31.5.4.2 The time equivalent (t_{eq}) needed to achieve ignition in one cubic metre can be calculated as follows:

$$t_{eq} = \frac{1000 \times \text{discharge time (s)}}{\text{Actual volume of drum } (dm^3)}$$

31.5.4.3 The deflagration density (D_{def}) needed to achieve ignition during the test may also be calculated as follows:

$$D_{def} = \frac{1000 \times \text{Amount of product dispensed (g)}}{\text{Actual volume of drum } (dm^3)}$$

31.5.4.4 An aerosol with a chemical heat of combustion less than 20 kJ/g for which no ignition occurs in the ignition distance test (see sub-section 31.4 of this Manual) is classified as flammable if the time equivalent is less than or equal to 300 s/m^3 or the deflagration density is less than or equal to 300 g/m^3. Otherwise the aerosol is classified as non-flammable.

31.6 Aerosol foam flammability test

31.6.1 *Introduction*

31.6.1.1 This test standard describes the method to determine the flammability of an aerosol spray emitted in the form of a foam, mousse, gel or paste. An aerosol, which emits a foam, mousse, gel or paste is sprayed (approx. 5 g) on a watchglass and an ignition source (candle, wax taper, match or lighter) is placed at the base of the watchglass to observe if ignition and sustained combustion of the foam, mousse, gel or paste occurs. Ignition is defined as a stable flame maintained for at least 2 s and a minimum 4 cm in height.

31.6.2 *Apparatus and material*

31.6.2.1 The following apparatus is required:

Graduated scale, support and clamp	(graduations in cm)
Fire-resistant watchglass roughly 150 mm in diameter	
Chronometer (stopwatch)	(accurate to ± 0.2 s)
Candle, wax taper, match or lighter	
Calibrated laboratory scales (balance)	(accurate to ± 0.1 g)
Water bath maintained at 20 °C	(accurate to ± 1 °C)
Thermometer	(accurate to ± 1 °C)
Hygrometer	(accurate to ± 5%)
Pressure gauge	(accurate to ± 0.1 bar)

31.6.2.2 The watchglass is placed on a fire-resistant surface within a draught-free area that may be ventilated after each test. The graduated scale is positioned exactly behind the watchglass and held vertically by means of a support and clamp.

31.6.2.3 The scale is positioned in such a way that its origin is on a level with the watchglass base in a horizontal plane.

31.6.3 *Procedure*

31.6.3.1 *General requirements*

31.6.3.1.1 Before testing, each aerosol dispenser shall be conditioned and then primed by discharging for approximately 1 second. The purpose of this action is to remove non-homogeneous material from the diptube.

31.6.3.1.2 The instructions of use shall be strictly followed, including whether the dispenser is intended to be used in the upright or inverted position. When shaking is required, shake immediately before testing.

31.6.3.1.3 The tests shall be carried out in a draught-free environment capable of ventilation, with the temperature controlled at 20 °C ± 5 °C and relative humidity in the range of 30–80%.

31.6.3.2 *Test procedure*

(a) A minimum of four full aerosol dispensers per product shall be conditioned to 20 °C ± 1 °C with at least 95% of the dispenser immersed in the water for at least 30 min before each test (if the aerosol is fully immersed, 30 min conditioning is sufficient);

(b) Comply with general requirements. Record the temperature and relative humidity of the environment;

(c) Determine the internal pressure at 20 °C ± 1 °C (to eliminate faulty or partly filled aerosol dispensers);

(d) Measure the discharge or flow rate of the aerosol product to be examined, so that the amount of test product dispensed can be more accurately gauged;

(e) Weigh one of the aerosol dispensers and note its mass;

(f) On the basis of the measured discharge or flow rate and following the manufacturer's instructions, release approximately 5 g of the product onto the centre of the clean watchglass with the aim of producing a mound no higher than 25 mm;

(g) Within 5 s of completion of discharge, apply the source of ignition to the edge of the sample at its base and at the same time start the chronometer (stopwatch). If necessary, the ignition source shall be removed from the edge of the sample after approximately two seconds, in order to clearly observe if ignition has occurred. If no ignition of the sample is apparent, the ignition source shall be reapplied to the edge of the sample;

(h) If ignition occurs note the following points:

    (i) The maximum height of the flame in cm above the base of the watchglass;

    (ii) The flame duration in s;

    (iii) Dry and re-weigh the aerosol dispenser and calculate the mass of the released product;

(i) Ventilate the test area immediately after each test;

(j) If ignition is not obtained and the released product remains in the form of a foam or paste throughout its period of use, steps (e) to (i) should be repeated. Allow the product to stand for 30 s, 1 min, 2 min or 4 min before applying the ignition source;

(k) Repeat the test procedure steps (e) to (j) twice more (a total of 3) for the same can;

(l) Repeat the test procedure steps (e) to (k) for another two aerosol cans (3 cans in total) of the same product.

31.6.4 ***Test criteria and method of assessing results***

31.6.4.1 A test report containing the following information shall be drawn up:

(a) Whether the product ignites;

(b) Maximum flame height in cm;

(c) Duration of flame in s;

(d) The mass of the product tested.

31.6.4.2 The aerosol product shall be classified as extremely flammable if the flame height is 20 cm or more and the flame duration is 2 s or more; or if the flame duration is 7 s or more and the flame height is 4 cm or more.

SECTION 32

CLASSIFICATION PROCEDURES, TEST METHODS AND CRITERIA RELATING TO LIQUID DESENSITIZED EXPLOSIVES AND FLAMMABLE LIQUIDS OF CLASS 3

32.1 Purpose

This section presents the United Nations scheme for the classification of liquid desensitized explosives and flammable liquids of Class 3 (see Chapter 2.3 of the Model Regulations). The text should be used in conjunction with the classification principles given in Chapter 2.3 of the Model Regulations and the test methods given in sub-sections 32.4 and 32.5 of this Manual.

32.2 Scope

32.2.1 Liquid desensitized explosives are explosive substances which are dissolved or suspended in water or other liquid substances, to form a homogeneous liquid mixture to suppress their explosive properties (see 2.3.1.4 of the Model Regulations).

32.2.2 Substances are listed in this class as flammable liquids only when their flash point is not more than 60 °C in a closed-cup test, or not more than 65.6 °C in an open-cup test or, in the case of substances transported or offered for transport at elevated temperatures, when they give off a flammable vapour at a temperature at or below the maximum transport temperature. However, liquids with a flash point of more than 35 °C, which do not sustain combustion, need not to be considered as flammable liquids for the purposes of the Model Regulations.

32.2.3 Flammable liquids listed (Chapter 3.2 of the Model Regulations) in this class by name should be regarded as chemically pure. In practice, goods shipped under the name of such substances are often commercial products which contain other added substances or impurities. Therefore, it may occur that liquids which are not included in the list because their flash point in their pure state is more than 60 °C in a closed-cup test, or more than 65.6 °C in an open-cup test, may be offered for transport as commercial products with a flash point at or below that limit. Moreover, liquids which would be listed, in their pure state, in packing group III could in fact be included in packing group II as commercial products because of the presence of added substances or impurities.

32.2.4 For these reasons caution should be exercised when using the lists, as they are only guides. In the event of doubt, the flash point of substances should be tested experimentally.

32.2.5 Liquids are considered to be unable to sustain combustion for the purposes of the Model Regulations (i.e. they do not sustain combustion under defined test conditions) if they have passed a suitable combustibility test (see section 32.5.2) or if their fire point, according to ISO 2592, is greater than 100 °C or if they are miscible solutions with a water content of more than 90% by mass.

32.3 Classification procedures

32.3.1 *Flammable liquids*

32.3.1.1 Table 32.1 should be used for the determination of the hazard grouping of a liquid that presents a risk due to flammability.

32.3.1.2 For liquids whose only risk is flammability, the packing group for the substance is shown in the hazard grouping shown in Table 32.1.

32.3.1.3 For a liquid with additional risk(s), the hazard group determined from Table 32.1 and the hazard group based on the severity of the additional risk(s) should be considered. In such cases, the table of precedence of hazard characteristics in Chapter 2.0, section 2.0.3 of the Model Regulations, should be used to determine the correct classification of the liquid. The hazard grouping indicating the highest degree of danger based on the different risks of a substance then becomes the packing group for the substance.

Table 32.1: HAZARD GROUPING BASED ON FLAMMABILITY

Packing Group	Flash Point (Closed-cup)	Initial Boiling Point
I	-	≤ 35 °C
II	< 23 °C	> 35 °C
III	≥ 23 °C, ≤ 60 °C	> 35 °C

32.3.2 ***Liquid desensitized explosives***

32.3.2.1 This sub-section presents the United Nations scheme for the classification of liquid desensitized explosives of Class 3 (see paragraph 2.3.1.4 of the Model Regulations). Liquid desensitized explosives are substances which are dissolved or suspended in water or other liquid substances to form a homogeneous liquid mixture to suppress their explosives properties.

32.3.2.2 New products which are thermally stable and have, or are suspected of having, explosive properties should first be considered for Class 1 and the Class 1 acceptance procedure and, if necessary, the assignment procedure applied.

32.3.2.3 Where a substance is assigned to Class 1 but is diluted to be exempted from Class 1 by test series 6 (see section 16), this diluted substance, when meeting the classification criteria or definition for another class or division, should be classified in that class or division at the highest concentration which exempts it from Class 1. When sufficiently diluted, such substances may be deemed to be non-dangerous (see also paragraph 2.1.3.5.3 of the Model Regulations).

32.3.2.4 The classification scheme of liquid desensitized explosives for supply and use (including storage) according to the Globally Harmonized System of Classification and Labelling of Chemicals (GHS) is referred to section 51.

32.4 Test methods used for determining flash point and viscosity

32.4.1 *Non-viscous flammable liquids*

The following methods for determining the flash point of flammable liquids may be used:

International standards:

ISO 1516
ISO 1523
ISO 2719
ISO 13736
ISO 3679
ISO 3680

National standards:

American Society for Testing Materials International, 100 Barr Harbor Drive, PO Box C700, West Conshohocken, Pennsylvania, USA 19428-2959:

ASTM D3828-07a, Standard Test Methods for Flash Point by Small Scale Closed Cup Tester

ASTM D56-05, Standard Test Method for Flash Point by Tag Closed Cup Tester

ASTM D3278-96(2004)e1, Standard Test Methods for Flash Point of Liquids by Small Scale Closed-Cup Apparatus

ASTM D93-08, Standard Test Methods for Flash Point by Pensky-Martens Closed Cup Tester

Association française de normalisation, AFNOR, 11, rue de Pressensé, 93571 La Plaine Saint-Denis Cedex:

French Standard NF M 07 - 019
French Standards NF M 07 - 011 / NF T 30 - 050 / NF T 66 - 009
French Standard NF M 07 - 036

Deutsches Institut für Normung, Burggrafenstr. 6, D-10787 Berlin:

Standard DIN 51755 (flash points below 65 °C)

State Committee of the Council of Ministers for Standardization, 113813, GSP, Moscow, M-49 Leninsky Prospect, 9:

GOST 12.1.044-84.

32.4.2 *Viscous flammable substances with a flash point of less than 23 °C*

32.4.2.1 The hazard group of paints, enamels, lacquers, varnishes, adhesives, polishes and other viscous flammable substances of Class 3 with a flash point of less than 23 °C is determined by reference to:

(a) The viscosity expressed as the flow-time in seconds (see 32.4.3);

(b) The closed-cup flash point (see 32.4.2.2); and

(c) A solvent separation test (see 32.5.1).

32.4.2.2 The closed-cup flash point is determined according to the ISO method ISO 1523:1983 for paints and varnishes. Where the temperature of the flash point is too low for the use of water in the water bath, the following modifications should be made:

(a) Use ethylene glycol in the water bath or other suitable similar container;

(b) Where appropriate, a refrigerator may be used to cool the sample and apparatus to below the temperature required by the method for the expected flash point. For lower temperatures, the sample and equipment should be cooled to a suitable temperature - for example, by adding of solid carbon dioxide slowly to the ethylene glycol, the sample being similarly cooled in a separate container of ethylene glycol;

(c) In order to obtain reliable flash points, it is important that the recommended rate of temperature rise for the sample during testing is not exceeded. Depending on the size of the water bath and the amount of ethylene glycol it contains, it may be necessary partially to insulate the water bath to achieve a sufficiently slow rate of temperature rise.

32.4.3 ***Viscosity test***

The flow-time in seconds is determined at 23 °C using the ISO standard cup with a 4 mm jet (ISO 2431:1984). Where the flow-time exceeds 100 seconds, a second test is carried out using the ISO standard cup with a 6 mm jet.

32.5 Test methods for determining solvent separation and sustained combustion

32.5.1 *Test L.1: Solvent separation test*

32.5.1.1 *Introduction*

This test is used to determine the extent of solvent separation in viscous liquids such as paints, enamels, varnishes, adhesives and polishes with a flash point of less than 23 °C.

32.5.1.2 *Apparatus and materials*

A stoppered 100 ml measuring cylinder is required of approximately 25 cm total height and of a uniform internal diameter of approximately 3 cm over the calibrated section.

32.5.1.3 *Procedure*

The paint should be stirred to obtain a uniform consistency and poured in up to the 100 ml mark. The stopper should be inserted and the cylinder left standing undisturbed for 24 hours. After 24 hours, the height of the upper separated layer should be measured.

32.5.1.4 *Test criteria and method of assessing results*

The height of the upper separated layer should be expressed as a percentage of the total height of the sample. If less than 3% of clear solvent separates out then the substance may be considered for inclusion in packing group III (see 32.3.1.6 and 32.3.1.7).

32.5.2 *Test L.2: Sustained combustibility test*

32.5.2.1 *Introduction*

This test is used to determine if a substance sustains combustion when heated under the test conditions and exposed to a flame. A metal block with a concave depression (sample well) is heated to a specified temperature. A specified volume of the substance under test is transferred to the well and its ability to sustain combustion is noted after application and subsequent removal of a standard flame under specified conditions.

32.5.2.2 *Apparatus and materials*

32.5.2.2.1 A combustibility tester is used consisting of a block of aluminium alloy or other corrosion resistant metal of high thermal conductivity. The block has a concave well and a pocket drilled to take a thermometer. A small gas jet assembly on a swivel is attached to the block. The handle and gas inlet for the gas jet may be fitted at any convenient angle to the gas jet. A suitable apparatus is shown in Figure 32.5.2.1 and the essential diagrams are given in figures 32.5.2.1 and 32.5.2.2. The following apparatus is also required:

(a) *Gauge*, for checking that the height of the centre of the gas jet above the top of the sample well is 2.2 mm (see Figure 32.5.2.1);

(b) *Thermometer*, mercury in glass, for horizontal operation, with a sensitivity not less than 1 mm/°C, or other measuring device of equivalent sensitivity permitting reading at 0.5 °C intervals. When in position in the block, the thermometer bulb should be surrounded with thermally conducting thermoplastic compound;

(c) *Hotplate*, fitted with a temperature-control device (Other types of apparatus with suitable temperature-control facilities may be employed to heat the metal block.);

(d) *Stopwatch*, or other suitable timing device;

(e) *Syringe*, capable of delivering 2.0 ml to an accuracy of ± 0.1 ml;

(f) *Fuel*, butane.

32.5.2.2.2 The sample should be representative of the substance to be tested and should be supplied and kept in a tightly closed container prior to the test. Because of the possibility of loss of volatile constituents, the sample should receive only the minimum treatment to ensure its homogeneity. After removing each test portion, the sample container should be immediately closed tightly to ensure that no volatile components escape from the container; if this closure is incomplete, an entirely new sample should be taken.

32.5.2.3 *Procedure*

32.5.2.3.1 It is essential that the apparatus is set up in a completely draught-free area[1] and in the absence of strong light to facilitate observation of flash, flame, etc.

32.5.2.3.2 Place the metal block on the hotplate or heat the metal block by other suitable means so that its temperature, as indicated by the thermometer placed in the metal block, is maintained at the specified temperature within a tolerance of ± 1 °C. The test temperature is 60.5 °C or 75 °C (see 32.5.2.3.8). Correct this temperature for the difference in barometric pressure from the standard atmospheric pressure (101.3 kPa) by raising the test temperature for a higher pressure, or lowering the test temperature for a lower pressure, by 1.0 °C for each 4 kPa difference. Ensure that the top of the metal block is exactly horizontal. Use the gauge to check that the jet is 2.2 mm above the top of the well when in the test position.

---

[1] ***Warning - do not carry out the test in a small confined area (for example, a glove box) because of the hazard of explosions.***

32.5.2.3.3 Light the butane with the jet away from the test position (i.e. in the "off" position, away from the well). Adjust the size of the flame so that it is 8 mm to 9 mm high and approximately 5 mm wide.

32.5.2.3.4 Using the syringe, take from the sample container at least 2 ml of the sample and rapidly transfer a test portion of 2.0 ml ± 0.1 ml to the well of the combustibility tester and immediately start the timing device.

32.5.2.3.5 After a heating time of 60 s, by which time the test portion is deemed to have reached its equilibrium temperature, and if the test fluid has not ignited, swing the test flame into the test position over the edge of the pool of liquid. Maintain it in this position for 15 s and then return it to the "off" position whilst observing the behaviour of the test portion. The test flame should remain alight throughout the test.

32.5.2.3.6 The test should be performed in triplicate. For each test observe and record:

(a) Whether there is ignition and sustained combustion or flashing, or neither, of the test portion before the test flame is moved into the test position; and

(b) Whether the test portion ignites while the test flame is in the test position, and, if so, how long combustion is sustained after the test flame is returned to the "off" position.

32.5.2.3.7 If sustained combustion, interpreted in accordance with 32.5.2.4, is not found, repeat the complete procedure with new test portions, but with a heating time of 30 s.

32.5.2.3.8 If sustained combustion, interpreted in accordance with 32.5.2.4, is not found at a test temperature of 60.5 °C, repeat the complete procedure with new test portions, but at a test temperature of 75 °C.

32.5.2.4 *Test criteria and method of assessing results*

The substance should be assessed either as not sustaining combustion or as sustaining combustion. Sustained combustion should be reported at either of the heating times or temperatures if one of the following occurs with any of the test portions:

(a) When the test flame is in the "off" position, the test portion ignites and sustains combustion; or

(b) The test portion ignites while the test flame is in the test position, maintained for 15 s, and sustains combustion for more than 15 s after the test flame has been returned to the "off" position.

Intermittent flashing should not be interpreted as sustained combustion. Normally, at the end of 15 s, the combustion has either clearly ceased or continues. In cases of doubt, the substance should be deemed to sustain combustion.

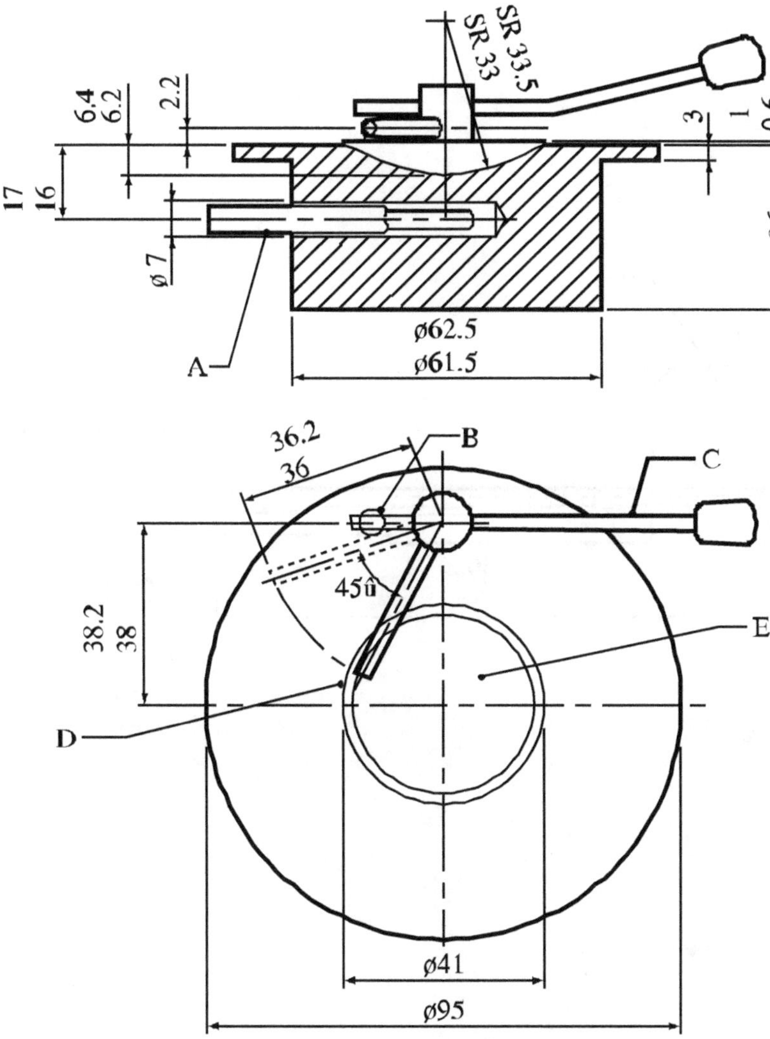

(A) Thermometer
(B) Stop
(C) Handle
(D) Test gas jet
(E) Sample well

Figure 32.5.2.1: COMBUSTIBILITY TESTER

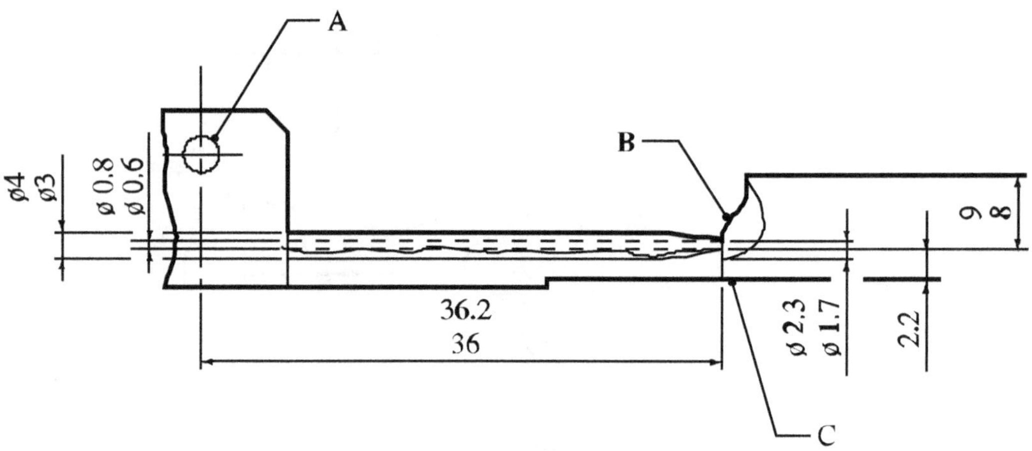

(A) Butane gas inlet
(B) Test flame
(C) Sample well

Figure 32.5.2.2: TEST GAS JET AND FLAME

32.6 Test methods used for determining initial boiling point

The following methods for determining the initial boiling point of flammable liquids may be used:

International standards:

ISO 3924
ISO 4626
ISO 3405

National standards:

American Society for Testing Materials International, 100 Barr Harbor Drive, PO Box C700, West Conshohocken, Pennsylvania, USA 19428-2959:

ASTM D86-07a, Standard Test Method for Distillation of Petroleum Products at Atmospheric Pressure

ASTM D1078-05, Standard Test Method for Distillation Range of Volatile Organic Liquids

Further acceptable methods:

Method A.2 as described in Part A of the Annex to Commission Regulation (EC) No 440/2008[2].

---

[2] *Commission Regulation (EC) No 440/2008 of 30 May 2008 laying down test methods pursuant to Regulation (EC) No 1907/2006 of the European Parliament and of the Council on the Registration, Evaluation, Authorisation and Restriction of Chemicals (REACH) (Official Journal of the European Union, No. L 142 of 31.05.2008, p.1-739 and No. L 143 of 03.06.2008, p.55).*

SECTION 33

CLASSIFICATION PROCEDURES, TEST METHODS AND CRITERIA RELATING TO CLASS 4

33.1 Introduction

This section of the Manual contains classification procedures, test methods and criteria relating to substances (except self-reactive substances of Division 4.1, see Part II) and articles of Class 4.

33.2 Division 4.1

33.2.1 *Flammable solids*

33.2.1.1 *Purpose*

33.2.1.1.1 Section 33.2.1.3 presents the United Nations scheme for the classification of readily combustible solids of Division 4.1 (see sub-section 2.4.2.2 of the Model Regulations). The text should be used in conjunction with the classification principles given in paragraphs 2.4.2.2.2 and 2.4.2.2.3 of the Model Regulations, the flow chart given in Figure 33.2.1.3 and the test prescription given here in 33.2.1.4.

33.2.1.1.2 In order to differentiate between substances that can be ignited and those which burn rapidly, or whose burning behaviour is particularly dangerous, only substances whose burning rate exceeds a certain limiting value are classified in Division 4.1.

33.2.1.1.3 The test procedures outlined here and in the Model Regulations adequately assess the relative hazard of readily combustible solids so that an appropriate classification for transport can be made.

33.2.1.2 *Scope*

33.2.1.2.1 Products offered for transport should be subjected to the classification procedures as set out in paragraphs 2.4.2.2.2 and 2.4.2.2.3 of the Model Regulations, unless it is impracticable (e.g. because of the physical form) to perform the tests. Substances or articles which cannot be tested should be classified by analogy with existing entries (see paragraph 2.4.2.2.2.2 of the Model Regulations). The classification procedure should be undertaken before a new product is offered for transport.

33.2.1.3 *Classification procedure for readily combustible solids*

33.2.1.3.1 A preliminary screening test is performed to determine if, on ignition by a gas flame, propagation by burning with flame or smouldering occurs. If propagation occurs within a specified time then the full test is carried out to determine the rate and vigour of burning.

33.2.1.3.2 The tests should only be applied to granular, paste-like or powdery substances. If in the screening test, the substance does not ignite and propagate combustion either by burning with flame or smouldering, it is not necessary to perform the complete burning rate test as the substance is not a readily combustible solid of Division 4.1. If propagation occurs and the burning time is less than a specified time, the full burning rate test should be performed. Whether the substance is a readily combustible solid of Division 4.1 and, if so, whether packing group II or III should be assigned is decided on the basis of the test result.

Figure 33.2.1.3: FLOW CHART FOR ASSIGNING READILY COMBUSTIBLE SOLIDS, EXCEPT METAL POWDERS, TO DIVISION 4.1

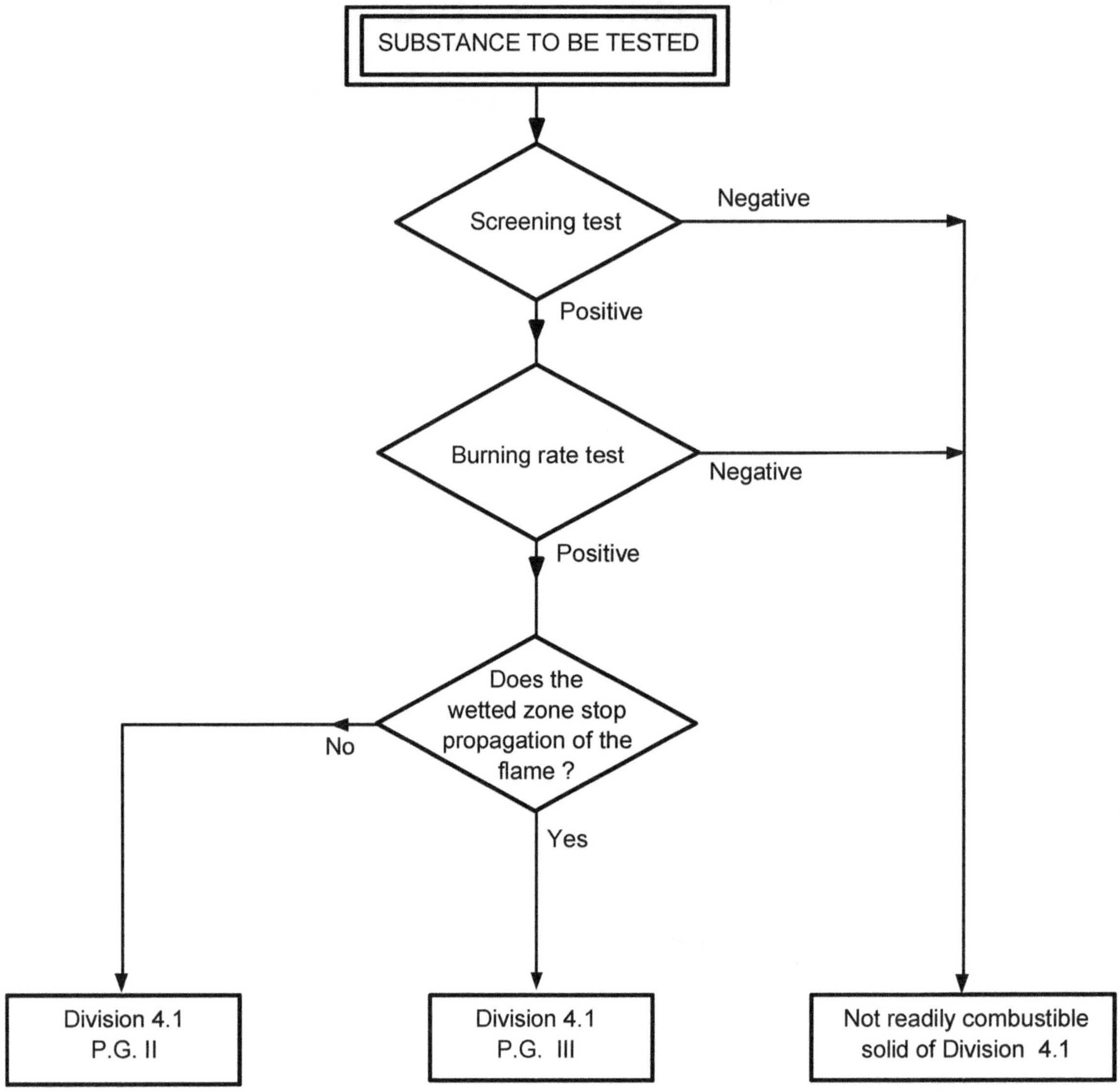

33.2.1.4 *Test N.1: Test method for readily combustible solids*

33.2.1.4.1 Introduction

The ability of a substance to propagate combustion is tested by igniting it and determining the burning time.

33.2.1.4.2 Apparatus and materials

A mould 250 mm long with a triangular cross-section of inner height 10 mm and width 20 mm is used to form the train for the burning rate test. On both sides of the mould, in the longitudinal direction, two metal sheets are mounted as lateral limitations which extend 2 mm beyond the upper edge of the triangular cross-section (Figure 33.2.1.4.1). An impervious, non-combustible, low heat-conducting plate is used to support the sample train.

33.2.1.4.3 Procedure

33.2.1.4.3.1 Preliminary screening test

The substance in its commercial form, should be formed into an unbroken strip or powder train about 250 mm long by 20 mm wide by 10 mm high on a cool, impervious, low heat-conducting base plate. A hot flame (minimum temperature 1 000 °C) from a gas burner (minimum diameter 5 mm) should be applied to one end of the powder train until the powder ignites or for a maximum of 2 minutes (5 minutes for powders of metals or metal-alloys). It should be noted whether combustion propagates along 200 mm of the train within the 2 minute test period (or 20 minutes for metal powders). If the substance does not ignite and propagate combustion either by burning with flame or smouldering along 200 mm of the powder train within the 2 minute (or 20 minute) test period, then the substance should not be classified as a flammable solid and no further testing is required. If the substance propagates burning of a 200 mm length of the powder train in less than 2 minutes or less than 20 minutes for metal powders, the full test programme in 33.2.1.4.3.2 should be carried out.

33.2.1.4.3.2 Burning rate test

33.2.1.4.3.2.1 The powdered or granular substance, in its commercial form, should be loosely filled into the mould. The mould is then dropped three times from a height of 20 mm onto a solid surface. The lateral limitations are then removed and the impervious, non-combustible, low heat-conducting plate is placed on top of the mould, the apparatus inverted and the mould removed. Pasty substances are spread on a non-combustible surface in the form of a rope 250 mm in length with a cross-section of about 100 mm^2. In the case of a moisture sensitive substance, the test should be carried out as quickly as possible after its removal from the container. The pile should be arranged across the draught in a fume cupboard. The air speed should be sufficient to prevent fumes escaping into the laboratory and should not be varied during the test. A draught screen may be erected around the apparatus.

33.2.1.4.3.2.2 For substances other than metal powders, 1 ml of a wetting solution should be added to the pile 30 - 40 mm beyond the 100 mm timing zone. Apply the wetting solution to the ridge drop by drop, ensuring the whole cross-section of the pile is wetted without loss of liquid from the sides. The liquid should be applied over the shortest possible length of the pile consistent with avoiding loss from the sides. With many substances, water rolls off the sides of the pile, so the addition of wetting agents may be necessary. Wetting agents used should be free from combustible diluents and the total active matter in the wetting solution should not exceed 1%. This liquid may be added to a hollow up to 3 mm deep and 5 mm in diameter in the top of the pile.

33.2.1.4.3.2.3 Any suitable ignition source such as a small flame or a hot wire of minimum temperature 1 000 °C is used to ignite the pile at one end. When the pile has burned a distance of 80 mm, measure the rate of burning over the next 100 mm. For substances other than metal powders, note whether or not the wetted zone stops propagation of the flame for at least 4 minutes. The test should be performed six times using a clean cool plate each time, unless a positive result is observed earlier.

33.2.1.4.4 Test criteria and method of assessing results

33.2.1.4.4.1 Powdered, granular or pasty substances should be classified in Division 4.1 when the time of burning of one or more of the test runs, in accordance with the test method described in 33.2.1.3.4.2, is less than 45 s or the rate of burning is more than 2.2 mm/s. Powders of metals or metal alloys should be classified when they can be ignited and the reaction spreads over the whole length of the sample in 10 minutes or less.

33.2.1.4.4.2 For readily combustible solids (other than metal powders), packing group II should be assigned if the burning time is less than 45 s and the flame passes the wetted zone. Packing group II should be assigned to powders of metal or metal alloys if the zone of reaction spreads over the whole length of the sample in five minutes or less.

33.2.1.4.4.3 For readily combustible solids (other than metal powders), packing group III should be assigned if the burning time is less than 45 s and the wetted zone stops the flame propagation for at least four minutes. Packing group III should be assigned to metal powders if the reaction spreads over the whole length of the sample in more than five minutes but not more than ten minutes.

33.2.1.4.5 Examples of results

Substance	**Prelimary burning time (s)**	**Burning time (s)**	**Delay time (s)**	**Result**
Manganese ethylene bis (dithiocarbamate) complex with zinc salt 88% (Mancozeb)	-	102	-	Not 4.1
Manganese ethylene bis (dithiocarbamate) complex with zinc salt 80% (Mancozeb)	-	145	-	Not 4.1
Manganese ethylene bis (dithiocarbamate) complex with zinc salt 75% (Mancozeb)	No ignition	-	-	Not 4.1

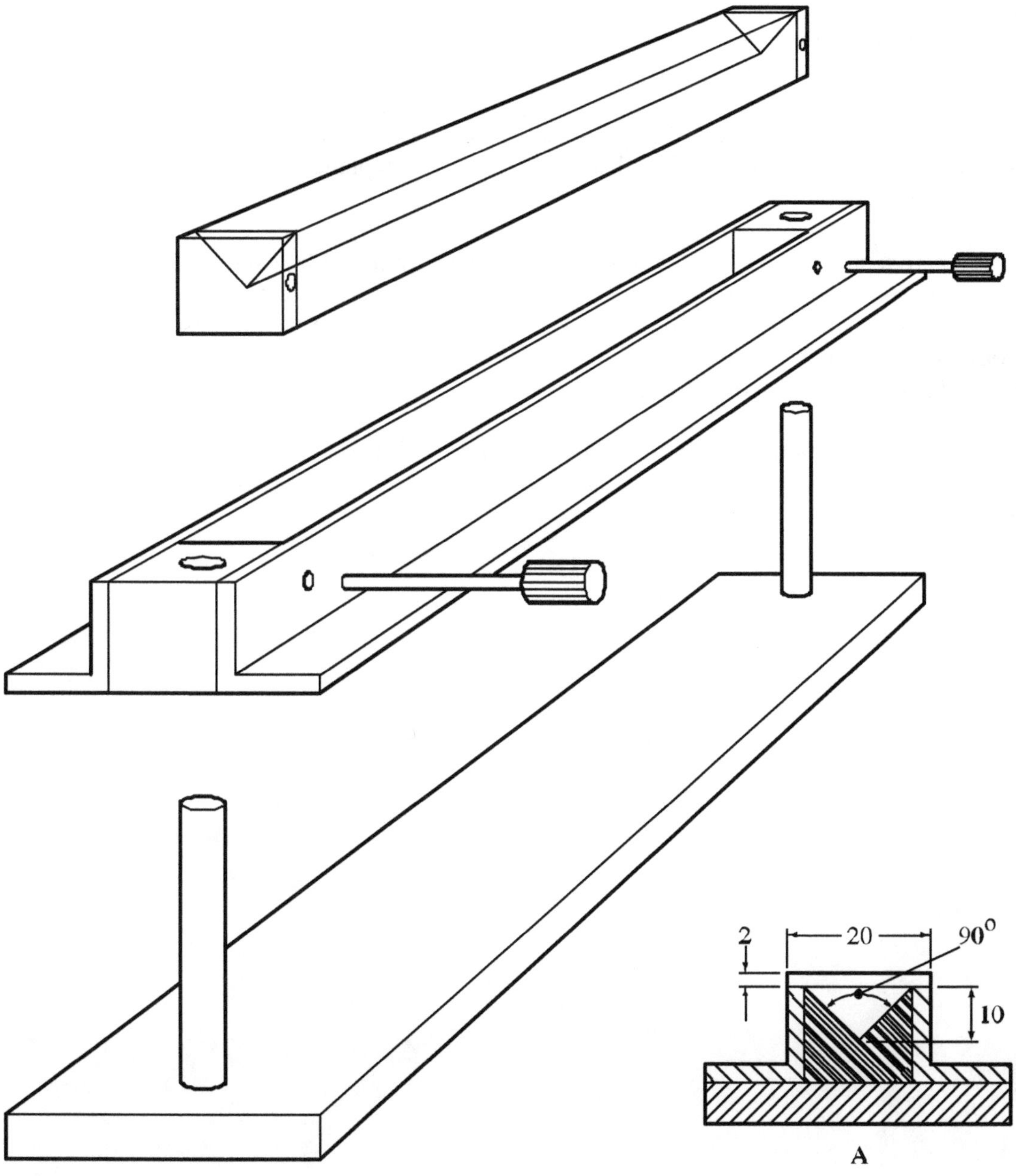

(A) Cross-section of 250 mm long mould

Figure 33.2.1.4.1: MOULD AND ACCESSORIES FOR THE PREPARATION OF THE PILE FOR THE BURNING RATE TEST

33.2.2 ***(Reserved)***

33.2.3 ***Solid desensitized explosives of Division 4.1***

33.2.3.1 This sub-section presents the United Nations scheme for the classification of desensitized explosives of Division 4.1 (see sub-section 2.4.2.4 of the Model Regulations). Solid desensitized explosives are substances which are wetted with water or alcohols or are diluted with other substances to form a homogeneous solid mixture to suppress their explosive properties.

33.2.3.2 New products which are thermally stable and have, or are suspected of having, explosive properties should first be considered for Class 1 and the Class 1 acceptance procedure and, if necessary, the assignment procedure applied.

33.2.3.3 Where a substance is assigned to Class 1 but is diluted to be exempted from Class 1 by test series 6 (see section 16), this diluted substance, when meeting the classification criteria or definition for another class or division, should be classified in that class or division at the highest concentration which exempts it from Class 1. When sufficiently diluted, such substances may be deemed to be non-dangerous (see also paragraph 2.1.3.5.3 of the Model Regulations).

33.2.3.4 The classification scheme of solid desensitized explosives for supply and use (including storage) according to the Globally Harmonized System of Classification and Labelling of Chemicals (GHS) is referred to section 51.

33.3 Division 4.2

33.3.1 *Substances liable to spontaneous combustion*

33.3.1.1 *Purpose*

33.3.1.1.1 This sub-section presents the United Nations scheme for the classification of substances liable to spontaneous combustion of Division 4.2 (see section 2.4.3 of the Model Regulations). The text should be used in conjunction with the classification principles given in sub-sections 2.4.3.2 and 2.4.3.3 of the Model Regulations and the test prescriptions given here in 33.3.1.4 to 33.3.1.6.

33.3.1.1.2 The test procedures are intended to identify two types of substances with spontaneous combustion properties:

(a) Substances, including mixtures and solutions (liquid or solid) which, even in small quantities, ignite within five minutes of coming in contact with air. These substances are the most liable to spontaneous combustion and are called pyrophoric substances;

(b) Other substances which, in contact with air and without an energy supply, are liable to self-heating. These substances will ignite only when in large amounts (kilograms) and after long periods of time (hours or days) and are called self-heating substances.

33.3.1.1.3 The test procedures outlined here adequately assess the relative hazard of substances liable to spontaneous combustion so that an appropriate classification for transport can be made.

33.3.1.2 *Scope*

33.3.1.2.1 Products offered for transport should be subjected to the classification procedures as set out in sub-sections 2.4.3.2 and 2.4.3.3 of the Model Regulations, unless it is impracticable (e.g. because of the physical properties) to perform the tests. The classification procedure should be undertaken before a product is offered for transport.

33.3.1.3 *Classification procedure for substances liable to spontaneous combustion*

33.3.1.3.1 Pyrophoric solids

A test is performed to determine if a solid ignites within five minutes of coming in contact with air. The recommended test method is given here, in 33.3.1.4. Whether a substance is a pyrophoric solid of Division 4.2 is decided on the basis of the test result. Packing group I is assigned to all pyrophoric solids.

33.3.1.3.2 Pyrophoric liquids

A test is performed to determine if a liquid ignites when added to an inert carrier and exposed to air for five minutes. If no ignition occurs then the second part of the test is performed to determine if it chars or ignites a filter paper. The recommended test method is given here in 33.3.1.5. Whether a substance is a pyrophoric liquid of Division 4.2 is decided on the basis of the test result. Packing group I is assigned to all pyrophoric liquids.

33.3.1.3.3 Self-heating substances

33.3.1.3.3.1 Tests are performed to determine if substances in a 25 mm or 100 mm sample cube, at test temperatures of 100 °C, 120 °C or 140 °C, undergo spontaneous ignition or dangerous self-heating, which is indicated by a 60 °C rise in temperature over the oven temperature within 24 hours. The classification scheme is illustrated in Figure 33.3.1.3.3.1. These criteria are based on the self-ignition temperature of charcoal, which is 50 °C for a sample cube of 27 m^3. Substances with a temperature of spontaneous combustion higher than 50 °C for a volume of 27 m^3 should not be assigned to Division 4.2. Substances with a spontaneous ignition temperature higher than 50 °C for a volume of 450 litres should not be assigned to packing group II of Division 4.2. The recommended test method is given here in 33.3.1.6.

33.3.1.3.3.2 If dangerous self-heating does not occur with the substance in a 100 mm sample cube at 140 °C then the substance is not a self-heating substance of Division 4.2.

33.3.1.3.3.3 If dangerous self-heating occurs with the substance in a 100 mm sample cube at 140 °C then a test with the substance in a 25 mm sample cube should be performed at 140 °C to determine if it should be assigned to packing group II.

33.3.1.3.3.4 If dangerous self-heating occurs at 140 °C with the substance in a 100 mm sample cube, but not a 25 mm sample cube, then a test with the substance in a 100 mm sample cube should be performed:

(a) At 120 °C if it is to be transported in packages of not more than 3 m^3 volume; or

(b) At 100 °C if the substance is to be transported in packages of not more than 450 litres volume.

Whether packing group III of Division 4.2 is assigned or the substance is not a self-heating substance of Division 4.2, in the packaging to be used, is decided on the basis of the test results.

33.3.1.3.3.5 Self-reactive substances type G which give a positive result with this test method may be classified in Division 4.2 (see 20.2.6).

Figure 33.3.1.3.3.1: CLASSIFICATION OF SELF-HEATING SUBSTANCES

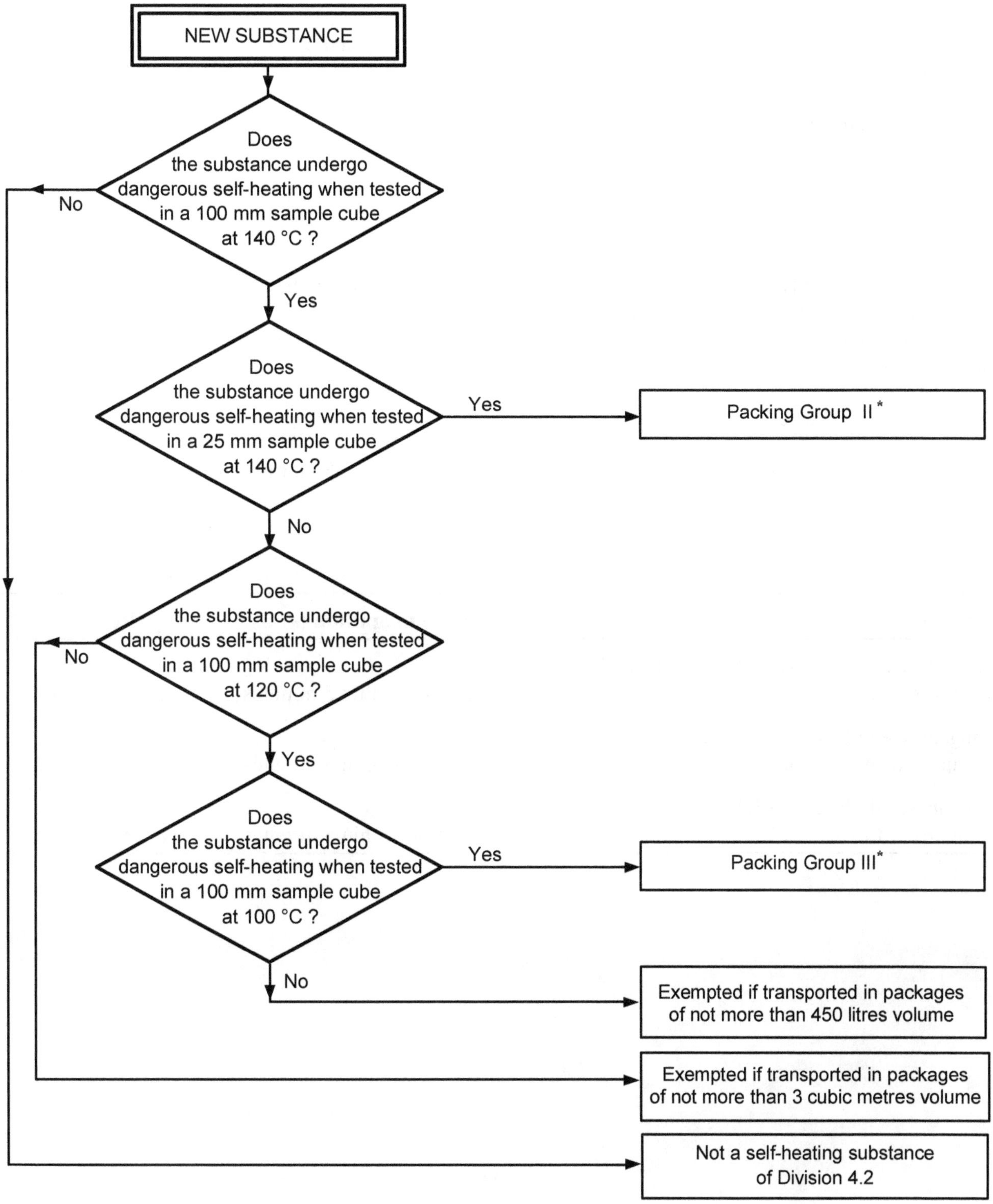

* *Substances with a temperature for spontaneous combustion higher than 50 °C for 27 m³ should not be classified in Division 4.2.*

33.3.1.4 *Test N.2: Test method for pyrophoric solids*

33.3.1.4.1 Introduction

The ability of a solid to ignite on contact with air is tested by exposure to air and determining the time to ignition.

33.3.1.4.2 Apparatus and materials

No special laboratory equipment is required.

33.3.1.4.3 Procedure

One to two ml of the powdery substance to be tested should be poured from about 1 m height onto a non-combustible surface and it is observed whether the substance ignites during dropping or within 5 minutes of settling. This procedure should be performed six times unless a positive result is obtained earlier.

33.3.1.4.4 Test criteria and method of assessing results

If the sample ignites in one of the tests, the substance should be considered pyrophoric and should be classified in packing group I of Division 4.2.

33.3.1.4.5 Examples of results

Substance	**Time to ignition (s)**	**Result**
Manganese ethylene bis (dithiocarbamate) complex with zinc salt 88% (Mancozeb)	No ignition within 5 minutes	Not PG I of 4.2
Manganese ethylene bis (dithiocarbamate) complex with zinc salt 80% (Mancozeb)	No ignition within 5 minutes	Not PG I of 4.2
Manganese ethylene bis (dithiocarbamate) complex with zinc salt 75% (Mancozeb)	No ignition within 5 minutes	Not PG I of 4.2

33.3.1.5 *Test N.3: Test method for pyrophoric liquids*

33.3.1.5.1 Introduction

The ability of a liquid to ignite when added to an inert carrier and exposed to air, or to char or ignite a filter paper on contact with air, is determined.

33.3.1.5.2 Apparatus and materials

A porcelain cup of about 100 mm diameter and some diatomaceous earth or silica gel is required for the first part of the test, and small pore size filter paper for the second part.

33.3.1.5.3 Procedure

33.3.1.5.3.1 A porcelain cup of about 100 mm diameter should be filled with diatomaceous earth or silica gel at room temperature to a height of about 5 mm. Approximately 5 ml of the liquid to be tested should be poured into the prepared porcelain cup and it is observed if the substance ignites within 5 minutes. This procedure should be performed six times unless a positive result is obtained earlier. If a negative result is obtained then the procedure in 33.3.1.5.3.2 should be followed.

33.3.1.5.3.2 A 0.5 ml test sample should be delivered from a syringe to an indented dry filter paper. The test should be conducted at 25 ± 2 °C and a relative humidity of 50 ± 5%. Observations are made to see if ignition or charring occurs on the filter paper within five minutes of addition of the liquid. This procedure should be performed three times using fresh filter paper each time unless a positive result is obtained earlier.

33.3.1.5.4 Test criteria and method of assessing results

If the liquid ignites in the first part of the test, or if it ignites or chars the filter paper, it should be considered to be pyrophoric and should be classified in packing group I of Division 4.2.

33.3.1.5.5 Examples of results

Substance	**Effect of exposure to air**	**Effect on filter paper**	**Result**
Diethyl aluminium chloride/isopentane (10/90)	No ignition	No charring	Not 4.2
Diethyl aluminium chloride/isopentane (15/85)	No ignition	Charring	4.2
Diethyl aluminium chloride/isopentane (95/5)	No ignition	Charring	4.2
Triethyl aluminium/heptane (10/90)	No ignition	No charring	Not 4.2
Triethyl aluminium/heptane (15/85)	No ignition	Charring	4.2
Triethyl aluminium/heptane (95/5)	No ignition	Charring	4.2

33.3.1.6 *Test N.4: Test method for self-heating substances*

33.3.1.6.1 Introduction

The ability of a substance to undergo oxidative self-heating is determined by exposure of it to air at temperatures of 100 °C, 120 °C or 140 °C in a 25 mm or 100 mm wire mesh cube.

33.3.1.6.2 Apparatus and materials

The following apparatus is required:

a) a hot-air circulating type of oven with an inner volume of more than 9 litres and capable of controlling the internal temperature at 100 °C, 120 °C or 140 °C ± 2 °C;

b) cubic sample containers of 25 mm and 100 mm side, made of stainless steel net with a mesh opening of 0.05 mm, with their top surface open; and

c) Chromel-Alumel thermocouples of 0.3 mm diameter; one placed in the centre of the sample and another between the sample container and the oven wall.

Each sample container should be housed in a cubic container cover made from a stainless steel net with a mesh opening of 0.60 mm, and slightly larger than the sample container. In order to avoid the effect of air circulation, this cover is installed in a second stainless steel cage, made from a net with a mesh size of 0.595 mm and 150 × 150 × 250 mm in size.

33.3.1.6.3 Procedure

The sample, powder or granular, in its commercial form, should be filled to the brim of the sample container and the container tapped several times. If the sample settles, more is added. If the sample is heaped it should be levelled to the brim. The container is housed in the cover and hung at the centre of the oven. The oven temperature should be raised to 140 °C and kept there for 24 hours. The temperature of the sample and of the oven should be recorded continuously. The first test[1] may be conducted with a 100 mm cube sample. A positive result is obtained if spontaneous ignition occurs or if the temperature of the sample exceeds the oven temperature by 60 °C. If a negative result is obtained, no further test is necessary. If a positive result is obtained, a second test should be conducted at 140 °C with a 25 mm cube sample to determine whether or not packing group II should be assigned. If a positive result is obtained at 140 °C with the substance in a 100 mm sample cube, but not a 25 mm sample cube, then an additional test with the substance in a 100 mm sample cube should be performed:

(a) At 120 °C if the substance is to be transported in packagings of more than 450 litre volume but not more than 3 m^3 volume; or

(b) At 100 °C if the substance is to be transported in packagings of not more than 450 litres volume.

33.3.1.6.4 Test criteria and method of assessing results

33.3.1.6.4.1 A positive result is obtained if spontaneous ignition occurs or if the temperature of the sample exceeds the oven temperature by 60 °C during the 24 hour testing time. Otherwise, the result is considered negative.

---

[1] *The tests may be performed in any order. For example, if it is expected that a positive result will be obtained using a 25 mm cube sample then, for safety and environmental protection, the first test may be performed with a 25 mm cube sample. If a positive result is obtained with a 25 mm cube sample then a test with a 100 mm cube sample is not necessary.*

33.3.1.6.4.2 A substance should not be classified in Division 4.2 if:

(a) A negative result is obtained in a test using a 100 mm cube sample at 140 °C;

(b) A positive result is obtained in a test using a 100 mm sample cube at 140 °C and a negative result is obtained in a test using a 25 mm cube sample at 140 °C, a negative result is obtained in a test using a 100 mm cube sample at 120 °C and the substance is to be transported in packages with a volume not more than 3 m^3;

(c) A positive result is obtained in a test using a 100 mm sample cube at 140 °C and a negative result is obtained in a test using a 25 mm cube sample at 140 °C, a negative result is obtained in a test using a 100 mm cube sample at 100 °C and the substance is to be transported in packages with a volume not more than 450 litres.

33.3.1.6.4.3 Packing group II should be assigned to self-heating substances which give a positive result in a test using a 25 mm sample cube at 140 °C.

33.3.1.6.4.4 Packing group III should be assigned to self-heating substances if:

(a) A positive result is obtained in a test using a 100 mm sample cube at 140 °C and a negative result is obtained in a test using a 25 mm cube sample at 140 °C and the substance is to be transported in packages with a volume of more than 3 m^3;

(b) A positive result is obtained in a test using a 100 mm sample cube at 140 °C and a negative result is obtained in a test using a 25 mm cube sample at 140 °C, a positive result is obtained in a test using a 100 mm cube sample at 120 °C and the substance is to be transported in packages with a volume of more than 450 litres;

(c) A positive result is obtained in a test using a 100 mm sample cube at 140 °C and a negative result is obtained in a test using a 25 mm cube sample at 140 °C and a positive result is obtained in a test using a 100 mm cube sample at 100 °C.

33.3.1.6.5 Examples of results

Substance	Oven temperature (°C)	Cube size (mm)	Maximum temperature reached (°C)	Result
Cobalt/molybdenum catalyst granules	140 140	100 25	>200 181	P.G. III of 4.2 [a]
Manganese ethylene bis (dithiocarbamate) 80% (Maneb)	140	25	> 200	P.G. II of 4.2
Manganese ethylene bis (dithiocarbamate) complex with zinc salt 75% (Mancozeb)	140	25	> 200	P.G. II of 4.2
Nickel catalyst granules with 70% hydrogenated oil	140	100	140	Not 4.2
Nickel catalyst granules with 50% white oil	140 140	100 25	> 200 140	P.G. III of 4.2 [a]
Nickel/molybdenum catalyst granules (spent)	140 140	100 25	> 200 150	P.G. III of 4.2 [a]
Nickel/molybdenum catalyst granules (passivated)	140	100	161	Not 4.2
Nickel/vanadium catalyst Granules	140	25	> 200	P.G. II of 4.2

[a] *Not tested at 100 °C or 120 °C.*

33.4 Division 4.3

33.4.1 *Substances which in contact with water emit flammable gases*

33.4.1.1 *Purpose*

33.4.1.1.1 This section of the Manual presents the United Nations scheme for the classification of substances of Division 4.3 which in contact with water emit flammable gases (see section 2.4.4 of the Model Regulations). The text should be used in conjunction with the classification principles given in sub-sections 2.4.4.2 and 2.4.4.3 of the Model Regulations, together with the test prescription given here in 33.4.1.4.

33.4.1.1.2 The test procedure is intended to determine whether the reaction of a substance with water leads to the development of a dangerous amount of gases which may be flammable.

33.4.1.1.3 The test procedures outlined here adequately assess the relative hazard of substances liable to emit flammable gases on contact with water (sometimes referred to as water reactive substances in the Model Regulations) so that an appropriate classification for transport can be made.

33.4.1.2 *Scope*

33.4.1.2.1 New products offered for transport should be subjected to the classification procedures as set out in sub-sections 2.4.4.2 and 2.4.4.3 of the Model Regulations. The classification procedure should be undertaken before a new product is offered for transport.

33.4.1.3 *Classification procedure for substances which in contact with water emit flammable gases*

33.4.1.3.1 The test method can be applied to solid and liquid substances. In case a pyrophoric substance is tested, the test should be executed under nitrogen atmosphere. The substance should be tested in its commercial form at ambient temperate (20 °C) by bringing it into contact with water. If during any stage of the test the gas emitted ignites then no further testing is necessary and the substance should be assigned to Division 4.3. If spontaneous ignition of the emitted gas does not occur then the final stage of the test should be performed to determine the rate of emission of flammable gas. The recommended test method, with possible results, is given here in 33.4.1.4. Whether a substance is a water-reactive substance of Division 4.3 and, if so, whether packing group I, II or III should be assigned is decided on the basis of the test result.

33.4.1.4 *Test N.5: Test method for substances which in contact with water emit flammable* gases

33.4.1.4.1 Introduction

The ability of a substance to emit flammable gases on contact with water is tested by bringing it into contact with water under a variety of conditions.

33.4.1.4.2 Apparatus and materials

No special laboratory apparatus is required.

33.4.1.4.3 Procedure

33.4.1.4.3.1 The substance should be tested according to the procedures described below; if spontaneous ignition occurs at any stage then no further testing is necessary. If it is known that the substance does not react violently with water then proceed to 33.4.1.4.3.5.

33.4.1.4.3.2 A small quantity (approximately 2 mm diameter) of the test substance should be placed in a trough of distilled water at 20 °C. It is noted:

(a) Whether any gas is evolved; and

(b) If spontaneous ignition of the gas occurs.

33.4.1.4.3.3 A small quantity of the test substance (approximately 2 mm diameter) should be placed on the centre of a filter paper which is floated flat on the surface of distilled water at 20 °C in a suitable vessel, e.g. a 100 mm diameter evaporating dish. The filter paper is to keep the substance in one place, under which condition the likelihood of spontaneous ignition of any gas is greatest. It is noted:

(a) Whether any gas is evolved; and

(b) If spontaneous ignition of the gas occurs.

33.4.1.4.3.4 The test substance should be made into a pile approximately 20 mm high and 30 mm diameter with a hollow in the top. A few drops of water are added to the hollow. It is noted whether:

(a) Any gas is evolved; and

(b) If spontaneous ignition of the gas occurs.

33.4.1.4.3.5 For solids, the package should be inspected for any particles of less than 500 μm diameter. If that powder constitutes more than 1% (mass) of the total, or if the substance is friable, then the whole of the sample should be ground to a powder before testing to allow for a reduction in particle size during handling and transport. Otherwise, as for liquids, the substance should be tested in its commercial state. This test should be performed three times at ambient temperature (20 °C) and atmospheric pressure. Water is put into the dropping funnel and enough of the substance (up to a maximum mass of 25 g) to produce between 100 ml and 250 ml of gas is weighed and placed in a conical flask. The tap of the dropping funnel is opened to let the water into the conical flask and a stop watch is started. The volume of gas evolved is measured by any suitable means. The time taken for all the gas to be evolved is noted and where possible, intermediate readings are taken. The rate of evolution of gas is calculated over 7 hours at 1 hour intervals. If the rate of evolution is erratic or is increasing after 7 hours, the measuring time should be extended to a maximum time of 5 days. The five day test may be stopped if the rate of evolution becomes steady or continually decreases and sufficient data has been established to assign a packing group to the substance or to determine that the substance should not be classified in Division 4.3. If the chemical identity of the gas is unknown, the gas should be tested for flammability.

33.4.1.4.4 Test criteria and method of assessing results

33.4.1.4.4.1 A substance should be classified in Division 4.3 if:

(a) Spontaneous ignition takes place in any step of the test procedure; or

(b) There is an evolution of a flammable gas at a rate greater than 1 litre per kilogram of the substance per hour.

33.4.1.4.4.2 Packing group I should be assigned to any substance which reacts vigorously with water at ambient temperatures and generally demonstrates a tendency for the gas produced to ignite spontaneously, or which reacts readily with water at ambient temperatures such that the rate of evolution of flammable gas is equal to or greater than 10 litres per kilogram of substance over any one minute period.

33.4.1.4.4.3 Packing group II should be assigned to any substance which reacts readily with water at ambient temperatures such that the maximum rate of evolution of flammable gas is equal to or greater than 20 litres per kilogram of substance per hour, and which does not meet the criteria for packing group I.

33.4.1.4.4.4 Packing group III should be assigned to any substance which reacts slowly with water at ambient temperatures such that the maximum rate of evolution of flammable gas is greater than 1 litre per kilogram of substance per hour, and which does not meet the criteria for packing groups I or II.

33.4.1.4.5 Examples of results

Substance	**Rate of gas emission (litre/kg.h)**	**Spontaneous ignition of gas (yes/no)**	**Result**
Manganese ethylene bis (dithiocarbamate) complex with zinc salt 88% (Mancozeb)	0	Not applicable	Not 4.3

SECTION 34

CLASSIFICATION PROCEDURES, TEST METHODS AND CRITERIA RELATING TO OXIDIZING SUBSTANCES OF DIVISION 5.1

34.1 Purpose

34.1.1 This section presents the United Nations scheme for the classification of oxidizing substances of Division 5.1 (see section 2.5.2 of the Model Regulations). The text should be used in conjunction with the classification principles given in paragraphs 2.5.2.2 and 2.5.2.3 of the Model Regulations and the test prescriptions given here in section 34.4.

34.2 Scope

34.2.1 New products offered for transport should be subjected to the classification procedures as set out in paragraphs 2.5.2.2.1 and 2.5.2.2.2 or 2.5.2.3.1 and 2.5.2.3.2 of the Model Regulations unless it is impracticable (e.g. because of the physical properties) to perform the tests. Substances which cannot be tested should be classified by analogy with existing entries. The classification procedure should be undertaken before a new product is offered for transport.

34.3 Classification procedure

The test procedures outlined here adequately assess the relative hazard of oxidizing substances so that an appropriate classification for transport can be made by the competent authority. In the event of divergence between test results and known experience, judgement based on known experience should take precedence over test results.

34.3.1 *Oxidizing solids*

A test is performed to determine the potential for a solid substance to increase the burning rate or burning intensity of a combustible substance when the two are thoroughly mixed. The recommended test methods are given here and in sub-section 2.5.2.2 of the Model Regulations. Whether a solid is an oxidizing substance of Division 5.1 and, if so, whether packing group I, II or III should be assigned, is decided on the basis of the test result (see also **Precedence of hazards characteristics** in section 2.0.3 of the Model Regulations). As the particle size has a significant effect on the result, the particle size of the substance tested should be stated in the test report.

34.3.2 *Oxidizing liquids*

A test is performed to determine the potential for a liquid substance to increase the burning rate or burning intensity of a combustible substance or for spontaneous ignition to occur when the two are thoroughly mixed. The recommended test method given here measures the pressure rise time during combustion. Whether a liquid is an oxidizing substance of Division 5.1 and, if so, whether packing group I, II or III should be assigned, is decided on the basis of the test result (see also **Precedence of hazards characteristics** in section 2.0.3 of the Model Regulations).

34.4 Test methods for oxidizing substances

34.4.1 *Test O.1: Test for oxidizing solids*

34.4.1.1 *Introduction*

This test method is designed to measure the potential for a solid substance to increase the burning rate or burning intensity of a combustible substance when the two are thoroughly mixed. Tests are conducted on the substance to be evaluated mixed with dry fibrous cellulose in mixing ratios of 1:1 and 4:1, by mass, of sample to cellulose. The burning characteristics of the mixtures are compared with the standard 3:7 mixture, by mass, of potassium bromate to cellulose. If the burning time is equal to or less than this standard mixture, the burning times should be compared with those from the packing group I or II reference standards, 3:2 and 2:3 ratios, by mass, of potassium bromate to cellulose respectively.

34.4.1.2 *Apparatus and materials*

34.4.1.2.1 Technically pure potassium bromate is required as a reference substance. It should be sieved, but not ground, and the fraction with nominal particle sizes in the range 0.15 to 0.30 mm used as the reference substance. The reference substance is dried at 65 °C to constant mass (for a minimum of 12 hours) and kept in a desiccator (with desiccant) until cool and required for use.

34.4.1.2.2 Dried fibrous cellulose[1], with a fibre length between 50 and 250 μm and a mean diameter of 25 μm, is used as the combustible material. It is dried in a layer no more than 25 mm thick at 105 °C to constant mass (for a minimum of 4 hours) and kept in a desiccator (with desiccant) until cool and required for use. The water content should be less than 0.5% by dry mass. If necessary, the drying time should be prolonged to achieve this.

34.4.1.2.3 An ignition source is required comprising an inert metal wire (e.g. nickel/chromium) connected to an electrical power source and with the following characteristics:

(a) Length = 30 ± 1 cm;

(b) Diameter = 0.6 ± 0.05 mm;

(c) Electrical resistance = 6.0 ± 0.5 Ω/m;

(d) Electrical power dissipated in the wire = 150 ± 7 W.

The wire should be shaped as in Figure 34.4.1.1.

34.4.1.2.4 A 60° glass funnel, sealed at the narrow end, with an internal diameter of 70 mm is required to form the mixtures into a truncated conical pile with base diameter of 70 mm on a cool, impervious, low heat conducting plate. A 150 mm by 150 mm plate with a thickness of 6 mm and a thermal conductivity (at a temperature of 0 °C) of 0.23 $W.m^{-1}.K^{-1}$ is suitable. Other plates with a similar conductivity may be used.

34.4.1.2.5 A fume cupboard or other kind of ventilated area is required in which there is some ventilation but with an air stream velocity of 0.5 m/s or less. ***The fume extraction system should be suitable for the capture of toxic fumes.***

34.4.1.2.6 The substance, in the form in which it will be transported, should be inspected for any particles less than 500 μm in diameter. If that powder constitutes more than 10% (mass) of the total, or if the substance is friable, then the whole of the test sample should be ground to a powder before testing to allow for a reduction in particle size during handling and transport.

---

1 *Source reference available from the national contact for test details in France (see Appendix 4).*

34.4.1.3 *Procedure*

34.4.1.3.1 30.0 g ± 0.1 g mixtures of the reference substance and cellulose are prepared in the potassium bromate to cellulose ratios of 3:7, 2:3 and 3:2, by mass. 30.0 g ± 0.1 g mixtures of the substance to be tested, in the particle size in which it will be transported (see 34.4.1.2.6), and cellulose are prepared in the oxidizer to cellulose ratios of 4:1 and 1:1, by mass. Each mixture should be mixed mechanically as thoroughly as possible without excessive stress. Each sample mixture should be made individually, used as soon as possible, and not taken from a batch.

34.4.1.3.2 Using the conical funnel, the mixture should be formed into a truncated conical pile, with a base diameter of 70 mm, covering the looped ignition wire resting on the low heat conducting plate. The plate should be placed in a ventilated area and the test performed at atmospheric pressure with the ambient temperature at 20 °C ± 5 °C.

34.4.1.3.3 Power is applied to the ignition wire and is maintained for the duration of the test or for three minutes if the mixture does not ignite and burn. The recorded burning time is taken from when the power is switched on to when the main reaction (e.g. flame, incandescence or glowing combustion) ends. Intermittent reaction, such as sparking or sputtering, after the main reaction should not be taken into account. If the heating wire breaks during the test then the test should be repeated unless breaking of the wire clearly does not affect the result. The test should be performed five times on the substance. Five tests should be performed with each reference mixture required to make the packing group assignment or to determine if the substance should not be classified in Division 5.1.

34.4.1.4 *Test criteria and method of assessing results*

34.4.1.4.1 The results are assessed on the basis of:

(a) The comparison of the mean burning time with those of the reference mixtures; and

(b) Whether the mixture of substance and cellulose ignites and burns.

34.4.1.4.2 The test criteria for determining oxidizing properties of the substance are:

Packing group I: Any substance which, in the 4:1 or 1:1 sample-to-cellulose ratio (by mass) tested, exhibits a mean burning time less than the mean burning time of a 3:2 mixture, by mass, of potassium bromate and cellulose.

Packing group II: Any substance which, in the 4:1 or 1:1 sample-to-cellulose ratio (by mass) tested, exhibits a mean burning time equal to or less than the mean burning time of a 2:3 mixture (by mass) of potassium bromate and cellulose; and which does not meet the criteria for packing group I.

Packing group III: Any substance which, in the 4:1 or 1:1 sample-to-cellulose ratio (by mass) tested, exhibits a mean burning time equal to are less than the mean burning time of a 3:7 mixture (by mass) of potassium bromate and cellulose; and which does not meet the criteria for packing groups I and II.

Not Division 5.1 Any substance which, in both the 4:1 and 1:1 sample-to-cellulose ratio (by mass) tested, does not ignite and burn, or exhibits mean burning times greater than that of a 3:7 mixture (by mass) of potassium bromate and cellulose.

For substances having other risks, e.g. toxicity or corrosivity, the requirements of section 2.0.3 of the Model Regulations should be met.

34.4.1.5 *Examples of results*

***NOTE:** The test results given should be used only as an illustration since the results obtained with a particular oxidizer will depend on the particle size etc.*

Substance	Mean burning times (s)		Results
	4:1	1:1	
Ammonium dichromate	55	189	PG III [a]
Ammonium nitrate (crystalline)	161	74	PG III
Calcium nitrate (anhydrous)	10	25	PG II [b]
Calcium nitrate (tetra hydrate)	268	142	Not 5.1
Ceric ammonium nitrate	10	36	PG II [c]
Chromium trioxide	3	33	PG I [a]
Cobalt nitrate (hexahydrate)	205	390	Not 5.1
Nickel nitrate	101	221	Not 5.1 [b]
Potassium nitrite	8	15	PG II
Potassium perchlorate	9	33	PG II
Potassium permanganate	17	51	PG II
Sodium chlorate	5	13	PG II
Sodium nitrite	15	22	PG II [b]
Sodium nitrate	56	39	PG II [b]
Strontium nitrate (anhydrous)	107	237	Not 5.1 [b]

Burning times for mixtures of reference substance to cellulose

3:7 Potassium bromate/cellulose	100 s
2:3 Potassium bromate/cellulose	54 s
3:2 Potassium bromate/cellulose	4 s

[a] *Currently PG II but borderline.*
[b] *Currently PG III.*
[c] *Not currently classified.*

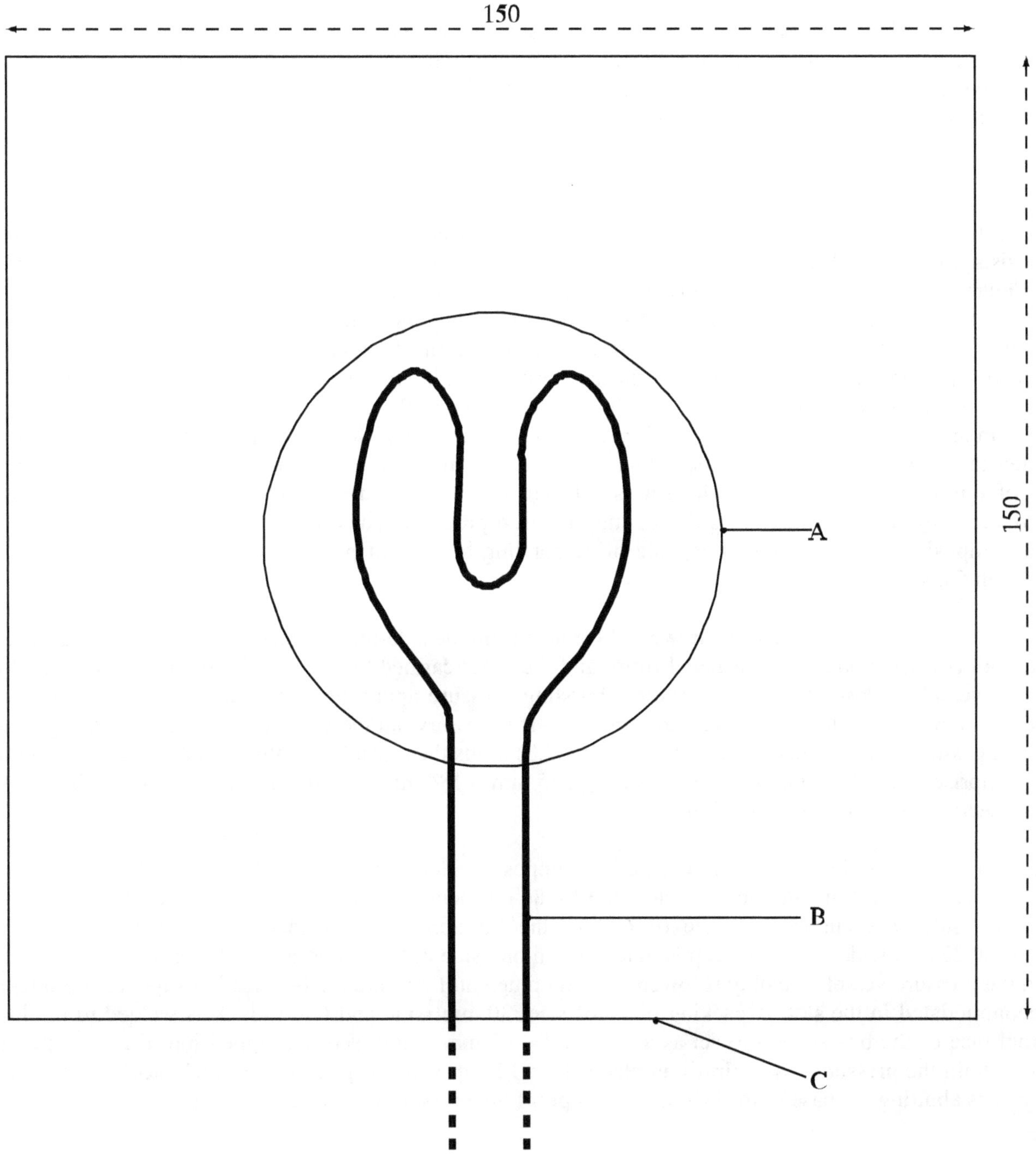

(A) Base of sample cone (70 mm diameter)
(B) Heating wire
(C) Low heat conducting plate

Figure 34.4.1.1: TEST PLATE AND IGNITION WIRE

34.4.2 ***Test O.2: Test for oxidizing liquids***

34.4.2.1 *Introduction*

This test method is designed to measure the potential for a liquid substance to increase the burning rate or burning intensity of a combustible substance when the two are thoroughly mixed or to form a mixture which spontaneously ignites. The liquid is mixed in a 1 to 1 ratio, by mass, with fibrous cellulose, the mixture heated in a pressure vessel and the rate of pressure rise determined[2].

34.4.2.2 *Apparatus and materials*

34.4.2.2.1 A pressure vessel, as used in the time/pressure test (see Part I, test 1 (c) (i)), is required. It consists of a cylindrical steel pressure vessel 89 mm in length and 60 mm in external diameter (see Figure 34.4.2.1). Two flats are machined on opposite sides (reducing the cross-section of the vessel to 50 mm) to facilitate holding whilst fitting up the firing plug and vent plug. The vessel, which has a bore of 20 mm diameter is internally rebated at either end to a depth of 19 mm and threaded to accept 1" British Standard Pipe (BSP). A pressure take-off, in the form of a side arm, is screwed into the curved face of the pressure vessel 35 mm from one end and at 90° to the machined flats. The socket for this is bored to a depth of 12 mm and threaded to accept the 1/2 " BSP thread on the end of the side-arm. If necessary, an inert seal is fitted to ensure a gastight seal. The side-arm extends 55 mm beyond the pressure vessel body and has a bore of 6 mm. The end of the side-arm is rebated and threaded to accept a diaphragm type pressure transducer. Any pressure-measuring device may be used provided that it is not affected by the hot gases or the decomposition products and is capable of responding to rates of pressure rise of 690-2 070 kPa in not more than 5 ms.

34.4.2.2.2 The end of the pressure vessel furthest from the side-arm is closed with a firing plug which is fitted with two electrodes, one insulated from, and the other earthed to, the plug body. The other end of the pressure vessel is closed by a bursting disc (bursting pressure approximately 2 200 kPa (320 psi)) held in place with a retaining plug which has a 20 mm bore. If necessary, an inert seal is used with the firing plug to ensure a gas-tight fit. A support stand (Figure 34.4.2.2) holds the assembly in the correct attitude during use. This comprises a mild steel base plate measuring 235 mm × 184 mm × 6 mm and a 185 mm length of square hollow section (S.-H.S.) 70 × 70 × 4 mm.

34.4.2.2.3 A section is cut from each of two opposite sides at one end of the length of S.H.S. so that a structure having two flat sided legs surmounted by 86 mm length of intact box section results. The ends of these flat sides are cut to an angle of 60° to the horizontal and welded to the base plate. A slot measuring 22 mm wide × 46 mm deep is machined in one side of the upper end of the base section such that when the pressure vessel assembly is lowered, firing plug end first, into the box section support, the side-arm is accommodated in the slot. A packing piece of steel 30 mm wide and 6 mm thick is welded to the lower internal face of the box section to act as a spacer. Two 7 mm thumb screws, tapped into the opposite face, serve to hold the pressure vessel firmly in place. Two 12 mm wide strips of 6 mm thick steel, welded to the side pieces abutting the base of the box section, support the pressure vessel from beneath.

---

[2] *In some cases, substances may generate a pressure rise (too high or too low), caused by chemical reactions not characterising the oxidizing properties of the substance. In these cases, it may be necessary to repeat the test with an inert substance, e.g. diatomite (kieselguhr), in place of the cellulose in order to clarify the nature of the reaction.*

34.4.2.2.4 The ignition system consists of a 25 cm long Ni/Cr wire with a diameter 0.6 mm and a resistance of 3.85 ohm/m. The wire is wound, using a 5 mm diameter rod, in the shape of a coil and is attached to the firing plug electrodes. The coil should have one of the configurations shown in Figure 34.4.2.3. The distance between the bottom of the vessel and the underside of the ignition coil should be 20 mm. If the electrodes are not adjustable, the ends of the ignition wire between the coil and the bottom of the vessel should be insulated by a ceramic sheath. The wire is heated by a constant current power supply able to deliver at least 10 A.

34.4.2.25 Dried, fibrous cellulose[3] with a fibre length between 50 and 250 μm and a mean diameter of 25 μm, is used as the combustible material. It is dried in a layer not more than 25 mm thick at 105 °C for 4 hours and kept in a desiccator, with desiccant, until cool and required for use. The water content of the dried cellulose should be less than 0.5% by dry mass. If necessary, the drying time should be prolonged to achieve this.

34.4.2.2.6 50% perchloric acid, 40% aqueous sodium chlorate solution and 65% aqueous nitric acid are required as reference substances.

34.4.2.2.7 The concentration of the substance tested should be specified in the report. If saturated solutions are tested, they should be prepared at 20 °C.

34.4.2.3 *Procedure*

34.4.2.3.1 The apparatus, assembled complete with pressure transducer and heating system but without the bursting disc in position, is supported firing plug end down. 2.5 g of the liquid to be tested is mixed with 2.5 g of dried cellulose in a glass beaker using a glass stirring rod. ***For safety, the mixing should be performed with a safety shield between the operator and mixture.*** (If the mixture ignites during mixing or filling, no further testing is necessary.) The mixture is added, in small portions with tapping, to the pressure vessel making sure that the mixture is packed around the ignition coil and is in good contact with it. It is important that the coil is not distorted during the packing process. The bursting disc is placed in position and the retaining plug is screwed in tightly. The charged vessel is transferred to the firing support stand, bursting disc uppermost, which should be located in a suitable, armoured fume cupboard or firing cell. The power supply is connected to the external terminals of the firing plug and 10 A applied. The time between the start of mixing and switching the power on should be about 10 minutes.

34.4.2.3.2 The signal produced by the pressure transducer is recorded on a suitable system which allows both evaluation and the generation of a permanent record of the time pressure profile obtained (e.g. a transient recorder coupled to a chart recorder). The mixture is heated until the bursting disc ruptures or until at least 60 s has elapsed. ***If the bursting disc does not rupture, the mixture should be allowed to cool before carefully dismantling the apparatus and precautions taken to allow for any pressurisation.*** Five trials are performed with the mixture and each of the reference substances. The time taken for the pressure rise from 690 kPa to 2 070 kPa above atmospheric is noted. The mean time interval should be used for classification.

34.4.2.4 *Test criteria and method of assessing results*

34.4.2.4.1 The test results are assessed on the basis of:

(a) Whether the mixture of substance and cellulose spontaneously ignites; and

(b) The comparison of the mean time taken for the pressure to rise from 690 kPa to 2 070 kPa gauge with those of the reference substances.

---

[3] *Source reference available from the national contact for test details in Sweden (see Appendix 4).*

34.4.2.4.2 The test criteria for determining the oxidizing properties of the substance are:

Packing group I:

- Any substance which, in the 1:1 mixture, by mass, of substance and cellulose tested, spontaneously ignites; or
- The mean pressure rise time of a 1:1 mixture, by mass, of substance and cellulose is less than that of a 1:1 mixture, by mass, of 50% perchloric acid and cellulose.

Packing group II:

- Any substance which, in the 1:1 mixture, by mass, of substance and cellulose tested, exhibits a mean pressure rise time less than or equal to the mean pressure rise time of a 1:1 mixture, by mass, of 40% aqueous sodium chlorate solution and cellulose; and
- The criteria for packing group I are not met.

Packing group III:

- Any substance which, in the 1:1 mixture, by mass, of substance and cellulose tested, exhibits a mean pressure rise time less than or equal to the mean pressure rise time of a 1:1 mixture, by mass, of 65% aqueous nitric acid and cellulose; and
- The criteria for packing groups I and II are not met.

Not Division 5.1:

- Any substance which, in the 1:1 mixture, by mass, of substance and cellulose tested, exhibits a pressure rise of less than 2 070 kPa gauge; or
- Exhibits a mean pressure rise time greater than the mean pressure rise time of a 1:1 mixture, by mass, of 65% aqueous nitric acid and cellulose.

For substances having additional risks, e.g. toxicity or corrosivity, the requirements of section 2.0.3 of the Model Regulations should be met.

34.4.2.5 *Examples of results*

Substance	Mean pressure rise time for a 1:1 mixture with cellulose (ms)	Results
Ammonium dichromate, saturated aqueous solution	20 800	Not Div. 5.1
Calcium nitrate, saturated aqueous solution	6 700	Not Div. 5.1
Ferric nitrate, saturated aqueous solution	4 133	PG III
Lithium perchlorate, saturated aqueous solution	1 686	PG II
Magnesium perchlorate, saturated aqueous solution	777	PG II
Nickel nitrate, saturated aqueous solution	6 250	Not Div. 5.1
Nitric acid, 65%	4 767 [a]	PG III [b]
Perchloric acid, 50%	121 [a]	PG II
Perchloric acid, 55%	59	PG I
Potassium nitrate, 30% aqueous solution	26 690	Not Div. 5.1
Silver nitrate, saturated aqueous solution	[c]	Not Div. 5.1
Sodium chlorate, 40% aqueous solution	2 555 [a]	PG II
Sodium nitrate, 45% aqueous solution	4 133	PG III
Inert substance		
Water:cellulose	[c]	

[a] *Mean value from inter-laboratory comparative trials.*
[b] *Packing group III by the test but Class 8 by the Precedence of Hazards table.*
[c] *Maximum pressure of 2 070 kPa not reached.*

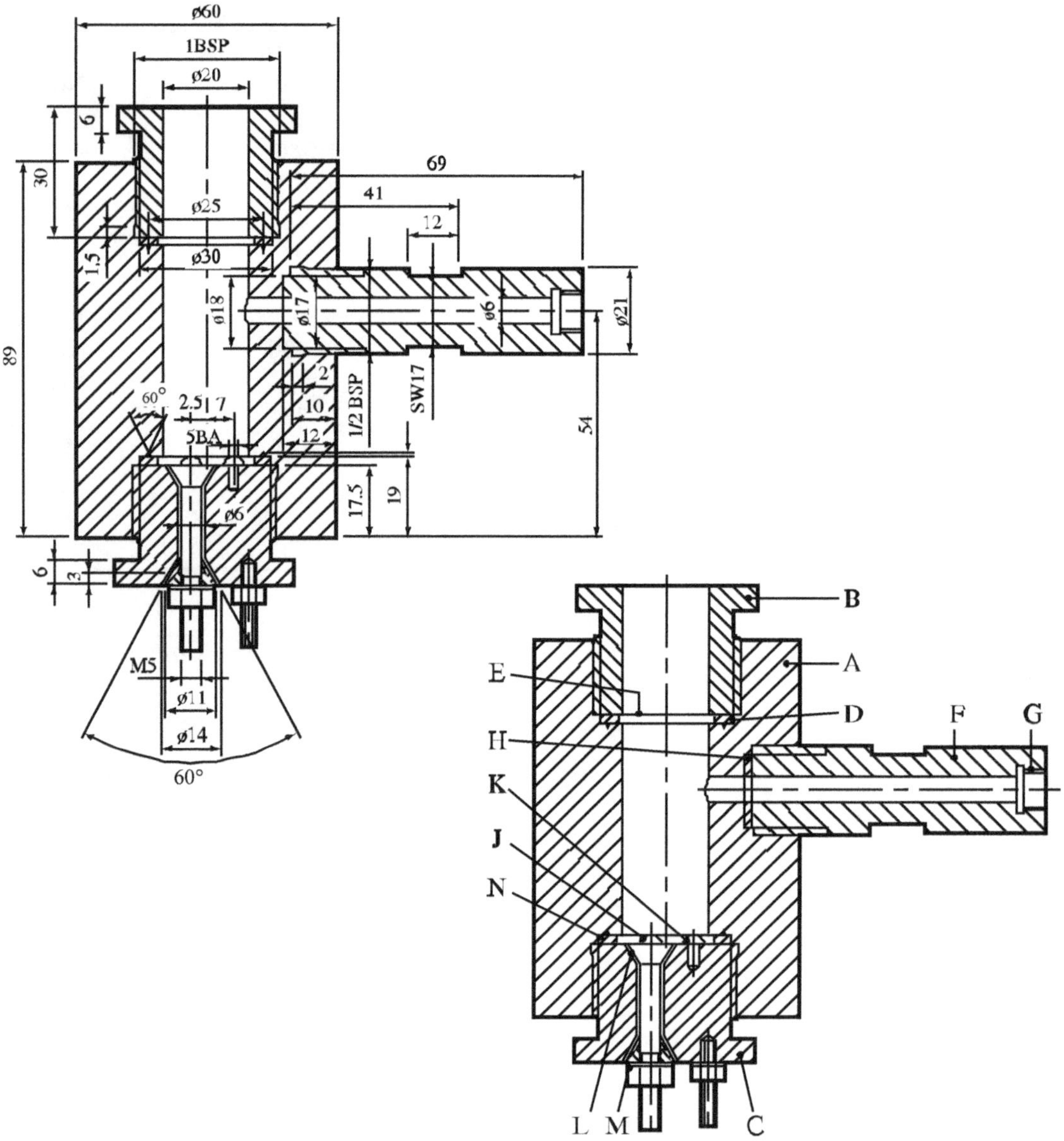

(A)	Pressure vessel body	(B)	Bursting disc retaining plug
(C)	Firing plug	(D)	Soft lead washer
(E)	Bursting disc	(F)	Side arm
(G)	Pressure transducer thread	(H)	Washer
(J)	Insulated electrode	(K)	Earthed electrode
(L)	Insulation	(M)	Steel cone
(N)	Washer distorting groove		

Figure 34.4.2.1: PRESSURE VESSEL

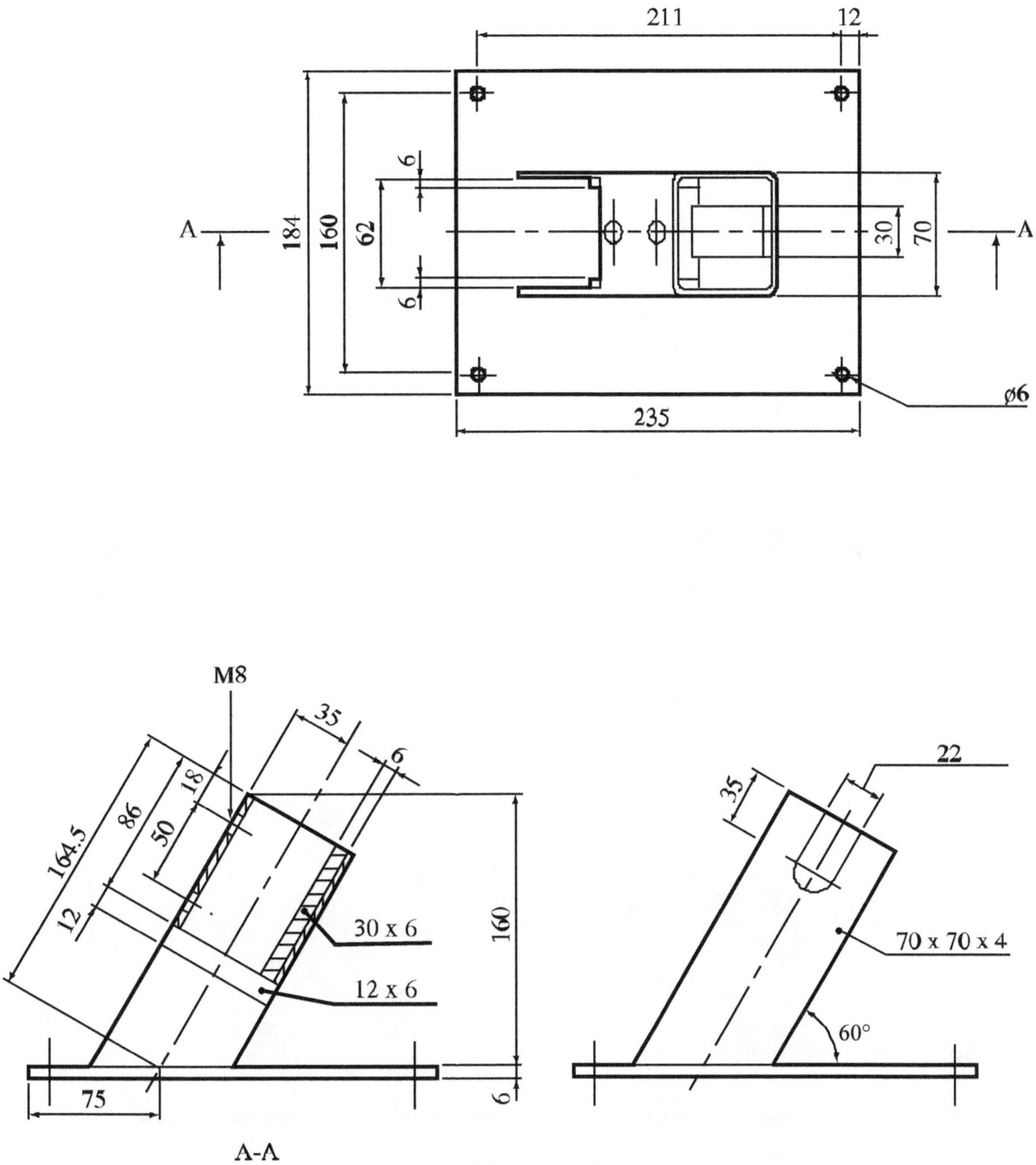

Figure 34.4.2.2: SUPPORT STAND

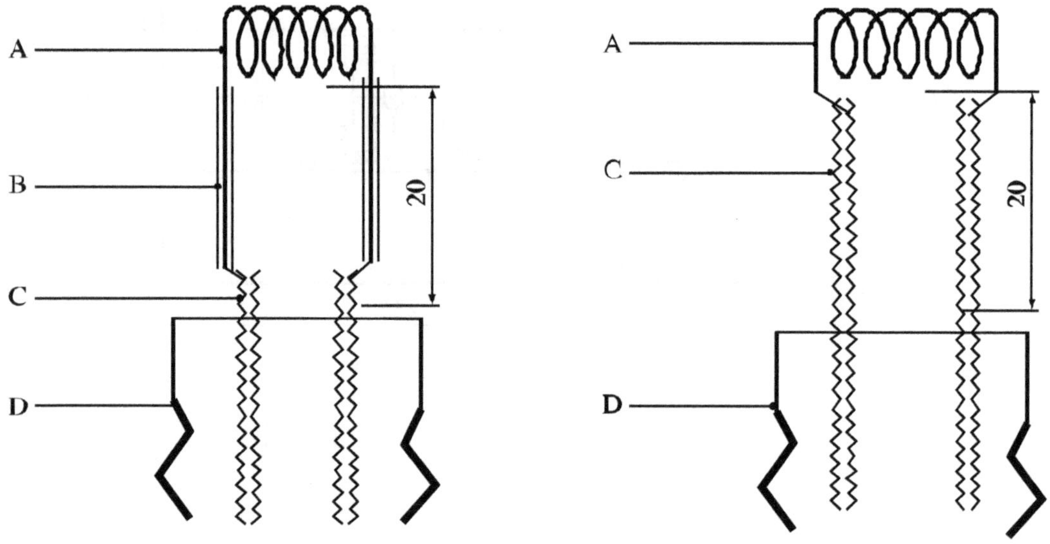

NOTE: Either of these configurations may be used.

(A) Ignition coil
(B) Insulation
(C) Electrodes
(D) Firing plug

Figure 34.4.2.3: IGNITION SYSTEM

34.4.3 *Test O.3: Gravimetric test for oxidizing solids*

34.4.3.1 *Introduction*

This test method is designed to measure the potential for a solid substance to increase the burning rate or burning intensity of a combustible substance when the two are thoroughly mixed. Tests are conducted on the test substance to be evaluated mixed with dry fibrous cellulose in mixing ratios of 1:1 and 4:1, by mass. The burning rates of these mixtures are compared to the reference mixture of calcium peroxide and cellulose in a mixing ratio of 1:2, by mass. The mass loss of the mixtures during combustion is determined using a balance connected to a suitable data recording system, and is recorded as a function of time. If the burning rate (g/s) is equal to or greater than the burning rate of the reference mixture for Packing Group III, it is then compared to the burning rate of the reference mixtures for Packing Group I or II (see table in 34.4.3.4).

The GHS also refers to this test for classification within the hazard class Oxidizing Solids. For the ranking according to oxidizing power, the GHS uses three categories which correspond exactly to the Packing Groups used within the transport of dangerous goods regime. The GHS categories 1, 2 and 3 thus correspond directly to Packing Groups I, II and III, respectively.

It is useful to have preliminary information on any potentially explosive properties of the substance before performing the test, screening procedures can be found in Appendix 6. This test is not applicable for substances which are explosive or flammable and also not for organic peroxides.

34.4.3.2 *Materials*

34.4.3.2.1 Technically pure, finely powdered calcium peroxide with a concentration of 75% ± 0.5% is required as the reference oxidizer. The concentration of impurities such as chlorides or compounds which generate water during combustion should be low, as they may impact on the burning behavior of the reference piles. Calcium peroxide meeting the specification below can be used without further pretreatment.[4]

CaO_2:	75% ± 0.5%
$Ca(OH)_2$:	20% to 25%
$CaCO_3$:	0% to 5%
Chloride:	max 500 ppm
Particle size:	min 99% < 75 µm, and thereof
	min 50% < 20 µm

34.4.3.2.2 Dried fibrous cellulose with a fibre length between 50 µm and 250 µm and a mean diameter of 25 µm is used as the combustible material. It is dried to constant mass (for a minimum of 4 hours), in a layer no more than 25 mm thick, at 105 °C and kept in a desiccator (with desiccant) until cool and required for use. The water content should be less than 0.5% by dry mass. If necessary, the drying time should be prolonged to achieve this. The bulk density of the cellulose used for the test should be such (sufficiently high) that the test mixture for Packing Group III (30.0 g ± 0.1 g) can be completely accommodated in the conical funnel.

34.4.3.2.3 The test-substance, in the form in which it will be presented, should be inspected for any particles less than 500 µm in diameter. If that powder constitutes more than 10% (by mass) of the total, or if the substance is friable, then the whole of the test sample should be ground to a powder of less than 500 µm before testing, to allow for a reduction in particle size during handling and transport. Because the particle size influences the oxidizing properties of a substance, a coarse oxidizer may be classified as Packing Group III, while a finer grade of the same material may show a faster burning rate and may result in different GHS classifications and assigned transport Packing Groups for different forms of the same material.

---

4 *Obtainable from the national contact for test details in Germany (see Appendix 4).*

34.4.3.3 *Apparatus*[5]

34.4.3.3.1 An appropriate balance is required, with a suitable range, accuracy and data transfer capability, equipped with an interface (e.g. USB or RS232) to allow data acquisition. The required data (time, mass) are recorded, preferably with a frequency of at least 5 data per second. Any suitable software capable of recording the output of the balance may be used. It is recommended to check the time frequency of the software and the monitoring of time by a stopwatch maybe useful.

34.4.3.3.2 An ignition source is required, comprising an inert metal wire connected to an electrical power source capable of maintaining the power dissipation specified below. The electrical resistance depends on the wire material. It is recommended to use Nickel/Chromium or AluChrom wire as follows:

(a) Length = 30 cm ± 1 cm;

(b) Diameter below 1 mm;

(c) Electrical power dissipated in the wire = 150 W ± 7 W.

The wire should be shaped as shown in figure 34.4.3.2. It is very important that the power connection to the heating wire is flexible enough to avoid any influence on the proper operation of the balance.

34.4.3.3.3 A 60° funnel, with an internal diameter of 70 mm and sealed at the narrow end, is required to form the mixtures into a truncated conical pile with a base diameter of 70 mm on a cool, impervious, low heat conducting plate.

34.4.3.3.4 An insulating test plate is needed to avoid loss of energy by thermal conduction. A 150 mm by 150 mm test plate with a thickness of at least 6 mm and a thermal conductivity at 0 °C of 0.23 $W*m^{-1}*K^{-1}$ or less is suitable. Other plates with a similar conductivity may be used. In order to increase the lifetime of the test plate, it may be covered with a thin ceramic sheet with the same low thermal conductivity.

34.4.3.3.5 A fume cupboard or other suitably ventilated area is required but the air stream velocity should be 0.5 m/s or less. ***The fume extraction system should be capable of capturing all toxic fumes.***

34.4.3.3.6 To construct the support for the burning trial on the balance a base plate of solid material (steel or other suitable material), a positioning plate of fire resistant material (the same material as test-plate is recommended) and guide bars are required.

34.4.3.3.7 The air stream of the ventilation must be prevented from affecting the balance and thereby altering the test result. A windshield protecting the complete test equipment against draughts from the ventilation or environment is the preferred means.

34.4.3.3.8 It is mandatory that the burning conical pile is always placed at the centre of the balance. It is also important to protect the balance from heat and burning particles during the test. To achieve this, the following general configuration is recommended (the letters in brackets refer to figure 34.4.3.1):

(a) Two plates are used to ensure always the same position on the balance and to protect the balance. The base plate (H) is larger than the balance and made of solid material. It is recommended to use some buffers fixed to the lower side to reduce vibrations from the environment. 2 to 4 metal guide bars (G) are fixed on the bottom plate as shown in figure 34.4.3.1 to ensure always the same positioning of the positioning plate (F) and test plate (C) on the balance during testing. The base plate should be rigid enough to keep the guide bars always in a stable position (e.g. 4 mm steel or

---

5 *Technical information about a suitable design and video training are obtainable from the national contact for test details in Germany (Appendix 4)*

16 mm polyamide). The balance has to be placed always at the same central position on the base plate;

(b) The positioning plate (F) is made of fire resistant low conductivity material, similar in properties to the test plate (see 34.4.3.3.4). The holes drilled in the positioning plate for the bars should be about 8 mm larger in diameter than the bars themselves;

(c) The bars always have to be positioned in the centre of the holes to prevent any contact between the positioning plate (F) and the bars, in order not to affect the operation of the balance. Some check marks should be fixed on the positioning plate (F) to position the test plate (C) in the correct position in the centre of the balance;

(d) The electrical connection between the power supply and the heated wire must be flexible enough to avoid interfering with free movement of the balance pan due to resistance or movement. This can be achieved by using a flexible wire and a support close to the test plate. Coiling of the wire between the support and the test plate will provide additional flexibility;

(e) The windshield (D) can be integral with the bottom plate or placed around the complete test equipment. The windshield should have no gaps at the bottom. The windshield itself should be closed and about 10 cm higher than the equipment to prevent air streams from the top.

34.4.3.4 *Procedure*

The following samples are required:

Sample	*Components*	*Mixing ratio by mass*
Test substance mixture 1:1	Test substance and cellulose	1:1
Test substance mixture 4:1	Test substance and cellulose	4:1
Reference mixture for Packing Group I	Reference substance and cellulose	3:1
Reference mixture for Packing Group II	Reference substance and cellulose	1:1
Reference mixture for Packing Group III	Reference substance and cellulose	1:2

34.4.3.4.1 30.0 g ± 0.1 g of each of the reference mixtures and 30.0 g ± 0.1 g of both mixtures of the test substance are prepared. Each mixture should be mixed mechanically as thoroughly as possible for at least 1 minute, but without excessive stress. Each mixture should be prepared individually, used as soon as possible, and not taken from a batch.

34.4.3.4.2 Using the conical funnel, the mixture should be formed into a truncated conical pile with a base diameter of 70 mm. Form the pile by tapping the funnel slightly after filling; cover the funnel with the test plate -including the sheet (if any)- and invert both. Tap slightly on the funnel before removing. The pile should now cover the looped ignition wire resting on test plate. The test should be performed at atmospheric pressure with an ambient temperature of 20 °C ± 5 °C and a relative humidity below 60%, to reduce absorption of moisture by the cellulose during handling.

34.4.3.4.3 The balance, placed in the ventilated area as mentioned above is set to zero. Power is applied to the ignition wire and is maintained for the duration of the test, or for three minutes if the mixture does not ignite and burn. Data collection should start a few seconds before the power supply is switched on, and should proceed until the reaction ends or the mass loss over 1 minute is less than 1 g. If the heating wire breaks, the test should be repeated to ensure that early breakage of the wire does not influence the test result.

34.4.3.5 *Test criteria and method of assessing results*

34.4.3.5.1 The results are assessed on the basis of:

(a) The comparison of the mean burning rate with those of the reference mixtures;

(b) Whether the mixture of the test substance and cellulose ignites and burns.

34.4.3.5.2 The combustion of the conical pile can be divided into three periods:

(a) Start period: between 0% and 20% of the total mass loss;

(b) Main combustion period: between 20% and 80% of the total mass loss; and

(c) End of the reaction: between 80% mass loss and the end of the reaction.

In the main combustion period the mass loss per unit time is quite constant. For this reason a linear regression (based on the least square approach) can be used to check the quality of the data collected.

34.4.3.5.3 The burning rate (BR) depends on the burning intensity and the amount of cellulose in the mixture. It is therefore defined as the quotient of 60% of the total amount of cellulose in the conical pile and the main combustion time $t_{20\text{-}80}$. The time $t_{20\text{-}80}$ is the time between 20% and 80% of the total mass loss. The total mass loss is the difference in mass before ignition and at the end of the main reaction, defined as a rate of mass loss rate of less than 1 g per minute.

The burning rate $BR_{20\text{-}80}$ is therefore calculated by the following equation:

$$BR_{20-80} = \frac{0.6 \times m_{cellulose}}{t_{20-80}}$$

where:

$BR_{20\text{-}80}$ = burning rate between 20% to 80% in g/s

$m_{cellulose}$ = mass of cellulose in mixture in gram

$t_{20\text{-}80}$ = burning time between 20% to 80% of total mass loss in seconds

The profile of each burning test has to be examined by plotting the mass loss as a function of time. The graph can also be used for decision making and should be used in case of doubt. The coefficient of correlation (R^2) of the mass curve of each burning test should be at least 0.95 between 20% to 80% mass loss, otherwise the burning trial has to be repeated. Five valid tests should be performed with each reference and test substance mixture. The standard deviation of the burning rates within these five tests should not exceed 10% in total.

34.4.3.5.4 The test criteria for determining transport Packing Groups on basis of the oxidizing properties of the substance are:

Packing Group I: Any substance which, in the 4:1 or 1:1 sample-to-cellulose ratio (by mass) tested, exhibits a mean burning rate greater than the mean burning rate of a 3:1 mixture, by mass, of calcium peroxide and cellulose.

Packing Group II: Any substance which, in the 4:1 or 1:1 sample-to-cellulose ratio (by mass) tested, exhibits a mean burning rate equal to or greater than the mean burning rate of a 1:1 mixture (by mass) of calcium peroxide and cellulose and the criteria for Packing Group I are not met.

Packing Group III: Any substance which, in the 4:1 or 1:1 sample-to-cellulose ratio (by mass) tested, exhibits a mean burning rate equal to or greater than the mean burning rate of a 1:2 mixture (by mass) of calcium peroxide and cellulose and the criteria for Packing Groups I and II are not met.

Not Division 5.1: Any substance which, in both the 4:1 and 1:1 sample-to-cellulose ratio (by mass) tested, does not ignite and burn, or exhibits mean burning rates less than that of a 1:2 mixture (by mass) of calcium peroxide and cellulose.

For substances having other risks, e.g. toxicity or corrosivity, section 2.0.3 of the Model Regulations applies.

In GHS terminology, the test criteria for determining the classification due to the oxidizing properties of a solid substance are analogous to those used in transport:

Category 1: Any substance which, in the 4:1 or 1:1 sample-to-cellulose ratio (by mass) tested, exhibits a mean burning rate greater than the mean burning rate of a 3:1 mixture (by mass) of calcium peroxide and cellulose.

Category 2: Any substance which, in the 4:1 or 1:1 sample-to-cellulose ratio (by mass) tested, exhibits a mean burning rate equal to or greater than the mean burning rate of a 1:1 mixture (by mass) of calcium peroxide and cellulose and the criteria for Category 1 are not met.

Category 3: Any substance which, in the 4:1 or 1:1 sample-to-cellulose ratio (by mass) tested, exhibits a mean burning rate equal to or greater than the mean burning rate of a 1:2 mixture (by mass) of calcium peroxide and cellulose and the criteria for Category 1 and 2 are not met.

Not an Oxidizing Solid: Any substance which, in both the 4:1 and 1:1 sample-to-cellulose ratio (by mass) tested, does not ignite and burn, or exhibits mean burning rates less than that of a 1:2 mixture (by mass) of calcium peroxide and cellulose.

34.4.3.5.5 If a false positive result is suspected, the results obtained when testing the substance mixed with an inert material and/or when testing under an inert atmosphere should also be considered when interpreting the results.

34.4.3.5.6 If the shape of the mass loss curve indicates the test is invalid, the mixing procedure or test set-up affecting free movement of the balance pan should be examined.

34.4.3.6 *Example of results*

NOTE: *The results given in this table should be understood as illustrative examples since the results obtained with a particular oxidizer will depend on the particle size etc. The test results of the examples are not intended to form the basis for classifications (see also 2.14.4.2.1 of Chapter 2.14 in Part 2 of the GHS and 2.5.2.1.1 of Chapter 2.5 in the Model Regulations, respectively).*

Substance	*Particle size, median value (D50 [μm])*	*Resulting Packing Group*
Ammonium dichromate	300	III[1]
Calcium nitrate (tetra hydrate)	1050	III[1]
Cobalt nitrate (hexahydrate)	1200	Not 5.1[1]
Nickel nitrate	1200	Not 5.1[1]
Potassium nitrite	200	I[2,a]
Potassium perchlorate	220	II[1]
Potassium perchlorate	30	II[2]
Potassium permanganate	200	I[1,2,a]
Sodium chlorate	220	I[2,a]
Sodium nitrite	320	II[1,b]
Sodium nitrate	200	II[1,b]
Strontium nitrate (anhydrous)	250	Not 5.1[1]

[a] Currently PG II

[b] Currently PG III

[1] Solvay

[2] BAM

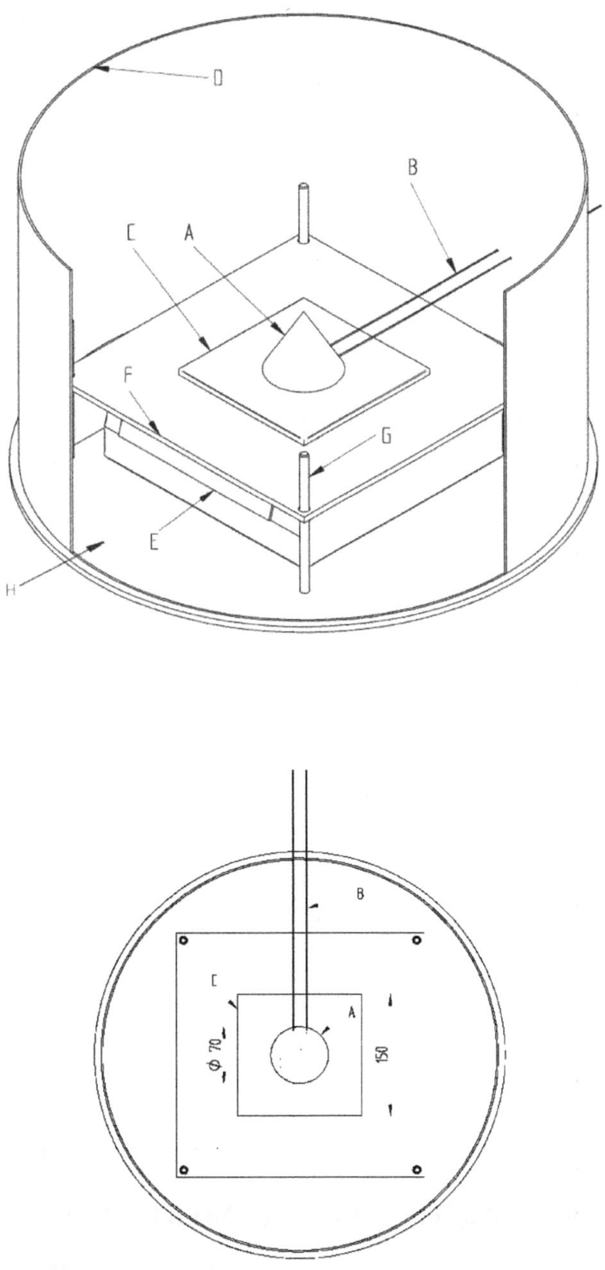

(A)	Conical pile
(B)	Heating wire
(C)	Test plate
(D)	Wind shield
(E)	Balance with interface
(F)	Positioning plate
(G)	Bars to fix positioning plate (F) and test plates (C) at defined location on balance
(H)	Base plate

Figure 34.4.3.1: Scheme of UN O.3 Test configuration

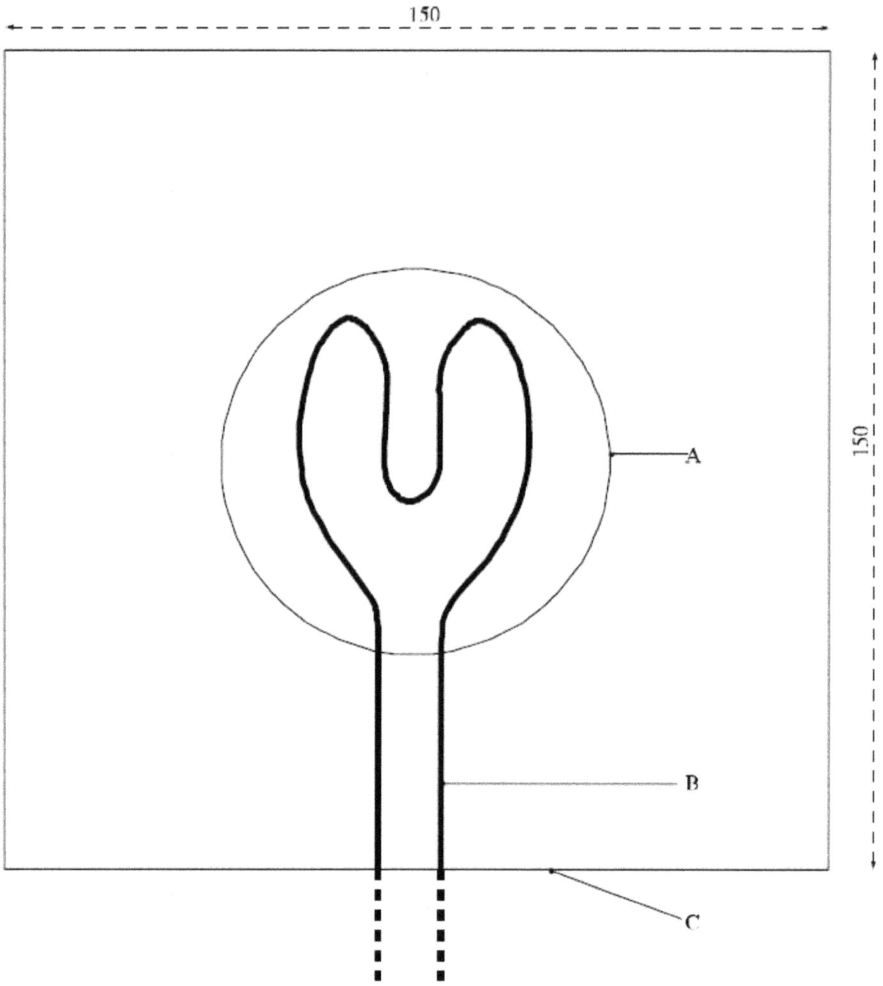

(A)	Conical pile (base)
(B)	Heating wire
(C)	Test plate

Figure 34.4.3.2: TEST PLATE AND IGNITION WIRE

SECTION 35

DETERMINATION OF CHEMICAL INSTABILITY OF GASES AND GAS MIXTURES

35.0 Introduction

This section presents the United Nations scheme for the classification of gases and gas mixtures as chemically unstable. The text should be used in conjunction with the classification principles given in Chapter 2.2 of the Globally Harmonized System of Classification and Labelling of Chemicals (GHS) and the test methods given in this section.

35.1 Purpose

35.1.1 This test method is used to determine the chemical instability of a gas or gas mixture by ignition tests in a closed vessel at ambient and elevated temperature and pressure.

35.1.2 For the purposes of this test method the following definitions apply:

Chemical instability means the propensity of a gas or gas mixture to react dangerously even in the absence of any reaction partner (e.g. air or oxygen) by decomposing and thereby causing a temperature and/or pressure increase;

Test gas means the gas or gas mixture to be evaluated by this test method;

Corresponding initial pressure means the pressure at which the test at 65 °C is carried out. For test gases that are completely gaseous the corresponding initial pressure is the pressure that a gas develops at 65 °C based on the maximum (filling) pressure at ambient temperature. For liquefied test gases the corresponding initial pressure is the vapour pressure at 65 °C.

35.2 Scope

35.2.1 The test method does not cover gas decomposition under process conditions in chemical plants and possible dangerous reactions between different gases in gas mixtures.

35.2.2 Mixtures of gases, where the components can react dangerously with each other, e.g. flammable and oxidizing gases, are not regarded as chemically unstable in the sense of this test method.

35.2.3 If the calculations in accordance with ISO 10156:2010 show that a gas mixture is not flammable it is not necessary to carry out the tests for determining chemical instability for classification purposes.

35.2.4 Expert judgement should be applied to decide whether a flammable gas or gas mixture is a candidate for classification as chemically unstable in order to avoid unnecessary testing of gases where there is no doubt that they are stable. Functional groups indicating chemical instability in gases are triple-bonds, adjacent or conjugated double-bonds, halogenated double-bonds and strained rings.

35.3 Concentration limits

35.3.1 *Generic concentration limits*

35.3.1.1 Gas mixtures containing only one chemically unstable gas are not considered as chemically unstable and therefore do not have to be tested for classification purposes if the concentration of the chemically unstable gas is below the higher of the following generic concentration limits:

(a) The lower explosion limit (LEL) of the chemically unstable gas; or

(b) 3 mole percent (mol %).

35.3.2 ***Specific concentration limits***

35.3.2.1 The following tables contain information about some gases with regard to their classification as chemically unstable. Specific concentration limits for their mixtures are given. Gas mixtures containing only one chemically unstable gas in concentrations below the specific concentration limit are not considered as chemically unstable and therefore do not have to be tested for classification purposes.

Table 35.1: Information about gases with regard to their chemical instability and concentration limits for their mixtures below which the mixtures are not classified as chemically unstable

Information about the pure gas					**Information about its mixtures**
Chemical Name	**Molecular formula**	**CAS No.**	**UN No.**	**Classification**	**Specific concentration limit (see Notes 1 and 2)**
Acetylene	C_2H_2	74-86-2	1001 3374	Chem. Unst. Cat. A	See Table 35.2 For other mixtures: Partial pressure of 1 bar abs
Bromotrifluoro-ethylene	C_2BrF_3	598-73-2	2419	Chem. Unst. Cat. B	8.4 mol % (LEL)
Butadiene-1,2	C_4H_6	590-19-2	1010	Not classified as chemically unstable	
Butadiene-1,3	C_4H_6	106-99-0	1010	Not classified as chemically unstable	
Butyne–1, Ethylacetylene	C_4H_6	107-00-6	2452	Chem. Unst. Cat. B	The specific concentration limits for acetylene may be applied, see Table 35.2. For other mixtures: Partial pressure of 1 bar abs
Chlorotrifluoro-ethylene	C_2ClF_3	79-38-9	1082	Chem. Unst. Cat. B	4.6 mol % (LEL)
Ethylene oxide	C_2H_4O	75-21-8	1040	Chem. Unst. Cat. A	15 mol % for mixtures containing rare gases. 30 mol % for other mixtures
Vinyl methyl ether	C_3H_6O	107-25-5	1087	Chem. Unst. Cat. B	3 mol %
Propadiene	C_3H_4	463-49-0	2200	Chem. Unst. Cat. B	The specific concentration limits for acetylene may be applied, see Table 35.2. For other mixtures: Partial pressure of 1 bar abs
Propyne	C_3H_4	74-99-7	3161	Chem. Unst. Cat. B	The specific concentration limits for acetylene may be applied, see Table 35.2. For other mixtures: Partial pressure of 1 bar abs
Tetrafluoro-ethylene	C_2F_4	116-14-3	1081	Chem. Unst. Cat. B	10.5 mol % (LEL)
Trifluoro-ethylene	C_2HF_3	359-11-5	1954	Chem. Unst. Cat. B	10.5 mol % (LEL)
Vinyl bromide	C_2H_3Br	593-60-2	1085	Chem. Unst. Cat. B	5.6 mol % (LEL)
Vinyl chloride	C_2H_3Cl	75-01-4	1086	Chem. Unst. Cat. B	3.8 mol % (LEL)
Vinyl fluoride	C_2H_3F	75-02-5	1860	Chem. Unst. Cat. B	3 mol %

NOTE 1: *The maximum pressure should be limited in order to avoid condensation.*

NOTE 2: *The test method is not applicable to liquefied gas mixtures. In case the gaseous phase above a liquefied gas mixture may become chemically unstable after withdrawal, this shall be communicated via the safety data sheet.*

Table 35.2: Specific concentration limits for binary mixtures with acetylene. These concentration limits may also be applied to butyne-1 (ethylacetylene), propadiene and propyne

Concentration limit for acetylene in mol %	Maximum (filling) pressure in bar for a mixture with						
	N_2	CO_2	NH_3	H_2	CH_4	C_3H_8	C_2H_4
3.0	200.0				200.0		
4.0	100.0						
5.0				40.0			40.0
6.0	80.0						
8.0	60.0						
10.0	50.0	38.0	5.6	20.0	100.0	6.0	20.0
15.0	30.0	30.0		10.0			10.0
20.0	25.0	20.0	6.2	5.0	50.0	6.6	7.5
25.0	20.0	15.0					5.0
30.0	10.0	10.0	6.9		25.0	7.3	
35.0			7.3				
40.0					15.0	8.2	
45.0							
50.0					5.0	9.3	
60.0						10.8	

35.4 Test method

35.4.1 *Introduction*

35.4.1.1 The propensity of a gas to decompose depends strongly on pressure, temperature and in case of gas mixtures on the concentration of the chemically unstable component. The possibility of decomposition reactions shall be evaluated at conditions which are relevant for handling and use and transport. Therefore two types of tests shall be performed:

(a) At ambient temperature and pressure,

(b) At 65 °C and the corresponding initial pressure.

35.4.2 Apparatus and material

35.4.2.1 The test apparatus (see Figure 35.1) consist of a pressure resistant test vessel (heatable) made of stainless steel; an ignition source; a measuring and recording system to record the pressure inside the ignition vessel; a gas supply; a venting system with bursting disc and additional piping, fitted with remote-controlled valves and cocks.

(a) Pressure resistant test vessel

The test vessel is a cylindrical stainless steel vessel with an inner volume of about 1 dm^3 and an inner diameter of 80 mm. An exploding wire ignition source is screwed into the bottom of the vessel. The vessel is equipped with a heating jacket which is connected to a temperature control unit that heats the outer vessel wall with an accuracy of ± 2 K. The test vessel is insulated with temperature resistant insulation material to avoid heat loss and temperature gradients. The test vessel shall be pressure resistant up to 500 bar (50 MPa).

(b) Exploding wire igniter

The ignition source is an exploding (fusing) wire igniter similar to that described in ASTM E 918 and EN 1839. The igniter consists of two insulated electrodes at 3 mm to 6 mm distance, holding a nickeline wire of 0.12 mm diameter at its ends. The ignition energy is provided by a 1.5 kVA/230 (115) V insulating transformer, which is switched for a short time period to this igniter. The wire melts and then an electrical arc burns between the electrodes for a period extending in maximum to half a period of the supply voltage (10 (8.3) ms). An electronic control unit allows switching different time periods of the mains voltage half wave to the igniter. The corresponding energy delivered shall be in the range of 15 J ± 3 J. The energy can be measured by recording the current and voltage during ignition.

(c) Pressure and temperature recording equipment

The pressure inside the ignition vessel shall be measured with a calibrated piezoresistive pressure transducer. The measuring range shall be 20 times higher than the initial pressure. The sensitivity shall be at least 0.1% of the full scale and the accuracy shall be better than 0.5% of the full scale.

The temperature of the ignition vessel shall be measured and controlled with a 3 mm thermocouple type "K" (NiCr/NiAl) mounted 50 mm below the top inside the autoclave.

After ignition the digitized pressure signal shall be recorded with a computer. The initial pressure (p0) and the highest pressure (pex) are derived from the raw data.

(d) Gas supply

Two different types of gas supply are necessary, one for test gases that are completely in the gaseous phase and one for liquefied test gases. Test gases in the gaseous phase are metered volumetric or by flow measurement and liquefied test gases are metered gravimetric.

(e) Bursting disc

The bursting disc is supposed to protect the test vessel. It is connected to a vent pipe for the exhaust gas. The free diameter of the bursting disc should be at least 10 mm, the inner diameter of the pipe at least 15 mm. The opening pressure of the bursting disc shall amount to 250 bar (25 MPa).

(f) Additional piping and valves

The piping and valves which are fitted directly to the test vessel, shall be pressure resistant up to 500 bar (50 MPa). The test apparatus shall be operated by remotely operated valves.

35.4.3 ***Test procedure***

35.4.3.1 The test gas is charged into a pressure resistant stainless steel vessel at controlled temperature and pressure. The vessel is equipped with a bursting disc. Ignition of the test gas is achieved with an exploding wire igniter. Whether a decomposition reaction has occurred is deduced from the pressure rise produced.

35.4.3.2 The tests shall be executed in the following sequence:

(a) Test at ambient temperature and pressure

For the tests at 20 °C and 1.01 bar (101.3 kPa) the exploding wire igniter shall be arranged in the middle of the test vessel. The test vessel and the piping are evacuated. The test gas is filled into the test vessel by using the remotely operated valves until ambient pressure (initial pressure) is reached. After closing the valves the igniter is fired. The ignition energy shall be about 15 J to avoid over-initiation in the test vessel at this relatively low pressure. Criterion for a reaction is a pressure rise of more than 20% after ignition (f = pex/p0 > 1.20). If no such pressure rise has occurred two further re-tests shall be carried out.

If the test gas shows a pressure rise of more than 20% in either of the tests it is to be classified as "chemically unstable at 20 °C and a standard pressure of 101.3 kPa". No further tests are required.

(b) Test at elevated temperature and pressure

If in the tests according to 35.4.3.2 (a) no pressure rise of more than 20% has occurred, further tests at 65 °C and the corresponding initial pressure shall be performed. The procedure is the same as in 35.3.3.2 (a) but care should be taken with potentially unstable gases under pressure. The ignition energy shall be about 15 J. If no pressure rise of more than 20% has occurred two further re-tests shall be carried out.

If the test gas shows a pressure rise of more than 20% in either of the tests it is to be classified as "chemically unstable at a temperature greater than 20 °C and/or a pressure greater than 101.3 kPa".

35.4.4 *Safety precautions*

35.4.4.1 Adequate shielding of the test apparatus must be provided to prevent injury in the event of equipment rupture. The apparatus is to be set up in such a way that the operator does not have to be in the same room as long as the vessel contains test gas. Alternatively, the test apparatus is separated by a blast proof barrier from the operator. Activation of the ignition source should be possible only from a position shielded from the test vessel.

35.4.4.2 The test vessel shall be fitted with a bursting disc connected to a vent pipe that allows venting the exhaust gas safely. Therefore it has to be taken into account that the exhaust gas can be hazardous itself (e.g. flammable or toxic).

35.4.4.3 The gas cylinder containing the test gas shall be equipped with a non-return valve and shall be separated from the test apparatus before the igniter is fired to avoid backfiring into the cylinder. The cylinder valve has to be closed immediately after finishing the filling.

35.4.4.4 Some chemically unstable gases can explode very violently especially at higher pressures. Therefore it is strongly recommended to start with the experiments at atmospheric pressure.

35.4.5 *Test criteria and method of assessing results*

35.4.5.1 Chemically unstable gases or gas mixtures shall be classified as "chemically unstable at 20 °C and a standard pressure of 101.3 kPa" or "chemically unstable at a temperature greater than 20 °C and/or a pressure greater than 101.3 kPa" according to the test results as follows:

(a) The gas is classified as "chemically unstable at 20 °C and a standard pressure of 101.3 kPa" if the test at 20 °C and 1.01 bar (101.3 kPa) shows a pressure rise of more than 20% of the initial absolute pressure;

(b) The gas is classified as "chemically unstable at a temperature greater than 20 °C and/or a pressure greater than 101.3 kPa" if the test at 65 °C and the corresponding initial pressure shows a pressure rise of more than 20% of the initial absolute pressure but no such pressure rise at 20 °C and 1.01 bar (101.3 kPa).

35.4.5.2 The gas is not classified according to this test method (i.e. it is chemically stable) if it does not show a pressure rise of more than 20% of the initial absolute pressure in either of the tests.

NOTE: *Chemically unstable gases not submitted to the classification procedure in this section should be classified as chemically unstable, Category A (see Chapter 2.2 of the GHS).*

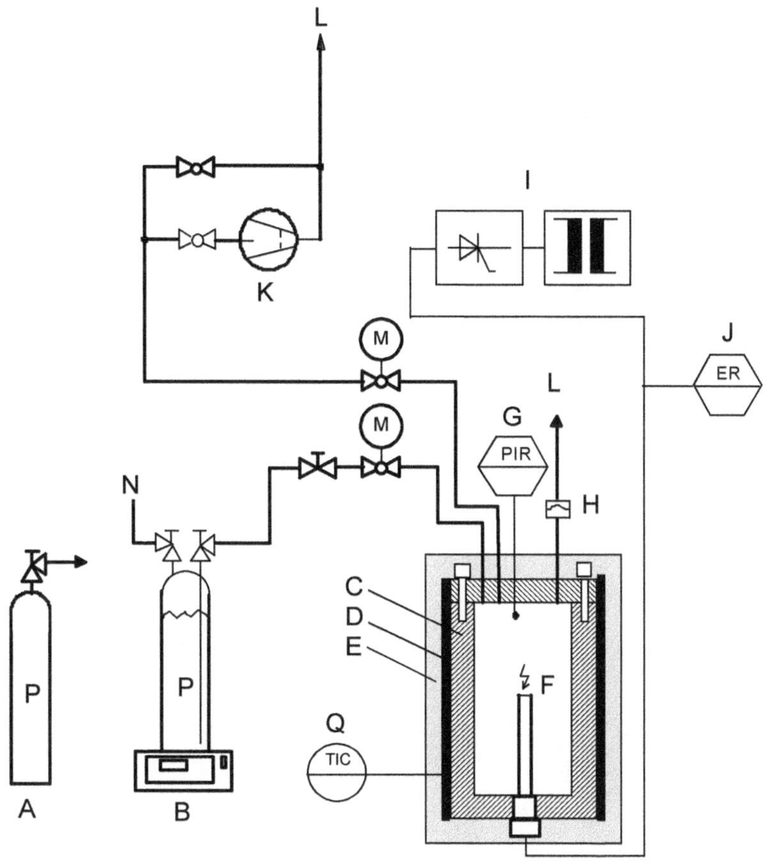

(A)	Test gas supply (gaseous)	(B)	Test gas supply (liquefied)
(C)	Pressure resistant test vessel	(D)	Regulated electric heater
(E)	Thermal insulation	(F)	Exploding wire igniter
(G)	Pressure sensor, pressure indication and registration (PIR)	(H)	Bursting disc
(I)	Electronic ignition device	(J)	Energy registration (ER)
(K)	Vacuum pump	(L)	Exhaust gas
(M)	Motor controlled valve	(N)	Pressurized helium
(P)	Test gas	(Q)	Temperature sensor, temperature indication and control (TIC)

Figure 35.1: TEST APPARATUS

[36. Reserved for classification procedures, test methods and criteria relating to Class 7]

SECTION 37

CLASSIFICATION PROCEDURES, TEST METHODS AND CRITERIA RELATING TO SUBSTANCES OF CLASS 8

37.1 Purpose

37.1.1 This section presents the United Nations scheme for the classification of corrosive substances of Class 8 (see sections 2.8.1 and 2.8.2 of the Model Regulations). The test method for corrosion is given in sub-section 37.4 of this Manual. The method for determining corrosion to skin is provided in OECD Guideline 404 and the criteria are provided in Chapter 2.8 of the Model Regulations. If a substance is shown to be corrosive to skin, then it is not necessary to conduct the tests for metal corrosion for the purposes of classification.

37.2 Scope

37.2.1 New products offered for transport shall be subjected to the classification procedures as set out in paragraph 2.8.2.5 (c) (ii) of the Model Regulations unless it is impracticable (e.g. because of the physical properties) to perform the tests. Substances which cannot be tested shall be classified by analogy with existing entries. The classification procedure shall be undertaken before a new product is offered for transport.

37.3 Classification procedure

The following test procedures are designed to assess the corrosion hazard for an appropriate classification for transport.

37.4 Test methods for corrosion to metals

37.4.1 *Introduction*

37.4.1.1 Test C.1: Test for determining the corrosive properties of liquids and solids that may become liquid during transport as dangerous goods of Class 8, packing group III.

37.4.2 *Apparatus and material*

For exposure to the medium being classified the specimens shall consist of 2 mm thick plates and shall be made of the following materials:

- Aluminium, non-clad types 7075-T6 or AZ5GU-T6 and

- Steel type, S235JR+CR (1.0037 resp. St 37-2), S275J2G3+CR (1.0144 resp. St 44-3), ISO 3574, Unified Numbering System (UNS) G10200 or SAE 1020 (see Figure 37.4.1).

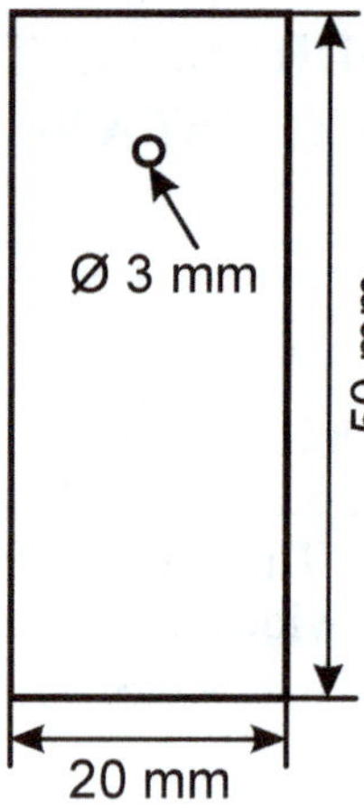

Figure 37.4.2.1: SPECIMEN

At least 3 sets of specimens shall be used for each metal (aluminium, steel). A cup-like reaction receptacle (of glass or PTFE) as shown in Figure 37.4.2.2 with three necks of suitable size (e.g. NS92/32 as well as one neck NS14) to accommodate the specimen as illustrated in Figure 37.4.2.1 and a fourth neck of sufficient size to accommodate a reflux condenser shall be used. The entrance of air into the receptacle shall be ensured. Aluminium and steel specimens shall be tested in different reaction receptacles. To prevent liquid loss a reflux condenser shall be attached (see Figure 37.4.2.2).

Figure 37.4.2.2: EXPOSURE RECEPTACLE WITH REFLUX CONDENSER

In order to carry out the test, the substance being classified shall have a minimum volume of 1.5 l to ensure enough reactive agent during the whole exposure time. Very long testing periods without changing the solution will sometimes give negative results. To get correct results and to avoid re-testing, the following items should be taken into account:

(a) Fresh solutions shall be provided during the course of the test;

(b) The volume should be large enough to avoid any appreciable change in its corrosivity during the test;

***NOTE**: If problems are expected, the composition should be checked by analysis at the end of the test to determine the extent of change in composition, such as might result from evaporation or depletion.*

37.4.3 *Procedure*

Metal sheets shall be polished with grinding paper of 120 grit. After removing the grinding remainings with alcohol in an ultrasound bath and degreasing with acetone, the metal specimens shall be weighed out to ± 0.0002 g. No chemical surface preparation (pickling, etching etc.) shall be performed to prevent surface "irritations" (inhibition, passivation). Specimens shall be fixed inside the receptacle by non-extruded PTFE-threads. Metal wire shall not be used. The test with the so prepared metals shall be initiated the same day to prevent reformation of oxide layer unless appropriate measures are taken to preserve the samples for further testing. For each test one metal specimen shall be dipped into the solution, another one only half way and a third one shall hang in the gas phase. The distance between the upper edge of the completely inserted specimen and the surface of the liquid shall be 10 mm. Losses of liquid shall be avoided.

The test temperature of 55 °C ± 1 shall be maintained throughout the test including the vapour phase as well.

Sheets shall be exposed at these stable conditions for at least one week (168 ± 1 hour).

After finishing the test, the metal specimens shall be rinsed off and cleaned with a brush with synthetic or natural bristles (no metal). For non-mechanically removable remainings (adherent corrosion product or depositions) inhibited pickling solutions should be used. In those cases an unexposed reference specimen needs to be treated in the same manner (time, temperature, concentration, surface preparation) to determine the mass loss caused by the pickling solution. This value needs to be subtracted before evaluating the corrosion rate. After final cleaning with alcohol and acetone in an ultrasound bath, and once dry, the metal samples shall be weighed. The resulting mass under consideration of the specific mass of the metal leads to the corrosion rate.

37.4.4 *Test criteria and method of assessing results*

Two types of corrosion behaviour need to be distinguished.

37.4. 4.1 *Test evaluation at uniform corrosion*

In case of uniform corrosion attack the mass loss of the most corroded sample shall be used. The test is considered positive if for any specimen the mass loss on the metal specimen is more than the amount stated in the following table:

Table 37.4.4.1: Minimum mass loss of specimens after different exposure times

exposure time	mass loss
7 days	13.5%
14 days	26.5%
21 days	39.2%
28 days	51.5%

NOTE: *These values are calculated based on a 6.25 mm/year corrosion rate.*

37.4.4.2 *Test evaluation at localised corrosion*

When localised corrosion occurs besides or instead of uniform corrosion attack of surface, the depth of the deepest hole respectively the strongest thickness reduction will be added or only be used to determine the intrusion. If the deepest intrusion (to be determined metallographically) exceeds the values shown in the following table, the result is considered positive.

Table 37.4.4.2: Minimum intrusion depths after exposure time

exposure time	min. intrusion depth
7 days	120 μm
14 days	240 μm
21 days	360 μm
28 days	480 μm

SECTION 38

CLASSIFICATION PROCEDURES, TEST METHODS AND CRITERIA RELATING TO CLASS 9

38.1 Introduction

This section contains classification procedures, test methods and criteria relating to substances and articles of Class 9.

38.2 Ammonium nitrate fertilizers capable of self-sustaining decomposition

38.2.1 *Purpose*

38.2.1.1 This section of the Manual presents the United Nations scheme for the classification of ammonium nitrate fertilizers of Class 9 (see UN 2071 and special provision 193 of the Model Regulations). The test procedure is designed to assess if an ammonium nitrate fertilizer is capable of undergoing self-sustaining decomposition.

38.2.2 *Scope*

New products offered for transport should be subjected to the classification procedure if the composition is covered by the definition for UN 2071. The classification procedure should be undertaken before a new product is offered for transport.

38.2.3 *Classification procedure*

38.2.3.1 The test method should be performed to determine if a decomposition initiated in a localised area will spread throughout the mass. The recommended test method is given in 38.2.4. Whether the substance is an ammonium nitrate fertilizer of Class 9, or not, is decided on the basis of the test result.

38.2.3.2 Packing group III is assigned to all Class 9 ammonium nitrate fertilisers.

38.2.3.3 Ammonium nitrate fertilizers with the composition given for UN 2071 may be regarded as not subject to the Model Regulations if shown not to be liable to self-sustaining decomposition and provided that they do not contain an excess of nitrate greater than 10% by mass (calculated as potassium nitrate).

38.2.4 *Test S.1: Trough test for determination of the self-sustaining exothermic decomposition of fertilizers containing nitrates*

38.2.4.1 *Introduction*

A fertilizer capable of self-sustaining decomposition is defined as one in which decomposition initiated in a localized area will spread throughout the mass. The tendency of a fertilizer, to be offered for transport, to undergo this type of decomposition can be determined by means of the Trough test. In this test localized decomposition is initiated in a bed of the fertilizer contained in a horizontally mounted trough. The amount of propagation, after removal of the initiating heat source, of decomposition through the mass is measured.

38.2.4.2 *Apparatus and materials*

38.2.4.2.1 The apparatus (Figure 38.2.4.1) consists of a trough of internal dimensions 150 × 150 × 500 mm, open at the top. The trough is constructed of square meshed gauze (preferably stainless steel) with a mesh width of about 1.5 mm and wire thickness of 1 mm, supported on a frame made from, for example, 15 mm wide, 2 mm thick steel bars. The gauze at each end of the trough may be replaced by 1.5 mm thick, 150 × 150 mm stainless steel plates. The trough should be rested on a suitable support. Fertilizers with a particle size distribution such that a significant amount falls through the mesh of the trough should be tested in a trough of smaller mesh gauze, or alternatively in a trough lined with gauze of a smaller mesh. During initiation, sufficient heat should be provided and maintained to establish a uniform decomposition front.

38.2.4.2.2 Two alternative heating methods are recommended. They are:

Electrical heating. An electrical heating element (capacity 250 watts) enclosed in a stainless steel box, placed inside and at one end of the trough (Figure 38.2.4.2). The dimensions of the stainless steel box are 145 × 145 × 10 mm, and the wall thickness about 3 mm. The side of the box which is not in contact with the fertilizer should be protected with a heat shield (insulation plate 5 mm thick). The heating side of the box may be protected with aluminium foil or a stainless steel plate.

Gas burners. A steel plate (thickness 1-3 mm) is placed inside one end of the trough and in contact with the wire gauze (Figure 38.2.4.1). The plate is heated by means of two burners which are fixed to the trough support and are capable of maintaining the plate at temperatures between 400-600 °C, i.e. dull red heat.

38.2.4.2.3 To prevent heat transport along the outside of the trough a heat shield consisting of a steel plate (2 mm thick) should be installed at about 5 cm from the end of the trough where the heating takes place.

38.2.4.2.4 The life of the apparatus may be prolonged if it is constructed of stainless steel throughout. This is particularly important in the case of the gauze trough.

38.2.4.2.5 Propagation may be measured using thermocouples in the substance and recording the time at which a sudden temperature rise occurs as the reaction front reaches the thermocouple.

38.2.4.3 *Procedure*

38.2.4.3.1 The apparatus should be set up under a fume hood to remove toxic decomposition gases or in an open area where the fumes can be readily dispersed. Although there is no explosion risk when performing the test, it is advisable to have a protective shield, e.g. of suitable transparent plastic, between the observer and the apparatus.

38.2.4.3.2 The trough is filled with the fertilizer in the form to be offered for transport and decomposition is initiated at one end, either electrically or by means of gas burners as described above. Heating should be continued until decomposition of the fertilizer is well established and propagation of the front (over approximately 3-5 cm) is observed. In the case of products of high thermal stability, it may be necessary to continue heating for two hours. If fertilizers show a tendency to melt, the heating should be done with care, i.e. using a small flame.

38.2.4.3.3 About 20 minutes after the heating has been discontinued, the position of the decomposition front is noted. The position of the reaction front can be determined by differences in colour, e.g. brown (undecomposed fertilizer) to white (decomposed fertilizer) or by the temperature indicated by adjacent pairs of thermocouples which bracket the reaction front. The rate of propagation may be determined by observation and timing or from thermocouple records. It should be noted whether there is no propagation after heating is discontinued or whether propagation occurs throughout the substance.

38.2.4.4 *Test criteria and method of assessing results*

38.2.4.4.1 If propagation of the decomposition continues throughout the substance, the fertilizer is considered capable of showing self-sustaining decomposition.

38.2.4.4.2 If propagation does not continue throughout the substance, the fertilizer is considered to be free from the hazard of self-sustaining decomposition.

38.2.4.4.3 Examples of results

NOTE: *The NPK proportions for a fertilizer should not be used as a guide to its ability to undergo self-sustaining decomposition as this depends on the chemical species present.*

Substance	Distance propagated (cm)	Result
Compound fertilizer NPK 17-11-22 [a]	50	+
Compound fertilizer NPK 15-11-8 [a]	10	-
Compound fertilizer NPK 14-14-14 [a]	10	-
Compound fertilizer NPK 21-14-14 [a]	10	-
Compound fertilizer NPK 12-12-18 [b]	50	+

[a] *Containing chloride.*
[b] *Containing traces of cobalt and copper but less than 1% chloride.*

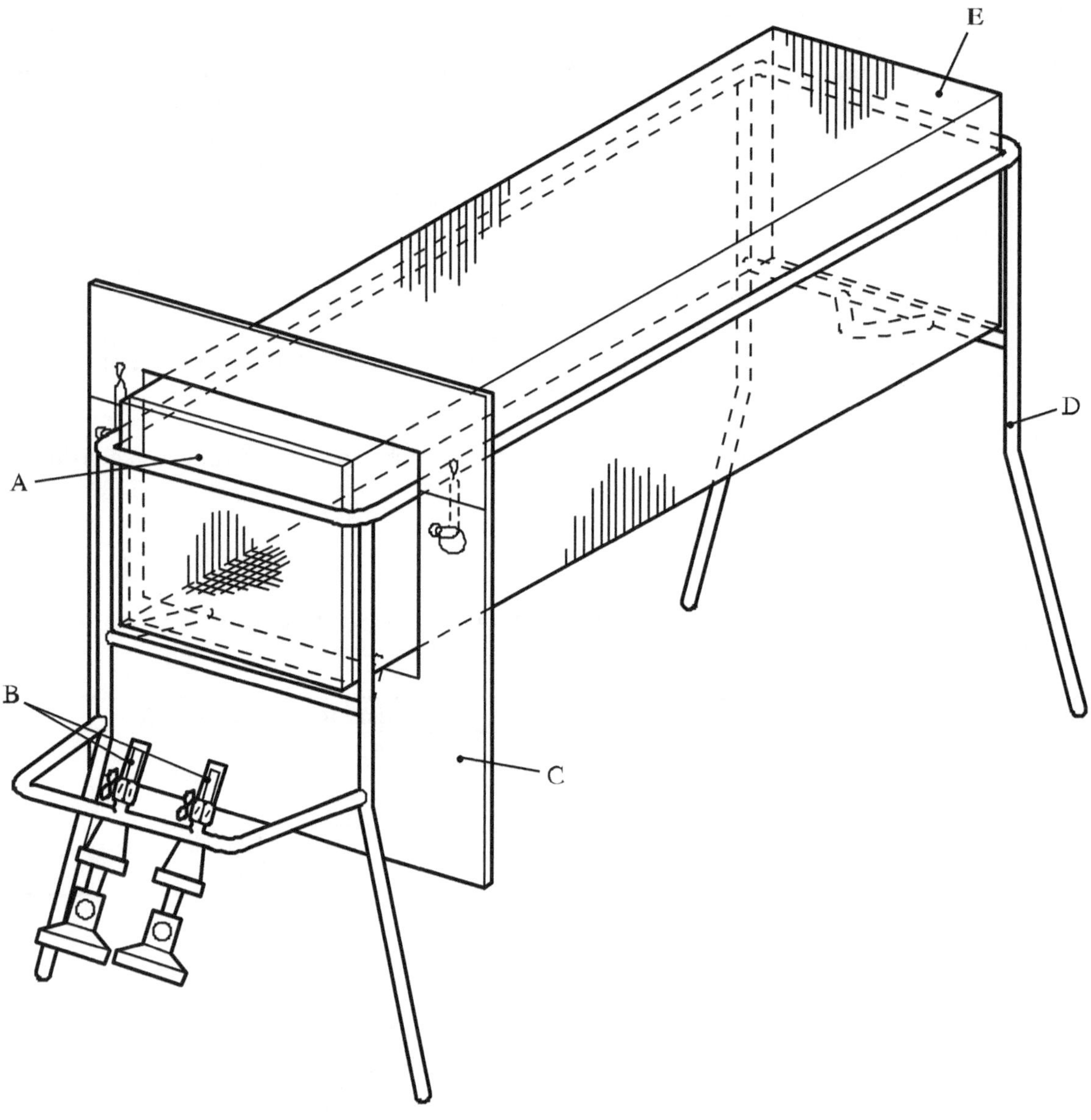

(A) Steel plate (150 × 150 mm and 1 to 3 mm thick)
(B) Gas burners (e.g. Teclu or Bunsen)
(C) Heat shield (2 mm thick)
(D) Stand (e.g. made from 15 mm wide, 2 mm thick steel bar)
(E) Gauze trough (150 × 150 × 500 mm)

Figure 38.2.4.1: GAUZE TROUGH WITH SUPPORT BURNERS

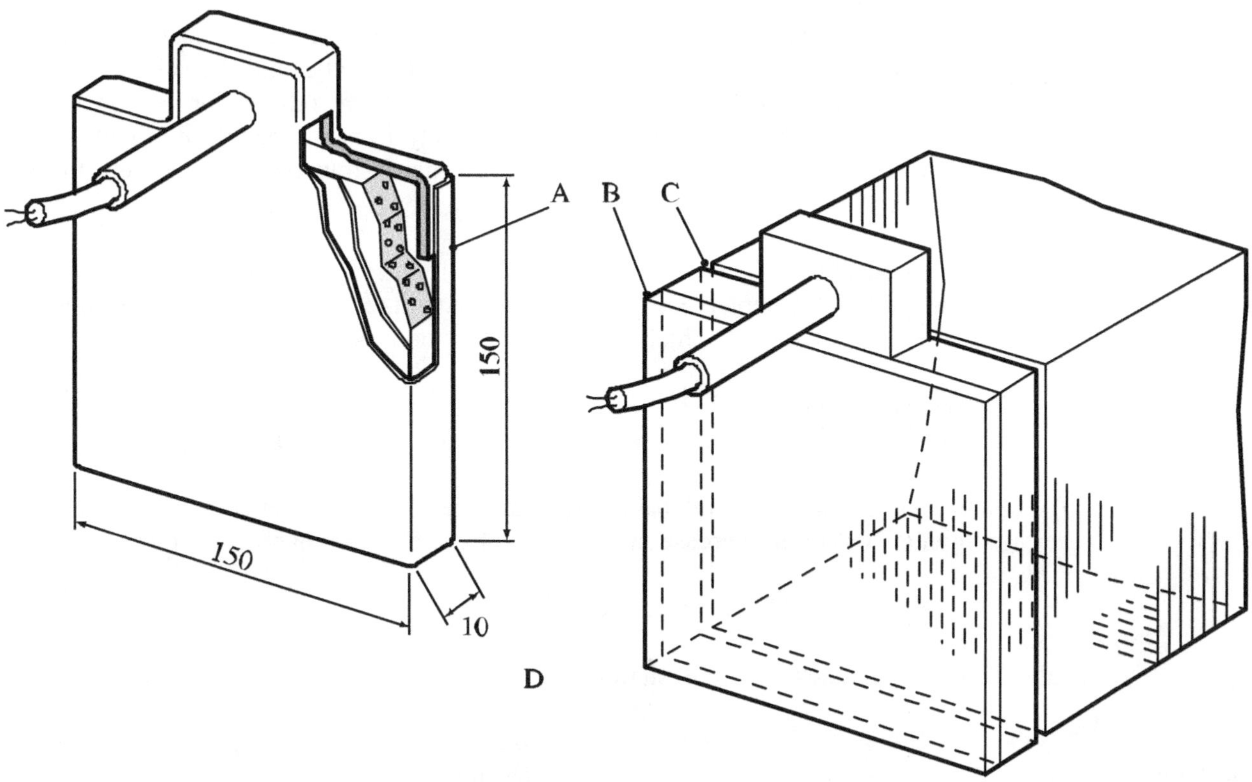

(A) Aluminium or stainless steel plate (wall thickness 3 mm)
(B) Insulating plate (wall thickness 5 mm)
(C) Aluminium or stainless steel plate (wall thickness 3 mm)
(D) Position of heating device in trough

Figure 38.2.4.2: ELECTRICAL HEATING DEVICE (CAPACITY 250 WATTS)

38.3 Lithium metal and lithium ion batteries

38.3.1 *Purpose*

This section presents the procedures to be followed for the classification of lithium metal and lithium ion cells and batteries (see UN Nos. 3090, 3091, 3480 and 3481, and the applicable special provisions of Chapter 3.3 of the Model Regulations).

38.3.2 *Scope*

38.3.2.1 All cell types shall be subjected to tests T.1 to T.6 and T.8. All non-rechargeable battery types, including those composed of previously tested cells, shall be subjected to tests T.1 to T.5. All rechargeable battery types, including those composed of previously tested cells, shall be subjected to tests T.1 to T.5 and T.7. In addition, rechargeable single cell batteries with overcharge protection shall be subjected to test T.7. A component cell that is not transported separately from the battery it is part of needs only to be tested according to tests T.6 and T.8. A component cell that is transported separately from the battery shall be subjected to tests T.1 to T.6 and T.8.

38.3.2.2 Lithium metal and lithium ion cells and batteries shall be subjected to the tests, as required by special provisions 188 and 230 of Chapter 3.3 of the Model Regulations prior to the transport of a particular cell or battery type. Cells or batteries which differ from a tested type by:

(a) For primary cells and batteries, a change of more than 0.1 g or 20% by mass, whichever is greater, to the cathode, to the anode, or to the electrolyte;

(b) For rechargeable cells and batteries, a change in nominal energy in Watt-hours of more than 20% or an increase in nominal voltage of more than 20%; or

(c) A change that would lead to failure of any of the tests,

shall be considered a new type and shall be subjected to the required tests.

***NOTE**:* *The type of change that might be considered to differ from a tested type, such that it might lead to failure of any of the test results, may include, but is not limited to:*

(a) A change in the material of the anode, the cathode, the separator or the electrolyte;

(b) A change of protective devices, including hardware and software;

(c) A change of safety design in cells or batteries, such as a venting valve;

(d) A change in the number of component cells;

(e) A change in connecting mode of component cells; and

(f) For batteries which are to be tested according to T.4 with a peak acceleration less than 150 g_n, a change in the mass which could adversely impact the result of the T.4 test and lead to a failure.

In the event that a cell or battery type does not meet one or more of the test requirements, steps shall be taken to correct the deficiency or deficiencies that caused the failure before such cell or battery type is retested.

38.3.2.3 For the purposes of classification, the following definitions apply:

Aggregate lithium content means the sum of the grams of lithium content contained by the cells comprising a battery.

Battery means two or more cells or batteries which are electrically connected together and fitted with devices necessary for use, for example, case, terminals, marking or protective devices. Units which have two or more cells that are commonly referred to as "battery packs", "modules" or "battery assemblies" having the primary function of providing a source of power to another piece of equipment are for the purposes of the Model Regulations and this Manual treated as batteries. See definitions for "cell" and "single cell battery".

Button cell or battery means a round small cell or battery when the overall height is less than the diameter.

Cell means a single encased electrochemical unit (one positive and one negative electrode) which exhibits a voltage differential across its two terminals, and may contain protective devices. See definitions for battery and single cell battery.

Component cell means a cell contained in a battery. A component cell is not to be considered a single cell battery.

Cycle means one sequence of fully charging and fully discharging a rechargeable cell or battery.

Disassembly means a vent or rupture where solid matter from any part of a cell or battery penetrates a wire mesh screen (annealed aluminium wire with a diameter of 0.25 mm and grid density of 6 to 7 wires per cm) placed 25 cm away from the cell or battery.

Effluent means a liquid or gas released when a cell or battery vents or leaks.

Fire means that flames are emitted from the test cell or battery.

First cycle means the initial cycle following completion of all manufacturing processes.

Fully charged means a rechargeable cell or battery which has been electrically charged to its design rated capacity.

Fully discharged means either:

a primary cell or battery which has been electrically discharged to remove 100% of its rated capacity; or

a rechargeable cell or battery which has been electrically discharged to its endpoint voltage as specified by the manufacturer.

Large battery means a lithium metal battery or lithium ion battery with a gross mass of more than 12 kg.

Large cell means a cell with a gross mass of more than 12 kg.

Leakage means the visible escape of electrolyte or other material from a cell or battery or the loss of material (except battery casing, handling devices or labels) from a cell or battery such that the loss of mass exceeds the values in Table 38.3.1.

Lithium content is applied to lithium metal and lithium alloy cells and batteries, and for a cell means the mass of lithium in the anode of a lithium metal or lithium alloy cell, which for a primary cell is measured when the cell is in an undischarged state and for a rechargeable cell is measured when the cell is fully charged. The lithium content of a battery equals the sum of the grams of lithium content contained in the component cells of the battery.

Lithium ion cell or battery means a rechargeable electrochemical cell or battery in which the positive and negative electrodes are both intercalation compounds (intercalated lithium exists in an ionic or quasi-atomic form with the lattice of the electrode material) constructed with no metallic lithium in either electrode. A lithium polymer cell or battery that uses lithium ion chemistries, as described herein, is regulated as a lithium ion cell or battery.

Mass loss means a loss of mass that exceeds the values in Table 38.3.2.2 below.

Table 38.3.1: Mass loss limit

Mass *M* of cell or battery	Mass loss limit
$M < 1$ g	0.5%
$1\text{ g} \leq M \leq 75$ g	0.2%
$M > 75$ g	0.1%

NOTE: *In order to quantify the mass loss, the following procedure is provided:*

$$\text{Mass loss (\%)} = \frac{(M_1 - M_2)}{M_1} \times 100$$

where M_1 is the mass before the test and M_2 is the mass after the test. When mass loss does not exceed the values in Table 38.3.1, it shall be considered as "no mass loss".

Nominal energy or Watt-hour rating, expressed in watt-hours, means the energy value of a cell or battery determined under specified conditions and declared by the manufacturer. The nominal energy is calculated by multiplying nominal voltage by rated capacity expressed in ampere-hours.

Nominal voltage means the approximate value of the voltage used to designate or identify a cell or battery.

Open circuit voltage means the voltage across the terminals of a cell or battery when no external current is flowing.

Primary cell or battery means a cell or battery which is not designed to be electrically charged or recharged.

Prismatic cell or battery means a cell or battery whose ends are similar, equal and parallel rectilinear figures, and whose sides are parallelograms.

Protective devices means devices such as fuses, diodes and current limiters which interrupt the current flow, block the current flow in one direction or limit the current flow in an electrical circuit.

Rated capacity means the capacity, in ampere-hours or milliampere-hours, of a cell or battery as measured by subjecting it to a load, temperature and voltage cut-off point specified by the manufacturer.

NOTE: *The following IEC standards provide guidance and methodology for determining the rated capacity:*

(1) IEC 61960 (First Edition 2003-12) : Secondary cells and batteries containing alkaline or other non-acid electrolytes – Secondary lithium cells and batteries for portable applications;

(2) IEC 62133 (First Edition 2002-10): Secondary cells and batteries containing alkaline or other non-acid electrolytes – Safety requirements for portable sealed secondary cells, and for batteries made from them, for use in portable applications;

(3) IEC 62660-1 (First Edition 2011-01): Secondary lithium-ion cells for the propulsion of electric road vehicles – Part 1: Performance testing.

Rechargeable cell or battery means a cell or battery which is designed to be electrically recharged.

Rupture means the mechanical failure of a cell container or battery case induced by an internal or external cause, resulting in exposure or spillage but not ejection of solid materials.

Short circuit means a direct connection between positive and negative terminals of a cell or battery that provides a virtual zero resistance path for current flow.

Single cell battery means a cell [externally] fitted with devices necessary for use in equipment or another battery which it is designed to power, for example protective devices. See definitions for cell and battery.

***NOTE**: A single cell battery is considered a "cell" and shall be tested according to the testing requirements for "cells" for the purposes of the Model Regulations and this Manual.*

Small battery means a lithium metal battery or lithium ion battery with a gross mass of not more than 12 kg.

Small cell means a cell with a gross mass of not more than 500 g.

Type means a particular electrochemical system and physical design of cells or batteries.

Undischarged means a primary cell or battery that has not been wholly or partly discharged.

Venting means the release of excessive internal pressure from a cell or battery in a manner intended by design to preclude rupture or disassembly.

Watt-hour rating, see *Nominal energy*.

38.3.3 When a cell or battery type is to be tested under this sub-section, the number and condition of cells and batteries of each type to be tested are as follows:

(a) When testing primary cells and batteries under tests T.1 to T.5 the following shall be tested in the quantity indicated:

(i) ten cells in undischarged states;

(ii) ten cells in fully discharged states;

(iii) four small batteries in undischarged states;

(iv) four small batteries in fully discharged states;

(v) four large batteries in undischarged states; and

(vi) four large batteries in fully discharged states.

(b) When testing rechargeable cells and batteries under tests T.1 to T.5 the following shall be tested in the quantity indicated:

(i) ten cells at first cycle, in fully charged states;

(ii) four small batteries at first cycle, in fully charged states;

(iii) four small batteries after 50 cycles ending in fully charged states;

(iv) two large batteries at first cycle, in fully charged states; and

(v) two large batteries after 25 cycles ending in fully charged states.

(c) When testing primary and rechargeable cells under test T.6, the following shall be tested in the quantity indicated:

(i) for primary cells, five cells in undischarged states and five cells in fully discharged states;

(ii) for component cells of primary batteries, five cells in undischarged states and five cells in fully discharged states;

(iii) for rechargeable cells, five cells at first cycle at 50% of the design rated capacity; and

(iv) for component cells of rechargeable batteries, five cells at first cycle at 50% of the design rated capacity.

(d) When testing rechargeable batteries or rechargeable single cell batteries under test T.7, the following shall be tested in the quantity indicated:

(i) four small batteries at first cycle, in fully charged states;

(ii) four small batteries after 50 cycles ending in fully charged states;

(iii) two large batteries at first cycle, in fully charged states; and

(iv) two large batteries after 25 cycles ending in fully charged states.

Batteries or single cell batteries not equipped with battery overcharge protection that are designed for use only as a component in another battery or in equipment, which affords such protection, are not subject to the requirements of this test.

(e) When testing primary and rechargeable cells and components under test T.8, the following shall be tested in the quantity indicated:

(i) ten primary cells in fully discharged states;

(ii) ten primary component cells in fully discharged states;

(iii) ten rechargeable cells, at first cycle in fully discharged states;

(iv) ten rechargeable component cells, at first cycle in fully discharged states;

(v) ten rechargeable cells after 50 cycles ending in fully discharged states; and

(vi) ten rechargeable component cells after 50 cycles ending in fully discharged states.

(f) When testing a battery assembly in which the aggregate lithium content of all anodes, when fully charged, is not more than 500 g, or in the case of a lithium ion battery, with a Watt-hour rating of not more than 6 200 Wh, that is assembled from batteries that have passed all applicable tests, one assembled battery in a fully charged state shall be tested under tests T.3, T.4 and T.5, and, in addition, test T.7 in the case of a rechargeable battery.

(g) When batteries that have passed all applicable tests are electrically connected to form a battery in which the aggregate lithium content of all anodes, when fully charged, is more than 500 g, or in the case of a lithium ion battery, with a Watt-hour rating of more than 6 200 Wh, the assembled battery does not need to be tested if the assembled battery is of a type that has been verified as preventing:

(i) Overcharge;

(ii) Short circuits; and

(iii) Over discharge between the batteries.

38.3.4 ***Procedure***

Tests T.1 to T.5 shall be conducted in sequence on the same cell or battery. Tests T.6 and T.8 shall be conducted using not otherwise tested cells or batteries. Test T.7 may be conducted using undamaged batteries previously used in Tests T.1 to T.5 for purposes of testing on cycled batteries.

38.3.4.1 *Test T.1: Altitude simulation*

38.3.4.1.1 Purpose

This test simulates air transport under low-pressure conditions.

38.3.4.1.2 Test procedure

Test cells and batteries shall be stored at a pressure of 11.6 kPa or less for at least six hours at ambient temperature (20 ± 5 °C).

38.3.4.1.3 Requirement

Cells and batteries meet this requirement if there is no leakage, no venting, no disassembly, no rupture and no fire and if the open circuit voltage of each test cell or battery after testing is not less than 90% of its voltage immediately prior to this procedure. The requirement relating to voltage is not applicable to test cells and batteries at fully discharged states.

38.3.4.2 *Test T.2: Thermal test*

38.3.4.2.1 Purpose

This test assesses cell and battery seal integrity and internal electrical connections. The test is conducted using rapid and extreme temperature changes.

38.3.4.2.2 Test procedure

Test cells and batteries are to be stored for at least six hours at a test temperature equal to 75 ± 2 °C, followed by storage for at least six hours at a test temperature equal to - 40 ± 2 °C. The maximum time interval between test temperature extremes is 30 minutes. This procedure is to be repeated until 10 total cycles are complete, after which all test cells and batteries are to be stored for 24 hours at ambient temperature (20 ± 5 °C). For large cells and batteries the duration of exposure to the test temperature extremes should be at least 12 hours.

38.3.4.2.3 Requirement

Cells and batteries meet this requirement if there is no leakage, no venting, no disassembly, no rupture and no fire and if the open circuit voltage of each test cell or battery after testing is not less than 90% of its voltage immediately prior to this procedure. The requirement relating to voltage is not applicable to test cells and batteries at fully discharged states.

38.3.4.3 *Test T.3: Vibration*

38.3.4.3.1 Purpose

This test simulates vibration during transport.

38.3.4.3.2 Test procedure

Cells and batteries are firmly secured to the platform of the vibration machine without distorting the cells in such a manner as to faithfully transmit the vibration. The vibration shall be a sinusoidal waveform with a logarithmic sweep between 7 Hz and 200 Hz and back to 7 Hz traversed in 15 minutes. This cycle shall be repeated 12 times for a total of 3 hours for each of three mutually perpendicular mounting positions of the cell. One of the directions of vibration must be perpendicular to the terminal face.

The logarithmic frequency sweep shall differ for cells and batteries with a gross mass of not more than 12 kg (cells and small batteries), and for batteries with a gross mass of more than 12 kg (large batteries).

For cells and small batteries: from 7 Hz a peak acceleration of 1 g_n is maintained until 18 Hz is reached. The amplitude is then maintained at 0.8 mm (1.6 mm total excursion) and the frequency increased until a peak acceleration of 8 g_n occurs (approximately 50 Hz). A peak acceleration of 8 g_n is then maintained until the frequency is increased to 200 Hz.

For large batteries: from 7 Hz to a peak acceleration of 1 g_n is maintained until 18 Hz is reached. The amplitude is then maintained at 0.8 mm (1.6 mm total excursion) and the frequency increased until a peak acceleration of 2 g_n occurs (approximately 25 Hz). A peak acceleration of 2 g_n is then maintained until the frequency is increased to 200 Hz.

38.3.4.3.3 Requirement

Cells and batteries meet this requirement if there is no leakage, no venting, no disassembly, no rupture and no fire during the test and after the test and if the open circuit voltage of each test cell or battery after testing in its perpendicular mounting position is not less than 90% of its voltage immediately prior to this procedure. The requirement relating to voltage is not applicable to test cells and batteries at fully discharged states.

38.3.4.4 *Test T.4: Shock*

38.3.4.4.1 Purpose

This test assesses the robustness of cells and batteries against cumulative shocks.

38.3.4.4.2 Test procedure

Test cells and batteries shall be secured to the testing machine by means of a rigid mount which will support all mounting surfaces of each test battery.

Each cell shall be subjected to a half-sine shock of peak acceleration of 150 g_n and pulse duration of 6 milliseconds. Alternatively, large cells may be subjected to a half-sine shock of peak acceleration of 50 g_n and pulse duration of 11 milliseconds.

Each battery shall be subjected to a half-sine shock of peak acceleration depending on the mass of the battery. The pulse duration shall be 6 milliseconds for small batteries and 11 milliseconds for large batteries. The formulas below are provided to calculate the appropriate minimum peak accelerations.

Battery	Minimum peak acceleration	Pulse duration
Small batteries	150 g_n or result of formula $Acceleration(g_n) = \sqrt{\left(\frac{100850}{mass*}\right)}$ whichever is smaller	6 ms
Large batteries	50 g_n or result of formula $Acceleration(g_n) = \sqrt{\left(\frac{30000}{mass*}\right)}$ whichever is smaller	11 ms

* Mass is expressed in kilograms.

***NOTE**: IEC Standard 60068-2-27 (Fourth Edition 2008-02): Environmental testing-Part 2-27: Tests – Test Ea and guidance: Shock provides guidance on tolerance for acceleration and pulse duration.*

The relationship between minimum peak acceleration and mass is illustrated in Figure 38.3.4.1 for small batteries and Figure 38.3.4.2 for large batteries.

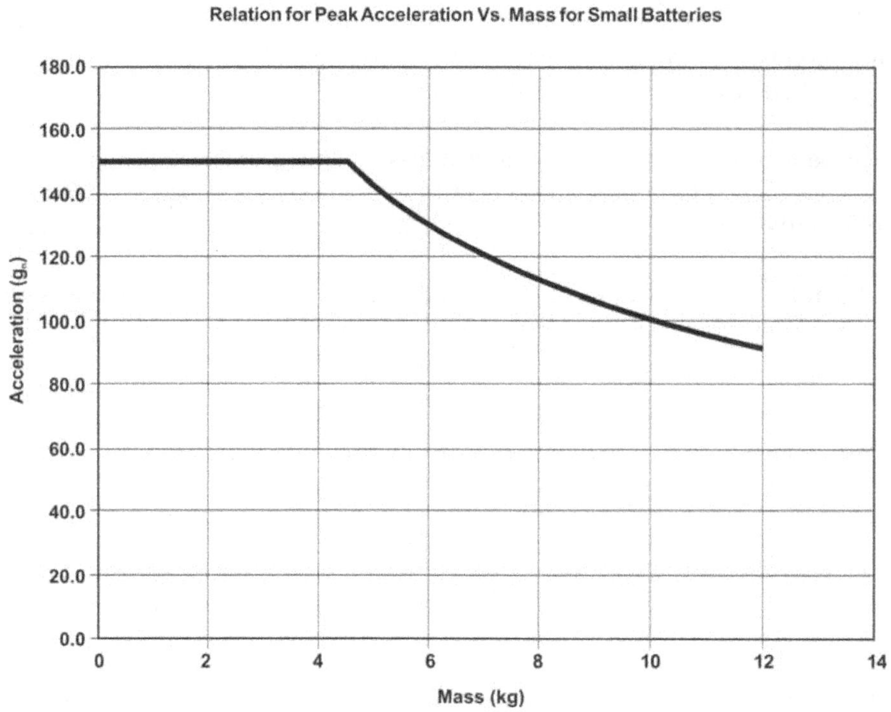

Figure 38.3.4.1: Relation between the Peak Acceleration and the Mass for small batteries (below 12.0 kg).

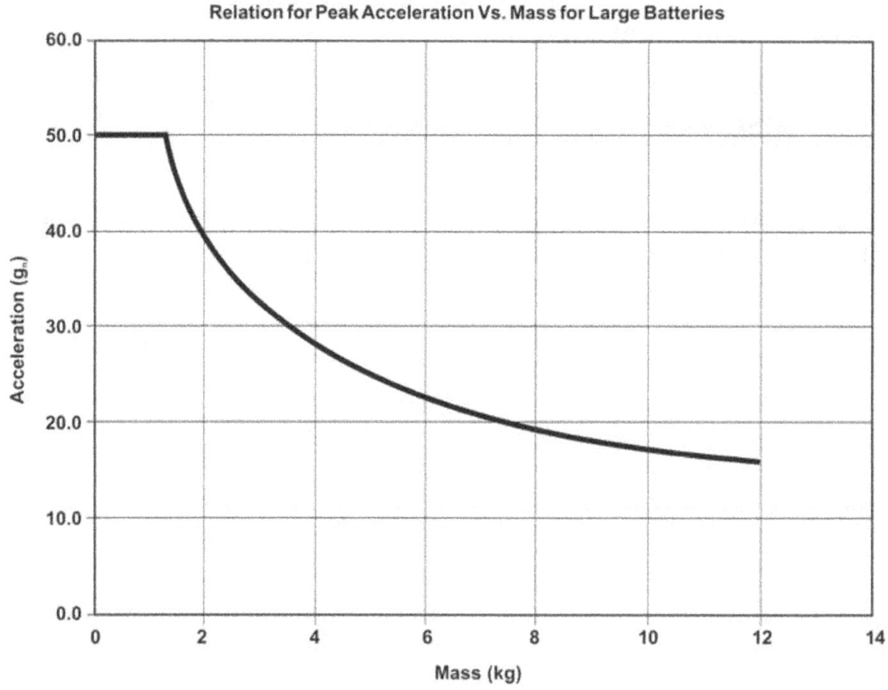

Figure 38.3.4.2: Relation between the Peak Acceleration and the Mass for large batteries (equal or above 12.0 kg).

Each cell or battery shall be subjected to three shocks in the positive direction and to three shocks in the negative direction in each of three mutually perpendicular mounting positions of the cell or battery for a total of 18 shocks.

38.3.4.4.3 Requirement

Cells and batteries meet this requirement if there is no leakage, no venting, no disassembly, no rupture and no fire and if the open circuit voltage of each test cell or battery after testing is not less than 90% of its voltage immediately prior to this procedure. The requirement relating to voltage is not applicable to test cells and batteries at fully discharged states.

38.3.4.5 *Test T.5: External short circuit*

38.3.4.5.1 Purpose

This test simulates an external short circuit.

38.3.4.5.2 Test procedure

The cell or battery to be tested shall be shall be heated for a period of time necessary to reach a homogeneous stabilized temperature of 57 ± 4 °C, measured on the external case. This period of time depends on the size and design of the cell or battery and should be assessed and documented. If this assessment is not feasible, the exposure time shall be at least 6 hours for small cells and small batteries, and 12 hours for large cells and large batteries. Then the cell or battery at 57 ± 4 °C shall be subjected to one short circuit condition with a total external resistance of less than 0.1 ohm.

This short circuit condition is continued for at least one hour after the cell or battery external case temperature has returned to 57 ± 4 °C, or in the case of the large batteries, has decreased by half of the maximum temperature increase observed during the test and remains below that value.

The short circuit and cooling down phases shall be conducted at least at ambient temperature.

38.3.4.5.3 Requirement

Cells and batteries meet this requirement if their external temperature does not exceed 170 °C and there is no disassembly, no rupture and no fire during the test and within six hours after the test.

38.3.4.6 *Test T.6: Impact/Crush*

38.3.4.6.1 Purpose

These tests simulate mechanical abuse from an impact or crush that may result in an internal short circuit.

38.3.4.6.2 Test procedure – Impact (applicable to cylindrical cells not less than 18.0 mm in diameter)

NOTE: *Diameter here refers to the design parameter (for example the diameter of 18 650 cells is 18.0 mm).*

The test sample cell or component cell is to be placed on a flat smooth surface. A 15.8 mm ± 0.1 mm diameter, at least 6 cm long, or the longest dimension of the cell, whichever is greater, Type 316 stainless steel bar is to be placed across the centre of the sample. A 9.1 kg ± 0.1kg mass is to be dropped from a height of 61 ± 2.5 cm at the intersection of the bar and sample in a controlled manner using a near frictionless, vertical sliding track or channel with minimal drag on the falling mass. The vertical track or channel used to guide the falling mass shall be oriented 90 degrees from the horizontal supporting surface.

The test sample is to be impacted with its longitudinal axis parallel to the flat surface and perpendicular to the longitudinal axis of the 15.8 mm ± 0.1 mm diameter curved surface lying across the centre of the test sample. Each sample is to be subjected to only a single impact.

38.3.4.6.3 Test Procedure – Crush (applicable to prismatic, pouch, coin/button cells and cylindrical cells less than 18.0 mm in diameter)

***NOTE**: Diameter here refers to the design parameter (for example the diameter of 18 650 cells is 18.0 mm).*

A cell or component cell is to be crushed between two flat surfaces. The crushing is to be gradual with a speed of approximately 1.5 cm/s at the first point of contact. The crushing is to be continued until the first of the three options below is reached.

(a) The applied force reaches 13 kN ± 0.78 kN;

Example: The force shall be applied by a hydraulic ram with a 32 mm diameter piston until a pressure of 17 MPa is reached on the hydraulic ram.

(b) The voltage of the cell drops by at least 100 mV; or

(c) The cell is deformed by 50% or more of its original thickness.

Once the maximum pressure has been obtained, the voltage drops by 100 mV or more, or the cell is deformed by at least 50% of its original thickness, the pressure shall be released.

A prismatic or pouch cell shall be crushed by applying the force to the widest side. A button/coin cell shall be crushed by applying the force on its flat surfaces. For cylindrical cells, the crush force shall be applied perpendicular to the longitudinal axis.

Each test cell or component cell is to be subjected to one crush only. The test sample shall be observed for a further 6 h. The test shall be conducted using test cells or component cells that have not previously been subjected to other tests.

38.3.4.6.4 Requirement

Cells and component cells meet this requirement if their external temperature does not exceed 170 °C and there is no disassembly and no fire during the test and within six hours after this test.

38.3.4.7 *Test T.7: Overcharge*

38.3.4.7.1 Purpose

This test evaluates the ability of a rechargeable battery or a single cell rechargeable battery to withstand an overcharge condition.

38.3.4.7.2 Test procedure

The charge current shall be twice the manufacturer's recommended maximum continuous charge current. The minimum voltage of the test shall be as follows:

(a) when the manufacturer's recommended charge voltage is not more than 18 V, the minimum voltage of the test shall be the lesser of two times the maximum charge voltage of the battery or 22 V.

(b) when the manufacturer's recommended charge voltage is more than 18 V, the minimum voltage of the test shall be 1.2 times the maximum charge voltage.

Tests are to be conducted at ambient temperature. The duration of the test shall be 24 hours.

38.3.4.7.3 Requirement

Rechargeable batteries meet this requirement if there is no disassembly and no fire during the test and within seven days after the test.

38.3.4.8 *Test T.8: Forced discharge*

38.3.4.8.1 Purpose

This test evaluates the ability of a primary or a rechargeable cell to withstand a forced discharge condition.

38.3.4.8.2 Test procedure

Each cell shall be forced discharged at ambient temperature by connecting it in series with a 12V D.C. power supply at an initial current equal to the maximum discharge current specified by the manufacturer.

The specified discharge current is to be obtained by connecting a resistive load of the appropriate size and rating in series with the test cell. Each cell shall be forced discharged for a time interval (in hours) equal to its rated capacity divided by the initial test current (in ampere).

38.3.4.8.3 Requirement

Primary or rechargeable cells meet this requirement if there is no disassembly and no fire during the test and within seven days after the test.

38.4 Substances evolving flammable vapour

38.4.1 *Purpose*

This section of the Manual presents the test procedure to determine whether during handling, transport and storage substances of Class 9 evolving flammable vapours (see UN No. 2211), are able to evolve a dangerous concentration of flammable vapours in closed containers resulting in the formation of a flammable atmosphere and, as a consequence, have to be classified or not.

38.4.2 *Scope*

The scope of the test method is to determine whether polymeric beads with encapsulated blowing agent, fulfilling the description of UN No. 2211, need not to be classified under these UN numbers.

38.4.3 *Classification procedure for substances liable to evolve flammable vapours*

Polymeric beads with encapsulated blowing agent shall be tested according to the procedures below to determine whether classification under UN No. 2211 is needed.

38.4.4 *Test U 1: Test method for substances liable to evolve flammable vapours*

38.4.4.1 *Introduction*

The ability to evolve flammable vapours is determined by placing the substance in a hermetically closed glass bottle, at a specified temperature for a prescribed period of time, and then, determine the identity and concentration of flammable vapours.

38.4.4.2 *Apparatus and materials*

A serum flask equipped with polytetrafluoroethylene septa with a volume of 50 ml to allow for enough samples to be analysed. A heating cabinet for storage of samples at prescribed time and temperature. A gas chromatographic (GC) apparatus and accompanying equipment, for analysis of flammable vapour concentration in the gas-phase.

38.4.4.3 *Procedure*

The substance as offered for transport should be put in a serum flask of 50 ml, with a degree of filling of 50% volume ratio and sealed with polytetrafluoroethylene septa. The sealed flask is put into a heating cabinet at a minimum of 50 °C for 14 days. Under these conditions analyse the gas twice by gas chromatography and calculate the average concentration of the flammable vapour. The test shall be performed on three samples of the same substance.

38.4.4.4 *Test criteria and method of assessing results*

Substances need not be classified as Polymeric beads, expandable if the concentration of the flammable vapours is less than or equal to 20% of the Lower Explosive Limit (LEL) of the flammable vapour in all of the three samples.

PART IV

TEST METHODS CONCERNING TRANSPORT EQUIPMENT

CONTENTS OF PART IV

SECTION 40

INTRODUCTION TO PART IV

40.1 Purpose

40.1.1 Part IV of the Manual presents the United Nations schemes for dynamic and longitudinal impact testing of portable tanks and MEGCs (see section 41 of this Manual and 6.7.2.19.1, 6.7.3.15.1, 6.7.4.14.1 and 6.7.5.12.1 of the Model Regulations).

40.2 Scope

40.2.1 The test methods of this Part should be applied when required by the Model Regulations.

SECTION 41

DYNAMIC LONGITUDINAL IMPACT TEST FOR PORTABLE TANKS AND MULTIPLE-ELEMENT GAS CONTAINERS (MEGCs)

41.1 General

41.1.1 This test method is intended to prove the ability of portable tanks and MEGCs to withstand the effects of a longitudinal impact, as required by 6.7.2.19.1, 6.7.3.15.1, 6.7.4.14.1 and 6.7.5.12.1 of the Model Regulations.

41.1.2 A representative prototype of each design of portable tank and MEGC meeting the definition of "container" under the International Convention for Safe Containers, 1972, as amended (CSC), shall be subjected to and shall satisfy the requirements of the dynamic longitudinal impact test. Testing shall be conducted by facilities approved for this purpose by the competent authority.

41.2 Permitted design variations

The following variations in container design from an already tested prototype are permitted without additional testing:

41.2.1 *Portable tanks*

(a) A reduction of no more than 10% or an increase of no more than 20% in capacity, resulting from variations in diameter and length;

(b) A decrease in maximum permissible gross mass;

(c) An increase in thickness, independent of design pressure and temperature;

(d) A change to the grade of material of construction provided that the permitted yield strength meets or exceeds that of the tested portable tank;

(e) A change in location of, or a modification to, nozzles and manholes.

41.2.2 *MEGCs*

(a) A decrease in the maximum design temperature, not affecting thickness;

(b) An increase in the minimum design temperature, not affecting thickness;

(c) A decrease in the maximum permissible gross mass;

(d) A decrease in the mass of each individual element and its lading or a decrease in the total mass of the elements and their lading;

(e) An increase of no more than 10% or a decrease of no more than 40% in the diameter of the elements;

(f) A change of no more than 10% in the length of the elements;

(g) A decrease of no more than 3.1 metres (10 feet) in the length of the MEGC framework;

(h) A decrease of no more than 50% in the height of the MEGC;

(i) A change of no more than 50% in the number of elements;

(j) An increase in the thickness of the materials of the framework provided the thickness stays within the range permitted by the welding procedure specifications;

(k) A change to the service equipment and manifold such that the total mass of the service equipment and manifold changes no more than 10% of the maximum permissible gross mass (but not resulting in an increase in the maximum permissible gross mass as compared to that of the already-tested prototype);

(l) The use of a different grade of the same type of material for the construction of the framework, provided that:

(i) The results of the design calculations for the different grade, using the most unfavourable specified values of mechanical properties for that grade, meet or exceed the results of the design calculation for the existing grade; and

(ii) The alternate grade is permitted by the welding procedure specifications.

NOTE: *For permitted MEGC design variations not requiring additional impact testing, the mounting apparatus attaching the elements to the framework must remain the same as that for the already-tested prototype MEGC design.*

41.3 Test apparatus

41.3.1 ***Test platform***

The test platform may be any suitable structure capable of sustaining without significant damage a shock of the prescribed severity with the container-under-test mounted securely in place. The test platform shall be:

(a) configured so as to allow the container-under-test to be mounted as close as possible to the impacting end;

(b) equipped with four devices, in good condition, for securing the container-under-test in accordance with ISO 1161:1984 (Series 1 Freight containers – Corner fittings – Specification); and

(c) equipped with a cushioning device to provide a suitable duration of impact.

41.3.2 ***Impact creation***

41.3.2.1 The impact shall be created by:

(a) the test platform striking a stationary mass; or

(b) the test platform being struck by a moving mass.

41.3.2.2 When the stationary mass consists of two or more railway vehicles connected together, each railway vehicle shall be equipped with cushioning devices. Free play between the vehicles shall be eliminated and the brakes on each of the railway vehicles shall be set.

41.3.3 ***Measuring and recording system***

41.3.3.1 Unless otherwise specified, the measuring and recording system shall comply with ISO 6487:2002 (Road vehicles – Measurement techniques in impact tests – Instrumentation).

41.3.3.2 The following equipment shall be available for the test:

(a) Two accelerometers with a minimum amplitude range of 200 g, a maximum lower frequency limit of 1 Hz and a minimum upper frequency limit of 3 000 Hz. Each accelerometer shall be rigidly attached to the container-under-test at the outer end or side face of the two adjacent bottom corner fittings closest to the impact source. The accelerometers shall be aligned so as to measure the acceleration in the longitudinal axis of the container. The preferred method is to attach each accelerometer to a flat mounting plate by means of bolting and to bond the mounting plates to the corner fittings;

(b) A means of measuring the velocity of the moving test platform or the moving mass at the moment of impact;

(c) An analogue-to-digital data acquisition system capable of recording the shock disturbance as an acceleration versus time history at a minimum sampling frequency of 1 000 Hz. The data acquisition system shall incorporate a low-pass anti-aliasing analogue filter with a corner frequency set to a minimum of 200 Hz and a maximum of 20% of the sampling rate, and a minimum roll off rate of 40 dB/octave; and

(d) A means of storing the acceleration versus time histories in electronic format so that they can be subsequently retrieved and analysed.

41.3.4 ***Procedure***

41.3.4.1 Filling the container-under-test may be undertaken before or after mounting on the test platform, as follows:

(a) Portable tanks: The tank shall be filled with water or any other non-pressurized substance to approximately 97% of the tank volumetric capacity. The tank shall not be pressurized during the test. If for reasons of overload it is not desirable to fill to 97% of capacity, the tank shall be filled so that the mass of the container-under test(tare and product) is as close as practicable to its maximum rated mass (R);

(b) MEGCs: Each element shall be filled with an equal quantity of water or any other non-pressurized substance. The MEGC shall be filled so that its mass is as close as practicable to its maximum rated mass (R) but in any event, to no more than 97% of its volumetric capacity. The MEGC shall not be pressurized during the test. Filling a MEGC is not required when its tare mass is equal to or higher than 90% of R.

41.3.4.2 The mass of the container, as tested, shall be measured and recorded.

41.3.4.3 The container-under-test shall be oriented in a manner that will result in the most severe test. The container shall be mounted on the test platform, as close as possible to the impacting end and secured using all four of its corner fittings so as to restrain its movement in all directions. Any clearance between the corner fittings of the container-under-test and the securing devices at the impacting end of the test platform shall be minimised. In particular, impacting masses shall be free to rebound after impact.

41.3.4.4 An impact shall be created (see 41.3.2) such that for a single impact the as tested Shock Response Spectrum (SRS, see 41.3.8.1) curve at both corner fittings at the impacting end equals or exceeds the minimum SRS curve shown in Figure 41.3.8.1 at all frequencies within the range from 3 Hz to 100 Hz. Repeated impacts may be required to achieve this result but the test results for each impact shall be considered individually;

41.3.4.5 Following an impact described in 41.3.4.4, the container-under-test shall be examined and the results recorded. To satisfy the test, the container shall show no leakage, permanent deformation or damage that would render it unsuitable for use, and shall be in conformity with the dimensional requirements regarding handling, securing and transfer from one means of transport to another.

41.3.5 *Processing and analysis of data*

41.3.5.1 *Data reduction system*

(a) The acceleration versus time history data from each channel shall be reduced to the shock response spectrum, ensuring that the spectra are presented in the form of equivalent static acceleration plotted as a function of frequency. The maximum absolute value acceleration peak shall be recorded for each of the specified frequency break points. The data reduction shall follow the following criteria:

(i) If required, the corrected impact acceleration versus time history data shall be scaled using the procedure outlined in section 41.3.5.2;

(ii) The acceleration versus time history data shall comprise the period commencing 0.05 seconds prior to the start of the impact event and the 2.0 seconds thereafter;

(iii) The analysis shall span the frequency range of 2 to 100 Hz and calculation of the shock response curve points shall be performed at a minimum of 30 frequency break points per octave. Each break point in the range shall constitute a natural frequency; and

(iv) A damping ratio of 5% shall be used in the analysis;

(b) Calculation of the test shock response curve points shall be made as described below. For each frequency break point:

(i) A matrix of relative displacement values shall be calculated using all data points from the shock input acceleration versus time history using the following equation:

$$\xi_i = -\frac{\Delta t}{\omega_d} \sum_{k=0}^{i} \ddot{X}_k e^{-\zeta\varpi_n \Delta t (i-k)} \sin\left[\omega_d \Delta t (i-k)\right]$$

where:

Δt = time interval between acceleration values;
ω_n = undamped natural frequency (in radians);
ω_d = damped natural frequency = $\omega_n\sqrt{1-\zeta^2}$;
$\ddot{x}_k$ = k_{th} value of acceleration input data;
ζ = damping ratio;
i = integer number, varies from 1 to the number of input acceleration data points;
k = parameter used in summation which varies from 0 to the current value of i.

(ii) A matrix of relative accelerations shall be calculated using the displacement values obtained in step i in the following equation:

$$\ddot{\xi}_i = 2\zeta\omega_n \Delta t \sum_{k=0}^{i} \ddot{x}_k e^{-\zeta\omega_n \Delta t (i-k)} \cos\left[\omega_d \Delta t (i-k)\right] + \omega_n^2 \left(2\zeta^2 - 1\right) \xi_i$$

(iii) The maximum absolute acceleration value from the matrix generated in step ii for the frequency break point under consideration shall be retained. This value becomes the SRS curve point for this particular frequency break point. Step i shall be repeated for each natural frequency until all natural frequency break points have been evaluated.

(iv) The test shock response spectrum curve shall be generated.

41.3.5.2 *Method for scaling measured acceleration versus time history values to compensate for under or over mass containers*

Where the sum of the as-tested payload mass plus tare mass of the container-under-test is not the maximum rated mass of the container-under-test, a scaling factor shall be applied to the measured acceleration versus time histories for the container-under-test as follows:

The corrected acceleration-time values, Acc(t) $_{(corrected)}$, shall be calculated from the measured acceleration versus time values using following formula:

$$Acc(t)_{(corrected)} = Acc(t)_{(measured)} \times \frac{1}{\sqrt{1+\dfrac{\Delta M}{M1+M2}}}$$

Where:

Acc(t) $_{(measured)}$ = actual measured-time value;

M1 = mass of the test platform, without the container-under-test;

M2 = actual test mass (including tare) of the container-under-test;

R = the maximum rated mass (including tare) of the container-under-test;

ΔM = R - M2;

The test SRS values shall be generated from the Acc(t) $_{(corrected)}$ values.

41.3.6 ***Defective instrumentation***

If the acquired signal from one accelerometer is faulty the test may be validated by the SRS from the functional accelerometer after three consecutive impacts provided that the SRS from each of the three impacts meets or exceeds the minimum SRS curve.

41.3.7 ***Alternate test severity validation method for portable tanks with frame length of 20 feet***

41.3.7.1 If the design of a tank container-under-test is significantly different from other containers successfully subjected to this test and the SRS curves obtained have correct features but remain below the minimum SRS curve, the test severity may be considered acceptable if three successive impacts are performed as follows:

(a) First impact at a speed higher than 90% of the critical speed referred to in 41.3.7.2; and

(b) Second and third impact at a speed higher than 95% of the critical speed referred to in 41.3.7.2.

41.3.7.2 The alternate validation method described in 41.3.7.1, shall be used only if the platform's "critical speed" had been determined beforehand. The critical speed is the speed where the platform's cushioning devices reach their maximum travel and energy absorption capacity beyond which the minimum SRS curve is normally obtained or exceeded. The critical speed shall have been determined from a minimum of five documented tests on five different tank containers. Each such test shall have been performed using the same equipment, measuring system and procedure.

41.3.8 ***Recording of data***

At least the following data shall be recorded in the application of this procedure:

(a) Date, time, ambient temperature, and location of test;

(b) Container tare mass, maximum rated mass, and as-tested payload mass;

(c) Container manufacturer, type, registration number if applicable, certified design codes and approvals if applicable;

(d) Test platform mass;

(e) Impact velocity;

(f) Direction of impact with respect to container; and

(g) For each impact, an acceleration versus time history for each instrumented corner fitting.

Figure 41.3.8.1: Minimum SRS Curve

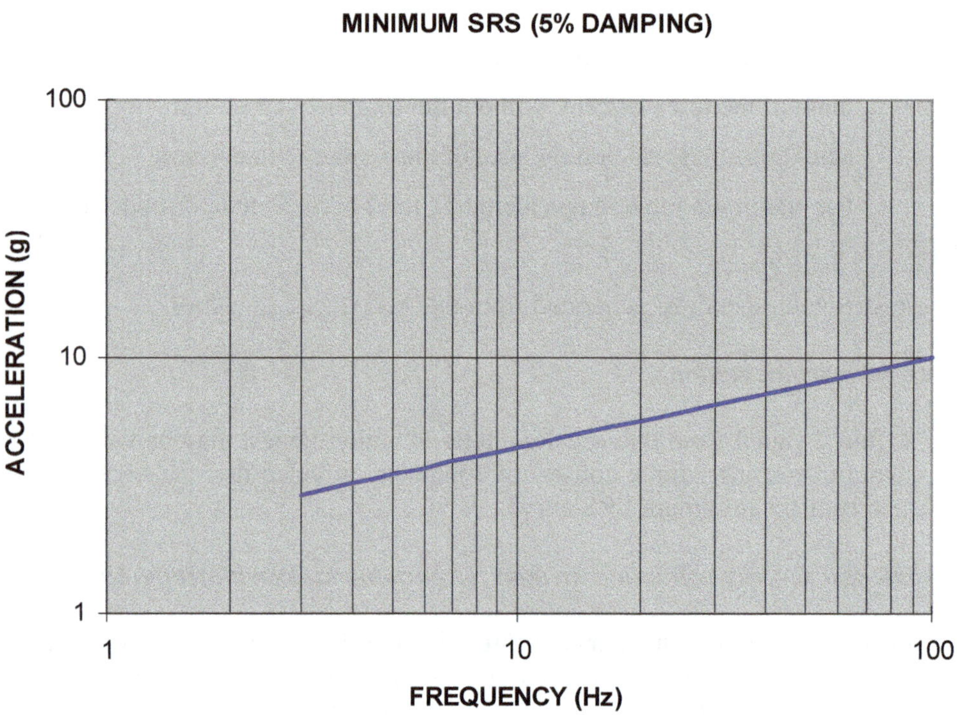

Equation for generating the above Minimum SRS Curve: $ACCEL = 1.95\ FREQ^{0.355}$

Table 41.3.8.1: Tabular representation of some data points for the minimum SRS curve above.

FREQUENCY (Hz)	ACCELERATION (g)
3	2.88
10	4.42
100	10.0

PART V

CLASSIFICATION PROCEDURES, TEST METHODS AND CRITERIA RELATING TO SECTORS OTHER THAN TRANSPORT

CONTENTS OF PART V

SECTION 50

INTRODUCTION TO PART V

50.1 Purpose

Part V of the Manual presents the United Nations schemes for the classification of desensitized explosives for supply and use (including storage) according to the GHS.

50.2 Scope

The test methods of this Part should be applied when required by the GHS.

SECTION 51

CLASSIFICATION PROCEDURES, TEST METHODS AND CRITERIA RELATING TO THE HAZARD CLASS DESENSITIZED EXPLOSIVES

51.1 Purpose

51.1.1 This section presents the United Nations scheme of the classification of liquid and solid desensitized explosives see Chapter 2.17 of the Globally Harmonized System of Classification and Labelling of Chemicals (GHS)). The text should be used in conjunction with the classification principles of Chapter 2.17 of the GHS and the test series given in sub-sections 16.4 and 16.5 of this Manual.

For testing of liquid desensitized explosives for transport purposes, refer to section 32, sub-section 32.3.2 of this Manual and to Chapter 2.3, sub-section 2.3.1.4 of the Model Regulations. Testing of solid desensitized explosives for transport purposes is addressed in section 33, sub-section 33.2.3 of this Manual and in Chapter 2.4, sub-section 2.4.2.4 of the Model Regulations.

51.2 Scope

51.2.1 Desensitized explosives are solid or liquid explosive substances or mixtures which are phlegmatized to suppress their explosive properties in such a manner that they may be excluded from the hazard class "Explosives" (Chapter 2.1 of GHS). Desensitized explosives, should be first tested according to the tests series 1 (type 1(a)), 2 and 6 (type (a) and (b), respectively) of this Manual[1] .

51.2.2 The appropriate classification procedures for desensitized explosives should be undertaken before they are offered for supply and use unless:

(a) They are manufactured with the view to producing a practical, explosive or pyrotechnic effect;

(b) They have a mass explosion hazard according to Test Series 6(a) or 6(b) or their corrected burning rate according to the burning rate test 51.4 is more than 1 200 kg/min;

(c) Their exothermic decomposition energy is less than 300 J/g[2].

51.3 Classification procedure

51.3.1 Before packaged substances or mixtures are subjected to the burning rate test, the test series 6 types 6 (a) and 6 (b) shall be performed in alphabetical order. The substances or mixtures should be tested first with a standard detonator (Appendix 1 of the Manual) and, if no explosion occurs, with an igniter just

---

[1] *Unstable explosives as defined in Chapter 2.1 of GHS can also be stabilized by desensitization and consequently may be classified as desensitized explosive, provided all criteria of Chapter 2.17 of GHS are met. In this case the desensitized explosive should be tested according to test series 3 (Part I of this Manual) because information about its sensitiveness to mechanical stimuli is likely to be important for determining conditions for safe handling and use. The results should be communicated in the safety data sheet.*

[2] *The exothermic decomposition energy should be determined using the explosive already desensitized (i.e.: the homogenous solid or liquids mixture formed by the explosive and the substance(s) used to suppress its explosive properties). The exothermic decomposition energy may be estimated using a suitable calorimetric technique (see Section 20, sub-section 20.3.3.3 in Part II of this Manual).*

sufficient (but not more than 30 g of black powder) to ensure ignition of the substance or mixture in the packaging. The initiation system giving a positive result in the 6 (a) test should be used for the 6 (b) test.

51.3.2 However, it is not always necessary to conduct tests of all types. Test type 6 (b) may be waived if in each type 6 (a) test:

(a) The exterior of the package is undamaged by internal detonation and/or ignition; or

(b) The contents of the package fail to explode, or explode so feebly as would exclude propagation of the explosive effect from one package to another in test type 6(b).

51.3.3 If a substance or mixture gives a negative result (no propagation of detonation) in the Series 1 type 1(a) test, the 6(a) test with a detonator may be waived[3]). If a substance or mixture gives a negative result (no or slow deflagration) in a Series 2 type 2(c) test, the 6 (a) test with an igniter may be waived.

51.3.4 The test for determination of the burning rate by large-scale test need not be performed if, in a test type 6 (b), there is practically instantaneous explosion of virtually the total contents of the stack. In such cases the product is assigned to Division 1.1.

51.4 Burning rate test (external fire)

51.4.1 *Introduction*

51.4.1.1 The test method for determination of the burning rate (10 000 kg scale burning rate) is to be used to determine the behaviour of substances or mixtures as packaged for storage and use if involved in an external fire. This test is performed with several packages of the substances or mixtures to determine:

(a) Whether there is a mass explosion hazard, a hazard from dangerous projection or a too violent burning,

(b) A burning rate (10 000 kg scaled), which depends on the total mass.

51.4.1.2 The burning rate is defined as the extrapolated burning rate for a mass of 10 000 kg packaged material. In practice, this burning rate is determined using both a single package and stacks of packages, following by an extrapolation procedure. The tests are performed with the substances or mixtures in the packages as provided for storage and use. All types of packages are subjected to the tests unless:

(a) A substance or mixture, as packed for supply and use, may be unambiguously assigned to a burning rate and category by a competent authority on the basis of results from other tests or of available information; or

(b) The substance or mixture, as packed for supply and use, is assigned to the hazard class "Explosives", Division 1.1.

51.4.1.3 The corrected burning rate (10 000 kg scaled) is to be used for classification into four different categories.

51.4.2 *Apparatus and materials*

51.4.2.1 The test should be applied to packages of substances or mixtures in the condition and form in which they are offered for supply and use (including storage). The following elements are needed:

(a) A number of 1, 6 and 10 packages, with a net mass of desensitized explosive of 25 kg in each package;

---

3 *If the type 1 (a) test is not carried out the Series 6 type 6(a) test cannot be waived.*

(b) A number of 1, 3 and 6 packages, with a net mass of desensitized explosive between 25 kg and 50 kg in each package;

(c) A number of 1 and up to six packages, with a net mass of desensitized explosive of more than 50 kg, the total net mass should not be greater than 500 kg;

(d) One or two trays with an adequate size and height to contain the wooden pallets and the packages and to protect the ground;

(e) Wooden pallets (e.g. according to DIN 15146), with wood-wool distributed between, under and above the packages;

(f) A suitable ignition source guaranteeing the ignition of the wooden pallets/wood-wool and consequently the tested packages (a mixture of gasoline and light fuel oil 10/90 evenly distributed over the packages and the wood-wool is recommended);

(g) Cine and/or video cameras and suitable equipment to measure the heat of radiation, e.g. infrared sensors and/or thermo cameras.

51.4.2.2 The number of tests and/or the total mass (whereas necessary) should be increased if the test results are ambiguous and the corresponding hazards cannot be clearly defined.

51.4.3 ***Procedure***

51.4.3.1 The tests start with a single package and then the number of packages are successively increased as mentioned under 51.4.2.1 (a), (b) or (c). Normally the burning rate test should be performed once for each number of packages. The required numbers of packages, in the condition and way in which they are offered for supply and use (including storage), are arranged in such way, that the most severe results are anticipated, on wooden and leveled pallets. The pallets are placed in one (or two, if necessary) trays. A tray must comprise at least one complete pallet including 10 cm open space all around the pallet. Flammable material (wood-wool, paper, etc.) is placed under and around the packages in such a way that an optimum ignition is guaranteed (see 51.4.2.1 (f)).

***NOTE**: A quantity of about 10 kg dry wood-wool is usually sufficient. The wooden pallets and the dry wood-wool shall be soaked with a liquid mixture of fuel (about 10 liter, see 51.4.2.1 (f)).*

51.4.3.2 The heat of radiation is measured during the test by suitable equipment, at least at three locations with three different distances from the seat of fire (the distances depend on the sensitivity of the equipment (sensors, thermo camera, etc.) and should be calculated before the test.

51.4.3.3 The signals are continuously recorded. The starting-point of the fire outbreak is defined as the moment when a reaction of the substance is detected. The end of the fire is determined from registered radiation curves.

51.4.3.4 If a mass explosion or individual explosions or metallic projections (fragments) are observed this should be noted in the test report.

51.4.4 ***Test criteria and method of assessing results***

51.4.4.1 The burning rates A and A_{10t} are determined as follows:

(a) The starting point of the fire is defined as the moment at which the substance or mixture reacts detectably. The end of the fire is characterized by a decrease in radiation level I (as caused by the fire) to less than 5% of the maximum level (I_{max}) (see Figure 51.4.1);

(b) The effect of either remainder or burning materials, if present, shall be taken into account in the evaluation;

(c) The burning time t is the time span between the starting point and the end of the fire;

(d) The burning rate A [kg/min] can be calculated for each tested quantity m [kg] and its corresponding burning time t [min] from the equation:

$$A = \frac{m}{t}$$

(e) log A is plotted against log m, where A is the determined burning rate, and m is the mass of substance or mixture used for the test. The observed test results are extrapolated by means of this graph to an uncorrected burning rate A_{10t} for a mass of 10 000 kg corresponding to the following function:

$$A_{10\,t} = \left(\frac{10000\ kg}{m}\right)^{\frac{2}{3}} A$$

51.4.4.2 The corrected burning rate A_C is determined as follows:

(a) The internal amount of energy of the substance is partially converted into radiation. The percent average radiation efficiency η at a distance from the fire is determined from the measured radiation level ($dose_{measured}$) and the theoretical maximum energy ($dose_{calculated}$);

$$\eta = \frac{dose_{\ measured}}{dose_{\ calculated}}$$

(b) The theoretical maximum energy is calculated by multiplying the individual mass of tested substance m [kg] with the heat of combustion H_v [kJ/kg][4]

$$dose_{\ calculated} = H_v \cdot m$$

(c) The amount of energy that in practice appears to be transferred by radiation is determined by integrating the area below the measured radiation curve;

$$dose_{\ measured} = f(t) = \left[\sum_{t\,=\,start}^{end} \frac{(I_{(t+\Delta t)} + I_t}{2} \cdot \Delta_t\right] \cdot 4\,\pi \cdot r^2$$

The numerical integration of the radiation intensities I_t [W/m^2] over the total burning time delivers $dose_{measured}$ [kJ] at the distance r [m].

(d) To this end a graph is made showing the radiation level I [kW/m^2] as a function of time. The complete radiation dose is calculated by integration of the smoothed and corrected curve down to 1% to 5% of I_{max};

[4] *Should be determined by a suitable technique e.g. combustion calorimeter.*

(e) $I_{relevant}$ is obtained from the maximum of the curve of heat radiation calculated as average value of the radiation by converting the integrated area in a rectangle of equal size during the same time span;

(f) The form factor f that must be taken into account during the maximum fire intensity can be averaged from the formula:

$$f = \frac{I_{relevant}}{I_{calculated}}$$

(g) The corrected burning rate A_c is calculated as follows:

$$A_c = A_{10r} \cdot \frac{H_v}{33\,500} \cdot \frac{\eta}{0.25} \cdot \frac{f}{2.78}$$

Where H_v is the heat of combustion of the substance [kJ/kg] (i.e. reaction enthalpy of the burning reaction); η is the radiation efficiency and f the form factor. A_C is the corrected burning rate [kg/min] for a quantity of 10 000 kg.

51.4.4.3 If a mass explosion or individual explosions or metallic projections (fragments) occur the substance or mixture is classified in the hazard class "explosives".

51.4.4.4 The test results are assessed on the basis of the corrected burning rate A_C for a quantity of 10 000 kg of the packaged substance or mixture.

51.4.4.5 The test criteria for determining the burning behavior of substances or mixtures are:

Category 1: Any substance or mixture with a corrected burning rate A_C equal to or greater than 300 kg/min but not more than 1200 kg/min;

Category 2: Any substance or mixture with a corrected burning rate A_C equal to or greater than 140 kg/min but less than 300 kg/min;

Category 3: Any substance or mixture with a corrected burning rate A_C equal to or greater than 60 kg/min but less than 140 kg/min;

Category 4: Any substance or mixture with a corrected burning rate A_C less than 60 kg/min.

Any substance or mixture with a corrected burning rate greater than 1200 kg/min is classified as an explosive (See Chapter 2.1 of the GHS).

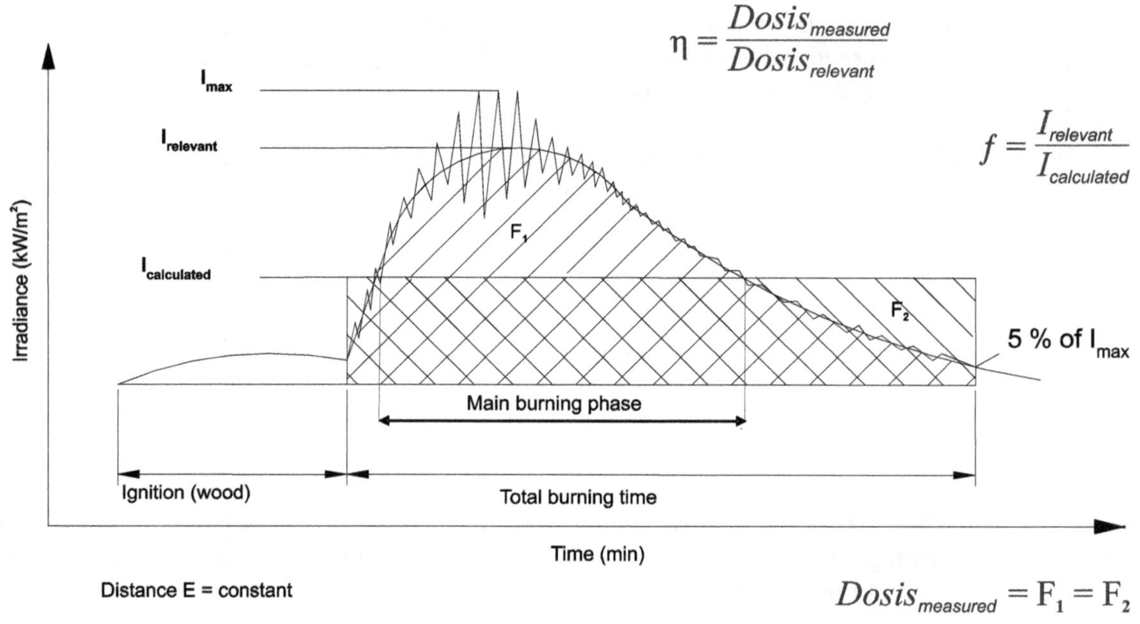

Figure 51.4.1: Measurement of radiation as a function of time

51.4.5 *Examples of results*

51.4.5.1 The nitrocellulose formulations are packed in fiber drums (1G) with a maximum mass of 140 kg and fiber board boxes (4G) with a maximum mass of 25 kg, assigned to categories as follows:

(a) Ester soluble (E-grades) nitrocellulose formulations with different phlegmatizers and a nitrogen content of 11.8% to 12.3%

NC-type	IPA 35%	IPA 30%	ETH 35%	ETH 30%	BUT 35%	BUT 30%	Water	Chips[a)]
12E	3	2	4	3	2	1 (330 kg/min)	4	1 (1115 kg/min)
22E	3	3	4	3	3	3	4	1 (1115 kg/min)
25E	3	3	4	3	3	3	3	1 (1115 kg/min)

IPA (Isopropanol), ETH (Ethanol), BUT (Butanol),
[a)] NC-Chips with 20% plasticizer

(b) Medium soluble (M-grades) nitrocellulose formulations with different phlegmatizers and a nitrogen content of 11.3% to 11.8%

NC-type	IPA 35%	IPA 30%	ETH 35%	ETH 30%	BUT 35%	BUT 30%	Water	Chips[a)]
15M	-	-	-	-	3	2	-	
27M	3	3	4	4	3	3	4	1 (1115 kg/min)
34M	3	3	4	4	4	-	-	1 (1115 kg/min)

IPA (Isopropanol), ETH (Ethanol), BUT (Butanol),
[a)] NC-Chips with 20% plasticizer

(c) Alcohol soluble (A-grades) nitrocellulose formulations with different phlegmatizers and a nitrogen content of 10.7% to 11.3%

NC-type	IPA 35%	IPA 30%	ETH 35%	ETH 30%	BUT 35%	BUT 30%	Water	Chips[a)]
15A	4	3	4	3	3	2	-	1 (1115 kg/min)
30A	4	3	4	4	3	3	4	1 (1115 kg/min)
32 A	4	3	4	4	4	3	-	-

IPA (Isopropanol), ETH (Ethanol), BUT (Butanol),
[a)] NC-Chips with 20% plasticizer

51.4.6 ***Example of a calculation***

NC-formulation (nitrogen content 10.7% to 11.2%) wetted with 30% isopropanol:

Mass of the tested NC formulation:	$m = 285$ kg
Burning time:	$t = 9.7$ min
Form factor:	$f = 3.73$
Radiation efficiency:	$\eta = 0.24$
Enthalpy of combustion:	$H_v = 15626$ kJ/kg

Calculation of the burning rate A:

$$A = \frac{m}{t} = \frac{285\ kg}{9.7\ \text{min}} = 29.4\frac{kg}{\text{min}}$$

Calculation of the burning rate A_{10t}:

$$A_{10_t} = \left(\frac{10\,000\ kg}{m}\right)^{\frac{2}{3}} \cdot A = \left(\frac{10\,000\ kg}{285\ kg}\right)^{\frac{2}{3}} \cdot 29.4\frac{kg}{\text{min}} = 315\frac{kg}{\text{min}}$$

Calculation of the corrected burning rate A_c:

$$A_t = A_{10t} \cdot \frac{H_v}{33\,500} \cdot \frac{\eta}{0.25} \cdot \frac{f}{2.78} = 315\frac{kg}{\text{min}} \cdot \frac{15\,626\frac{kJ}{kg}}{33\,500\frac{kJ}{kg}} \cdot \frac{0.24}{0.25} \cdot \frac{3.73}{2.78} = 189\frac{kg}{\text{min}}$$

The desensitized explosive is classified in category 2.

References

[1] German "Guideline for the assignment of substances which may show explosive properties to Storage Groups (SprengLR011)"

[2] Thermal radiation hazards from organic peroxides, Roberts, T.A. and Merrifield, R., J. Loss. Prev. Process Ind. 1990, 3, 244.

[3] Thermal radiation hazard and separation distances for industrial cellulose nitrate, Roberts, T.A. and Merrifield, R., J. Loss. Prev. Process Ind. 1992, 5,311.

[4] Storage of Organic Peroxides, Publication Series on Dangerous Substances 8 (PGS 8), Ministries of Social Affairs and of the Interior, The State Secretary of Housing, Spatial Planning and Environment (VROM), The Netherlands 2006.

[5] The storage and handling of organic peroxides, Guidance Note CS21, Health and Safety Executive, 1998, United Kingdom

APPENDICES

CONTENTS OF APPENDICES

APPENDIX 1

SPECIFICATIONS OF STANDARD DETONATORS

1. Description of the standard electric blasting cap 0.6 g PETN

Drawing No.	Part No.	Part	Description	Remarks
I	A	Blasting cap		
	B	Fusehead	It must not undergo compression. Recommended amount of the pyrotechnic substance forming the bead: 20 mg to 50 mg	Example: Electrical fusehead Fa. DNAG, Germany, T 10 - U - with aluminium coating
II	A	Tube	Hollow-drawn tube of pure copper, (5% zinc) or of other alloys whose compositions are in the range between that of the above alloy and pure copper. The dimensions of the tube are shown in the figure. If required, the tubes for making standard detonators should be selected by checking the exact dimensions of each tube to be used.	
	B	(a) Secondary charge	Base charge: 0.40 g (± 0.01 g) PETN; compressed at 440 bar;	The PETN may contain up to 0.5% of carbonaceous matter to prevent electrostatic charges during handling and to improve the flow properties.
	C		Intermediate charge 0.20 g (± 0.01 g) PETN; compressed at 20 bar.	Total height of secondary charge 12.3 mm (± 0.6 mm)
	D	Priming charge (initiation charge)	Free choice of the substance and of its quantity. However, at least twice the minimum quantity required for initiation should be used. The total oxygen balance of the priming charge plus the secondary charge should not be more negative than -9.5% O_2	Example: 0.30 ± 0.01 g dextrinated lead azide with a purity of 88% compressed at 440 bar
	E	Inner cup (pierced)	-	It is not required to use a pierced inner cup. Pressing the priming charge onto a highly compressed part of the secondary charge is excluded.
III	A	Fusehead	-	Example: Electrical fusehead Fa. DNAG, Germany T 10 - U - with aluminium coating
	B	Closing plug	-	No special requirements. It should however provide an absolutely tight seal (to avoid the formation of cuprous azide and to ensure the desired initiating strength). The usual commercial design is satisfactory.
	C	Wire	-	Free choice, providing the electrical risks (static electricity, stray currents) are taken into account. However, the use of an insulating sheath of plastic material inside the detonator tube is not allowed.
	D	Short-circuiting tube	Plastic tube	
	E	Label		

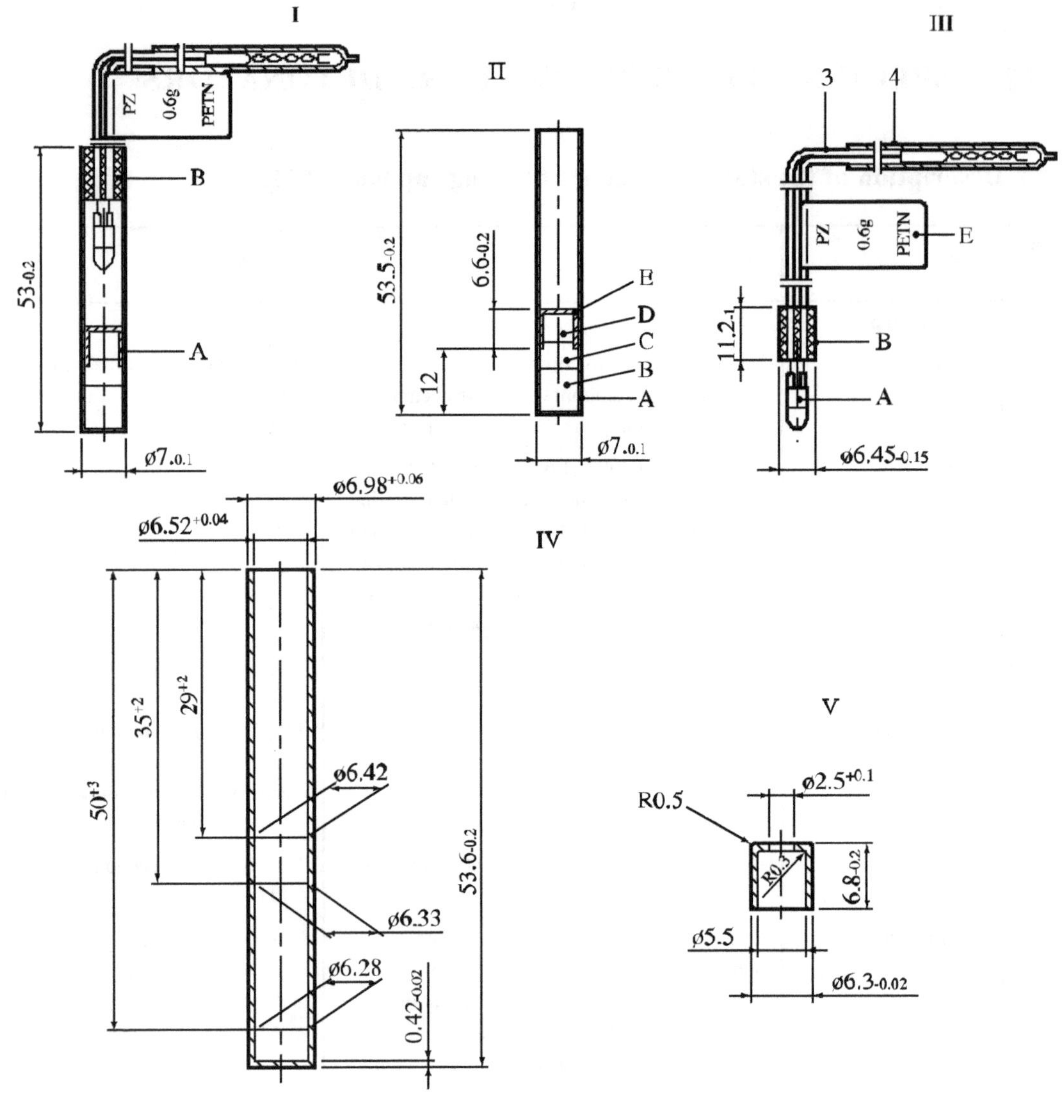

I	Electric blasting cap (standard detonator)	II	Blasting cap (standard detonator)
III	Fusehead	IV	Tube
V	Inner cup		
(A)	Blasting cap	(B)	Fusehead
(C)	Intermediate charge	(D)	Priming charge
(E)	Inner cup		

Figure A1.1: STANDARD DETONATOR (EUROPEAN)

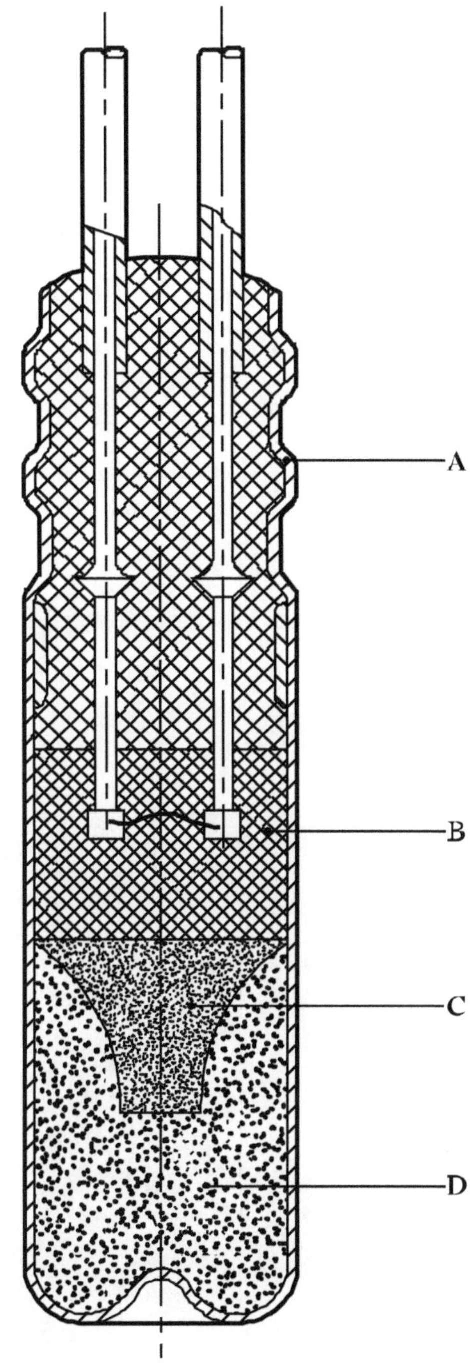

(A) Aluminium shell (material - 5052 aluminium alloy; length 31.8 mm; outer diameter 7.06 mm; wall thickness 0.19 mm)
(B) Brigewire and ignition charge
(C) Primer charge (0.195 g dextrinated lead azide)
(D) Base charge (0.447 g PETN pressed at 28 MPa)

Figure A1.2: No. 8 (USA) DETONATOR

APPENDIX 2

BRUCETON AND SAMPLE COMPARISON METHODS

1. Bruceton method

Introduction: The Bruceton method is used to determine the level of stimulus at which there is a 50% probability of obtaining a positive result.

Procedure: The method involves applying different levels of stimulus and determining whether or not a positive reaction occurs. Performance of the trials is concentrated around the critical region by decreasing the stimulus by one level in the next trial if a positive result is obtained and increasing the stimulus by one level if a negative result is obtained. Usually about 5 preliminary trials are performed to find a starting level in approximately the right region and then at least 25 trials are performed to provide the data for the calculations.

Calculation of results: In determining the level at which the probability of obtaining a positive result is 50% (H_{50}), only the positives (+) or only the negatives (-) are used, depending on which has the smaller total. If the numbers are equal, either may be used. The data are recorded in a table (e.g. as in Table A2.1) and summarised as shown in Table A2.2. Column 1 of Table A2.2 contains the drop heights, in ascending order, starting with the lowest level for which a test result is recorded. In column 2, 'i' is a number corresponding to the number of equal increments above the base or zero line. Column 3 contains the number of positives (n(+)) or negatives (n(-)) for each drop height. The fourth column tabulates the result of multiplying 'i' times 'n' and the fifth column tabulates the results of multiplying the square of 'i' times 'n'. A mean is calculated from the following equation:

$$H_{50} = c + d \times \left(\frac{A}{N_S} \pm 0.5 \right)$$

where $N_S = \sum n_i$, $A = \sum (i \times n_i)$, c = lowest drop height and d = height interval.

If negative results are used, the sign inside the brackets is positive; it is negative if positive results are used. The standard deviation, s, may be estimated using:

$$s = 1.62 \times d \times \left(\frac{N_S \times B - A^2}{{N_S}^2} + 0.029 \right)$$

where $B = \sum (i^2 \times n_i)$.

Example of results: Using the data from Table A2.2, lowest drop height 10 cm, height interval 5 cm, sum of i.n(-) 16, sum of i².n(-) 30 and sum of n(-) 12; the mean height is given as:

$$H_{50} = 10 + 5 \times \left(\frac{16}{12} + 0.5 \right) = 19.2 \text{ cm}$$

and the standard deviation as:

$$s = 1.62 \times 5 \times \left(\frac{12 \times 30 - 16^2}{12^2} + 0.029 \right) = 6.1$$

Reference: *W.J. Dixon and F.V. Massey, Jr. "Introduction to Statistical Analysis, McGraw-Hill Book Co., Toronto, 1969.*

Table A2.1: RECORDING DATA

Drop height (cm)	DROP RESULT																									FREQUENCY	
	1	2	3	4	5	6	7	8	9	10	11	12	13	14	15	16	17	18	19	20	21	22	23	24	25	+	-
30								+																		1	
25							-		+				+				+		+							4	1
20				+		-				+		-		+		-		-		+		+				5	4
15	+		-		-						-				-						-		+		+	3	5
10		-																						-			2
																										13	12

Table A2.2: SUMMARISING DATA

	CALCULATIONS USING NEGATIVES			
Height (cm)	i(-)	n(-)	i(-).n(-)	i^2(-).n(-)
25	3	1	3	9
20	2	4	8	16
15	1	5	5	5
10	0	2	0	0
TOTALS		$N_s = 12$	A = 16	B = 30

2. Sample comparison method

Introduction: This technique can be applied to any test where the Bruceton method is used. The Sample Comparison Test (SCT) is a non-parametric procedure designed to afford a high degree of confidence in any difference in sensitiveness in situations where the mean values given by the Bruceton method are close to one another.

Procedure: Samples of explosive A are tested following a normal Bruceton method, but are tested alternately with those of sample B. However, instead of following their own up-and-down programme, each sample of explosive B is subjected to the same level of stimulus as in the immediately preceding trial with sample A. Thus, at each level of stimulus as the test proceeds, one trial is performed with sample A and one with sample B. If both react or both do not react then the result is ignored for the appraisal. Only pairs of results which have afforded different responses are used for the appraisal.

Calculation of results: If there are **n** pairs of results which have afforded different responses and **x** is the number of positive reactions of the least sensitive sample from these pairs, i.e. $x < (n - x)$, then the confidence, **K**%, that this sample is really less sensitive is calculated using Bernoullian statistics. **K** can be estimated by:

$$K = 100 \times \left(1 - 2^{-n} \times \left(\sum_{i=0}^{x} \frac{n!}{i! \times (n-i)!}\right)\right)$$

Various illustrative values of K are given in the table below for a series of values of x and n.

x \ n	**15**	**20**	**25**	**30**
2	99			
3	98	99		
4	94	99		
5	85	98	99	
6	70	94	99	
7		87	98	99
8		75	95	99
9		59	89	98
10			79	95

Where there is no real difference in two samples, the proportion of occasions where the pairs of results are the same increases and, at the same time, (n - 2x) does not show a general tendency to increase as the testing proceeds.

Examples of results: HMX admixed with 0.01% of 45-63 μm airborne grit, compared with unadulterated HMX, gave x = 3 at n = 13 indicating that the former was more sensitive at the confidence level of:

$$K = 100 \times \left(1 - 2^{-13} \times \left(\sum_{i=0}^{3} \frac{13!}{i! \times (13 - i)!}\right)\right)$$

$$= 100 \times \left(1 - \frac{1 + 13 + 78 + 286}{8192}\right) = 95.4\%$$

Comparing a suspect sample of milled HMX with a normal sample gave x = 6 at n = 11 indicating that the former was more sensitive at the confidence level of:

$$K = 100 \times \left(1 - 2^{-11} \times \left(\sum_{i=0}^{6} \frac{11!}{i! \times (11 - i)!}\right)\right)$$

$$= 100 \times \left(1 - \frac{1 + 11 + 55 + 165 + 330 + 462 + 462}{2048}\right) = 27.4\%$$

showing no evidence that the suspect sample was other than normal.

NOTE: *The simplest way to estimate K is to use* $K = 100 \times \{0.5 + G(z)\}$ *where G(z) is the Gaussian area between the centre ordinate and the ordinate at abscissa z where* $z = n^{0.5} - (2x+1)/n^{0.5}$. *For example, where n = 13 and x = 3, z = 1.6641, G(z) = 0.452 and K = 95.2%.*

Reference: *H J Scullion, Journal of Applied Chemistry and Biotechnology, 1975, <u>25</u>, pp. 503 - 508.*

APPENDIX 3

CAVITATION OF SAMPLES

1. German method

When a liquid is to be tested in the cavitated state, the cavitation may be achieved by passing a constant stream of gas bubbles through it. The test method is modified (see Figure A3.1) as follows:

The bottom of the tube (extended by 100 mm) is closed with a screw cap and PTFE gasket instead of the normal welded plate. A short steel tube of approximately 5 mm inner diameter is welded into a central hole made in this cap. A porous glass filter is attached to the inner end of the tube by means of a flexible plastics tube so that it is positioned centrally and as close as possible to the bottom of the cap. The porous disc should have a diameter of at least 35 mm and have a pore size in the range 10 to 16 μm (porosity 4). The air, oxygen or nitrogen flow rate should be 28 ± 5 litres/hour. In order to prevent build-up of pressure the top cap should have four additional 10 mm diameter holes drilled through it.

2. USA method

The apparatus for the detonation tests with cavitated liquids is the same as that for solids and uncavitated liquids except that a method of injecting bubbles into the liquid sample is provided. An example of the experimental set-up is given in Figure A3.2. The bubbles are injected by means of a 23.5 mm diameter loop of vinyl plastic tubing of the type used for medical catherisation with an outer diameter of 1.8 mm and a wall thickness of 0.4 mm located at the bottom of the sample. This loop is perforated with two rows of holes diametrically opposite to each other with the holes in each row spaced 3.2 mm apart. The holes are made by inserting a 1.3 mm diameter needle through the wall of the tubing. Because of the elastic nature of the tubing the holes contract almost completely when the needle is withdrawn, so the actual hole diameter is much smaller than 1 mm. The tubing is sealed at one end of the loop with epoxy cement and a length of the tubing from the other hand of the loop is led outside to the air supply through a hole in the steel tubing, which is sealed with epoxy cement. Air is supplied at a pressure of 30 to 100 kPa to obtain a flow rate of 1.2 litres/minute.

3. French method

This method uses glass micro-balloons (hollow closed spheres) which are commonly used to sensitize emulsion explosives e.g. soda lime borosilicate glass bubbles, apparent density 0.15, mean diameter 50 μm, maximum diameter 200 μm and with 25% having a diameter less than 30 μm. It is applicable to liquids and pastes. Glass micro-balloons are added, if necessary with the aid of a small quantity of dispersant which is compatible with the test substance, in the ratio of 500 mg to one litre of test substance. The mixture is agitated until formed into a homogeneous, stable dispersion and is then loaded into the firing tube.

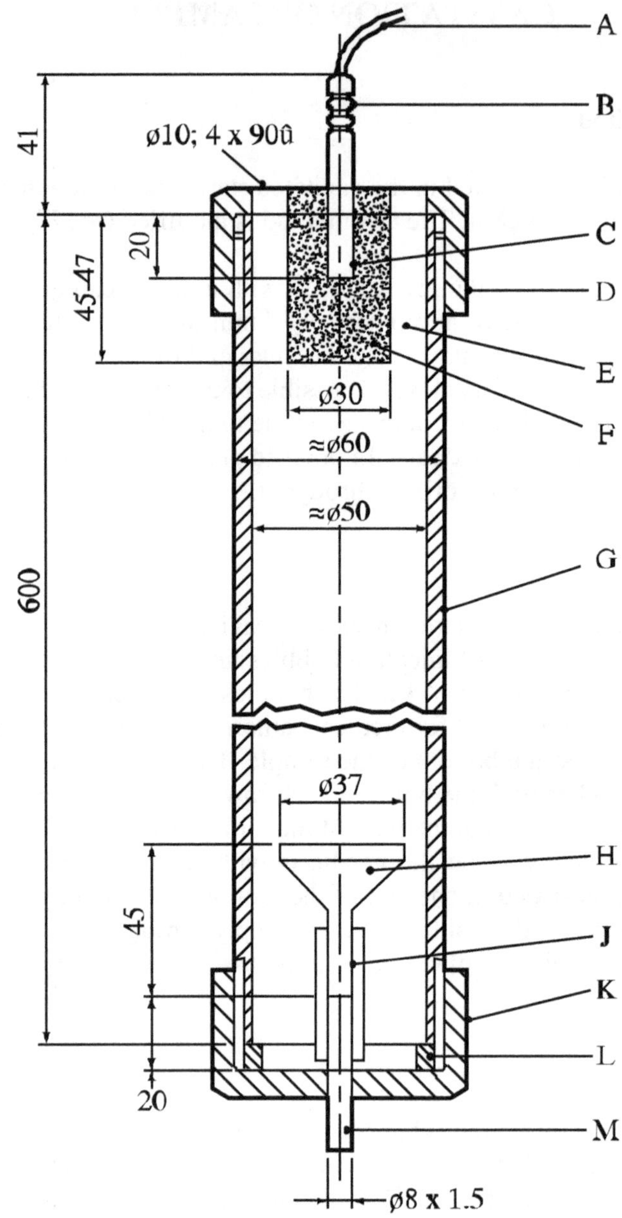

(A)	Lead wires	(B)	Electric igniter
(C)	Detonator	(D)	Screw cap of malleable cast iron
(E)	Substance under test	(F)	Booster charge of RDX/wax (95/5)
(G)	Steel tube to DIN 2441 specification material St 37 to DIN 1629 specification sheet 3	(H)	Porous glass filter
(J)	Flexible plastics tube	(K)	Screw cap of steel St 35
(L)	PTFE gasket	(M)	Small steel tube

Figure A3.1: GERMAN METHOD OF CAVITATION

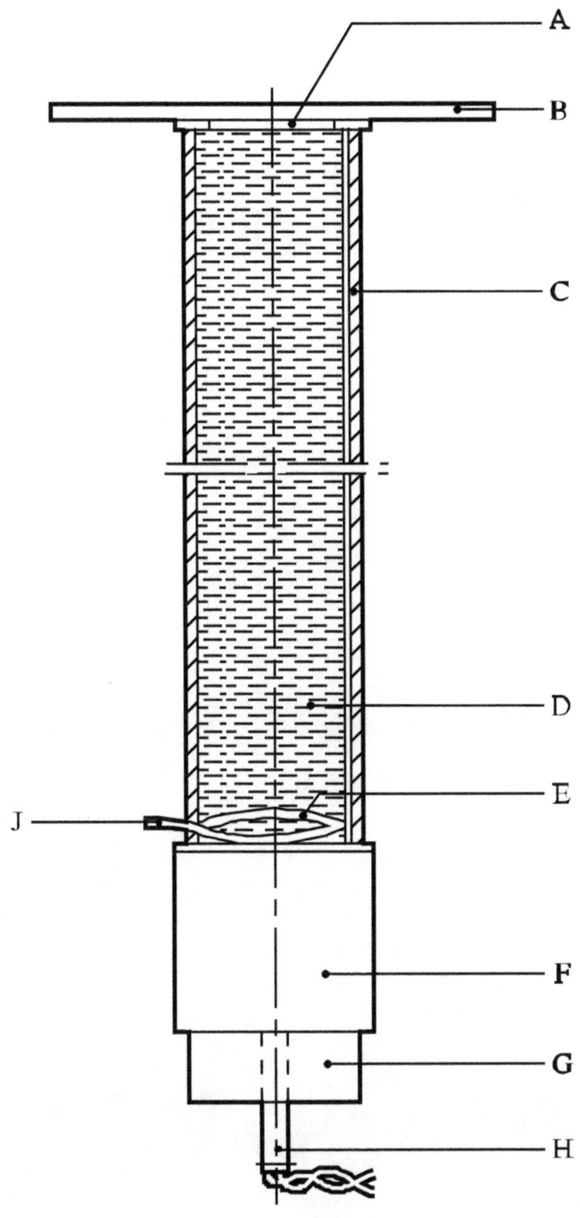

(A)	Spacers	(B)	Witness plate
(C)	Steel tube	(D)	Substance under test
(E)	Bubbler	(F)	Pentolite pellet
(G)	Detonator holder	(H)	Detonator
(J)	Air supply		

Figure A3.2: USA METHOD OF CAVITATION

APPENDIX 4

NATIONAL CONTACTS FOR TEST DETAILS

Country	Code	Address
CANADA	C	Canadian Explosives Research Laboratory Department of Natural Resources CANMET Complex, Bells Corners Ontario, Canada K1A 0G1
FRANCE	F	INERIS/LSE Parc Technologique ALATA B.P. 2 60550 Verneuil-en-Halatte France
GERMANY	D	Abteilung II Bundesanstalt für Materialforschung und -prüfung Unter den Eichen 87 D - Berlin 12205 Germany
NETHERLANDS	NL	TNO Prins Maurits Laboratory P.O. Box 45 2280 AA Rijswijk The Netherlands
JAPAN	J	Technology and Safety Division Transport Policy Bureau Ministry of Transport 2-1-3 Kasumigaseki Chiyoda-ku Tokyo 100, Japan
POLAND	PL	Institute of Industrial Organic Chemistry Laboratory of Dangerous Properties of Materials 6, Annopol Street 03 - 236 Warsaw Poland
RUSSIAN FEDERATION	RUS	The State Committee of the Russian Federation on Defensive Branches of Industry Central Scientific and Design Bureau 20 Goncharnaya Street Moscow, 109240 Russian Federation
SPAIN	E	Laboratorio Oficial Madariaga (LOM) Alenza, 1 Madrid 28002 Spain
SWEDEN	S	Saab Bofors Dynamics AB Research and Development Explosives S-691 80 Karlskoga Sweden

NATIONAL CONTACTS FOR TEST DETAILS (continued)

Country	Code	Address
SWITZERLAND	CH	Eidg. Gefahrgutinspektorat Richtistrasse 15 CH-8304 Wallisellen Switzerland
UNITED KINGDOM	GB	HSE, Health and Safety Laboratory Harpur Hill, Buxton Derbyshire SK17 9JN United Kingdom
UNITED STATES OF AMERICA	USA	Associate Director for Hazardous Materials Safety RSPA/DOT Washington D.C. USA 20590

APPENDIX 5

EXAMPLE OF A TEST METHOD FOR VENT SIZING

1. Introduction

This example of a method for vent sizing is used to determine the required emergency vent capacity to be fitted to a specific IBC or tank for a particular organic peroxide Type F, or self-reactive substance Type F, or formulations thereof. The method is based on experimental data which indicates that, for organic peroxide or self-reactive substance formulations, the ratio of the minimum emergency vent area to the capacity of the IBC or tank is constant and can be determined using a reduced scale tank with a 10 litre capacity. In the tests, the reduced scale tank is heated at rates equivalent to that given by complete fire engulfment or, in the case of insulated IBC or tanks, the heat transfer through the insulation with the assumption that 1% of the insulation is missing (see 4.2.1.13.8 and 4.2.1.13.9 of the Model Regulations). Others methods may be used provided that they adequately size the emergency relief device(s) on an IBC or a tank to vent all the material evolved during self-accelerating decomposition or a period of not less than one hour of complete fire-engulfment.

Warning: The method does not take into account the possibility of initiation of deflagration. If this is a possibility, particularly if initiation in the vapour phase can propagate to the liquid phase, then tests should be performed which take this into account.

2. Apparatus and materials

The reduced scale tank consists of a stainless steel test vessel with a gross volume of 10 *l*. The top of the tank is provided with either a 1 mm opening which simulates the pressure relief valve (PRV) of the IBC or tank or a real PRV of a diameter which is scaled using the vent area to vessel volume ratio. A second opening simulates the emergency vent opening and is closed by a bursting disc. The diameter of this vent opening can be varied by using orifice plates with different apertures. The bursting pressure of the disc fixed to the 10 *l* vessel should be equal to the maximum rupture pressure of the bursting discs to be fitted to the IBC or tank. This pressure should be lower than the test pressure of the tank involved. Usually, the bursting pressure is set at a level that can cope with the pressures encountered during normal transport conditions such as hydrostatic pressure from the liquid due to turn over of the tank, slopping of the contents, etc. The 10 *l* vessel should be provided with a bursting disc with a set pressure in the range of the disc(s) fitted on the tank or IBC as to be used in transport. For safety, it is recommended to provide the test vessel with an extra bursting disc (bursting pressure approximately 80% of the design pressure of the 10 *l* test vessel) with a large opening for additional emergency venting of the test vessel in the event that the chosen orifice diameter is too small.

The outer surface of the test vessel, below the liquid level, is provided with an electrical heating coil or cartridge heaters connected to a power supply. Vessel contents should be heated at a constant rate independent of the heat being generated by the organic peroxide or self-reactive substance. The resistance of the heating coil should be such that, with the power available, the calculated heating rate (see section 3) can be achieved. The whole vessel is insulated with rock wool, cellular glass or ceramic fibre.

The temperature inside the tank is measured by means of three thermocouples, two located in the liquid phase (near the top and bottom) and one in the gas phase. Two thermocouples are used in the liquid phase to check the homogeneity of the heating. The pressure is recorded by a pressure transducer(s) capable of recording slow and fast (at least 1 000 points/sec.) changes of pressure. Examples of test vessels are illustrated in Figure A5.1. Additional information may be obtained if the tank is mounted in a tray designed to collect any solids or liquids ejected.

The tests should be performed at a test site with suitable safety distances. Alternatively, the test can be performed in a bunker provided with sufficient ventilation and vent openings to prevent pressure build-up in it. Explosion-proof electrical equipment should be used in such a bunker to minimise the risk of ignition. ***However, the tests should be performed on the assumption that the decomposition products will ignite.***

3. Calculation of the heating rate to be used in the test

If an IBC or tank is non-insulated, a heat load of the shell as given in 4.2.1.13.8 of the Model Regulations is required. For an insulated IBC or tank, the Model Regulations require that the heat load to the shell be equivalent to the heat transfer through the insulation plus the heat load to the shell on the assumption that 1% of the insulation is missing.

The following information on the IBC or tank and organic peroxide or self-reactive substance is needed for the heating rate calculation:

F_r	=	fraction of tank directly heated (1 if non-insulated, 0.01 if insulated)	[–]
M_t	=	total mass of organic peroxide or self-reactive substance and diluent	[kg]
K	=	heat conductivity of the insulation layer	$[W.m^{-1}.K^{-1}]$
L	=	thickness of insulation layer	[m]
U	=	K/L= heat transfer coefficient	$[W.m^{-2}.K^{-1}]$
A	=	wetted area of IBC or tank	$[m^2]$
C_p	=	specific heat of the organic peroxide or self-reactive substance formulation	$[J.kg^{-1}.K^{-1}]$
T_{po}	=	temperature of organic peroxide or self-reactive substance formulation at relieving conditions	[K]
q_i	=	indirectly exposed heat	[W]
q_d	=	directly exposed heat	[W]
F	=	insulation factor	[–]

Heat input, q_i (W), via <u>indirectly</u> exposed surface (insulated part) is calculated by equations (1) and (2):

$$q_i = 70961 \times F \times [(1 - F_r) \times A]^{0.82} \qquad (1)$$

where: F = insulation factor;
F = 1 for non-insulated shells, or

$$F = 2 \times \frac{U(923 - T_{PO})}{47032} \quad \text{for insulated shells} \qquad (2)$$

In the calculation of F a multiplication factor of 2 is introduced to take into account a 50% loss in insulation efficiency in an incident.

Heat input, q_d (W), via the <u>directly</u> exposed surface (non-insulated part) is calculated by equation (3):

$$q_d = 70961 \times F \times [F_r \times A]^{0.82} \qquad (3)$$

where: F = insulation factor = 1 (non-insulated)

The overall heating rate, dT/dt (K/min), due to fire engulfment is calculated by equation (4):

$$dT/dt = \frac{(q_i + q_d)}{M_t C_P} 60 \qquad (4)$$

Example 1: insulated tank

For a typical 20 m^3 insulated tank:

F_r	=	fraction of tank directly heated	=	0.01
M_t	=	total mass of organic peroxide or self-reactive substance and diluent	=	16 268 kg
K	=	heat conductivity of the insulation layer	=	0.031 W.m^{-1}.K^{-1}
L	=	thickness of the insulation layer	=	0.075 m
U	=	heat transfer coefficient	=	0.4 W.m^{-2}.K^{-1}
A	=	wetted area of tank	=	40 m^2
C_p	=	specific heat of the organic peroxide form	=	2 000 J.kg^{-1}.K^{-1}
T_{po}	=	temperature of peroxide at relieving conditions	=	100 °C

and

$$q_i = 70961 \times 2 \times \frac{0.4 \times (923 - 373)}{47032} \times [(1 - 0.01) \times 40]^{0.82} = 13558 \text{ W}$$

$$q_d = 70961 \times 1 \times [0.01 \times 40]^{0.82} = 33474 \text{ W}$$

$$\frac{dT}{dt} = \frac{(13558 + 33474)}{16268 \times 2000} \times 60 = 0{,}086 \text{ K} \cdot \text{min}^{-1}$$

Example 2: non-insulated IBC

For a typical 1.2 m^3 non-insulated stainless steel IBC (only direct heat input, q_d):

F_r	=	fraction of rank directly heated	=	1
M_t	=	total mass of organic peroxide and diluent	=	1 012 kg
A	=	wetted area of IBC	=	5.04 m^2
C_p	=	specific heat of the organic peroxide form	=	2 190 J.kg^{-1}.K^{-1}

and

$$q_d = 70961 \times 1 \times [1 \times 5.04]^{0.82} = 267308 \text{ W}$$

$$q_i = 0$$

$$\frac{dT}{dt} = \frac{(0 + 267308)}{1012 \times 2190} \times 60 = 7.2 \text{ K} \cdot \text{min}^{-1}$$

4. Procedure

Fill the test vessel shell with the amount of organic peroxide or self-reactive substance required to give the same degree of fill (by volume of the shell) as to be used in the tank (maximum degree of fill 90%, by volume) and then install the required orifice plate[1] and bursting disc. For example, it is common practice to fit four 250 mm diameter bursting discs to a 20 ton tank. This corresponds to a test vessel orifice diameter of about 11 mm.

The vessel is heated at the desired rate by applying power to the heating coil. A higher than calculated heating rate may be applied initially until a temperature 5 °C above the self-accelerating decomposition temperature (for a 50 kg package) of the organic peroxide or self-reactive substance is reached. The calculated heating rate should be applied once this temperature is reached. The temperature and

[1] *It is recommended that either small-scale vent experiments (100 - 200 ml scale) or experiments using a very strong vessel (>100 bar) be performed prior to the performance of the 10 l vent test in order to obtain information on the maximum pressure effect from the test substance and on the required orifice diameter to be used in the first 10 l scale vent test.*

pressure in the test vessel are recorded during the entire experiment. After rupture of the bursting disc, the heating should be continued for approximately 30 minutes more to be sure that all dangerous effects are measured. ***Keep distance during the execution of the test and afterwards the vessel should not be approached until the contents have cooled.***

The diameter of the orifice should be varied (if necessary) until a suitable opening is determined at which the maximum recorded pressure does not exceed the pressure as specified in Section 5, Test criteria and method of assessing the results. The step size used should be related to the options available in practice for the tank, i.e. larger vent sizes or more vents. If necessary the concentration of the organic peroxide or self-reactive substances can be lowered. The test should be performed in duplicate at the level at which the total vent area has sufficient capacity.

5. Test criteria and method of assessing the results

The minimum or suitable (if it is acceptable to use a vent size larger than the minimum vent size) IBC or tank vent area, A_{IBC} or A_{tank} (m^2), can be calculated using the minimum or suitable orifice vent area as tested in the 10 litre test at which the maximum pressure during venting is:

- for tanks not more than test pressure of the tank (according to 4.2.1.13.4 of the Model Regulations, tank shall be designed for a test pressure of at least 0.4 MPa),
- for IBC not more than 200 kPa gauge pressure, as tested according to 6.5.6.8.4 of the Model Regulations, or higher under an approval granted by the competent authority,

and the volumes of the test vessel and IBC or tank.

The minimum total IBC or tank vent area is given by:

For IBCs:

$$A_{IBC} = V_{IBC} \times \left(A_{test\ vessel} / V_{test\ vessel} \right)$$

For tanks:

$$A_{tank} = V_{tank} \times \left(A_{test\ vessel} / V_{test\ vessel} \right)$$

where:

$A_{test\ vessel}$	=	Area of venting of 10 litre test vessel	[m^2]
A_{IBC}	=	Area of venting of IBC	[m^2]
A_{tank}	=	Area of venting of tank	[m^2]
$V_{test\ vessel}$	=	Volume of 10 litre test vessel	[m^3]
V_{IBC}	=	Volume of IBC	[m^3]
V_{tank}	=	Volume of tank	[m^3]

Example:

For a typical organic peroxide in a 20 m^3 insulated tank:

$A_{test\ vessel}$	=	Minimum suitable orifice area found in test	=	$9.5 \times 10^{-5}\ m^2$
V_{tank}	=	Volume of tank	=	$20\ m^3$
$V_{test\ vessel}$	=	Volume of test vessel	=	$0.01\ m^3$

$$A_{tank} = 20 \times \frac{9.5 \times 10^{-5}}{0.01} = 0.19\ m^2$$

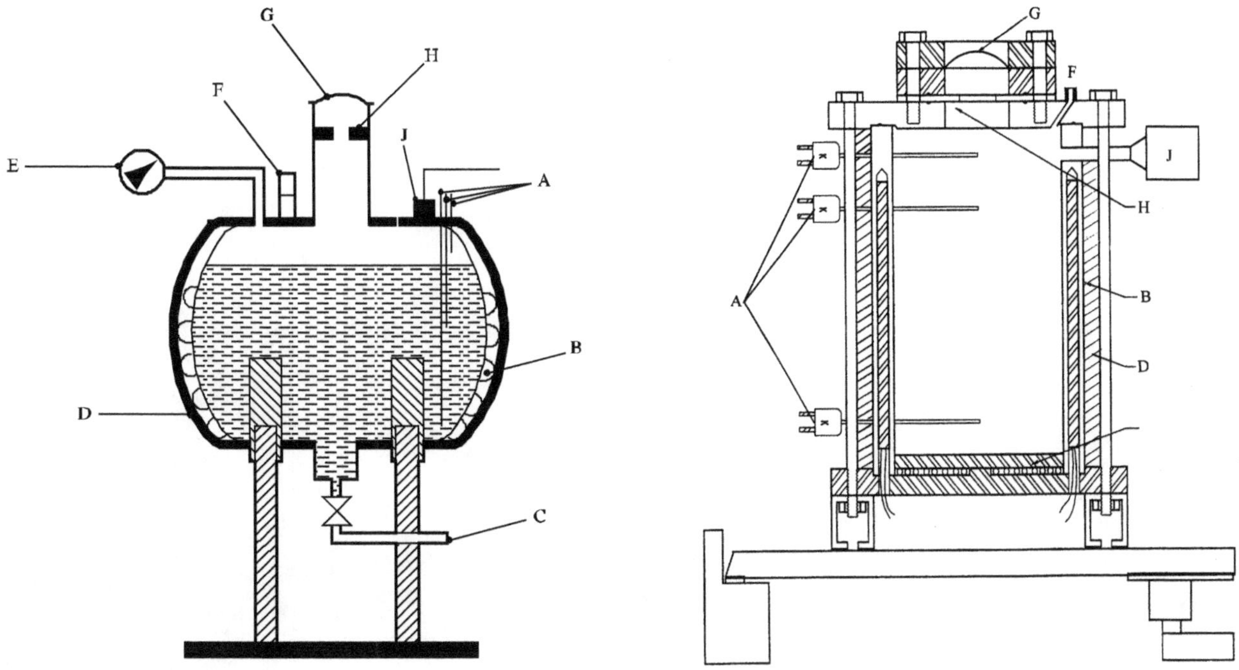

(A) Thermocouples (two in liquid and one in vapour space)
(B) Heating coil/heating cartridge
(C) Drain line, optional
(D) Insulation
(E) Manometer, optional
(F) Pressure relief valve, optional
(G) Bursting disc
(H) Orifice plate
(J) Pressure transducer or pressure relief valve and transducer on tee

Figure A5.1: 10 LITRE VESSELS FOR VENTING TESTS

APPENDIX 6

SCREENING PROCEDURES

1. Purpose

1.1 Industry uses screening procedures to identify the hazard potential of raw materials, reactive mixtures and intermediates, products and by-products. The use of such procedures is essential to ensure safety during research and development and to ensure that new products and processes are as safe as possible. These procedures usually consist of a combination of a theoretical appraisal and small-scale tests and, in many cases, enable an adequate hazard evaluation to be carried out without the need for larger scale classification tests. This reduces the quantity of material required, lessens any detrimental effect on the environment and minimizes the amount of unnecessary testing.

1.2 The purpose of this Appendix is to present example screening procedures. It should be used in conjunction with any screening procedures given in the introductions to the relevant test series. With the specified safety margin, the results from the screening procedures adequately predict when it is not necessary to perform the classification test as a negative result would be obtained. They are presented for guidance and their use is not compulsory. Other screening procedures may be used provided that adequate correlation has been obtained with the classification tests on a representative range of substances and there is a suitable safety margin.

2. Scope

2.1 A hazard evaluation for a new substance should be undertaken before it is offered for transport. Initially this evaluation can use the screening procedures given in this Appendix. If the screening procedure indicates that there is a hazard, the full classification procedure should be applied.

2.2 The screening procedures are only applicable to substances and stable, homogeneous mixtures of substances. If a mixture can separate out during transport, the screening procedure should also be performed on each reactive component of the mixture in addition to the mixture.

2.3 The remarks 1.1.2 from section 1 "General introduction" are emphasized that competence on the part of the testing authority is assumed and responsibility for classification is left with them.

3. Screening procedures for substances which may have explosive properties

3.1 The screening procedure may be used for new substances which are suspected of having explosive properties. When considering the explosive properties of self-reactive substances of Division 4.1 or organic peroxides of Division 5.2, refer to Part II of this Manual and section 5.1 of this Appendix. It should not be used for substances manufactured with the intention of producing a practical explosive or pyrotechnic effect.

3.2 Explosive properties are associated with the presence of certain chemical groups in a molecule which can react to produce very rapid increases in temperature or pressure. The screening procedure is aimed at identifying the presence of such reactive groups and the potential for rapid energy release. If the screening procedure identifies the material to be a potential explosive, the Class 1 Acceptance Procedure (see 10.3) should be applied.

***NOTE**: Neither a Series 1 type (a) propagation of detonation test nor a Series 2 type (a) test of sensitivity to detonative shock is required if the exothermic decomposition energy of organic materials is less than 800 J/g. For organic substances and mixtures of organic substances with a decomposition energy of 800 J/g or more, tests 1 (a) and 2 (a) need not be performed if the outcome of the ballistic mortar Mk.IIId test (F.1), or the ballistic mortar test (F.2) or the BAM Trauzl test (F.3) with initiation by a standard No. 8 detonator (see Appendix 1) is "No". In this case, the results of test 1 (a) and 2 (a) are deemed to be "-".*

3.3 The acceptance procedure for Class 1 explosives need not be applied:

(a) When there are no chemical groups associated with explosive properties present in the molecule. Examples of groups which may indicate explosive properties are given in Table A6.1; or

Table A6.1 EXAMPLES OF CHEMICAL GROUPS INDICATING EXPLOSIVE PROPERTIES IN ORGANIC MATERIALS

Structural feature	Examples
C-C unsaturation	Acetylenes, acetylides, 1,2-dienes
C-Metal, N-Metal	Grignard reagents, organo-lithium compounds
Contiguous nitrogen atoms	Azides, aliphatic azo compounds, diazonium salts, hydrazines, sulphonylhydrazides
Contiguous oxygen atoms	Peroxides, ozonides
N-O	Hydroxylamines, nitrates, nitro compounds, nitroso compounds, N-oxides, 1,2-oxazoles
N-halogen	Chloramines, fluoroamines
O-halogen	Chlorates, perchlorates, iodosyl compounds

(b) When the substance contains chemical groups associated with explosive properties which include oxygen and the calculated oxygen balance is less than -200.

The oxygen balance is calculated for the chemical reaction:

$$C_xH_yO_z + [x + (y/4)-(z/2)]\ O_2 \leftrightarrows x\ CO_2 + (y/2)\ H_2O$$

using the formula:

$$\text{oxygen balance} = -1600 \times \frac{\left(2x + \frac{y}{2} - z\right)}{\text{molecular weight}}; \text{ or}$$

(c) For the organic substance or a homogenous mixture of organic substances containing chemical group (or groups) associated with explosive properties:

- when the exothermic decomposition energy is less than 500 J/g, or

- when the onset of exothermic decomposition is 500 °C or above as indicated by Table A6.2.

Table A6.2 DECISION TO APPLY THE ACCEPTANCE PROCEDURE FOR CLASS 1 FOR AN ORGANIC SUBSTANCE OR A HOMOGENOUS MIXTURE OF ORGANIC SUBSTANCES

Decomposition energy (J/g)	Decomposition onset temperature (°C)	Apply acceptance procedure for Class 1? (Yes/No)
< 500	< 500	No
< 500	≥ 500	No
≥ 500	< 500	Yes
≥ 500	≥ 500	No

The exothermic decomposition energy may be determined using a suitable calorimetric technique (see 20.3.3.3); or

(d) For mixtures of inorganic oxidizing substances of Division 5.1 with organic material(s), the concentration of the inorganic oxidizing substance is:

- less than 15%, by mass, if assigned to packing group I (high hazard) or II (medium hazard);

- less than 30%, by mass, if assigned to packing group III (low hazard).

3.4 When the substance is a mixture containing any known explosives, the class 1 acceptance procedure should be applied.

4. Screening procedures for mixtures which may be flammable liquids (Class 3)

4.1 The procedure only applies to possible flammable mixtures[1] containing known flammable liquids in defined concentrations although they may contain non-volatile components e.g. polymers, additives etc. The flash point of these mixtures need not be determined experimentally if the calculated flash point of the mixture, using the method given in 4.2, is at least 5 °C[2] greater than the relevant classification criterion (23 °C and 60 °C, respectively) and provided that:

(a) The composition of the mixture is accurately known (if the material has a specified range of composition the composition with the lowest calculated flash point should be selected for assessment);

(b) The lower explosion limit of each component is known (an appropriate correlation has to be applied when these data are extrapolated to other temperatures than test conditions) as well as a method for calculating the lower explosion limit of the mixture;

(c) The temperature dependence of the saturated vapour pressure and of the activity coefficient is known for each component as present in the mixture;

(d) The liquid phase is homogeneous.

---

[1] *Up to now, the calculation method is validated for mixtures containing up to six volatile components. These components may be flammable liquids like hydrocarbons, ethers, alcohols, esters (except acrylates), and water. It is however not yet validated for mixtures containing halogenated, sulphurous, and/or phosphoric compounds as well as reactive acrylates.*

[2] *If the calculated flash point is less than 5 °C greater than the relevant classification criterion, the calculation method may not be used and the flash point should be determined experimentally.*

4.2 A suitable method is described by Gmehling and Rasmussen (Ind. Eng. Chem. Fundament, 21, 186, (1982)). For a mixture containing non-volatile components, e.g. polymers or additives, the flash point is calculated from the volatile components. It is considered that a non-volatile component only slightly decreases the partial pressure of the solvents and the calculated flash point is only slightly below the measured value.

5. Screening procedures for substances which may be flammable solids (Class 4)

5.1 Substances which may be self-reactive substances (Division 4.1)

The classification procedures (see section 20.4) for self-reactive substances need not be applied if:

(a) There are no chemical groups present in the molecule associated with explosive or self-reactive properties; examples of such groups are given in Tables A6.1 and A6.3; or

Table A6.3: EXAMPLES OF CHEMICAL GROUPS INDICATING SELF-REACTIVE PROPERTIES IN ORGANIC MATERIALS

Structural feature	Examples
Mutually reactive groups	Aminonitriles, haloanilines, organic salts of oxidizing acids
S=O	Sulphonyl halides, sulphonyl cyanides, sulphonyl hydrazides
P-O	Phosphites
Strained rings	Epoxides, aziridines
Unsaturation	Olefins, cyanates

(b) For a single organic substance or a homogeneous mixture of organic substances, the estimated SADT is greater than 75 °C or the exothermic decomposition energy is less than 300 J/g. The onset temperature and decomposition energy may be estimated using a suitable calorimetric technique (see 20.3.3.3).

5.2 Substances which may be liable to spontaneous combustion (Division 4.2)

5.2.1 The classification procedure for *pyrophoric solids and liquids* need not be applied when experience, in production or handling, shows that the substance do not ignite spontaneously on coming into contact with air at normal temperatures (i.e. the substance is known to be stable at room temperature for prolonged periods of time (days)).

5.2.2 The classification procedure for *self-heating substances* need not be applied if the results of a screening test can be adequately correlated with the classification test and an appropriate safety margin is applied. Examples of screening tests are:

(a) The Grewer Oven test (VDI guideline 2263, part 1, 1990, *Test methods for the Determination of the Safety Characteristics of Dusts*) with an onset temperature 80 K above the reference temperature for a volume of 1 *l* (33.3.1.6);

(b) The Bulk Powder Screening Test (Gibson, N. Harper, D. J. Rogers, R. *Evaluation of the fire and explosion risks in drying powders,* Plant Operations Progress, 4 (3), 181 - 189, 1985) with an onset temperature 60 K above the reference temperature for a volume of 1 *l* (33.3.1.6).

5.3 ***Substances which in contact with water may react to emit flammable gases (Division 4.3)***

The classification procedure for substances which may react with water to emit flammable gases need not be applied if:

(a) The chemical structure of the substance does not contain metals or metalloids; or

(b) Experience in production or handling shows that the substance does not react with water, e.g. the substance is manufactured in water or washed with water; or

(c) The substance is known to be soluble in water to form a stable mixture.

6. Screening procedures for substances which may be oxidizing substances and those which may be organic peroxides (Class 5)

6.1 ***Substances which may be oxidizing substances (Division 5.1)***

6.1.1 For *organic compounds*, the classification procedure for oxidizing substances of Division 5.1 need not be applied if:

(a) The compound does not contain oxygen, fluorine or chlorine; or

(b) The compound contains oxygen, fluorine or chlorine and these elements are chemically bonded only to carbon or hydrogen.

6.1.2 For *inorganic substances*, the test procedure in Section 34 need not be applied if the substance does not contain any oxygen or halogen atoms.

6.2 ***Substances which may be organic peroxides (Division 5.2)***

6.2.1 Organic peroxides are classified by definition based on their chemical structure and on the available oxygen and hydrogen peroxide content of formulations (see 20.2.2).

APPENDIX 7

HSL FLASH COMPOSITION TEST

1. Introduction

This test is used to determine whether pyrotechnic substances in powder form or as pyrotechnic units as presented in the fireworks, that are used to produce an aural effect, or used as a bursting charge or lifting charge, are considered to be flash compositions for the purposes of determining the classification of fireworks using the UN default fireworks classification table in 2.1.3.5.5 of the Model Regulations.

2. Apparatus and materials

2.1 The time/pressure apparatus (Figure A7.2) consists of a cylindrical steel pressure vessel 89 mm in length and 60 mm in external diameter. Two flats are machined on opposite sides (reducing the cross-section of the vessel to 50 mm) to facilitate holding whilst fitting the cone in firing plug and vent plug. The vessel, which has a bore of 20 mm diameter, is internally rebated at either end to a depth of 19 mm and threaded to accept 1" British Standard Pipe (BSP). A pressure take-off, in the form of a side-arm, is screwed into the curved face of the pressure vessel 35 mm from one end and at 90° to the machined flats. The socket for this is bored to a depth of 12 mm and threaded to accept the 1/2" BSP thread on the end of the side-arm. A washer is fitted to ensure a gastight seal. The side-arm extends 55 mm beyond the pressure vessel body and has a bore of 6 mm. The end of the side-arm is rebated and threaded to accept a diaphragm type pressure transducer. Any pressure-measuring device may be used provided that it is not affected by the hot gases or decomposition products and is capable of responding to rates of pressure rise of 690-2 070 kPa in not more than 1 ms.

2.2 The end of the pressure vessel furthest from the side-arm is closed with a cone in firing plug which is fitted with two electrodes, one insulated from, and the other earthed to, the plug body. The other end of the pressure vessel is closed by an aluminium bursting disc 0.2 mm thick (bursting pressure approximately 2 200 kPa) held in place with a retaining plug which has a 20 mm bore. A soft lead washer is used with both plugs to ensure a good seal.

2.3 A support stand (Figure A7.8) holds the assembly in the correct attitude during use. This comprises a mild steel base plate measuring 235 mm × 184 mm × 6 mm and a 185 mm length of square hollow section (S.H.S.) 70 × 70 × 4 mm. A section is cut from each of two opposite sides at one end of the length of S.H.S. so that a structure having two flat sided legs surmounted by an 86 mm length of intact box section results. The ends of these flat sides are cut to an angle of 60° to the horizontal and welded to the base plate.

2.4 A slot measuring 22 mm wide × 46 mm deep is machined in one side of the upper end of the base section such that when the pressure vessel assembly is lowered, firing plug end first, into the box section support, the side-arm is accommodated in this slot. A packing piece of steel 30 mm wide and 6 mm thick is welded to the lower internal face of the box section to act as a spacer. Two 7 mm thumb screws, tapped into the opposite face, serve to hold the pressure vessel firmly in place. Two 12 mm wide strips of 6 mm thick steel, welded to the side pieces abutting the base of the box section, support the pressure vessel from beneath.

2.5 The ignition system consists of a Vulcan electric fusehead, with lead wires, of the type commonly used for igniting pyrotechnic substances. Fuseheads with equivalent properties may be used.

2.6 The wires of the fusehead are cut to such a length that the fusehead sits 10 mm above the top of the cone of the firing plug (see Figure A7.1). The fusehead leads are held in position using the grub screws (see Figure A7.3).

3. Procedure

3.1 The apparatus, assembled complete with pressure transducer but without the aluminium bursting disc in position, is supported firing plug end down. 0.5 g of the substance is introduced into the cone of the firing plug. Where the pyrotechnic substance is in consolidated form greater than 0.5 g it should be broken to produce a piece as close to 0.5 g as possible. Where the pyrotechnic substance is in consolidated form less than 0.5 g then whole and broken units should be chosen to give 0.5 g pyrotechnic substance. The lead washer and brass or aluminium bursting disc are placed in position and the retaining plug is screwed in tightly. The charged vessel is transferred to the firing support stand, bursting disc uppermost, which should be contained in a suitable, armoured fume cupboard or firing cell. An exploder dynamo is connected to the external terminals of the firing plug and the charge is fired. The signal produced by the pressure transducer is recorded on a suitable system which allows both evaluation and a permanent record of the time/pressure profile to be achieved (e.g. transient recorder coupled to a chart-recorder).

3.2 The test is carried out three times. The time taken for the pressure to rise from 690 kPa to 2 070 kPa above atmospheric is noted. The shortest interval of three firings should be used for classification.

4. Test criteria and method of assessing results

The test results are interpreted in terms of whether a gauge pressure of 2 070 kPa is reached and, if so, the time taken for the pressure to rise from 690 kPa to 2 070 kPa gauge. The result is considered positive "+" and the pyrotechnic substances in powder form or as pyrotechnic units as presented in the fireworks, that are used in waterfalls, or used as a bursting charge or lifting charge, is to be considered as flash composition if the minimum time taken for the pressure rise is shown to be less than, or equal to, 6 ms for 0.5 g of pyrotechnic substance.

Examples of results:

Substance	Maximum pressure rise (kPa)	Mean time for a pressure rise from 690 to 2 070 kPa (ms)	Result
1	> 2 070	0.70	Flash composition
2	> 2 070	4.98	Flash composition
4	> 2 070	1.51	Flash composition
5	> 2 070	0.84	Flash composition
6	> 2 070	11.98	Not flash composition

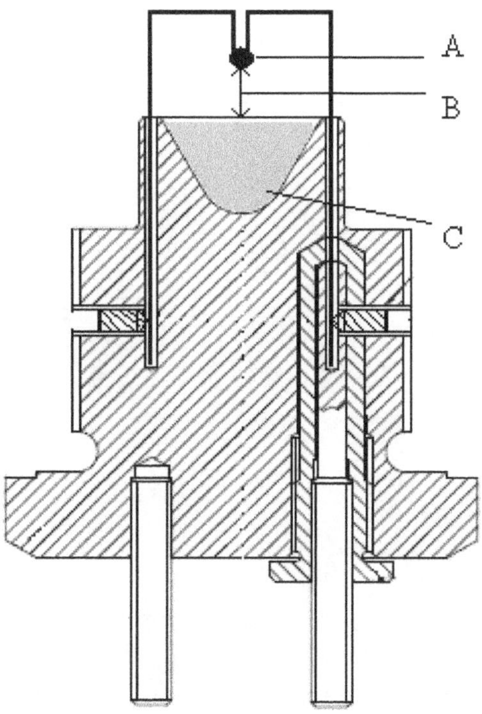

(A) Fusehead
(B) 10 mm gap
(C) Substance under test

Figure A7.1: SAMPLE SETUP

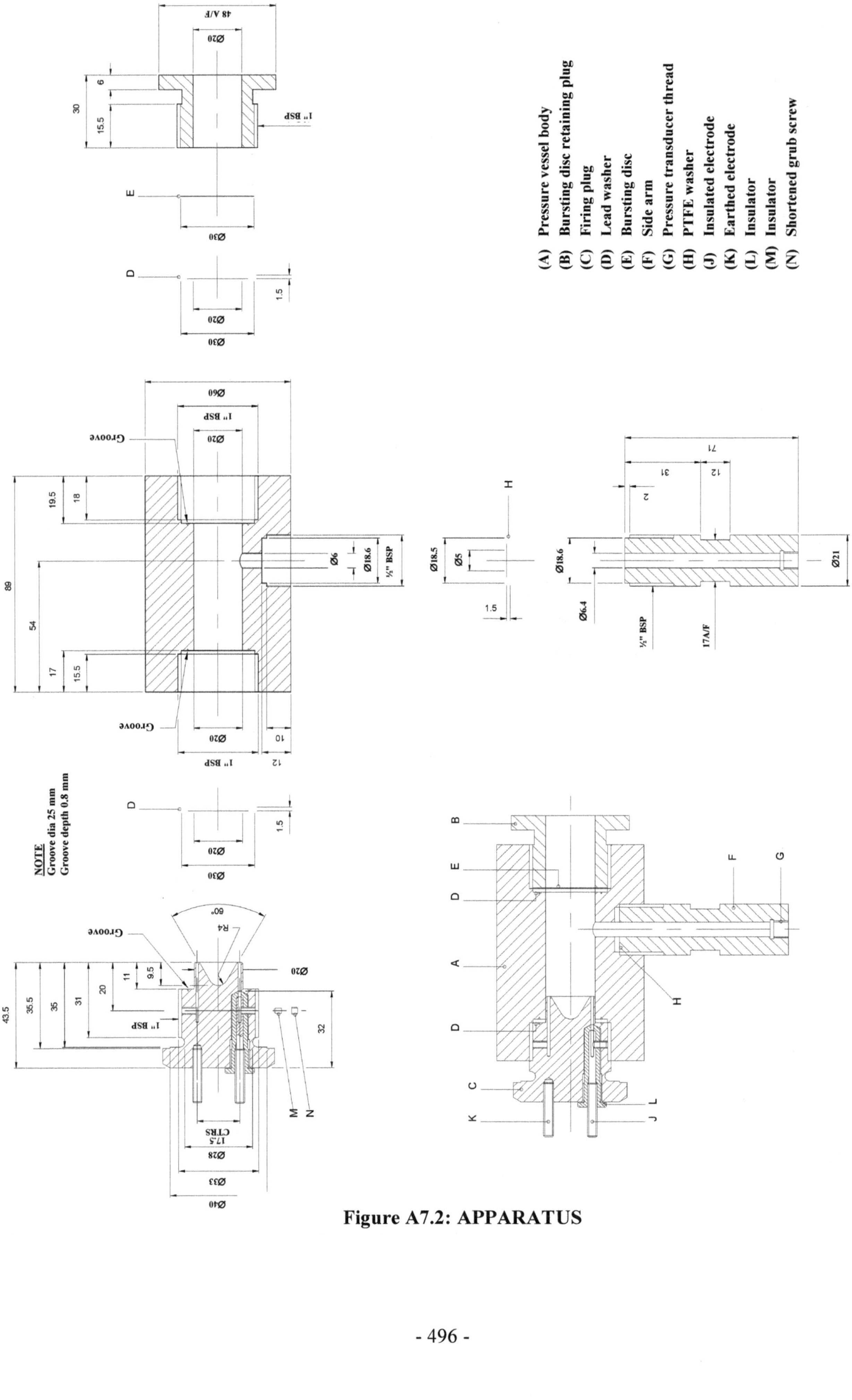

Figure A7.2: APPARATUS

MACHINING/ASSEMBLY SEQUENCE

1. SCREW JN0003490:B2 INTO PRESSURE PLUG BODY
2. SCREW JN0003490:A2 INTO JN0003490:B2
3. DRILL & TAP M3 * 0.5P * 7 DEEP HOLE
4. SCREWCUT 1" BSP PARALLEL THREAD ON PRESSURE PLUG BODY

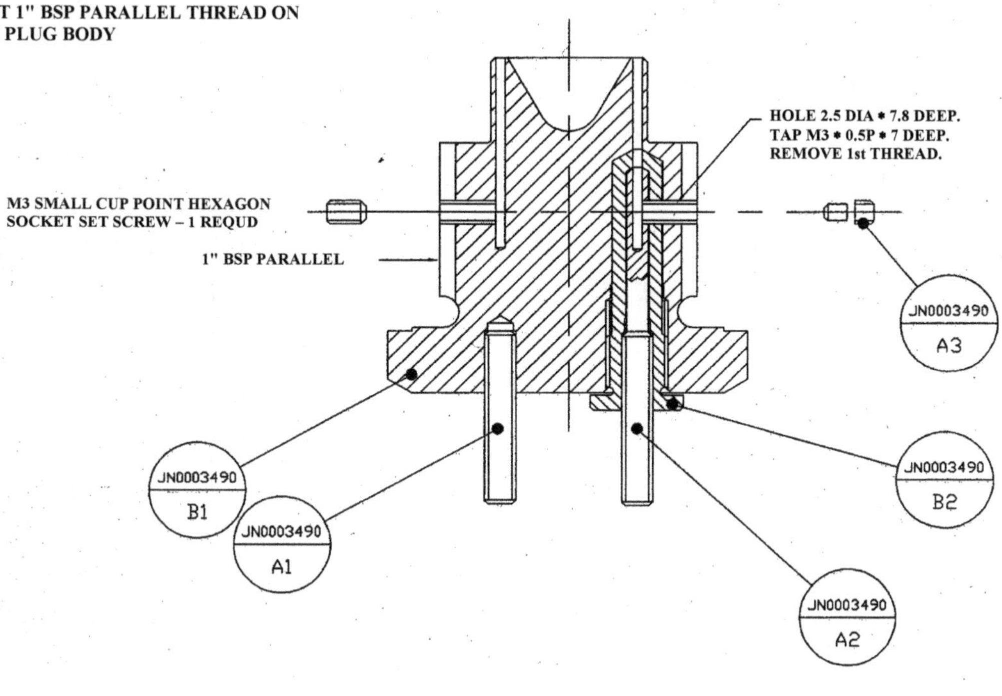

Figure A7.3: ASSEMBLY

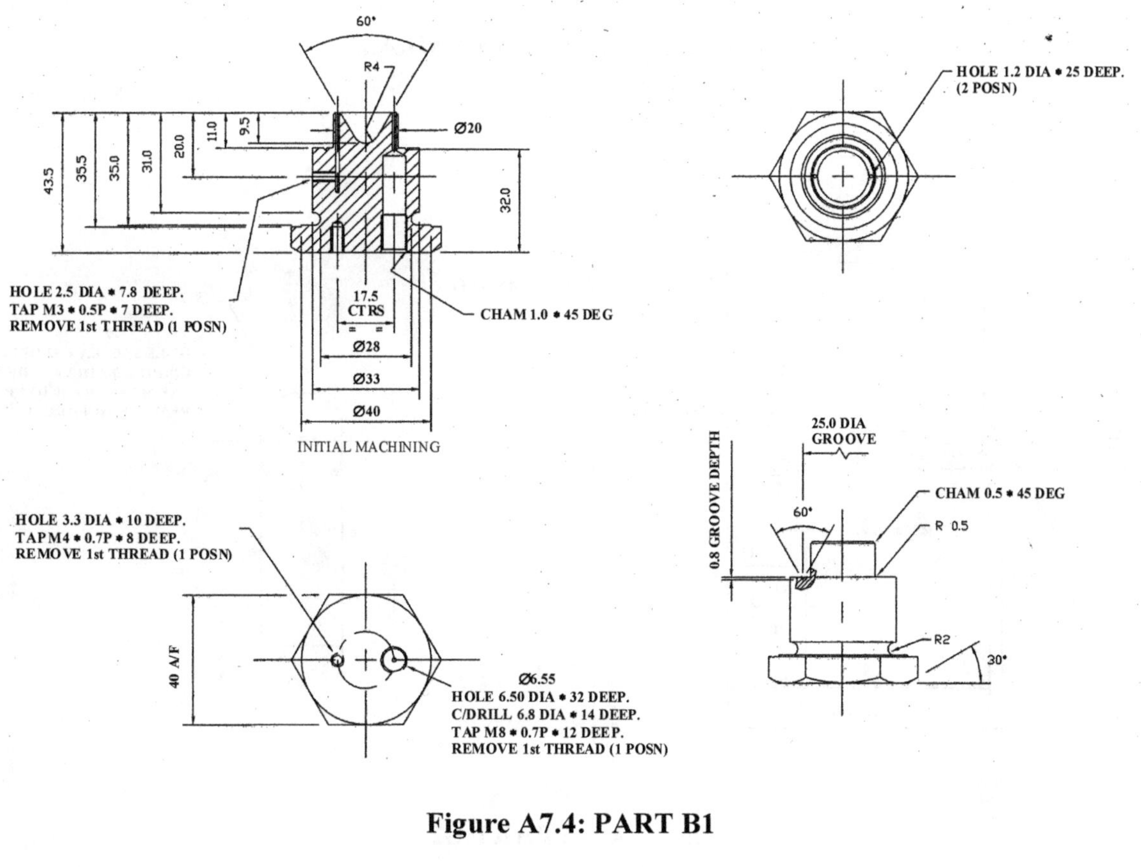

Figure A7.4: PART B1

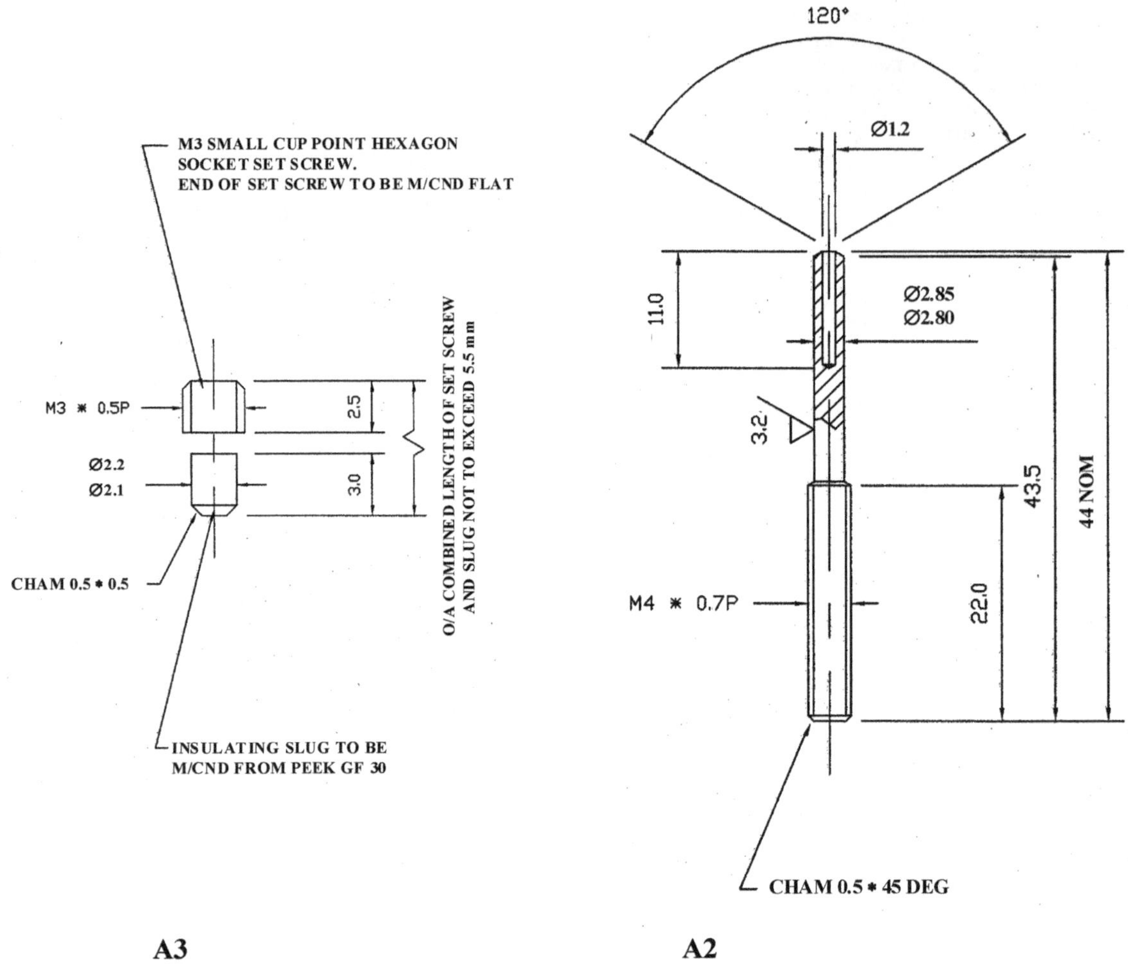

Figure A7.5: PARTS A3 AND A2

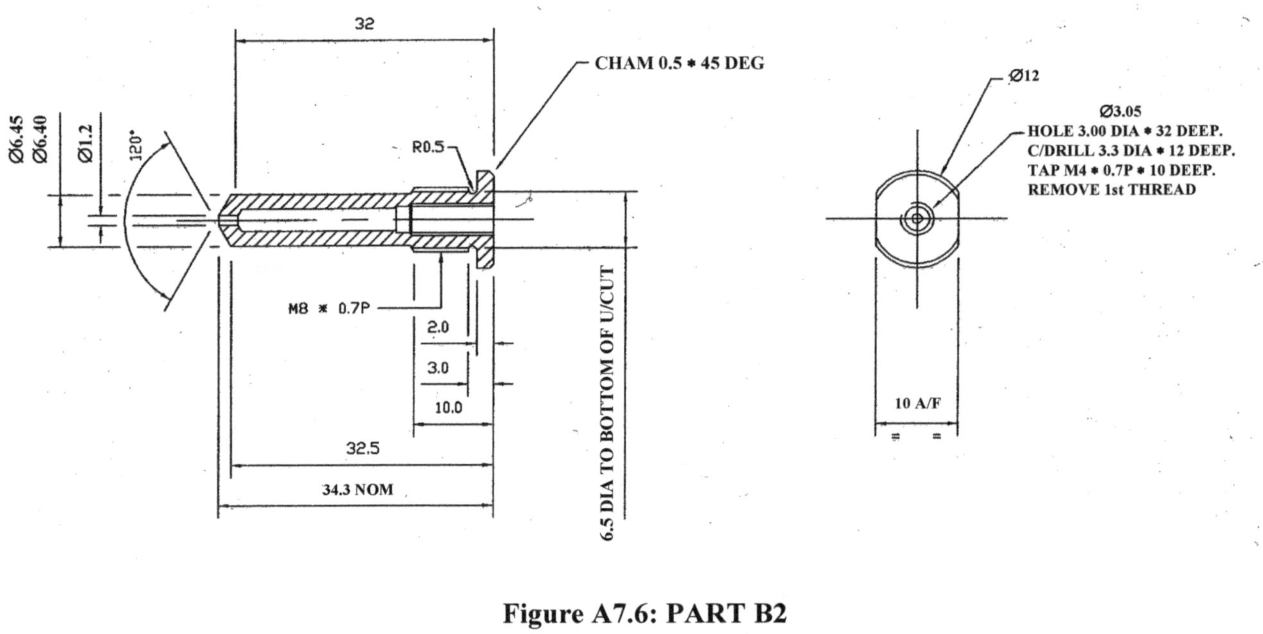

Figure A7.6: PART B2

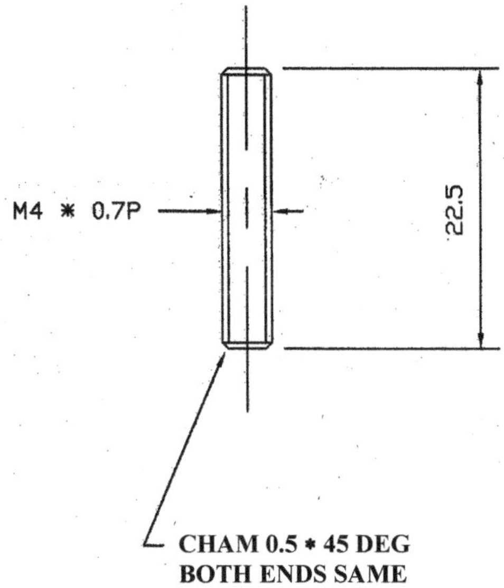

Figure A7.7: PART A1

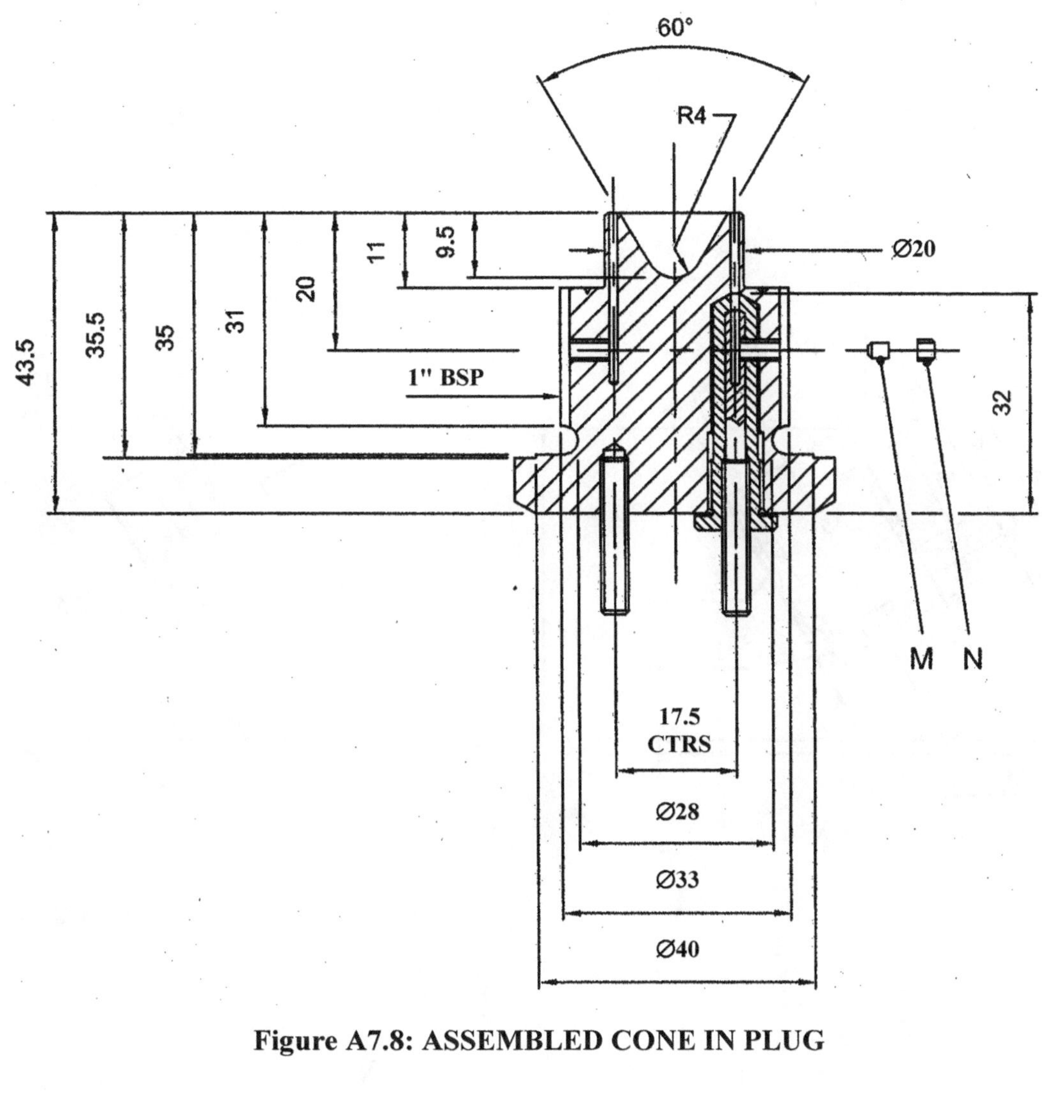

Figure A7.8: ASSEMBLED CONE IN PLUG

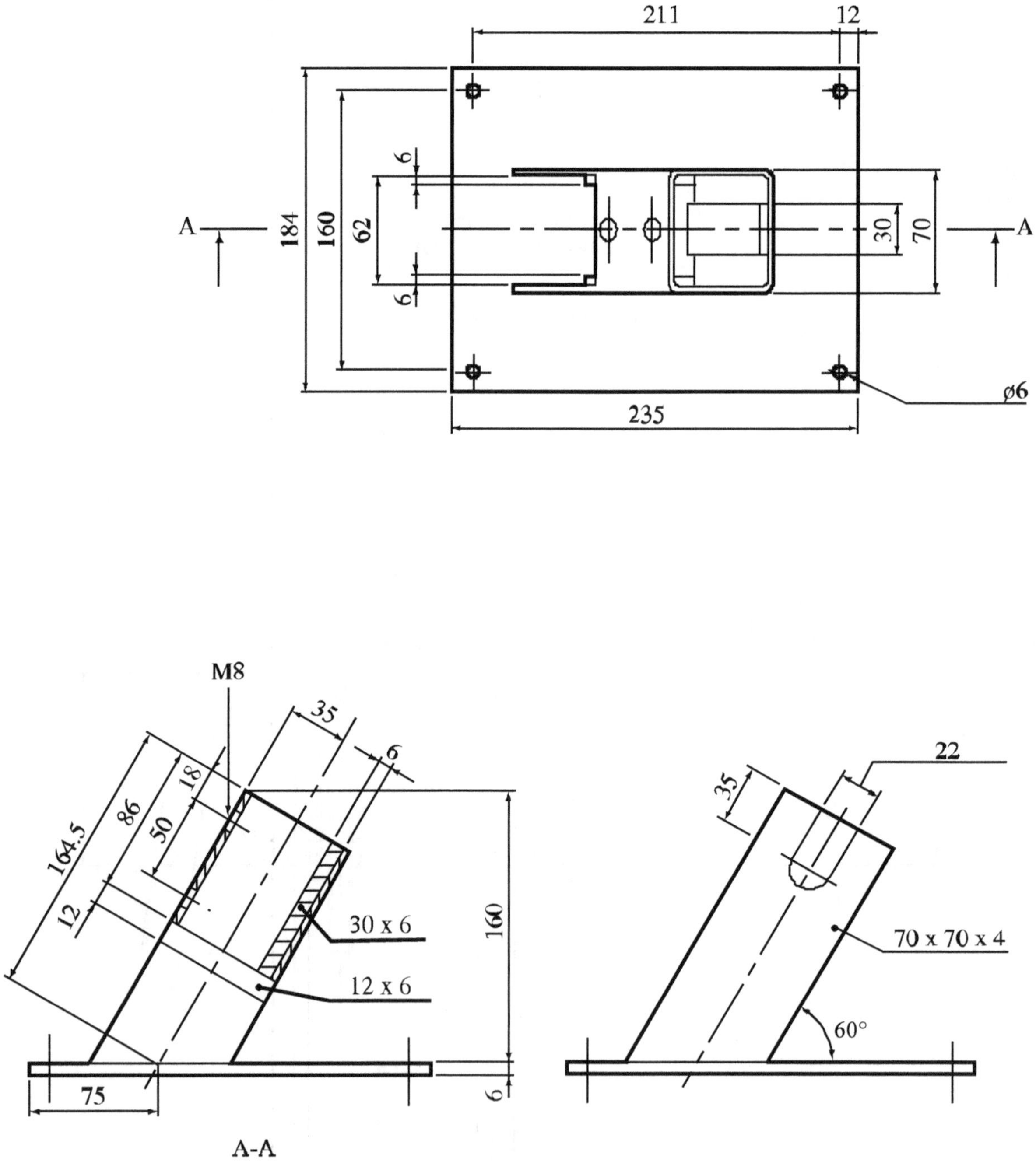

Figure A7.9: SUPPORT STAND

APPENDIX 8

RESPONSE DESCRIPTORS

These response descriptors are to be used for the purposes of Test Series 7 criteria and designed to be used by the competent authority to determine the response type of articles. For example, articles vary greatly in size, type, packaging and explosive substances; these differences need to be taken into account. For a reaction to be judged a particular type, the primary evidence (denoted P in the table below) for that type would need to be present. The entire (both primary and secondary) body of evidence must be weighed carefully and used in its entirety by the competent authority to assess the reaction. The secondary evidence provides other indicators that may be present.

Response level	Observed or measured effects				
	Explosive Substances (ES)	Case	Blast	Fragment or ES projection	Other
Detonation	Prompt consumption of all ES once the reaction starts	(P) Rapid plastic deformation of the metal casing contacting the ES with extensive high shear rate fragmentation	(P) Shock wave with magnitude & timescale = to a calculated value or measured value from a calibration test	Perforation, fragmentation and/or plastic deformation of witness plates	Ground craters of a size corresponding to the amount of ES in the article
Partial detonation		(P) Rapid plastic deformation of some, but not all, of the metal casing contacting the ES with extensive high shear rate fragmentation	(P) Shock wave with magnitude & timescale < that of a calculated value or measured value from a calibration test Damage to neighboring structures	Perforation, plastic deformation and/or fragmentation of adjacent witness plates. Scattered burned or unburned ES.	Ground craters of a size corresponding to the amount of ES that detonated.
Explosion	(P) Rapid combustion of some or all of the ES once the article reaction starts	(P) Extensive fracture of metal casings with no evidence of high shear rate fragmentation resulting in larger and fewer fragments than observed from purposely detonated calibration tests ¤	Observation or measurement of a pressure wave throughout the test arena with peak magnitude << and significantly longer duration that of a measured value from a calibration test	Witness plate damage. Significant long distance scattering of burning or unburned ES.	Ground craters.
Deflagration	(P) Combustion of some or all of the ES	(P) Rupture of casings resulting in a few large pieces that might include enclosures or attachments. *	Some evidence of pressure in the test arena which may vary in time or space.	(P) At least one piece (casing, enclosure or attachment) travels beyond 15m with an energy level > 20J based on the distance/mass relationship of Figure 16.6.1.1. Significant scattered burning or unburned ES, generally beyond 15 m.	(P) There is no primary evidence of a more severe reaction and there is evidence of thrust capable of propelling the article beyond 15m. Longer reaction time than would be expected in an explosion reaction.

Response level	Observed or measured effects				
	Explosive Substances(ES)	Case	Blast	Fragment or ES projection	Other
Burn	(P) Low pressure burn of some or all of the ES	(P) The casing may rupture resulting in a few large pieces that might include enclosures or attachments. *	Some evidence of insignificant pressure in the test arena.	(P) No item (casing, enclosure, attachment or ES) travels beyond 15m with an energy level > 20J based on the distance/mass relationship detailed at Figure 16.6.1.1. (P) A small amount of burning or unburned ES relative to the total amount in the article may be scattered, generally within 15m but no farther than 30m.	(P) No evidence of thrust capable of propelling the article beyond 15m. For a rocket motor a significantly longer reaction time than if initiated in its design mode.
No Reaction	(P) No reaction of the ES without a continued external stimulus. (P) Recovery of all or most of the unreacted ES with no indication of a sustained combustion.	(P) No fragmentation of the casing or packaging greater than that from a comparable inert test item. *	None	None	None

** Note: Mechanical threats will directly induce damage causing disruption of the article or even a pneumatic response resulting in parts, particularly closures, being projected. This evidence can be misinterpreted as being driven by the reaction of the explosive substance contained in the article, which may result in a more severe response descriptor being assigned. Comparison of observed evidence with that of a corresponding inert article can be useful in helping to determine the article's response.*

APPENDIX 9

BALLISTIC PROJECTION ENERGY TEST FOR CARTRIDGES, SMALL ARMS (UN No. 0012)

1. Introduction

This test is conducted with candidates for Cartridges, small arms (UN No. 0012) with individual cartridges and is used to determine the maximum possible energy of a projection that could be generated upon functioning in transport. The test takes worst-case conditions into account, since no packaging attenuates the energy of the projectile and the cartridge is supported by a fixed anvil block. It is not necessary to reverse the test set-up to a situation where the cartridge is propelled, because experimentation shows that energy transfer from the propellant to the bullet is equal or more than that to the case.

2. Apparatus and materials

The following items are required:

(a) A suitable actuator to initiate ammunition and

(b) A ballistic pendulum with an interception device for the projectile for determining the energy, or a high-speed camera and a background with a scale to determine the velocity of the projectile.

3. Procedure

The test is performed on single cartridges. The cartridge is actuated as designed by means of the primer cap and a firing pin. The cartridge, actuator and measuring device are arranged along the flight path in such a way that angle errors are minimized. The test is performed three times.

4. Test criteria and method of assessing the results

The energy of the projectile is calculated either from the maximum displacement of the ballistic pendulum or from the velocity (v) determined by the high-speed camera taking the mass (m) of the projectile into account. The value of energy (E) can be calculated from the equation:

$$E = \frac{1}{2}mv^2$$

If the energy of the projectile does not exceed 8 J in any of the test runs, the article, in the appropriate packaging in accordance with Chapter 3.2 of the Model Regulations, may be assigned to Cartridges, small arms (UN No. 0012).

www.ingramcontent.com/pod-product-compliance
Lightning Source LLC
Chambersburg PA
CBHW082347230926
51846CB00004B/937
* 9 7 8 9 2 1 1 3 9 1 5 5 8 *